Handbook of
ANALYSIS OF SYNTHETIC POLYMERS AND PLASTICS

Ellis Horwood Series in Analytical Chemistry

Consultant Editor: Dr. R. A. Chalmers

University of Aberdeen

Founded as a library of fundamental books on important and growing subject areas in analytical chemistry, this series will serve chemists in industrial work and research, and in teaching or advanced study

Published or in active preparation:

Handbook of Process Stream Analysis
K. J. Clevett, *Crest Engineering (U.K.) Inc.*

Automatic Methods in Chemical Analysis
J. K. Foreman, P. B. Stockwell, *Laboratory of the Government Chemist, London*

Particle Size Analysis
Z. K. Jelínek, *Organic Synthesis Research Institute, Pardubice*

Gradient Liquid Chromatography
C. Liteanu & S. Gocan, *University of Cluj*

Theoretical Foundations of Chemical Electroanalysis
Z. Galus, *Warsaw University*

Electroanalytical Chemistry
G. F. Reynolds, *University of Reading*

Analysis of Organic Solvents
V. Šedivec, J. Flek, *Institute of Hygiene and Epidemiology, Prague*

Methods of Catalytic Analysis
G. Svehla, *Queen's University of Belfast*
H. Thompson, *University of New York*

Handbook of Organic Reagents in Inorganic Analysis
Z. Holzbecher *et al.*, *Institute of Chemical Technology, Prague*

Analysis of Synthetic Polymers
J. Urbański *et al.*, *Warsaw Technical University*

Spectrophotometric Determination of Elements
Z. Marczenko, *Warsaw Technical University*

Operational Amplifiers in Chemical Instrumentation
R. Kalvoda, *Heyrovský Institute of Polarography, Prague*

Analytical Applications of Complex Equilibria
J. Inczédy, *University of Chemical Engineering, Veszprém*

Electrochemical Stripping Analysis
F. Vydra, K. Štulik, E. Juláková, *Heyrovský Institute of Polarography, Prague*; *Charles University, Prague*

Handbook of ANALYSIS OF SYNTHETIC POLYMERS AND PLASTICS

J. URBAŃSKI, W. CZERWIŃSKI, K. JANICKA,
F. MAJEWSKA and H. ZOWALL
Warsaw Technical University
and Warsaw Institute of Plastics

Translation Editor:
G. GORDON CAMERON
University of Aberdeen

ELLIS HORWOOD LIMITED
Publisher Chichester

Halsted Press: a division of
JOHN WILEY & SONS INC.
New York • London • Sydney • Toronto

English Edition first published in 1977 by
ELLIS HORWOOD Ltd., Coll House, Westergate, Chichester, Sussex, England
in co-edition with
WYDAWNICTWA NAUKOWO-TECHNICZNE, Warsaw, Poland

Distributed in:

Australia, New Zealand, South-east Asia by
JOHN WILEY & SONS AUSTRALASIA PTY LIMITED
1-7, Waterloo Road, North Ryde, N.S.W., Australia

Canada by
JOHN WILEY & SONS CANADA LIMITED
22 Worcester Road, Rexdale, Ontario, Canada

Europe, Africa by
JOHN WILEY & SONS LIMITED
Baffins Lane, Chichester, Sussex, England

North and South America and the rest of the world by
Halsted Press a division of
JOHN WILEY & SONS INC.
605 Third Avenue, New York, N.Y. 10016, U.S.A.

Translated by Andrzej Skup from the Polish
Analiza polimerów syntetycznych
Published by Wydawnictwa Naukowo-Techniczne, Warsaw

Library of Congress Cataloging in Publication Data:

Main entry under title:
Handbook of analysis of synthetic polymers and plastics
Translation of Analiza polimerów syntetycznych
1. Plastics—Analysis. 2. Polymers and polymerization—Analysis.
I. Urbański, Jerzy
TP1140. A4713 668.4'19 76-5880
ISBN 0-470-15081-5 (Halsted Press)
ISBN 85312-020-X (Ellis Horwood, Publisher).

PRINTED IN POLAND

AUTHORS' PREFACE

THE book has been compiled by a team of authors from the Warsaw Institute of Plastics, concerned in their everyday work with experimental problems connected with analysis of synthetic polymers. It consists of two separate parts.

Part I of the book (Chapters 1–5) gives a general survey of classic and instrumental techniques most frequently used in the analysis of polymers and plastics, as for instance infrared and ultraviolet spectroscopy, NMR spectroscopy, chromatography and polarography. Each chapter is devoted to a particular technique and gives a concise description of the underlying theoretical foundations (in the case of instrumental techniques), the experimental procedure, and most important applications.

In the present English edition major changes have been made in the chapters on chromatography and NMR spectroscopy. The applications of these methods to polymer analysis have been significantly extended at the cost of a considerable reduction in the theoretical foundations.

Part II gives detailed descriptions of the analysis of particular types of polymers. Each chapter, dealing with a polymer or group of polymers, opens by providing the principal information about the properties, production and application of the relevant materials. Chemical structure is considered from those aspects of interest to analytical chemists. The main text of each chapter then goes on to the important analytical procedures for determining the chemical composition of the plastics under discussion. Account is taken of all experimental techniques of any practical value, including techniques not considered in Part I. In view of the observed dynamic development of thermoplastics, the chapters dealing with these materials have been expanded in Part II of the English edition.

The book gives an exhaustive survey of the world literature in the field, including numerous references to works from East European countries and particularly Poland, Czechoslovakia and the USSR. The personal experiences and researches of the authors make an important contribution.

The book is intended primarily for chemists working in those industries producing and fabricating synthetic polymers, and should also prove useful to workers in universities and research institutes where problems in polymer analysis are encountered.

JERZY URBAŃSKI
Warsaw Technical University

FOREWORD TO THE ENGLISH EDITION

ONE of the major problems in science today is that of communication, and in spite of the technical ease of transmission of information, the language barrier remains the rate-determining step in dissemination of knowledge. In this respect translations are particularly valuable when they are the result of the linguistic and editorial efforts of those actively working in the field, who are thus able to bring personal knowledge and experience to bear on questions of usage and terminology, and ensure a high degree of accuracy. Further, translations make available to a wide readership an informed survey, set in context, of a whole range of literature often inaccessible to many readers, except in the form of abstracts. When the book itself is topical and compiled by experts, it deserves as wide reading as possible. We therefore welcome the opportunity to publish the book in this series, and so help to transcend these problems.

R. A. CHALMERS
University of Aberdeen

January, 1977

CONTENTS

Part I

GENERAL METHODS OF ANALYSIS

Chapter 1†

CHEMICAL METHODS

1.1 PREPARATION OF TEST SAMPLE AND SEPARATION OF POLYMER

Sampling of the polymer for analysis is based on general principles so as to obtain an average sample representative of the whole material. Preparation of samples for preliminary tests or of samples of polymers of simple homogenous structure is carried out by reducing the size of the material, for example by cutting with scissors or a knife. In some cases a small fragment of the material is sufficient for qualitative analysis. Usually for quantitative purposes powdering of the material is required. This is essential in the investigation of plastics of complex composition, when the polymers must be separated from other constituents of the material. Sometimes the material can be crushed in a mortar, in other cases it can be pulverised by filing or in a mill. During pulverisation the material should not become hot, otherwise undesirable decomposition reactions may occur.

The methods of separating polymer from other constituents of the plastic material (plasticisers, inorganic fillers, antioxidants, etc.) depend on the composition of the sample being analysed. Details of the procedures can be found in subsequent chapters on methods of testing for individual types of plastics. It is useful to mention here only the general principles of methods for polymer separation.

In the case of polymeric materials containing only the polymer and plasticiser there are three basic procedures: a method using a solvent and a non-solvent for precipitation, extraction in a Soxhlet-type apparatus, and vacuum distillation.

The most commonly used is the first method. It comprises solution of the polymer sample in a solvent followed by precipitation by addition of a non-solvent. The non-solvent should be miscible with the solvent used; it should also dissolve the plasticiser completely. Very often petroleum ether is used as a non-solvent. After removal of the polymer by filtration or centrifugation the separated compounding ingredients can be recovered by evaporation of the solvent.

In the Soxhlet-type extraction method the pulverised sample of the material is extracted by means of a suitable solvent such as ethyl ether or other low boiling liquid. Efforts should be made to select a solvent that will remove only the plasticiser. Solvents which react with the polymer, partially dissolve it or are strongly absorbed, must be avoided.

† References to this chapter are given at the end of each main section.

According to the vacuum distillation method a disc of the test material (about 5 cm in diameter and 4 mm thick) is hung in a heated chamber which is connected to a vacuum line through a glass cooler and a receiver. Relatively low boiling plasticiser is collected in the receiver.

The inorganic fillers or pigments can be separated by dissolving the polymer (with chemical modification such as hydrolysis, if necessary) followed by centrifuging. The polymer is preferably dissolved by extraction. The polymer may be recovered from the extract by precipitation with a non-solvent or by distilling off the solvent under diminished pressure. The best solvent is selected by trial and error or on the basis of some prior knowledge of the composition of the material.

For the quantitative determination of the polymer it is necessary to dry the separated material, preferably under high vacuum, and then to weigh the residue.

The isolated polymer may contain some low molecular weight constituents such as residual catalysts, activators, modifiers, emulsifiers, stabilisers, etc. Many of these impurities can be removed by extraction with an appropriate solvent.

In order to purify a polymer by the fractional precipitation method the following procedure has been suggested [1].

A 1–2% solution of the polymer in a relatively volatile solvent is passed several times through a glass filter and then the polymer is precipitated with an appropriate non-solvent. For polystyrene for instance, benzene or toluene is used as a solvent and isopropanol as non-solvent. The precipitated polymer is separated by decantation, filtration or centrifugation and washed with a hot non-solvent. The polymer is dissolved in another solvent (methyl ethyl ketone in the case of polystyrene) and precipitated with a low boiling non-solvent (methanol in our example). The precipitate is washed again and the procedure of alternate dissolving and precipitating repeated. Finally, the precipitate is slowly dried under vacuum at room temperature. Drying is the most critical step in the whole procedure. It may cause irreversible changes in the polymer such as crosslinking and degradation reactions. Therefore it is essential to use in the last precipitation a very volatile non-solvent which can be quantitatively removed under vacuum at low temperature.

Polymer emulsions must first be broken before their ingredients can be separated. This can be accomplished by coagulation of the polymer with non-solvents or inorganic salts. Occasionally coagulation can be accomplished by cooling below the freezing point of water.

1.2 IDENTIFICATION TESTS

The aim of the identification tests is to determine the type of polymer which is the main constituent of the plastic material analysed. One of the possible ways consists in carrying out systematic tests according to suitable identification schemes. Some such schemes are not completely satisfactory, because the methods usually cover a limited number of plastic materials. In the case of newer plastics and mixtures of plastic materials as well as complex compounding ingredients, misleading results can be obtained from these schemes.

One brief general scheme for identification includes detailed tables of properties of various plastic materials [2]. Another scheme [3], presented in the form of an identification chart, is based on solubility tests, elemental composition and the burning and odour characteristics of polymers.

A fairly complete scheme for identification has been worked out by Shaw [4]. This scheme is based on previous separation of pure polymer by removing compounding ingredients (plasticisers, fillers or solvents) using appropriate means such as the solvent–non-solvent procedure, extraction, etc. This is followed by determination of elemental composition (nitrogen, sulphur, chlorine), saponification number and acetyl number. From this information the material can be classed in one of several groups, and the polymer identified by characteristics such as certain physical properties (density, refractive index, etc.), behaviour on burning, solubility in specific solvents and chemical classification tests.

Methods of separating plasticisers from plastic materials as well as methods for their qualitative and quantitative analysis, including chromatographic and spectrophotometric procedures, have been presented in a monograph [5].

In some cases simpler schemes are sufficient for the identification of polymers. They include such tests as thermal decomposition, behaviour on ignition, colour chemical tests or examination of infrared spectra. The spectrophotometric methods are now widely used for identification of plastic materials [6], along with differential thermal analysis and nuclear magnetic resonance [7].

There are also identification schemes and qualitative tests based on pyrolysis of plastics followed by spectroscopy and gas chromatography [8–10].

Reviews of works published from 1958 on the analysis of plastics and compounding ingredients appear in Kunststoff Rundschau [11] each year. These are very comprehensive and include chemical and physical tests.

Comparative tests carried out using both unknown samples and appropriate known plastics are generally useful. Such tests confirm the identity of the examined material. Comparison with the standard and skill in evaluation of the results of simple tests often permit identification without laborious and time-consuming complete systematic examinations.

Identification by Thermal Decomposition

A simple preliminary identification test consists in observing the burning characteristics of the unknown polymeric material. A small sample placed on a spatula is slowly brought towards the edge of a clear Bunsen flame. Ease of ignition, self-extinguishing characteristics and the odour of any vapours and fumes are noted. The identification of odour is a matter of experience and comparison with a known specimen is very helpful. The use of a micro-burner is preferred. If the flame is too large the decomposition is too fast for accurate observation. Polymers that decompose into aromatic hydrocarbons burn with a yellow, sooty flame. Less soot is formed when aliphatic hydrocarbons are evolved. With increasing oxygen content in the decomposition products the flame turns more blue.

Detailed tables containing a summary of the behaviour of common plastics subjected to the ignition test have been worked out. There is also an identification scheme [12] with a rough qualitative classification of the polymer. The results of the combustion and pyrolysis tests are not always synonymous, and the same sample can be classified in more than one position in the scheme.

The modification of ignition tests has been published [13] and a corresponding scheme worked out [14]. The latter includes also solubility tests in hydrochloric acid for nylon-6,6 and in nitric and hydrochloric acids (1:3) for polyfluorocarbons. Detailed tables are also available that describe the results of heating and burning tests [15]. The scheme below has been restricted to include only the more commonly encountered plastics.

The sample on the spatula should be held at the edge of the Bunsen flame. If it does not ignite instantly it should be held in the flame for a few seconds and withdrawn. The odour should be noted after each withdrawal.

A. The sample fails to ignite, or ignites only with difficulty; it retains its shape or slowly chars, evolving formaldehyde odour in three cases:
    I. No additional odour: urea-formaldehyde resin
    II. Odour resembling that of burnt fish: melamine-formaldehyde resin
    III. Phenol odour: phenol-formaldehyde resin with inorganic filler
    IV. Pungent acidic odour on intense heating: polyfluorocarbon

B. The sample extinguishes itself on removal from the flame (it should be held just at the edge of the flame):
    I. Green zone of the flame is formed:
        1. Odour of burnt rubber and hydrogen chloride, green colour overshadowed by yellow: polychloroprene
        2. Odour acrid (but not that of rubber): poly(vinyl chloride) and copolymers
        3. Odour sweet and acrid, black heavy ash: poly(vinylidene chloride)
    II. Odour of phenol and of burning wood or paper: phenolic resin with organic filler
    III. Odour of burnt protein: casein plastics
    IV. Acetic odour, sparks from the flame: cellulose acetate

C. The sample burns freely after removal from the flame (note the colour of the flame in the first moments of ignition):

I. Burns very rapidly with white flame: cellulose nitrate

II. Burns slowly, black smoke, pungent phenol-like odour: epoxy resin

III. Blue colour with yellow-white tip of the flame:

1. Strong fruity odour: polymethacrylate
2. Odour reminiscent of burning vegetation or fresh celery, sample melts to clear liquid which can be drawn into a fibre: nylon polymers
3. Weak sweetish odour, flame more yellow: poly(vinyl formal)
4. Odour partially of formaldehyde: polyformaldehyde
5. Odour of rancid butter or cheese:
   a. Sparks from the yellowish flame: cellulose acetate-butyrate
   b. No sparks: poly(vinyl butyral)
6. Odour of burning candle, sample melts with drops burning: polyethylene, polypropylene

IV. Flame surrounded by a purple mantle, sparks, odour of acetic acid: poly(vinyl acetal)

V. Flame luminous, yellow:

1. Odour of cyanide and burning wood: polyacrylonitrile and copolymers
2. Odour floral, sweetish, flame dark, smoky: poly(ethylene terephthalate)
3. Sweet floral odour, styrene monomer, flame smoky: polystyrene
4. Odour of burnt paper, flame smoky: regenerated cellulose
5. Odour of burnt rubber, flame smoky: natural rubber and polyisoprene

VI. Flame with yellow-blue mantle:

1. Burns with difficulty, sparks, odour acetic: cellulose acetate
2. Burns readily after ignition, faint sweet odour, the melted plastic darkens when allowed to drop into water: ethylcellulose
3. Sample cracks and splits: polyester resin

Preliminary identification can also be based on a simple test for some easily detectable pyrolysis products [16].

Place a few mg of the sample in the bottom of an ignition tube ca. 10 cm long carrying a cork with a side arm. Heat on a low flame and collect the pyrolysis vapours in another test tube cooled with ice-water. Test the condensate with a moistened indicator paper.

Neutral vapours are evolved from hydrocarbon polymers, silicones and some polyesters. Acid vapours may come from carbohydrate polymers and their derivatives. Highly acidic vapours (hydrochloric acid confirmed by detection of chlorides—precipitate or cloudiness with silver nitrate) indicate the presence of such polymers as poly(vinyl chloride) and its copolymers, poly(vinylidene chloride), natural and synthetic chlorinated rubber or rubber hydrochloride, chloroprene, chlorinated polyethylene, chlorinated polyethers, etc.

The detection of sulphates (precipitate or haze with barium chloride) may indicate polysulphones, polysulphides, factice (rubber substitute), vulcanised natural or synthetic rubber.

Alkaline vapours almost always indicate the presence of such nitrogen-containing plastics as polyamides, polyurethanes, proteins, and amino-formaldehyde resins.

The detection of nitric acid (formation of a precipitate with nitron [17]) confirms the presence of nitrocellulose. Other acids, organic or inorganic, indicate the presence of corresponding cellulose or poly(vinyl esters) as well as some polyesters (e.g. acetic acid from acetates, phthalic anhydride from alkyds) or polyfluorocarbons (volatile fluoro derivatives giving a strongly acidic reaction in the presence of moisture).

Polymers may be identified by depolymerisation or decomposition, using simple dry distillation in a test tube, or heating with alkali, as for example in testing odour on carbonate fusion [4], or heating with strong acid.

Odour on heating with 25% sulphuric acid is a valuable test for the following polymers:

Odour	Plastic material
Acetic acid	poly(vinyl acetate) or cellulose acetate
Butyric acid	cellulose butyrate
Butyraldehyde	poly(vinyl butyral)
Benzaldehyde	benzylcellulose
Formaldehyde	urea resins, phenolics, poly(vinyl formal)
Camphor	cellulose nitrate plastic

Heating with 80% sulphuric acid decomposes such polymers as natural rubber, styrene-butadiene rubber, polychloroprene, acrylonitrile, polysulphides, silicone rubber, and polyester, whereas others, including polyethylene, polyisobutylene, poly(vinyl chloride) and polychlorotrifluoroethylene remain unaffected.

Solubility Tests

Solubility tests form a basis of some older identification schemes for main types of plastic materials [18]. However, in many cases solubility varies considerably for different samples of the same resin and it is difficult to interpret the results. Solubility tests form a part of a comprehensive analytical scheme for identification of plastics [4]. Detailed tables are available [19] which contain solvents and non-solvents for a wide range of plastic materials. They may prove valuable in conjunction with other tests.

In one of the newest identification schemes the separation procedure has been based on the solubility of the plastic material [20]. The procedure starts with the investigation of the solubility of the sample, using five solvents: water, tetrahydrofuran, boiling xylene, dimethylformamide and formic acid. On this basis, the plastics are divided into two groups: soluble and insoluble. Pyrolysis of the sample then leads to three subgroups according to pH value of the pyrolysate: acid, neutral or alkaline. The identification within the subgroups is made by systematic separation procedures of pyrolysates and hydrolysates using thin layer chromatography as well as by simple chemical tests and colour reactions.

Solubility tests are helpful in the selection of a suitable solvent or mixture of solvents for the separation of polymer from other constituents of a plastic material. Of practical importance is the solubility in organic solvents, water and solutions of acids and bases (Table 1.1).

Table 1.1 Solubility of Plastic Materials

Plastic material	*Soluble in*	*Insoluble in*
Alkyd resin	Chlorinated hydrocarbons, lower alcohols, esters	Hydrocarbons
Amino-formaldehyde resin, cured	Benzylamine (160°C), ammonia	
Cellulose, regenerated	Schweizer's reagent	Organic solvents
Cellulose ethers		
Methylcellulose	Water, dil. sodium hydroxide, ethylene chlorohydrin, methylene chloride, methanol	Acetone, ethanol, etc.
Ethylcellulose	Methanol, methylene chloride, formic acid, acetic acid, pyridine	Aliphatic and aromatic hydrocarbons, water
Benzylcellulose	Acetone, ethyl acetate, benzene, butanol	Aliphatic hydrocarbons, lower alcohols, water
Cellulose esters	Ketones, esters	Aliphatic hydrocarbons, water
Cellulose nitrate	Lower alcohols, acetic esters, ketones, ether-alcohol (3:1)	Ether, benzene, chlorinated hydrocarbons
Chlorine-containing polymers		
Chlorinated rubber	Esters, ketones, carbon tetrachloride, linseed oil (at 80–100°C), tetrahydrofuran	Aliphatic hydrocarbons
Chloroprene rubber	Toluene, chlorinated hydrocarbons	Alcohols
Rubber hydrochloride	Ketones	Aliphatic hydrocarbons, cerbon tetrachloride
Chlorinated polyethers	Cyclohexanone	Ethyl acetate, dimethylformamide, toluene
Polychlorotrifluoroethylene	Hot fluorinated solvents (e.g. 2,5-dichloro-α-trifluorotoluene at 130°C)	All common solvents
Poly(vinyl chloride)	Dimethylformamide, tetrahydrofuran, cyclohexanone	Alcohols, butyl acetate, hydrocarbons, dioxan
Chlorinated poly(vinyl chloride)	Methylene chloride, cyclohexane, benzene, tetrachloroethylene	
Poly(vinylidene chloride)	Tetrahydrofuran, ketones, butyl acetate, dimethylformamide (hot), chlorobenzene (hot)	Alcohols, hydrocarbons

Table 1.1 *continued*

Plastic material	*Soluble in*	*Insoluble in*
Copolymers		
Acrylonitrile/butadiene/styrene	Methylene chloride	Alcohols, aliphatic hydrocarbons, water
Styrene/butadiene	Ethyl acetate, benzene, methylene chloride	Alcohols, water
Vinyl chloride/vinyl acetate	Methylene chloride, cyclohexanone, tetrahydrofuran	Alcohols, hydrocarbons
Coumarone/indene resin	Aromatic and chlorinated hydrocarbons, ketones, esters, pyridine, drying oil	Alcohols, water
Epoxy resin		
Intermediates	Alcohols, dioxan, ketones, esters	Hydrocarbons, water
Cured	Practically insoluble	
Fluorine-containing polymers		
Polytetrafluoroethylene	Fluorocarbon oil e.g. $C_{21}F_{44}$ (hot)	All solvents, boiling conc. sulphuric acid
Poly(vinyl fluoride)	Above 110°C: cyclohexanone, propylene carbonate, dimethylsulphoxide, dimethylformamide	
Poly(vinylidene fluoride)	Dimethylsulphoxide, dioxan	
Natural rubber	Chlorinated and aromatic hydrocarbons	Alcohols, acetone, ethyl acetate
Phenolic resin, cured	Benzylamine (200°C), hot alkalis	
Phenolic resin, uncured	Alcohol, ketones	Chlorinated hydrocarbons, aliphatic hydrocarbons
Poly(acrylic acid) derivatives		
Polyacrylamide	Water	Alcohols, esters, hydrocarbons
Polyacrylonitrile	Dimethylformamide, butyrolactone, nitrophenol, mineral acids, dimethylsulphoxide, aqueous solutions of some inorganic salts	Alcohols, esters, ketones, formic acid, hydrocarbons
Poly(acrylic acid esters)	Aromatic hydrocarbons, esters, chlorinated hydrocarbons, acetone, tetrahydrofuran	Aliphatic hydrocarbons
Poly(methacrylic acid esters)	Aromatic hydrocarbons, dioxan, chlorinated hydrocarbons, esters, ketones	Ether, alcohols, aliphatic hydrocarbons
Polyamides	Phenols, formic acid, tetrafluoropropanol, conc. mineral acids	Alcohols, esters, hydrocarbons
Polybutadiene	Aromatic hydrocarbons, cyclohexane, dibutyl ether	Alcohols, esters
Polycarbonates	Chlorinated hydrocarbons, dioxan, cyclohexanone	Alcohols, aliphatic hydrocarbons, water
Polyesters, unsaturated uncured	Ketones, styrene, acrylic esters	Aliphatic hydrocarbons

Table 1.1 *continued*

Plastic material	*Soluble in*	*Insoluble in*
Poly(ethylene terephthalate)	Cresol, conc. sulphuric acid, chlorophenol	
Polyethylene	Dichloroethylene, tetralin, hot hydrocarbons	Polar solvents: alcohols, esters, etc.
Poly(ethylene glycol)	Chlorinated hydrocarbons, alcohols, water	Aliphatic hydrocarbons
Polyformaldehyde	Hot solvents: phenols, benzyl alcohol, dimethylformamide	Alcohols, ketones, aromatic hydrocarbons
Polyisoprene	Benzene	Alcohols, ketones, esters
Polyoxymethylene	Dimethylformamide (150°C), dimethylsulphoxide	Hydrocarbons, alcohols
Polypropylene	At elevated temp.: aromatic and chlorinated hydrocarbons, tetralin	Alcohols, esters, cyclohexanone
Polystyrene	Aromatic and chlorinated hydrocarbons, pyridine, ethyl acetate, methyl ethyl ketone, dioxan, tetralin	Alcohols, water, aliphatic hydrocarbons
Polyurethane	Tetrahydrofuran, pyridine, dimethylformamide, formic acid, dimethylsulphoxide	Ether, alcohols, benzene, water, hydrochloric acid (6*N*)
Poly(vinyl acetal)	Esters, ketones, tetrahydrofuran	Aliphatic hydrocarbons, methanol
Poly(vinyl formal)	Ethylene dichloride, dioxan, glacial acetic acid, phenols	Aliphatic hydrocarbons
Poly(vinyl acetate)	Aromatic and chlorinated hydrocarbons, acetone, methanol, esters	Aliphatic hydrocarbons
Poly(vinyl ethers)		
Poly(vinyl methyl ether)	Methanol, water, benzene	Alkalis, soluble salts, aliphatic hydrocarbons
Poly(vinyl ethyl ether)	Aromatic and chlorinated hydrocarbons, esters, alcohols, ketones	Water
Poly(vinyl butyl ether)	Aliphatic, aromatic and chlorinated hydrocarbons, ketones	Alcohols
Poly(vinyl alcohol)	Formamide, water	Ether, alcohols, aliphatic and aromatic hydrocarbons, esters, ketones
Poly(vinyl carbazole)	Aromatic and chlorinated hydrocarbons, tetrahydrofuran	Ether, alcohols, esters, aliphatic hydrocarbons, ketones, carbon tetrachloride

Preparation of Schweizer's reagent (ammoniacal cupric hydroxide solution). To a solution of 10 g of cupric sulphate in 100 ml of water, add 5 g of sodium hydroxide dissolved in 50 ml of water.
Filter the precipitate, wash with water and place in a beaker.
Then add 20 ml of 25% ammonia and allow to stand overnight.

Many kinds of plastics are insoluble in common solvents. Phenolic, urea-formaldehyde, melamine-formaldehyde and aniline-formaldehyde resins, as well as nylon, polyethylene and polytetrafluoroethylene, are not readily soluble. Phenolic plastics are slowly attacked by molten 1- or 2-naphthol or resorcinol. Nylon is soluble in cresol and 60% hydrochloric acid. With some exceptions, such as poly(acrylic acid), poly(vinyl alcohol), poly(vinyl methyl ether), methylcellulose and starch, polymers are insoluble in water.

Solubility of plastics may vary according to the grade or to whether or not other constituents are present in the sample, and this should not be considered as an absolute criterion for identification. The data in Table 1.1 refer in principle to pure polymers, although even those polymers may exhibit differences in solubility. The ease of solution or swelling is affected by the length of polymer molecules, their arrangement, tacticity and crystallinity. Crystallinity in a linear polymer restricts solubility, but sometimes this can be improved by copolymerisation. The presence of crosslinks prevents solubility and swelling. In the third column of Table 1.1 are mentioned substances which in many cases act as precipitants for individual polymers. This can sometimes be used as a further qualitative indication.

The solubility test should be carried out directly in a test tube. To about 100 mg of a powdered sample add 10 ml of solvent, mix, occasionally shake the contents of the test tube and observe for a few hours. Often swelling may occur before complete dissolution of the polymer.

Qualitative Elemental Analysis

The detection of additional elements, i.e. elements other than carbon, hydrogen and oxygen, is one of the most important preliminary tests for the identification of unknown polymers or polymer compositions. Long and comprehensive tables have been published [21], in which plastics are classified according to the elements present.

For the detection of chlorine an easy and simple Beilstein copper wire test can be employed. It consists in pyrolysis of the sample placed on a copper wire in the flame of a burner. A green colour in the flame indicates the presence of chlorine [22], bromine or iodine.

Heat one end of a stout copper wire, ca. 2 mm in diameter, in an oxidising Bunsen flame until it ceases to impart any colour. While the wire is still hot push it into, or rub it on, the sample to be tested, then return it to the flame. A distinct green or green-blue colour of the flame appears for a time depending on the quantity of halogen. A yellow colour caused sometimes by traces of sodium can mask the green colour. In such cases a cobalt glass should be used.

Fluorine does not behave in the above way. It can be detected by thermal decomposition followed by heating with sulphuric acid. The sample, ca. 0·5 g, is heated in a test tube to complete decomposition of the organic compound. After cooling, a few ml of conc. sulphuric acid is added. A typical uneven wetting of the glass surface by conc. sulphuric acid (the liquid flows down in separate drops) indicates the presence of fluorine.

Elements such as nitrogen, sulphur, phosphorus and halogen may be detected by the sodium or potassium fusion test [22, 23].

About 0·05 g of dry polymer is placed in a small, soft glass test tube. Add a piece of freshly cut sodium or potassium (size of a pea) and heat for a few minutes to almost dull red. Cool and add carefully 1 ml of ethanol to react with the remaining sodium. Heat the test tube slowly to evaporate the ethanol, then heat strongly to a dull red heat. Drop the hot test tube into a small beaker half filled with water. Stir the crushed test tube with a glass rod and filter the solution. Apply the following tests to separate portions of the filtrate.

Sulphur. Acidify ca. 1 ml of the filtrate with acetic acid and add a few drops of lead acetate solution. A black precipitate indicates sulphur. To another portion add 1–2 drops of a fresh solution of sodium nitroprusside. In the presence of sulphur a distinct violet colour is produced.

Chlorine. Boil a portion of the filtrate with dilute nitric acid, add silver nitrate solution. A white precipitate, soluble in ammonium hydroxide indicates chlorine (bromine or iodine yields a yellow precipitate).

Fluorine. Acidify 1 ml of the filtrate with acetic acid. Boil the solution, cool and add 2 drops of saturated calcium chloride solution. A gelatinous precipitate forms on standing if fluorine is present. Another test for fluorine consists in using zirconium-alizarin test paper [22]. A yellow colour on the red paper indicates fluorine.

Nitrogen. Add to the filtrate two drops of freshly prepared saturated ferrous sulphate solution and boil for one minute. In the presence of sulphur filter off the precipitated iron sulphide. Cool the filtrate, add 1 drop of 5% ferric sulphate solution and acidify with dilute hydrochloric acid until the iron hydroxide is just dissolved. In the presence of nitrogen a green-blue colour develops, turning after some time into a Prussian blue precipitate.

Another detection test for nitrogen is based on pyrolytic oxidation in the presence of manganese dioxide. Nitrogen oxides evolved are detected in the gas phase by reaction with the Griess reagent [24]. Mix the powdered sample with about an equal amount of manganese dioxide and place in a micro test tube. Cover the mouth of the test tube with a filter paper moistened with the Griess reagent. Heat the bottom of the tube with a micro flame for about 2 min. A red or pink circle on the paper indicates the presence of nitrogen. Prepare the Griess reagent immediately before use by mixing equal volumes of 1% sulphanilic acid solution in 30% acetic acid and 1% of α-naphthylamine† in 30% acetic acid.

Phosphorus. Acidify a portion of the filtrate with conc. nitric acid and add a few drops of ammonium molybdate solution and heat to boiling. In the presence of phosphorus (casein resin) a yellow precipitate is formed.

† Caution: α-naphthylamine is a potent carcinogen. The 7-sulphonic acid is preferred.

An alternative procedure for detecting small amounts of additional elements depends on combustion of a small sample in oxygen, followed by specific colorimetric reactions for the elements [25].

Identification by Colour Tests

The identification technique by spot tests on a micro or semimicro scale has been introduced by Feigl in the qualitative analysis of some polymers [26]. The amounts of analysed sample are very small, a few mg is sufficient in most cases. Suitable apparatus should be used, such as micro test tubes, micro crucibles, special pipettes for measuring small amounts of liquids and solutions. In spot tests colour chemical reactions are mostly employed. They can be specific for one type of polymer or have a more general selective character giving a positive result for a group of materials of similar type.

In general, plasticisers, fillers and modifying agents do not participate in the reactions used. However, their presence reduces the sensitivity of tests and in many cases they must be separated from the polymer. To avoid misinterpretation of results it is recommended that simultaneous tests with a material of known composition are carried out.

The tests given below are mainly of a selective character. Specific identification tests may be found in the following chapters dealing with individual polymer types.

Liebermann–Storch–Morawski Test

The Liebermann–Storch–Morawski test [4,27] belongs to classical colour reactions and it has a number of modifications used for the identification of polymers.

A small fragment of plastic material is placed on a spot plate and covered with a few drops of acetic anhydride. One drop of conc. sulphuric acid is added and the colours are noted as they appear, both in the liquid and on the surface of the sample. The temperature and concentration of reagents must be stable, otherwise the colours observed may differ for the same polymer type. The results are not always conclusive, so the Liebermann test may be considered as a confirmation of other identification tests. Colours obtained with various polymers are given in a collective work edited by Kline [28].

Detection of Formaldehyde

Formaldehyde obtained from a polymer on pyrolysis, decomposition reaction or distillation with sulphuric acid may be detected in the liquid or vapour phase by the chromotropic acid colour test.

A small fragment of the test material is placed in a test tube, mixed with 1 ml of 0·1% chromotropic acid in 72% sulphuric acid and warmed for 10 min at 60–70°C. A blank test is also carried out. The colour is observed after standing one hour at room temperature. A violet colour indicates the presence of formaldehyde.

Formaldehyde may be released from the following materials: amino- and phenol-formaldehyde resins, polyformaldehyde, poly(vinyl alcohol) (especially from acetalised fibres), poly(vinyl formal), poly(methyl methacrylate) (on pyrolysis only), casein-formaldehyde resin, methoxymethyl polyamides and formal-derived polysulphide rubber. Phenol-formaldehyde modified rosin may give a negative result. Some other polymers also give colour reactions, thus interfering with formaldehyde detection. A red colour may appear in the presence of cellulose nitrate, poly(vinyl acetate), high acetyl cellulose acetate, poly(vinyl butyral), some poly(vinyl acetals) and rosin. Polysulphones develop a violet colour, rosin-modified coumarone polymer gives an orange colour. Detection limit is about 0·15 mg formaldehyde in the sample. Therefore detection of formaldehyde in the gaseous phase is recommended as being more selective [29].

For this purpose a small test tube with a glass stopper is used. A little of the test sample is placed in the bottom of the test tube and 1–2 drops of conc. sulphuric acid are introduced. A knob at the lower end of the stopper is charged with a drop of a freshly prepared solution of chromotropic acid in sulphuric acid and the stopper is put in place. The end of the test tube is immersed in a bath previously heated to 170°C. After 1–10 min, depending on the amount of formaldehyde evolved, a more or less intense violet colour appears in the suspended drop. If the drop is brought onto a white spot plate, even slight colour is easily visible.

Detection of Phenolic Resins

Phenols are released from phenolic resins on dry distillation, on fusion with alkali or heating with soda-lime and sometimes on boiling with water or aqueous alkali. It should be noted that phenols may also result from aryl ester plasticisers, such as tritolyl phosphate.

For the detection of phenols a modified Gibbs indophenol test is recommended [30]. A small fragment of the test material is heated in a micro test tube over an open flame. The mouth of the test tube is covered with a disc of filter paper impregnated with 2,6-dichloroquinone-4-chloroimine (the discs are prepared by immersion of the filter paper in a benzene solution of the reagent and drying). During the heating, the evolving vapours make contact with the reagent for several minutes. The reagent disc is then exposed to ammonia vapour. A blue stain indicates the presence of phenol. The test is based on the condensation of phenols with 2,6-dichloroquinone-4-chloroimine resulting in the formation of coloured indophenols. Free indophenols are yellow to brown, their ammonium salts are blue.

Epoxy resins of the bisphenol type and epoxy resin esters, as well as coumarone-indene resins and some alkyds, also give a positive indophenol test.

In the case of uncured lacquer resins, partially soluble in benzene or butyl acetate, the detection of phenolic resin may be confirmed by a test described by Swann [31].

About 50 mg of the sample is added to a 50 ml cylinder, with ground glass stopper, containing 40 ml of benzene. If the sample is insoluble in benzene, it should be dissolved in 20 ml of butyl acetate and then 20 ml of benzene added. A volume of 10 ml of 3·6*N* sulphuric acid is added followed by 2 drops of freshly prepared 20% sodium nitrate solution. The cylinder is stoppered and the contents are mixed vigorously for at least 10 sec. After the two layers separate, an intense yellow colour in the benzene layer indicates the presence of phenolic resin. Bisphenol epoxy resins and alkyds do not interfere with this test.

Detection of Phthalic Acid

The production of phthalic anhydride, for example as a sublimate from pyrolysis, is characteristic of polyester (alkyd resins). One of the most selective tests for phthalates is the fluorescein test.

A small sample is heated with 2–3 times its amount of resorcinol and a trace of sulphuric acid for 5–10 min at 120°C. After cooling, the residue is extracted with water, then with ethyl ether; the ether extract is washed with water and finally with dilute sodium hydroxide solution. The appearance in the aqueous solution of a pink to red colour and characteristic greenish yellow fluorescence of fluorescein indicates the presence of phthalates [31, 32]. Overheating is to be avoided in this test and it is advisable to run blank controls. The test may also be applied to the sublimate obtained by pyrolysis of the resin.

Phthalate plasticisers also respond to the test [33]. Some other organic compounds give fluorescence with resorcinol. These are: citric acid, tartaric acid, maleic acid and succinamide. With isophthalates and terephthalates similar colours are observed in daylight, but without fluorescence in the light from a quartz lamp.

To confirm the identity of the resin, other tests are advisable, such as the phenolphthalein test described on page 227.

Another confirmatory test for alkyd resins is the detection of glycol or glycerol. The sample fused with potassium hydrogen sulphate yields acetaldehyde (characteristic odour) in the case of glycol esters and acrolein (pungent odour) in the case of resins or other compounds containing glycerol.

Detection of Carboxylic Esters

Carboxylic acid esters, mostly acetates in the case of polymers, react with hydroxylamine to form hydroxamic acids and alcohols. All hydroxamic acids react with ferric chloride to form coloured (usually violet) complex salts [34].

Treat a small piece of resin in a clean test tube with 1 ml of 6% alcoholic sodium hydroxide solution. Add 1 drop of a saturated solution of hydroxylamine hydrochloride. Mix well and allow to stand for 5 min. Heat to boiling for about 30 sec and add without cooling 1 drop of 1% ferric chloride solution. Acidify with dilute hydrochloric acid to dissolve the ferric hydroxide precipitate with only a small excess of acid; too much acid may destroy the colour. A violet or reddish-wine colour indicates carboxylic acid esters. If the colour is too intense dilute the solution with water.

Some esters such as poly(vinyl chloride-acetate) copolymers, poly(methyl acrylate) and methacrylate give a negative test. With poly(vinyl acetals) and alkyds, as well as some amides and imides, the results are positive.

Detection of Styrene

Polystyrene or its monomer forms a mononitrobenzene compound when heated and evaporated with fuming nitric acid. These nitro compounds release phenol on pyrolysis as do other oxygen-containing aromatics. Phenol can be detected by the indophenol colour reaction according to Gibbs, as described above. Copolymers of styrene and of methylstyrene also give a positive reaction.

Phenolic and crosslinked epoxy resins which are oxygen-containing aromatics also give phenol on pyrolysis. However, they are converted into polynitrophenols when treated with fuming nitric acid, yield no phenol and give a negative test.

Styrene may also be detected by nitration to *p*-nitrobenzoic acid, followed by reduction to *p*-aminobenzoic acid. The latter is diazotised and coupled with α-naphthol in alkaline solution, producing a red colour [35].

An acidic extract from aniline-formaldehyde resin will give similar results but without requiring nitration and reduction.

Detection of Epoxy Resins

Both cured epoxy resins and intermediates split off acetaldehyde on heating in a test tube at 240–250°C in a glycerol bath. The acetaldehyde can be detected in the gas phase by the colour reaction with sodium nitroprusside and morpholine [30]. A blue spot develops on a filter paper disc moistened with the aqueous reagent.

On stronger heating in the flame, pyrolysis products evolving from cured resin contain phenol, which can easily be detected by using the Gibbs indophenol reaction as above. This test allows differentiation of uncured resins from epoxy resins crosslinked with aromatic amines.

Detection of Melamine- and Urea-Formaldehyde Resins

Melamine resins can be detected by the presence of melamine chloride and the colour reaction with thiosulphate [31, 36].

Place a small fragment of the sample in a micro test tube and treat with one drop of conc. hydrochloric acid. Heat the mixture gradually to 190–200°C in a glycerol bath until all hydrogen chloride is removed (no colour change of Congo red paper in contact with the vapour). Cool the residue and add 50 mg or so of sodium thiosulphate. Prepare a disc of

Congo red paper moistened with 3% hydrogen peroxide. Cover the mouth of the test tube with the disc and heat at 140°C. In the presence of melamine a blue colour appears on the indicator paper, due to evolution of sulphur dioxide. Non-volatile chlorides of weak bases, such as melamine chloride, behave as weak acids in these conditions.

Urea-formaldehyde resins give a negative test, because urea formed during the decomposition is further hydrolysed to ammonium hydroxide, which does not react with thiosulphate.

The test for detection of urea resin consists in the conversion of urea to diphenylcarbazide on heating with an excess of phenylhydrazine [26].

Treat a few mg of the sample in a micro test tube with one drop of conc. hydrochloric acid and evaporate to dryness at 110°C. Cool, add one drop of phenylhydrazine and heat the mixture to 195°C for 5 min in an oil bath. After cooling stir the mixture with three drops of dil. ammonia (1:1) and five drops of 10% nickel sulphate solution. Shake with 10–12 drops of chloroform. The violet colour of nickel diphenylcarbazide in the chloroform layer indicates the presence of urea resin. The detection limit of the test is 10 μg of urea and 800 μg of thiourea.

Another test for urea resin is based on hydrolysis with dil. sulphuric acid and reaction with xanthydrol [35]. After addition of a few drops of a saturated alcoholic solution of xanthydrol to the hydrolysate, a copious white precipitate of dixanthylurea (m.p. 256°C) slowly forms. Melamine is without effect, except on long standing. Thiourea resins do not interfere with the detection.

Detection of Cellulose and its Derivatives

The following tests distinguish cellulose and its derivatives, and related carbohydrates, from practically all other polymers.

One of the tests consists in thermal decomposition of cellulose and its hydrolysis to a volatile hydroxymethylfurfural [26] which can be detected in the gaseous phase in the same manner as furfural, i.e. by means of the colour test with aniline acetate [37]. To detect cellulose derivatives such as cellulose acetate, methyl- and ethylcellulose, phosphoric acid must be used to hydrolyse these materials to cellulose.

A few mg of the material to be tested are placed in a micro crucible and one drop of syrupy phosphoric acid is added. The mixture is cautiously heated to 40°C. The mouth of the crucible is covered with filter paper moistened with aniline acetate solution (10% solution of aniline in 10% acetic acid). The bottom of the crucible is cautiously heated for 30–60 sec with a micro burner. A pink to red stain indicates the presence of cellulose derivatives. The test allows detection of about 5 μg of methyl-, ethyl- or acetylcellulose.

Cellulose nitrate gives a negative result in the test. It can be detected by means of the reaction with diphenylamine [38].

A small piece of the material is placed on a spot plate and treated with

1–2 drops of diphenylamine solution in ca. 70% sulphuric acid. An intense blue colour appears when cellulose nitrate is present. Coumarone resins, some plasticisers and oxidising agents must be removed before the test is made.

Another colour test for cellulose and its derivatives is the Molisch test [37] based on the reaction of hydrolysed carbohydrates with α-naphthol.

Shake a ca. 5 mg sample with 1 ml of water and 2 drops of 10% solution of α-naphthol in chloroform. Slowly add about 2 ml of conc. sulphuric acid to form a lower layer beneath the aqueous one without mixing with it. In the presence of carbohydrates a violet ring is produced at the interface within a few seconds. On standing or careful mixing a purple solution is obtained which yields a violet precipitate after dilution with water. The precipitate will change to rusty yellowish brown on addition of an excess of ammonia. The test is also positive if the cellulosic materials include starches, gums or glycoproteins. Cellulose nitrate gives a green colour.

Identification of Synthetic Elastomers

Specific tests are described [39] for the identification and differentiation of polymer types in the main group of elastomers. The tests include: the azo-dye reaction for styrene (see page 31), detection of isobutylene by mercuration characteristic of polyisobutylene or butyl rubber (preparation, from the pyrolysate, of methoxyisobutylmercuriacetate and determination of its melting point), the Weber reaction for natural rubber (bromination and colour reaction with phenol), pyridine-caustic soda test for PVC (see page 357), morpholine test for poly(vinylidene chloride) (see page 363), colour tests for polyurethanes from the formation of hydroxamic acid and its reaction with ferric chloride (see page 321) and the reaction with *p*-dimethylaminobenzaldehyde (see page 320), as well as the colour test for chlorosulphonated polyethylene [40] based on the reaction with 2-aminofluorene in pyridine solution.

Another identification method for synthetic elastomers [41] includes colour tests with pyrolysis products, colour spot tests and chemical confirmatory tests. The tables with the specification of colours obtained in these tests refer to the following elastomers: chloroprene, styrene, nitrile and polyisobutylene rubbers, thioplasts, polyvinyl plastics and natural rubber.

References

1. Mark, H., *Anal. Chem.*, **20**, 104 (1948).
2. *Testing and Analysis of Plastics, Part I, The Identification of Plastics*, London, The Plastics Institute, 1955.
3. Gruben, A., Leiner, H. H., *Modern Plastics*, **37**, 88 (1960).
4. Shaw, T. P. G., *Ind. Eng. Chem., Anal. Ed.*, **16**, 541 (1944).
5. Wandel, M., Tengler, H., Ostromow, H., *Die Analyse von Weichmachern*, Springer–Verlag, Heidelberg, 1967.

6. Haslam, J., Willis, H. A., *Identification and Analysis of Plastics*, 1st Ed., Iliffe Books, London, 1965; 2nd Ed., 1972.
7. Ke, B., *Newer Methods of Polymer Characterisation, Polymer Reviews No 6*, Wiley, New York, 1964.
8. Lindley, G., *Lab. Practice (London)*, **14**, 826 (1965).
9. Nelson, D. F., Yee, J. I., Kirk, P. L., *Microchim. J.*, **6**, 225 (1962).
10. Levy, R. L., Lederer, M., *Chromatographic Reviews*, **8**, 48 (1966).
11. Ledwoch, K. D., *Kunstoff Rundschau*, Nos. **11–12**, 1959–1972.
12. Nechamkin, H., *Ind. Eng. Chem., Anal. Ed.* **15**, 40 (1943); *J. Chem. Ed.*, **28**, 97 (1951).
13. Shemilt, H. R., *Electronic Components*, **8**, 1397 (1967).
14. Domsch, G. H., *Kunststoffe*, **61**, 669 (1971).
15. Krause, A., Lange, A., *Kunststoff-Bestimmungsmöglichkeiten*, 2nd Ed., Hanser Verlag, Munich, 1970.
16. Mark, H., Raff R., *High Polymeric Reactions: Their Theory and Practice*, Interscience, New York, 1941, p. 40.
17. Scott, W. W., *Standard Methods of Chemical Analysis*, 5th Ed., Van Nostrand, Princeton, 1950, p. 639.
18. Fischer, E. J., *Laboratoriumsbuch für die organischen Plastischen Massen*, Knapp, Halle, 1938.
19. Krause, A., Lange, A., *Introduction to the Chemical Analysis of Plastics*, Iliffe Books, London, 1969.
20. Braun, D., *Farbe u. Lack*, **76**, 651 (1970).
21. Saunders, K. J., *The Identification of Plastics and Rubbers*, Chapman and Hall, London, 1966.
22. Shriner, R. L., Fuson, R. C., Curtin, D. Y., *The Systematic Identification of Organic Compounds*, Wiley, New York, 1965.
23. Cheronis, N. D., Entrikin J. B., *Semimicro Qualitative Organic Analysis*, Interscience, New York, 1957, p. 173.
24. Feigl, F., Amaral, J. R., *Anal. Chem.*, **39**, 1148 (1958).
25. Haslam, J., Hamilton, J. B., Squirrel, D. C. M., *Analyst*, **86**, 239 (1961).
26. Feigl, F., Anger, V., *Spot Tests in Organic Analysis*, 7th Ed., Elsevier, London, 1966.
27. Nortz, M., Dupuy, P., Rabate, P., *Peintures, Pigments, Vernis*, **30**, 568 (1954).
28. *Analytical Chemistry of Polymers, Part III, Identification Procedures and Chemical Analysis*, Kline, G. M. (Ed.), Interscience, New York, 1962.
29. Feigl, F., Hainberger, L., *Chim. Anal. (Paris)*, **44**, 47 (1955).
30. Feigl, F., Anger, V., *Modern Plastics*, **37**, 151 (1960).
31. Lucchesi, C. A., Tessari, D. J., *Official Digest Fed. Soc. for Paint Tech.*, **34**, 387 (1962).
32. Mattiello, J. J., *Protective and Decorative Coatings, Vol. 5.* Wiley, New York, 1946, p. 105.
33. *Chemical Analysis of Resin-Based Coating Materials*, Kappelmeier, C.P.A. (Ed.), Interscience, New York, 1959, p. 167.
34. Buckles, R. E., Thielen C. J., *Anal. Chem.*, **22**, 676 (1950).
35. Ford, J. E., Roff, W. J., *J. Textile Inst.*, **45**, T 580 (1954).
36. Feigl, F., *Spot Tests in Inorganic Analysis, 5th Ed.*, Elsevier, New York, 1958, p. 307.
37. Struszyński, M., *Jakościowa Analiza Organiczna (Qualitative Organic Analysis)*, PWN, Warsaw, 1960.
38. Roberts, A. G., *Anal. Chem.*, **21**, 813 (1949).
39. Wake, W. C., *The Analysis of Rubber and Rubber-Like Polymers*, Elsevier, London, 1969.
40. Esposito, G. G., *Off. Dig. Fed. Paint Varn. Prod. Clubs*, **32**, 67 (1960).
41. *American Society for Testing Materials. Tentative methods of identification anp quantitative analysis of synthetic elastomers, D833–56T, A.S.T.M. Standards, Pt. 9*, Philadelphia, 1958.

1.3 ELEMENTAL ANALYSIS

Carbon and Hydrogen

In the majority of polymers carbon and hydrogen are determined by conventional methods described in text books and monographs of organic analysis. It is not surprising therefore that few studies of these analyses as applied to polymers are to be found in the pertinent literature. The studies that are available are largely concerned with polymers whose behaviour deviates in some degree from the usual behaviour of organic compounds during the combustion process. Most of the studies deal with determination of carbon or hydrogen in polymeric silicones which are known to undergo incomplete combustion with the formation of highly oxidation-resistant silicon carbide (cf. Chapter 20). Organophosphorus compounds also offer certain problems in combustion analysis, as the ash produced consists of phosphorus oxides and various amounts of carbon. In this connection the residue must be strongly ignited, otherwise the results obtained are too low [1]. Likewise, the combustion of fluorine-containing polymers necessitates the use of special conditions.

According to Haslam and Willis [2], in the combustion analysis of fluorine-containing hydrocarbons and polymers, the lead dioxide at the beak end of the combustion tube should be replaced by granular sodium fluoride to absorb silicon tetrafluoride produced as a result of reaction of hydrogen fluoride with the silica of the combustion tube. The temperature of the sodium fluoride section is maintained at about 180°C with thermostatic control, but sometimes a higher temperature is recommended (270±20°C). For the majority of compounds the usual temperature of 750±20°C in the combustion zone of the tube is sufficient. However, the fluorinated hydrocarbons most resistant to oxidation require a temperature of 950°C. Under such conditions accurate results are also obtained for polymers like Teflon (polytetrafluoroethylene) or polychlorotrifluoroethylene. Silver metavanadate deposited on pumice as the combustion tube packing and a pre-combustion of the material at a temperature of 900°C are recommended by Lebedeva *et al.* [3] for the carbon and hydrogen determination on a micro scale (3–6 mg of material) in fluoroorganic monomers and polymers.

Somewhat different reaction conditions are needed in the determination of carbon and hydrogen in polymers of high hydrogen content such as polyethylene. To avoid water vapour condensation between the tube end and the water absorption vessel, an increased temperature of the lead oxidefilled tube section from 176 to 190°C is suggested by Pickhardt *et al.* [4]. According to Bartels [5], for polymers containing halogens and sulphur in larger quantities, good analysis results are achieved with ceric oxide packing in the combustion tube.

Finally, special conditions must be realised in the determination of the composition of copolymers when the parent monomers differ only slightly in their carbon to hydrogen ratio. Here a very high accuracy and precision is required. This aim was attained by Wood *et al.* [6] in the analysis of butadiene-styrene copolymers by using rather large samples (0·5–2·0 g). The materials tested were burnt in Pyrex test tubes plugged with platinum wire gauze. According to these authors, if a test tube is used in place of a boat, the combustion rate may be more easily controlled. This is of particular significance when the styrene content is small.

Polymers can also be analysed by a rapid "empty tube" technique of combustion analysis [7]. If the material under test contains other elements besides carbon, hydrogen and oxygen, their combustion products are absorbed by suitable absorbants placed in the combustion boat or the tube (e.g. sulphur and halogens are fixed by silver gauze).

Nitrogen

Nitrogen in polymers is usually determined by the Kjeldahl and Dumas methods on the semimicro and micro scale.

The Kjeldahl Method

The Kjeldahl method can be used for almost all polymers [1]. It will be recalled here that the method consists in heating a material under test in sulphuric acid, usually in the presence of mercury or copper salts as catalysts. Under such conditions organic materials are decomposed and oxidised, and the nitrogen turns into ammonium ion in the form of ammonium hydrogen sulphate. By treatment with concentrated alkalis ammonia is liberated from the salt and is distilled off with steam and absorbed in a standard acid solution.

In a method developed for butadiene-acrylonitrile copolymers, recommended also for other polymers, Haslam and Willis [2] use copper sulphate (0·3 g per 0·5 g of the polymer) as a catalyst for the decomposition reaction with dilute sulphuric acid (10 ml of water per 20 ml of conc. sulphuric acid). In another procedure designed for rapid determination of nitrogen in the same copolymers the authors recommend the same catalyst. The ammonia is not, however, isolated from the reaction mixture by steam-distillation but is determined directly in the mixture by a method based on a hypobromite oxidation reaction:

$$3\,NaOBr + 2\,NH_3 = N_2 + 3\,NaBr + 3\,H_2O$$

Unreacted sodium hypobromite is determined iodometrically

$$NaOBr + 2\,KI + H_2SO_4 = NaBr + I_2 + K_2SO_4 + H_2O$$

REAGENTS

Sulphuric Acid-Catalyst Mixture. One part of water (by vol.) is treated with 1 part (by vol.) of conc. sulphuric acid. After cooling of the mixture

8 g of sodium sulphate and 1 g of crystalline copper sulphate are added per 100 ml of the acid, and the whole is heated to dissolve the solids.

Neutralising Solution. Boric acid (30·9 g) is dissolved in 1 l. of 4*N* sodium hydroxide.

Solution A. Potassium bromide (8 g) is dissolved in water, 1 ml of bromine is added and the solution is diluted with water to a volume of 400 ml.

Solution B. Crystalline boric acid (24·7 g) is dissolved in 100 ml of 4*N* sodium hydroxide and the resulting solution is diluted with water to a volume of 1 l.

Hypobromite Solution. Two volumes of solution A are mixed with 5 volumes of solution B.

Sodium Sulphate. A 20% aqueous solution.

ANALYTICAL PROCEDURE. About 0·05 g of the polymer is heated in a 100 ml Kjeldahl flask with 10 ml of the acid–catalyst mixture and 10 ml of 20% aqueous sodium sulphate until sulphuric acid fumes appear and the mixture assumes a clear blue-green colour.

The solution is then heated for another 10 min, cooled and 10 ml of water is added. The mixture is reheated to boiling. After cooling, the contents of the flask are transferred to a 100 ml measuring flask and the volume is made up to the mark with water.

Twenty five ml of the solution thus obtained is transferred with a pipette to a bottle with a tight stopper and the neutralising solution is added until the appearance of a blue precipitate, then 0·1 ml more of the neutralising solution is added. On cooling, 25 ml of sodium hypobromite solution is added, and the contents of the bottle are swirled and allowed to stand for 2 min. Subsequently 10 drops of 50% potassium iodide solution and 3·0 ml of 3*M* sulphuric acid are added. The liberated iodine is titrated with a 0·01*N* sodium thiosulphate solution in the presence of starch.

A blank determination is run separately.

Polymers containing nitrogen in a ring decompose less readily than other nitrogen-containing polymers. In the case of the former polymers, Cole and Parks [8] suggest the use as catalyst of a mixture of anhydrous potassium sulphate with selenium and mercuric oxide, recommending longer heating. These authors have determined in this way nitrogen in butadiene– or chloroprene–2-vinyl-pyridine copolymers or in butadiene–acrylonitrile copolymers. A catalyst of similar composition was also used by Haslam and Squirrell [9]. A catalyst composed of potassium sulphate and selenium is used by Lebedeva and Novozhilova [10] in their method of nitrogen determination in low molecular weight compounds. The determination itself is carried out directly in the reaction mixture by a method involving a colour reaction with the Nessler reagent.

In determining nitrogen in a variety of polymers Škoda and Schurz [11] used copper and mercury salts as catalysts. The determination was carried out by heating a sample of 0·3–0·4 g of polymer for 1 hr with 40 ml of conc. sulphuric acid, 1 g of crystalline copper sulphate, 0·7 g of mercuric oxide, 0·5–0·7 g of mercury and 9 g of sodium sulphate. The reaction mixture was neutralised with 40% sodium hydroxide solution containing 7 g of sodium thiosulphate. The usual procedure was followed thereafter.

Compounds containing nitro, nitroso or azo groups should be qantitatively reduced to amines prior to determination by the method described above.

THE DUMAS METHOD

Determination of nitrogen by the Dumas method is carried out by pyrolysis in a stream of carbon dioxide in a combustion tube. Any nitrogen oxides formed in the oxidation of the compound are reduced to free nitrogen by hot copper, and the nitrogen is displaced from the combustion tube in a carbon dioxide stream and collected over potassium hydroxide solution in a nitrometer.

According to the original Dumas procedure, the sample is oxidised with cupric oxide powder, and the combustion tube is connected to the nitrometer via a stopcock. As further modified by Zimmermann [12] this method involves combustion of the sample in a tube equipped with stopcocks at both ends. These prevent atmospheric gases from entering the tube and giving too high results. One step further from the original Dumas method is the combustion of the sample in a stream of carbon dioxide and oxygen as in the determination of carbon and hydrogen, followed by removal of the excess of oxygen. This technique is reported by Ingram [13] and Alford [14], and has the advantage of being faster.

Nitrogen-containing polymers, which are decomposed with difficulty, yield too low results whether analysed by the Kjeldhal or the Dumas method. For polymers of this kind accurate results were obtained with the use of vanadium pentoxide as oxidising agent and nickel oxide NiO as packing of the tube. In this way the combustion temperature could be raised to 1000°C (cf. Pippel and Römer [15]).

The polymer combustion products may also be determined by gas chromatography. Using an analyser of special design Gautheret *et al.* [16] determined in this way minute quantities of nitrogen (0·05–0·5%) in polyethylene.

Chlorine

The problem of determination of chlorine in polymers has been given much more attention than the determination of any other element. This point is evidenced not only by the considerable number of published papers but also by the remarkable quantity of existing techniques for determination of this element. These are: wet combustion (the Carius method and the related techniques), dry combustion (in the tube and by the Schöniger method), fusion (with or without added oxidising agents), splitting in solution with alkalis and metal alkoxides, and pyrolysis.

Finally chlorine may be determined in polymers like PVC by X-ray excitation [17], that is, without the necessity of previous chemical decomposition of samples under analysis.

Wet Combustion

The oldest known method of this type is the Carius method which consists in oxidising a sample in a sealed tube with fuming nitric acid in the presence of silver nitrate at 280–300°C. The chlorine present in the polymer precipitates in the form of insoluble silver chloride which can be determined gravimetrically or titrimetrically.

Several methods are known in which the hydrogen halides and free halogens produced in the decomposition of the compound analysed in the absence of silver nitrate are removed from the reaction mixture with a gas stream (oxygen or carbon dioxide), and determined by absorption in a suitable absorbing solution.

Specifically, by the technique reported by Hofmeier and Schröder [18], the material analysed is decomposed with a conc. sulphuric acid-potassium nitrate mixture. The hydrogen chloride formed is removed with a carbon dioxide stream, absorbed in 0·05*N* silver nitrate solution, and the chlorine is determined gravimetrically in the form of silver chloride.

According to the ASTM D 1156–52 method reported in the Kline monograph [1], fuming sulphuric acid (20% SO_3) with potassium persulphate is used as oxidant for the analysed material s (vinyl chloride polymers and copolymers). Evolving chlorine is absorbed in alkaline sodium arsenite, where it undergoes reduction to chloride ions.

By the wet combustion method, followed by separation of chlorine from the reaction mixture, this element was also determined in polymers by Rao and Shah [19] (in the presence of nitrogen and sulphur), and also by Bather [20].

Analysis by Combustion in the Combustion Tube

Polymers are analysed by combustion in the combustion tube in an oxygen stream just like common organic compounds.

According to the method developed by Bogina and Martyukhina [21] total chlorine and bromine in rubbers and vulcanisates is determined by combustion with an addition of vanadium pentoxide. The process is run at a temperature of 1100°C. The reaction products are absorbed in a sodium hydroxide solution containing hydrogen peroxide. Free halogens undergo reduction to halide ions, e.g.:

$$Cl_2 + 2OH^- + H_2O_2 = 2Cl^- + 2H_2O + O_2$$

Halide ions are directly determined in the absorbing solution by titration with 0·05*N* mercuric nitrate solution with diphenylcarbazone as indicator. For the determination of trace quantities of chlorine in monomers and polymers Gurvich *et al.* [22] employed potentiometric titration.

COMBUSTION ANALYSIS BY THE SCHÖNIGER METHOD

Determination of chlorine in polymers by this method, which involves combustion of a sample in an atmosphere of oxygen in a sealed flask, is carried out in a manner similar to that used with common organic compounds.

According to Haslam *et al.* [2, 23, 24] the combustion products are absorbed in a sodium hydroxide solution containing sodium sulphite. The chloride ions in the solution thus obtained are determined by potentiometric titration with 0·01*N* silver nitrate solution [2].

Chlorine and bromine when simultaneously present were determined jointly in polymers by coulometric titration of a solution of the combustion products in dilute potassium hydroxide by Cassani [25]. Truscott [26] determined chlorine in PVC by examining a solution of the combustion products with the help of atomic absorption spectrometry.

In determining chlorine present in large amounts (e.g. in hexachlorobutadiene) Lebedev *et al.* [27] advise an addition of 20–30 mg of EDTA to the analysed sample of 40–50 mg.

A procedure for determination of very small quantities of chlorine, sulphur, or phosphorus in large samples (up to 8 g) worked out by Farley and Winkler [28] consists in combustion of the sample in an oxygen stream at a flow rate of 45 l. per min in a 2 litre vessel. The reaction products are taken up in a cylindrical absorber with a sintered glass bottom. The determination is finished in the usual manner.

The Schöniger method is of the same accuracy as the combustion analysis in the combustion tube.

FUSION

A substantial number of procedures are available for determination of chlorine in polymers by fusion. The most accurate results are obtained in the method involving the Parr bomb. Described below is a semimicro technique developed by Haslam and Hall [2,29].

The Reference Electrode Solution. To 64 ml of a saturated sodium sulphate solution are added 6·0 ml of 2*N* nitric acid, 30 ml of 0·1*N* sodium chloride solution and 30·1 ml of 0·1*N* silver nitrate solution. The mixture is allowed to stand until the silver chloride precipitate settles to the bottom of the vessel. The supernatant solution is used to fill the reference electrode.

Electrodes. Silver wire electrode as the indicator electrode. As a reference electrode a silver wire dipped in the reference electrode solution filling a small glass tube provided with an opening.

PROCEDURE. A 20 mg sample of a polymer is weighed into the cup of a semimicro Parr bomb, and 0·06 g of dry starch and 1·0 g of sodium peroxide are added. The contents of the cup are thoroughly mixed with a thin glass rod, and the bomb is well sealed with a screw-on lid. Next, the electric current is turned on or the cup is heated from the bottom with a micro-burner flame 10 mm high for 50 sec. After the fusion the bomb is disassembled by removing the lid, then, along with the cup, it is immersed in 10 ml of distilled water contained in a 50 ml beaker. The beaker is heated until the melt is completely dissolved, then the lid and the cup are removed from the solution and rinsed into the beaker thoroughly with as little distilled water as possible. The liquid is heated to boiling for 15 min to decompose the peroxides, then it is cooled, made acidic with conc. nitric acid against methyl red indicator (0·05% alcoholic solution), cooled again, a few more drops of the indicator are added and the resulting solution is made alkaline with 10% sodium hydroxide solution. The solution is neutralised with 2*N* nitric acid and an excess of 0·25 ml of this acid is added. The total volume of the liquid should amount to about 30 ml.

The electrodes are now dipped in the solution and the chloride ions present are titrated with 0·02*N* silver nitrate solution. The silver nitrate solution is standardised against 20 mg of sodium chloride (which has been previously dried at 270±10°C for 4–5 hr and subsequently fused with 1·0 g of sodium peroxide and 0·06 g of starch). The titration is carried out following exactly the same procedure as with the unknown sample.

In routine analyses in which high accuracy is not required, the sample may be fused in an open crucible or even in a test tube. Thus Stoeckhert [30] and Quenum *et al.* [31] fused the polymer in an open steel crucible with a mixture composed of 1 part by wt. of anhydrous sodium carbonate and 2 or 3 parts by wt. of sodium peroxide. Ubaldini and Capizzi-Maitan [32] fused a polymer sample with a mixture of 2 parts by wt. of magnesia and 1 part by wt. of sodium carbonate. Khoroshaya *et al.* [33] fused the polymer (PVC) in a test tube with a mixture of sodium and potassium carbonates and magnesia (1:1:1), whereas Kojima and Ueno [34] determined chlorine and fluorine in fluorocarbon polymers by fusion with potassium carbonate in a porcelain crucible. In the Piria–Schiff method the polymers are fused with calcium oxide [7].

Pyrolytic Method

Poly(vinyl chloride) and related polymers decompose on heating and release hydrogen chloride. At elevated temperatures the reaction is almost quantitative. Advantage was taken of this effect for quantitative determination of chlorine in these polymers.

According to Parlashkevich *et al.* [35] poly(vinyl chloride) is heated in a nitrogen stream in a glass flask, and the hydrogen chloride produced is taken up in water. The chloride ion content in the absorbing solution is determined conductometrically.

Minor quantities of chlorine in polypropylene were determined by Maltese *et al.* [36], who pyrolysed the polymer at a temperature below 350°, with subsequent combustion of the pyrolysate in oxygen in an apparatus of special design. After absorption in sodium carbonate solution, the combustion products were analysed by potentiometric titration or spectrophotometrically.

Pyrolysis gas chromatography was used for the analysis of chlorine in polymers by Tsuge *et al.* [37]. The chlorine content was calculated from the relative percentages of the characteristic pyrolysis products.

Cleavage of Chlorine with Chemical Agents

Petukhov and Guseva [38] determined the chlorine content in PVC and perchlorovinyl resins by the Stepanov method, in which a polymer sample dissolved in a non-aqueous solvent is treated with alkoxides. A quantity of 20 mg of polymer is dissolved in 20 ml of dioxan (or tetrahydrofuran), then 15 ml of the monomethyl or monoethyl ether of ethylene glycol is added. The mixture is heated again to boiling, and 1·3–1·8 g of sodium is added in small portions over a period of 40 min. The chloride ions are determined in the reaction mixture, on acidification with nitric acid, by titration with 0·1*N* silver nitrate by the Volhard method.

Nadalin and Theodore [39] determined chlorine in polypropylene after removal of this element from the polymer by refluxing in 0·1*N* methanolic sodium hydroxide.

Halogens present in very low quantities in copolymers of styrene, methyl methacrylate or vinyl acetate with vinyl chloride or with tetrachloroethylene were determined by Kumarsaha *et al.* [40] by a technique which involves converting the halogens into the corresponding quaternary ammonium (pyridinium) salts. The salts were subsequently determined colorimetrically as a colour complex with disulphine blue.

Fluorine

The exceptional position of fluorine in the halogen group is reflected in the behaviour of fluorine-containing polymers during their elemental analysis. Satisfactory analyses are achieved only when very vigorous conditions for the decomposition of the material are adopted, specifically by such methods as fusion in the Parr bomb, fusion in an oxygen-hydrogen flame, and decomposition with potassium. This is undoubtedly due to the considerably higher stability of the C-F compared with the other carbon-halogen bonds.

The fluoride ions are usually determined in solutions of the pyrolysate of fluoropolymers by titration with thorium nitrate, but not until the interfering compounds that are formed during combustion or reduction of the material under analysis are removed. The removal is accomplished either by precipitation with suitable reagents or with the aid of ion-exchange resins [41]. Alternatively the interfering compounds may be got rid of by stripping off fluorine with steam as fluorosilicic acid.

Combustion in a Combustion Tube

In general the method yields low results for materials of high fluorine content. Freier *et al.* [42] succeeded in obtaining fairly accurate results for low molecular weight compounds only by combustion at 1150°C in an atmosphere of moist oxygen, but in the analysis of polymers this method is of no major significance. For analysis of polymers of high fluorine content Selig [43] recommends the combustion process to be run in a mixture with excess paraffin.

Fusion with Alkaline Reagents

Accurate results in analysis of fluoropolymers cannot be achieved by fusion with such alkaline compounds as sodium carbonate, calcium oxide, or calcium hydroxide. On the other hand, according to Kojima and Ueno [34] good results in the analysis of fluorocarbon polymers are provided by fusion with potassium carbonate at 500°C.

Fusion with Sodium Peroxide

This method gives excellent results in the fluorine analysis of polytetrafluoroethylene, as reported by Haslam and Whettem [44] (76·07 and 75·98% F against the theoretical value of 76·0%). In the opinion of these authors the method may well be used for the analysis of other fluoropolymers. Occasionally, however, this method also fails to provide accurate results [2].

PROCEDURE. Thoroughly dried starch (0·7 g) is weighed into the bottom of the metal bomb cup, 15 g of sodium peroxide is added, and the contents are intimately mixed. About 2/3 of the contents are poured out into a dry 50 ml beaker, and a polymer sample (0·12–0·14 g) is weighed out exactly into the cup. The beaker contents are added back to the cup and the whole is thoroughly mixed. The lid is replaced and the charge is electrically fired. After 5 min the bomb is allowed to cool, then it is opened and the melt is dissolved in water as described previously (see p. 41). The solution is freed from the peroxides by boiling for 10 min, transferred to a 100 ml volumetric flask, and the volume made up to the mark with water. A 50 ml aliquot of the solution is placed in a 250 ml Claisen flask (Fig. 1.1), which is connected to a steam generator and to a reflux condenser. Into the receiver (a 500 ml beaker) 50 ml of water is added, and 50 ml of 60% perchloric acid is added to the flask.

The steam distillation is conducted with the temperature of the contents maintained in the range 130–140°C until 250 ml of the distillate is collected. The contents of the receiver are first neutralised with 1*N* sodium hydroxide solution to pH 6·0, then made acidic with 0·1*N* hydrochloric acid to pH 3·3. The solution is titrated with 0·05*N* thorium nitrate with 1 ml of 0·1% aqueous alizarine red S added as indicator. During the titration the pH of the titrated solution is maintained at 3·0 by adding 0·1*N* sodium hydroxide. Titration is carried on until the colour of the titrated solution is the same as that of the comparison colour solution. The comparison colour solution is prepared by adding 15 ml of 1% cobaltous sulphate and 1·0 ml of 0·1% potassium dichromate solutions to 150 ml of water. After this, a standard sodium fluoride solution is added in a quantity sufficient

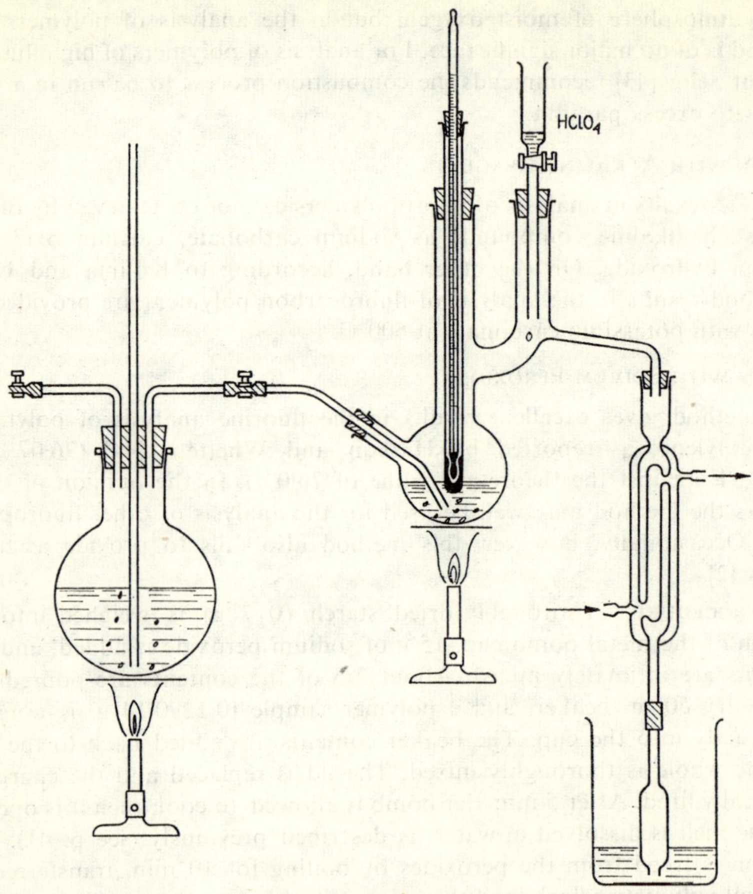

Fig. 1.1 Steam-distillation apparatus; determination of fluorine in polytetrafluoroethylene (by permission of the copyright holders, ref. [44])

to give the expected quantity of fluorine, and then combined with an equivalent quantity of thorium nitrate. The whole is diluted with water to 300 ml.

The thorium nitrate solution is standardised against recrystallised sodium fluoride under identical conditions as in the analysis itself (including the distillation).

SODIUM FUSION [45]

A 0·15 g sample is decomposed by fusion with sodium at 300°C for 90 min in a nickel crucible equipped with a screw-on lid. The resulting melt is dissolved in water and the aqueous solution is passed through a cation-exchange bed. In the eluate, hydrofluoric acid is determined by titration with 0·1*N* sodium hydroxide solution against a mixed indicator (ethanolic solution of 0·125 g of methyl red and 0·08 g of methylene blue in 100 ml of the alcohol). In the same solution chloride may be determined by the Volhard method.

COMBUSTION IN AN OXYGEN-HYDROGEN FLAME

In this technique [46] the partial combustion products of the unknown are passed through an oxygen-hydrogen flame where they undergo complete decomposition to carbon dioxide and hydrogen fluoride. These products are absorbed in dilute sodium hydroxide, and in the resultant solution fluoride ions are determined by the thorium nitrate titration as described above.

According to Haslam and Willis [2] the method yields slightly low results (by 0·5% in case of polytetrafluoroethylene). The method is unsuited to determination of fluorine in compounds containing inorganic material, as here inorganic fluorides may be produced and remain in the ash. Organic sulphur compounds interfere with the determination so that the results are too high.

Sulphur

Determination of sulphur in polymers offers no special problems, and can be achieved by the usual routine techniques used in analysis of common organic compounds. These techniques include the Schöniger method, combustion in a combustion tube, fusion with sodium peroxide in the Parr bomb, or with alkaline compounds in an open crucible, and oxidation with nitric acid in a sealed tube (the Carius method).

In the oxidation products the sulphate ions produced are determined by a variety of procedures depending on the oxidation process followed and on the elemental composition of the compound under analysis.

The sulphuric acid produced in the combustion tube process, and also in the analysis by the Schöniger method, can be determined in the absence of nitrogen, halogens, or phosphorus by acidimetric titration. When these acid-forming elements are present in the sample, sulphur is best determined as barium sulphate. Sulphuric acid may also be precipitated with 4-amino-4′-chlorobiphenyl, the precipitate filtered off, and the sulphuric acid content in the precipitate determined by acidimetric titration [47]. For other polymer oxidation procedures sulphate ions are most commonly determined as barium sulphate (volumetrically or gravimetrically), and also as benzidine sulphate [48].

The micro Schöniger determination of sulphur in polymers is treated in considerable detail in the literature. Minor quantities of sulphur in poly(methyl methacrylate) were determined by Haslam and Squirrell [49] by combustion of a sample in a 2 litre flask. The combustion products were taken up in an alkaline hydrogen peroxide solution, and sulphate determined turbidimetrically as barium sulphate.

Very detailed studies on the determination of sulphur in fibre-forming polymers by the Schöniger combustion technique were made by Majewska [50]. The sulphate ions were determined in the combustion products by precipitation titration both visually and conductometrically, and, after conversion to the sulphide form, colorimetrically on the basis of the formation of methylene blue. The latter procedure proved to be best for the analysis of trace quantities of sulphur.

Trace quantities of sulphur in monomers and polymers by the combustion tube technique at 900°C were determined by Gurvich *et al.* [22]. The combustion products were absorbed in alkaline hydrogen peroxide solution, and the sulphate ions produced determined by conductometric titration with 0·01*N* standard barium nitrate solution.

The sodium peroxide oxidation in a Parr micro bomb is reported by Lincoln *et al.* [51]. Cheney [52] uses a mixture of nitric acid and zinc oxide, and also bromine or potassium chlorate for oxidation of the polymer, while Recchia and Carraroli [48] in the analysis of synthetic rubbers employ a potassium hydroxide-potassium nitrate mixture for this purpose.

Minor quantities of sulphur and phosphorus in polypropylene are determined according to Takeuchi *et al.* [53] by ashing a sample with an addition of 10% by wt. of sodium carbonate.

Heslop and Tay [54] analysed combined or impurity sulphur in fluoropolymers by neutron activation.

Phosphorus

Phosphorus in polymers is usually determined by such methods as wet ashing with conc. sulphuric acid or a sulphuric acid-nitric acid mixture (in a Kjeldahl flask or in a conical flask under reflux), by fusion with sodium peroxide in the Parr micro bomb, or by the Schöniger method [1].

The phosphates formed are determined gravimetrically as ammonium molybdophosphate or indirectly by titration with EDTA after precipitation of magnesium ammonium phosphate. Minor quantities of phosphates after wet ashing with conc. sulphuric acid or decomposition by the Schöniger procedure can be determined colorimetrically through the formation of phosphomolybdenum blue [55,56].

Recently Malyukova and Zaitseva [57] made a critical estimation of the available procedures for the determination of phosphorus in low- and high-molecular weight organophosphorus compounds and concluded that the sodium peroxide fusion is the best.

References

1. Brauer, G. M., Horowitz, E., in *High Polymers, Vol. XII, Analytical Chemistry of Polymers*, Kline, G. M. (Ed.), *Part III, Identification Procedures and Chemical Analysis*, Interscience, New York, 1962.
2. Haslam, J., Willis, H. A., *Identification and Analysis of Plastics*, 1st Ed., Iliffe Books, London, 1965; 2nd Ed., 1972.
3. Lebedeva, A. I., Nikolaeva, N. A., Orestova, V. A., *Izv. Akad. Nauk SSSR, Otd. Khim. Nauk*, 1961, 1350.

4. Pickhardt, W. P., Safranski, L. W., Mitchell, J., Jr., *Anal. Chem.*, **30**, 1298 (1958).
5. Bartels, U., *Chem. Tech.*, **19**, 696 (1967).
6. Wood, L. A., Madorsky, I., Paulson, R. A., *J. Res. Nat. Bur. Stand.*, **64A**, 157 (1960).
7. Kasterina, G. N., Kalinina, L. S., *Khimicheskie Metody Issledovaniya Sinteticheskikh Smol i Plasticheskikh Mass* (*Chemical Methods of Analysis of Synthetic Polymers and Plastics*), Goskhimizdat, Moscow, 1963.
8. Cole, J. O., Parks, C. R., *Ind. Eng. Chem., Anal. Ed.*, **18**, 61 (1946).
9. Haslam, J., Squirrell, D. C. M., *Analyst*, **82**, 511 (1957).
10. Lebedeva, A. I., Novozhilova, I. V., *Zh. Analit. Khim.*, **22**, 914 (1967).
11. Škoda, W., Schurz, J., *Z. Anal. Chem.*, **162**, 259 (1958).
12. Zimmermann, W., *Mikrochemie ver. Mikrochim. Acta*, **31**, 42 (1943).
13. Ingram, G., *Mikrochim. Acta*, **1953**, 131.
14. Alford, W. C., *Anal. Chem.*, **24**, 881 (1952).
15. Pippel, G., Römer, S., *Chem. Techn.*, **15**, 173 (1963).
16. Gautheret, J. M., Kubiak, S., Nicco, A., *Chim. Anal.* (*Paris*), **52**, 283 (1970).
17. Leuteritz, F., Brunner, G., *Plaste u. Kautschuk*, **14**, 887 (1967).
18. Hofmeier, H., Schröder, W., *Kunststoffe*, **34**, 104 (1944).
19. Rao, D. S., Shah, G. D., *Mikrochemie ver. Mikrochim. Acta*, **40**, 254 (1953).
20. Bather, J. M., *Analyst*, **81**, 536 (1956).
21. Bogina, L. L., Martyukhina, S. P., *Kauch. Rezina*, **18**, 55 (1959).
22. Gurvich, D. B., Balandina, V. A., Paikina, L. M., Yur'evsakaya, I. M., Noskova, M. P., *Zavodsk. Lab.*, **32**, 409 (1966).
23. Haslam, J., Hamilton, J. B., Squirrell, D. C. M., *Analyst*, **85**, 556 (1960).
24. Haslam, J., Hamilton, J. B., Squirrell, D. C. M., *J. Appl. Chem.*, **10**, 97 (1960).
25. Cassani, F., *Chim. Ind.* (*Milan*), **51**, 167 (1969).
26. Truscott, E. D., *Anal. Chem.*, **42**, 1657 (1970).
27. Lebedev, D. D., Korobkina, T. V., Vereshchinskii, I. V., *Zavodsk. Lab.*, **32**, 530 (1966).
28. Farley, L. L., Winkler, R. A., *Anal. Chem.*, **35**, 772 (1963).
29. Haslam, J., Hall, J. I., *Analyst*, **83**, 196 (1958).
30. Stoeckhert, K., *Kunststoffe*, **37**, 53 (1947).
31. Quenum, B. M., Grandaud, J. L., Berticat, P., Vallet, G., *Chim. Anal.* (*Paris*), **53**, 629 (1971).
32. Ubaldini, I., Capizzi-Maitan, F., *Chim. Ind.* (*Milan*), **37**, 779 (1955).
33. Khoroshaya, E. S., Kovrigina, G. I., Eliseeva, L. I., *Zavodsk. Lab.*, **30**, 1450 (1964).
34. Kojima, R., Ueno, H., *J. Kogyo Kagaku Zasshi*, **61**, 270 (1958).
35. Parlashkevich, N. Ya., Gribkova, R. N., Bil'dina, V. P., Ivanovskaya, T. S., *Plast. Massy*, **1961**, [6], 55.
36. Maltese, P., Clementini, L., Mori, A., *Chim. Ind.* (*Milan*), **49**, 1070 (1967).
37. Tsuge, S., Okumoto, T., Takeuchi, T., *Bull. Chem. Soc. Japan*, **42**, 2870 (1969).
38. Petukhov, G. G., Guseva, T. V., *Zavodsk. Lab.*, **29**, 806 (1963).
39. Nadalin, R. J., Theodore, M. L., *Anal. Chim. Acta*, **53**, 135 (1971).
40. Kumarsaha, M., Ghosh, P., Palit, S. R., *J. Polymer Sci., Pt. A*, **2**, 1365 (1964).
41. Gurvich, D. B., Balandina, V. A., *Plast. Massy*, **1962**, [6], 54.
42. Freier, H. E., Nippoldt, B. W., Olson, P. B., Weiblen, D. G., *Anal. Chem.*, **27**, 146 (1955).
43. Selig, W., *Z. Anal. Chem.*, **234**, 261 (1968).
44. Haslam, J., Whettem, S. M. A., *J. Appl. Chem.*, **2**, 339 (1952).
45. Schröder, E., Waurick, U., *Plaste u. Kautschuk*, **7**, 9 (1960).
46. Sweetser, P. B., *Anal. Chem.*, **28**, 1766 (1956).
47. Belcher, R., Nutten, A. J., Stephen, W. I., *Mikrochim. Acta*, **1953**, 51.
48. Recchia, S., Carraroli, D., *Chim. Ind.* (*Milan*), **24**, 203 (1942).
49. Haslam, J., Squirrell, D. C. M., *J. Appl. Chem.*, **11**, 244 (1961).
50. Majewska, J., *Chem. Anal.* (*Warsaw*), **13**, 29 (1968).

51. Lincoln, R. M., Carney, A. S., Wagner, E. C., *Ind. Eng. Chem., Anal. Ed.*, **13**, 358 (1941).
52. Cheney, L. V. E., *Ind. Eng. Chem., Anal. Ed.*, **15**, 164 (1943).
53. Takeuchi, T., Suzuki, M., Yamamoto, K., *Japan. Analyst*, **12**, 752 (1963).
54. Heslop, R. B., Tay, S. K., *Anal. Chim. Acta*, **47**, 183 (1969).
55. Fleischer, K. D., Southworth, B. C., Hodecker, J. H., Tuckerman, M. M., *Anal. Chem.*, **30**, 152 (1958).
56. Yu, H. Y., Sha, I. H., *Hua Hsueh Tung Pao*, **1965**, 557.
57. Malyukova, F. S., Zaitseva, A. D., *Plast. Massy*, **1966**, [4], 57.

1.4 ESTIMATION OF CHEMICAL CHARACTERISTICS

Acid Number

Acid numbers of polymers are determined by titration of their solutions in inert organic solvents such as ethanol (either absolute or in mixtures with water), acetone, benzene, dioxan, or some other suitable solvent, with 0·1*N* sodium or potassium hydroxide, using phenolphthalein indicator. The determination is conducted at room temperature.

For strongly coloured materials the titration end point is difficult or sometimes impossible to establish, unless carried out by potentiometry or by the Coburn procedure [1], which involves the use of a two-phase titration system. The upper phase is an alcohol-benzene mixture, in which the dark organic material is dissolved. The lower phase is water or more exactly a saturated aqueous solution of sodium chloride (sodium chloride is added to produce a more distinct interface); the lower layer contains the titration indicator.

The method for determination of acid number by visual and potentiometric methods is described below. The values of acid numbers of a variety of polymers are listed in Table 1.2 [2].

Visual Titration

A resin sample (10–50 g, depending on the expected acid number) is dissolved in 100 ml of ethanol, previously neutralised to phenolphthalein, and titrated with 0·1*N* aqueous sodium hydroxide (phenolphthalein indicator) until a pink colour persists. Ethanol-insoluble materials are dissolved in one of the solvents or mixtures mentioned above, and titrated with 0·1*N* alcoholic potassium hydroxide (according to the German Standard DIN 53,402).

The acid number (in mg KOH per g) is found from the formula

$$\text{Acid number} = 56{\cdot}1vn/m$$

where:

v = the volume (ml) of sodium hydroxide titrant consumed in titrating the resin sample,
n = normality of the titrant,
m = weight of resin sample, g.

Table 1.2 Acid Numbers of Polymers
(by permission, from *Kunststoff–Bestimmungsmöglichkeiten*, Carl Hanser Verlag, München, 1970)

Acid No.	*Polymer*
0	Cellulose nitrate, Regenerated cellulose, Vinyl chloride-vinyl acetate copolymers, Styrene-butadiene copolymers, Cellulose acetate, Poly(vinyl chloride), Aniline resins, Melamine resins, Urea resins, Thiourea resins.
0–200	Polyesters, non-crosslinked.
2	Cellulose acetate propionate.
<3	Benzylcellulose, Poly(ethylene glycols), Methylcellulose.
4	Poly(methyl methacrylate).
4–5	Vinyl acetate-fumarate (or maleate) copolymers.
<5	Poly(vinyl acetals), Polystyrene, Coumarone or coumarone-indene resins.
5	Polyisobutylene.
7	Polyindene.
<10	Ethylcellulose, Polyacrylates, Poly(vinyl alcohol), Poly(vinyl acetate).
10–30	Alkyd resins, oil- and phenol-modified.
<20	Phenolic resins.
20–50	Alkyd resins, unmodified.
20–100	Alkyd resins, oil-modified.
<50	Polyesters, unsaturated.

POTENTIOMETRIC TITRATION

The sample to be analysed is dissolved in ethanol or in a (1:1) ethanol-water mixture (water should be made neutral to pH 7·0). The titration, using a conventional electrode system, is also taken to pH 7·0. Ethanol may be replaced by acetone.

Saponification Value

The adopted conditions, including reaction medium, hydrolytic agent, and temperature for determination of saponification value, depend on the ease with which the ester groups in the polymer undergo hydrolysis. Among the most readily hydrolysable esters are those which produce hydroxyl groups directly attached to the polymer chain carbon atoms, notable examples being acetylcellulose and poly(vinyl acetate). Esters in which the carboxyl groups are directly linked to the polymer chain, as in polyacrylates are much more resistant to hydrolysis. Polymethacrylates are especially difficult to hydrolyse because of the presence of the α-methyl group. Phosphates are also difficult to hydrolyse. The rate of hydrolysis is also affected by the configuration of the polymer molecules [3].

It should be emphasised that the result from the determination of the saponification value also includes the acid number, since in this determination no distinction is made between free acid groups and those combined in the form of esters. This must be taken into account in the calculation of the saponification value.

The saponification value of purified resins free of acidic, low molecular weight contaminants constitutes an important characteristic of considerable assistance in the identification of polymers.

Readily Saponifiable Esters

A 2–3 g sample is weighed into a 250 ml conical flask equipped with a ground-glass stopper. 25 ml of ethanol and 50 ml of benzene are added, and the contents are agitated to accelerate dissolution of the material. When dissolution is complete 25 ml of 2*N* alcoholic potash is added, the flask is connected to a reflux condenser (equipped with a soda-lime tube) and the whole is heated to the boil for 1 hr. After this time the contents are allowed to cool and the condenser is well rinsed with alcohol. The hydroxide excess is titrated with 1*N* hydrochloric acid (phenolphthalein indicator). For very dark solutions a special "universal indicator" is of assistance*.

If the resin is not completely dissolved the solution must be vigorously shaken during titration to neutralise any alkali present in the bulk of the swollen polymer.

The saponification value (mg KOH per g) is found from the formula

$$\text{Saponification number} = 56{\cdot}1(n_1 v_1 - n_2 v_2)/m$$

where:

n_1 = normality of the potassium hydroxide,
v_1 = volume (ml) of the potassium hydroxide added,
n_2 = normality of the hydrochloric acid,
v_2 = volume (ml) of hydrochloric acid consumed in titration of the resin solution,
m = weight of polymer sample, g.

Poorly Saponifiable Esters

Ten ml of the hydrolysing reagent (1*N* potassium hydroxide in ethylene glycol) is pipetted into a 100 ml Quickfit conical flask and 0·4–0·6 g of the resin is added. The flask is stoppered and heated on an oil bath so that the temperature rises to 70–80°C within 2–3 min. The flask is then removed from the bath for a moment, the contents are vigorously shaken, the stopper is removed for a time, then the restoppered flask is put back in the bath and heated to 120–130°C. After 3 min heating the flask is allowed to cool to 80–90°C, and the stopper is removed and rinsed with distilled water. About 15 ml of water is added, the contents are shaken, and the potassium hydroxide excess is neutralised with standard 0·25*N* hydrochloric acid, using phenolphthalein indicator.

For esters that are saponified only with great difficulty (like polymethacrylates) ethanolamine has proved useful.

An ingenious procedure reported by Heitler [4] for determining sapon-

* Frequently the universal indicator used is that due to Bogen. This indicator comprises 0·1 g phenolphthalein, 0·2 g methyl red, 0·3 g dimethylazobenzene, 0·4 g bromothymol blue, and 0·5 g thymol blue in 500 ml anhydrous ethanol.

ification value consists in ethanolysis of the material in an ebulliometer and calculation of the results on the basis of the boiling point depression of the solvent.

Iodine Value

Iodine values of polymers are substantially affected by the double bond reactivity which in turn depends on the system containing the double bond and the type of its neighbouring groups. It is also affected by the nature of the iodinating reagent used. Moreover, the iodine value depends on the presence of other groupings that can react with iodine, and also on the reaction temperature and reaction time, the kind of solvent used and the intensity of light during the iodination process.

The iodine value of polymers is commonly determined by Wijs method, in which the sample is treated with iodine monochloride in a non-aqueous solvent at room temperature. The reagent reacts with olefinic double bonds of vinyl and allyl esters, allyl ethers, styrene double bonds, and also isolated and conjugated double bonds of unsaturated fatty acids. Double bonds in maleic and fumaric acids and their esters, as well as acrylate and methacrylate double bonds, are unreactive.

In addition to the Wijs reagent certain other reagents may be used for the determination of iodine value. These include the Hanus reagent, which is a solution of BrI in glacial acetic acid, and the Kaufmann reagent, which comprises a solution of bromine and sodium bromide in methanol. However, the results obtained by the Hanus procedure are 4–5% lower than those from the corresponding analysis by the Wijs method.

Iodine values of certain polymers are collected in Table 1.3.

Table 1.3 Iodine Values of Polymers
(by permission, from *Kunststoff–Bestimmungsmöglichkeiten*, Carl Hanser Verlag, München, 1970)

Iodine value	*Polymer*	*Iodine value*	*Polymer*
<1	Polyisobutylene	340–360	Styrene-butadiene copolymer (21:79)
<5	Butyl rubber	345–375	Natural rubber
ca. 140	Styrene-butadiene copolymer (66:34)	385–440	Polybutadiene
ca. 290	Styrene-butadiene copolymer (36:64)		

The methods available for the determination of olefinic bonds are reviewed by Polgár and Jungnickel [5].

THE WIJS METHOD

The Wijs Reagent. Iodine (9 g) and iodine trichloride (8 g) are dissolved in 1 l. of glacial acetic acid. The resultant solution is filtered into an amber coloured bottle and stored in the dark. The reagent must be used within 30 days.

PROCEDURE [6]. A 0·2–1·0 g polymer sample is weighed into a 200–300 ml stoppered conical flask and dissolved in a suitable solvent such as chloroform or *p*-dichlorobenzene; 20 ml of 0·2*N* Wijs solution is added to the flask, which is then stoppered and allowed to stand in the dark for 30 min. After that time 100 ml of 10% aqueous potassium iodide is added. The resultant two-phase mixture is titrated with 0·1*N* sodium thiosulphate to decolorise the aqueous layer. Then 5 ml of 0·5% starch solution is added, the contents are well shaken and the titration is resumed until the iodine is completely removed from the organic layer. A blank is run separately, taking the same volumes of chloroform and Wijs reagent.

The iodine value, in g of iodine per 100 g of the substance, is calculated from the formula

$$\text{Iodine value} = (v_1 - v_2)\,12{\cdot}69\; n/m$$

where:

v_2 and v_1 = the volumes (ml) of 0·1*N* sodium thiosulphate used for titration of the sample and the blank, respectively,

n = normality of sodium thiosulphate titrant,

m = weight of polymer sample, g.

Hydroxyl Value

The hydroxyl value is usually determined by acetylation with acetic anhydride in pyridine solution. The excess anhydride is decomposed with water, and the resulting acetic acid, which is formed both in the hydrolysis and in the acetylation process, is titrated with a standard alkali solution, using phenolphthalein as indicator.

Hydroxyl values of a number of polymers are listed in Table 1.4.

ACETYLATION

REAGENT. Anhydrous pyridine (880 ml) is mixed with 120 ml of acetic anhydride. Best results are achieved with a freshly prepared reagent.

PROCEDURE [6]. About 2 g of polymer is weighed into a 250 ml Quickfit conical flask and 20 ml of the acetylating mixture is added. The flask is connected to a reflux condenser and the contents are heated on a boiling water bath with occasional shaking for an hour or more to effect complete dissolution of the material. When solution is complete, 25 ml of dry dichloroethane or benzene is added, the flask is stoppered and vigorously shaken. Unless dissolution was complete during heating, the solvent addition should help in dissolving or at least in finely dividing the material. Subsequently, 100–150 ml of water is poured in and the solution, on vigorous shaking, is titrated with 0·5*N* potassium hydroxide, with phenolphthalein added in just about double the quantity usually taken, until a pink colour persists (for a minute).

Table 1.4 Hydroxyl Values of Polymers
(by permission, from *Kunststoff–Bestimmungsmöglichkeiten*, Carl Hanser Verlag, München, 1970)

Hydroxyl Value	*Polymer*
0	Polyesters of maleic acid,
	Styrene-butadiene copolymers
	Vinyl chloride-vinylidene chloride copolymers
	Vinyl chloride-vinyl acetate copolymers
	Poly(vinyl chloride)
	Polyindene
	Polyisobutylene
	Poly(methyl methacrylate)
	Poly(vinyl acetate)
	Polystyrene
	Cellulose triacetate*
	Cellulose tributyrate*
	Cellulose trinitrate*
	Cellulose tripropionate*
	Coumarone and coumarone-indene resins
	Melamine resins
	Urea resins
	Thiourea resins
0–350	Polyesters (unsaturated)
10–70	Alkyd resins, oil-modified
20–100	Alkyd resins, unmodified
20–250	Cellulose nitrate
40–215	Poly(ethylene glycol)
65–250	Ethylcellulose
	Poly(vinyl acetals)
100	Benzylcellulose
105	Cellulose 2,5-acetate*
120	Methylcellulose
125–450	Phenolic resins
200–300	Regenerated cellulose
1000–1270	Poly(vinyl alcohol)
1038	Cellulose*
1080–1270	Poly(vinyl acetate) of a lower acetyl content

* Theoretical values.

Hydroxyl number, in mg KOH per g, is found from the formula

$$\text{Hydroxyl number} = 56{\cdot}1(v_1 - v_2)\ n/m$$

where:

v_2 and v_1 = the volumes (ml) of 0·5*N* potassium hydroxide used for titration of the sample and the blank, respectively,

n = normality of the potassium hydroxide,

m = weight of polymer sample, g.

The acid value of the material under analysis must be taken into account if it is higher than 25.

The procedure described is equally good for the determination of alcoholic hydroxyl as well as phenolic hydroxyl groups such as those found in the phenolic resins. When alcoholic hydroxyl groups are to be determined separately in the presence of the phenolic groups, phthalic anhydride should be used in place of acetic anhydride. The use of this anhydride has also proved advantageous in the presence of aldehydes.

PHTHALOYLATION

A polymer sample which contains about 20 mequiv of hydroxyl groups is weighed into a 500 ml Quickfit conical flask and 100 ml of 1*N* phthalic anhydride solution in anhydrous pyridine is added. The flask is stoppered with the stopper moistened with pyridine, and heated for 1 hr on a boiling water bath. Both the stopper and the flask walls are then rinsed with 10 ml of water and the flask is reheated for 3 min. On cooling, the reaction mixture is titrated with 1*N* aqueous potassium hydroxide (phenolphthalein indicator). The hydroxyl value is calculated from the formula reported above.

Recently certain other procedures have come to be used for determining hydroxyl values (see Chapter 8).

References

1. Coburn, H. H., *Ind. Eng. Chem., Anal. Ed.*, **2**, 181 (1930).
2. Krause, A., Lange, A., *Kunststoff-Bestimmungsmöglichkeiten*, Carl Hanser Verlag, Munich, 1970.
3. Glavis, F. J., *J. Polymer Sci.*, **36**, 547 (1959).
4. Heitler, C., *Talanta*, **11**, 1081 (1964).
5. Polgár, A., Jungnickel, J. L., in Mitchell, J., Jr., Kolthoff, I. M., Proskauer, E. S., Weissberger, A., *Organic Analysis, Vol. III*, Interscience, New York, 1956.
6. Brauer, G. M., Horowitz, E., in *High Polymers, Vol. XII, Analytical Chemistry of Polymers*, Kline, G. M. (Ed.), *Part III, Identification Procedures and Chemical Analysis*, Interscience, New York, 1962.

1.5 DETERMINATION OF WATER

Too high a water content may adversely affect polymer properties such as material strength, mechanical, electrical or optical characteristics, and may result in difficult fabrication. For these reasons water is usually very thoroughly removed from both polymeric materials and the additives used in fabricating processes. In this connection analytical control of the drying process is required.

Of the great number of available literature methods for determination of moisture in polymers and in their derivative materials, the following have gained considerable popularity: drying, azeotropic distillation with water-immiscible solvents, and direct or indirect tiration by the Karl Fischer method. In addition, the manometric technique is quite popular, and occasionally also such instrumental techniques as infrared spectrophotometry and NMR spectroscopy are used.

None of the methods mentioned is, however, very versatile. Their range of applicability is limited on the one hand by the peculiar properties of these techniques, and on the other, by the properties of the polymers and plastics analysed. Thus for each individual case the choice of method requires careful consideration.

Drying

The method of drying, especially convenient because of its simplicity, is not quite suitable for determination of water in a number of polymers, as it provides results which are usually too high. This refers, in the first place, to polymeric resins which always contain unpolymerised monomers in some quantity, and in general, to polymers in which water is not the only volatile component. Moreover, as the drying process is commonly carried out at elevated temperatures, polymers of low heat resistance may either decompose during the process to produce volatile matter or they may undergo polycondensation reactions attended by formation of water. For instance, with resol-type phenolics and amino resins a constant weight cannot be achieved by conventional drying in a drier and the results obtained are erroneous and too high. The interfering condensation reactions in these resins can actually be considerably suppressed by drying under reduced pressure, over drying agents such as phosphorus pentoxide or conc. sulphuric acid, but this makes the determination much longer.

The drying method is also of no use for a number of plastic materials which occur in a compact resinous form. Despite prolonged heating at higher temperatures, water fails to separate completely from the polymer because of the very slow transport from the bulk of material to the surface layers.

Until recently the drying was usually effected in a common electrically heated drier. Lately, infrared radiation has been increasingly adopted in drying polymers [1].

The procedure followed with infrared radiation is given below. A sample, accommodated in an aluminium vessel of diameter 80 mm, is heated with radiation of wavelength 1–16 μm, using a 500 W infrared lamp mounted at a distance of 5 cm from the sample. The vessel is placed on an asbestos disc in the middle of the radiation and dried to constant weight. The drying time is found experimentally. It is usually very short, 3–10 min.

By the drying methods discussed above the moisture content can be analysed in such polymers as cellulose esters and ethers, fluoroethylene derivatives, polyacrylonitrile, emulsion poly(vinyl chloride) [2], amino resins [3] and certain other polymers. In the case of suspension poly(vinyl chloride) [2] as well as polystyrene, polyacrylates and polymethacrylates, monomer also escapes along with moisture.

Azeotropic Distillation

Azeotropic distillation by the Dean–Stark method consists in removal of water by distillation together with water-immiscible solvents which form azeotropes with water. Water separates from the distillate as a lower layer in a graduated receiver which makes possible the measurement of the volume of water collected. Xylene is usually used as solvent, but the lower boiling compounds toluene and benzene are also used. Occasionally a benzene-ethanol mixture (b.p. of the water azeotrope 64·9°C) or a carbon tetrachloride-ethanol mixture (b.p. of water azeotrope 61·8°C) is employed.

The Dean–Stark method is used for the determination of water in polymers containing more than 2% water. Specifically, it is used for such polymers as furan resins, butadiene-styrene copolymers, and also in phenolics soluble in the solvents used in the Dean–Stark method. For amino resins the method is unsuitable for the same reasons as apply to the drying at higher temperatures.

The Karl Fischer Method

For determination of water this is the most widely used method that can be followed successfully in the analysis of a large variety of both organic and inorganic materials. The method is also extensively used in analysis of polymers. The high specificity of this technique is unquestionably a major factor in its success. It will be recalled that with the help of the Karl Fischer reagent moisture can be determined even in compounds that themselves react with iodine, such as mixtures containing olefins [4,5], paraformaldehyde and α-polyoxymethylene [6]. The Karl Fischer method is distinguished by a high sensitivity, which makes it possible to analyse polymers where hundredths of a percent of water are involved. An additional advantage of this reagent is its power in dissolving, or at least swelling, a great number of polymers.

The preparative procedures for the Karl Fischer reagent, including the methods of titration with this reagent and the reaction mechanisms, are treated at great length in monographs by Mitchell and Smith [7], and Eberius [8], as well as in numerous articles and monographs on analysis of polymers. In this book some information on the available variants of the Fischer reagent, which are essential from the viewpoint of the analysis of polymers, will be given.

The Karl Fischer reagent in its classical and most used form comprises a methanolic solution of iodine, sulphur dioxide and pyridine.

Of practical significance is the classical variant of the Karl Fischer reagent suggested by Mitchell and Smith [7]. This reagent contains in a volume of 1 litre about 84 g of iodine, 64 g of (45 ml) sulphur dioxide, 270 ml of pyridine and 670 ml of methanol.

A number of other variants of the Karl Fischer reagent are known. The modifications primarily consist in the replacement of methanol, in part or entirely, by other solvents.

To increase stability of the Karl Fischer reagent Peters and Jungnickel [9] replaced methanol completely with the monomethyl ether of ethylene glycol (methyl Cellosolve). The Peters–Jungnickel reagent, which comprises a mixture of 133 g of iodine, 70 ml of sulphur dioxide, 425 ml of pyridine and 425 ml of methyl Cellosolve, is not only more stable but also less volatile. Because of this the reagent is particularly advantageous for determination of moisture in gases. The gas under analysis is passed through the reagent to absorb moisture present.

A number of compounds interfere with the determination of water by the conventional Karl Fischer reagent. These are primarily compounds that react with methanol to yield water during titration. This group of compounds includes substances with active carbonyl groups, such as aldehydes and ketones, and also those incorporating hydroxymethyl groups linked to the nitrogen of amino groups, as in amino-formaldehyde resins. Water is produced from the reactions of these substances with methanol according to the equations:

$$RCHO + 2CH_3OH \rightarrow RCH(OCH_3)_2 + H_2O$$

$$R_2CO + 2CH_3OH \rightarrow R_2C(OCH_3)_2 + H_2O$$

$$\rangle N{-}CH_2OH + CH_3OH \rightarrow \rangle N{-}CH_2OCH_3 + H_2O$$

For this reason substitution of methanol in the Karl Fischer reagent by the less reactive methyl Cellosolve, as introduced by Peters and Jungnickel, makes this method suitable also for determination of water in such compounds. A similar result was attained by Mitchell and Smith [7], who replaced part of the methanol by pyridine.

Undesirable side reactions of the kind mentioned above may also be prevented by using certain solvents other than methanol in preparing solutions of substances under analysis. Specifically, Smith and Kellum [10] in determining water in silanols and silicones containing silanol groups that react with methanol, recommend using alcohols containing more than eight carbon atoms in the molecule for the preparation of the polymer solutions. For urea- and melamine-formaldehyde resins dimethylformamide was used for this purpose [8,11].

Direct Methods

In the usual manner, that is, by dissolving a polymer sample in a suitable solvent and titrating the resulting solution with the Karl Fischer reagent, water can be determined only in polymers readily soluble in the reagent and in solvents not interfering with the analysis. Thus, by direct titration with the Fischer reagent water can be analysed in such polymers as phenolic resins, acrylic resins, poly(vinyl alcohol) and even polyamides [12].

For amino resins the method, although the most suitable of all available techniques for determination of water, yields too high results. The reason for this is the reaction of free hydroxymethyl groups to produce water. By operating at low temperature (−40°C) and in dimethylformamide solution, according to Mitchell and Smith [7,13], the error due to this reaction can be so decreased that for urea- and melamine-formaldehyde resins satisfactory results are achieved. As demonstrated by the results of more recent studies by Bertz *et al.* [14], by working at a sufficiently low temperature good results can be achieved for the analysis of melamine resins even in methanolic solutions. According to Vašta and Seidl [15] on the other hand, the Fischer method, even in the modification developed by Mitchell and Smith, is unsuitable for analysis of moisture in urea resins. This is due to the fact that at −35°C the reaction of water with the Karl Fischer reagent in a non-aqueous medium fails to go to completion, whereas at higher temperatures (above −25°C) the side reactions mentioned above occur.

A problem in determination of moisture in some sparingly soluble polymers is too slow a diffusion of moisture from the polymer sample to the reagent during titration, which is conducted at a temperature that cannot be higher than room temperature. To accelerate this process Cornish [16] refluxed such polymers in methanol previously titrated with the Karl Fischer reagent. When allowed to cool to room temperature the mixture was titrated with the Karl Fischer reagent by a modified dead-stop titration technique, worked out by Cornish.

In the case of insoluble polymers the methods of direct determination of water are good only for the moisture adsorbed on the surface of the polymer under analysis.

WATER RELEASE METHODS

These methods are also applicable to the determination of water contained in the bulk of polymers that are resinous, hard or horny materials, barely soluble or practically insoluble in any of the available solvents.

According to Haslam and Clasper [17] total moisture content in nylon-6,6 is determined by heating a sample under diminished pressure. The liberated water, collected in a cold trap, is titrated with the Karl Fischer reagent. Accurate results, however, are difficult to achieve by this method especially if the water content of the polymer is low.

The method of Haslam and Clasper, improved by Reid and Turner [18], involves heating a polymer sample in a dry nitrogen stream at 120°C. The vapours entrained by nitrogen are absorbed in an absorbing vessel, the contents of which are titrated continuously throughout the analysis. A detailed procedure is reported below.

REAGENTS AND SOLUTIONS

The Karl Fischer Reagent. After Peters and Jungnickel (see p. 56).

Nitrogen. From a gas cylinder, dried over magnesium perchlorate.

APPARATUS. The set-up as shown in Fig. 1.2. A "dead-stop" end point indicator [17].

PROCEDURE. An oil-bath is heated to 120°C. Absorber *4* is filled with 20 ml of the Karl Fischer reagent containing a slight excess of iodine. The apparatus train is completely dried by passing nitrogen through it at a rate of 0·5 l./min. Throughout the drying period the absorbing liquid should always contain a slight excess of the Karl Fischer reagent, which is achieved by portionwise addition of the reagent in 0·1 ml volumes as the iodine is exhausted.

The drying is carried on until the meter of the dead-stop assembly ceases indicating variations within 15 min after the last 0·1 ml portion of the reagent was added. The indicator very slowly drops to the "dead-stop" end point. When the indicator reaches this point, the side neck of vessel *8* immersed in the oil-bath is opened, a 20 g polymer sample is quickly dropped in, and the neck is immediately closed with a tightly fitting stopper. At that moment the level of the liquid in burette *2* is noted and at once 0·1 ml of the reagent is added to the vessel. Thereafter, as water is absorbed, 0·1 ml portions of the reagent are introduced when the indicator returns

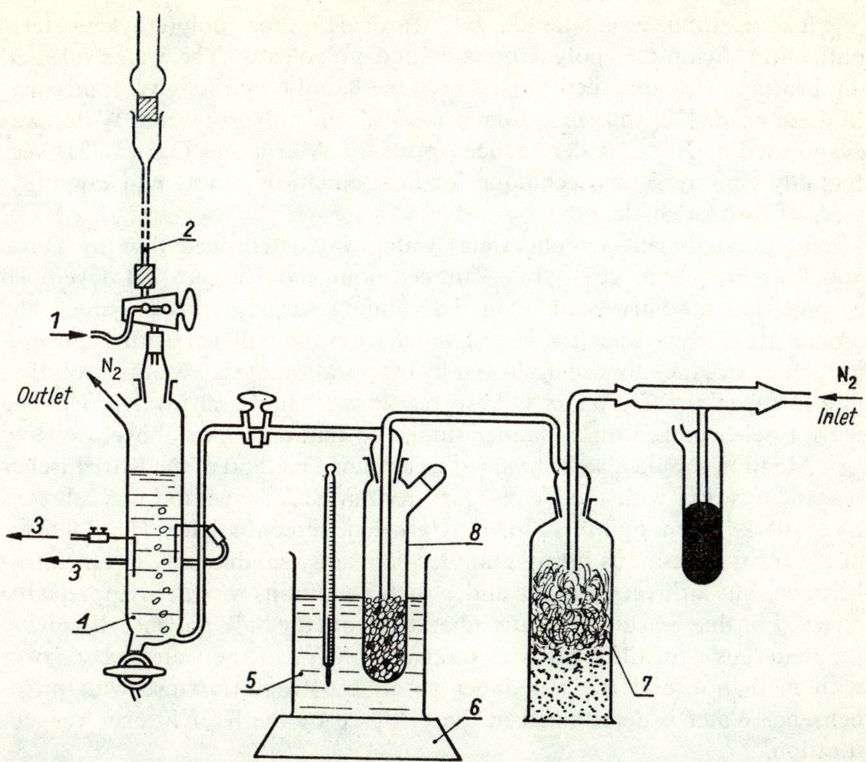

Fig. 1.2 Apparatus for determination of water in polymers; *1*—the Karl Fischer reagent supply, *2*—microburette, of capacity 2 ml, *3*—connections to the dead-stop titrimeter, *4*—titration vessel-absorber, *5*—oil-bath at 120°C, *6*—heating plate, *7*—nitrogen drying bottle, *8*—vessel with sample under analysis (from [18] by permission of the copyright holders)

to the dead-stop end point. When water is totally removed from the polymer, the flow of nitrogen is continued until the pointer ceases showing deflections after the last 0·1 ml portion of the reagent was added, as was done in the preliminary stages of the analysis.

The moisture content in the polymer ($x\%$) is found from the formula

$$x = \frac{vT}{1000m} 100 = \frac{vT}{10m}$$

where:

v = volume (ml) of the Karl Fischer reagent used in titration of water liberated from polymer,

T = titre of the reagent (mg per ml),

m = weight of sample, g.

The method was originally developed for carbon-black pigmented polyethylene, but the authors also used this method to advantage for other polymers. This method may therefore be regarded as a general one for determination of water in polymers.

The method was adapted by Muroi [19] for poly(ethylene terephthalate), nylon-6,6, polycarbonates and polyolefins. The water released on heating was absorbed in a (1:1) methanol-ethylene glycol mixture. Maltese *et al.* [20] thus determined moisture in polypropylene. Water was evaporated at 200°C under reduced pressure. Muroi and Ogawa [21] successfully employed this technique for determination of water in low polymers of formaldehyde.

In poly(ethylene terephthalate) water was determined thus by Dinse and Praeger (see p. 267) while Zimmermann and Hoyme [22] developed a somewhat modified method in determining water in this polymer. The modification involves extraction of water from the polymer with a phenol-tetrachloroethane mixture, followed by evaporation of the resulting solution in a nitrogen stream. Water is absorbed in methanol and titrated with the Karl Fischer reagent in a manner similar to that described above.

Mention should also be made of a combined method of the Karl Fischer reagent titration with the Dean–Stark technique. The method was adopted by Lardera [2] among others for poly(ethylene terephthalate). The polymer, in the form of resinous horny granules, is readily soluble only in hot nitrobenzene, but a direct titration under such conditions results in appreciable errors. For that reason, the water liberated from the bulk polymer dissolving the materials is distilled off with methanol. Most of the water passes over with methanol, and the remainder as an azeotropic mixture with nitrobenzene. Water is determined in the distillate by the Karl Fischer reagent titration.

Manometric Method

In the manometric method use is made of the volatility of water contained in a polymer. The method consists in heating a polymer sample under high vacuum to a temperature higher than the boiling point of water at the pressure existing in the apparatus. Owing to the evaporation of water in the apparatus the pressure rises and this rise is measured in an oil-filled manometer calibrated with small quantities of water.

The manometric method is not, however, specific only for water, as the pressure rise may result from other volatile polymer components, such as volatile monomers or polymer degradation products. The method has therefore similar limitations to the drying method with the one difference that it is much more sensitive than the latter. Thus accurate results in determination of moisture can only be obtained when the polymer does not contain volatile components in major quantities and when it does not decompose appreciably at the temperature of the measurement.

The fundamentals of the manometric method have been described by Gmelin [23]. The polymer sample is heated to 195–210°C [24]. The method has been found equally good for determination of volatile monomers, oligomers, plasticisers, and other volatile components of polymers.

David *et al.* [25] determined total volatile matter in polycarbonates, polyesters, polyurethanes, and other polymers by this method, using samples about 10 g in size. Recently Illing and Hobe [26] employed this technique for determination of water (0·01–2%) in polyamides and polycarbonates, and Schmalz [27] used it for water in polyamides and poly(ethylene terephthalate).

Gas Chromatography

It follows from the foregoing discussion, that the majority of methods available for determination of water in polymers involve its liberation from the sample by heating. The released water is subsequently determined by a variety of procedures. Being a volatile compound, water may also be analysed by gas chromatography. Okubo *et al.* [28] thus determined the moisture content in PVC, using the gas chromatography method developed by Nadeau and Oaks [29]. Keister and Harrington [30] developed a rapid gas chromatographic determination procedure for pigmented polyethylene.

Gas chromatography may also be used in analysis of water evolved from polymer by the method based on the reaction with calcium carbide.

Infrared Spectroscopy

Infrared spectrophotometric studies are conducted at room temperature and reagents are unnecessary. For these reasons the method may prove valuable in determining water in polymers when other procedures fail because of negligible solubility of the material under analysis, or in view of the reactions, occurring with the Karl Fischer reagent, which apparently interfere with the analysis, or else on account of interfering chemical reactions occurring at the elevated temperatures used in some of the methods.

In the infrared spectrophotometric determination of water in polymers the characteristic bands due to water in the near infrared region are commonly used. In hydrocarbon polymers water may sometimes be determined in the same manner as in liquid hydrocarbons by measuring absorbance at 2·69 μm [31]. In polyamides water was determined [24] by measuring absorbance at 1·89 μm in a 5% polymer solution in a (1:1) phenol-tetrachloroethane mixture. In polyacrolein water was determined on the basis of absorbance measured at 6·1 μm [32].

Analyses of water in polymers which exhibit considerable absorption within the analytical bands for water offer serious problems. According to Langbein and Seufert [33] water may be determined in polymers such as poly(ethylene terephthalate), but the analysis is based in this case on measurements at as many as three different wave numbers, namely 3630, 4080, and 7350 cm^{-1}, with a fairly complex mathematical treatment in the calculation of the results.

NMR Spectroscopy

Water in polymers may be determined by nuclear magnetic resonance spectroscopy (see also p. 152). In addition to the rapidity and convenience of the analysis, the NMR method has the extra asset of being suitable for continuous control of moisture content in materials. Obviously owing to the latter feature, the method has aroused considerable interest industrially. In highly industrialised countries hygrometers operated on the NMR principle are now commercially available.

The NMR spectrum of solid polymers containing moisture is complex in nature. The narrow signal of the water protons is superimposed on the broad line of the solid. The experimental conditions can, however, be so arranged that the solid broad signal is very weak and the narrow line due to the water protons predominates.

The best results with the NMR technique are achieved for a moisture content $\geqslant 10\%$, as then the relation between the NMR signal height and the moisture content is linear. In the range of low moisture content this relation deviates from linearity because of interactions between the absorbed water and the solid resulting in a broadening of the NMR signal.

Slonim, Urman, and Konovalov [34] used the NMR technique in the analysis of such polymers as polyamides, aminoplasts and wood flour, and found that the method is viable for the analysis of water in the range 3–17%. For lower water contents the broadened signal of the water protons coalesces with the signal for the solid, while for higher quantities of water deviations from linearity of the calibration curve are observed.

Other Methods

The water liberated from polymers on heating was determined by Salzer and Geldermann [35] with the help of a coulometric hygrometer (after Keidel).

Barabanov *et al.* [36] determined water in nitrocellulose solutions in acetic acid by conductometric titration with acetic anhydride.

In the analysis of polymers physical methods such as hygrometry or psychrometry are also used for determination of water. Electrical conductance of the substance or the measurement of its dielectric constant [37] has also been employed.

References

1. Balandina, V. A., Gurvich, D. B., Kleshcheva, M. S., Nikitina, V. A., Nikolaeva, A. P., Novikova, E. M., *Analiz polimerizatsionnykh plastmass* (*Analysis of Polymerisation Plastics*), Izd. Khimiya, Leningrad, 1967.
2. Lardera, M. R., *Mat. Plastische*, **23**, 247 (1957).
3. Rybnikář, F., Ditrych, Z., Klácel, Z., Ordelt, O., *Analýza a zkoušeni plastickych hmot* (*Analysis and Testing of Plastic Materials*), SNTL, Prague, 1965.
4. Martin, A. R., Lloyd, A. C., *J. Am. Oil Chem. Soc.*, **30**, 594 (1953).

5. Sneed, R. W., Altman, R. W., Mosteller, J. C., *Anal. Chem.*, **26**, 1018 (1954).
6. Lyubomilov, V. I., Usevich, T. D., *Plast. Massy*, **1961**, [2], 67.
7. Mitchell, J., Jr., Smith, D. M., *Aquametry*, Interscience, New York, 1948.
8. Eberius, E., *Wasserbestimmung mit Karl-Fischer-Lösung*, Verlag Chemie, Weinheim, 1958.
9. Peters, E. D., Jungnickel, J. L., *Anal. Chem.*, **27**, 450 (1955).
10. Smith, R. C., Kellum, G. E., *Anal. Chem.*, **38**, 67 (1966).
11. Weltzien, W., Achwal, W., *Die Bestimmung das Wassergehaltes mit Hilfe der Karl-Fischer-Methode in Harnstoff-Formaldehyd-Kunstharzen sowie in unbehandelten und in mit diesen Kunstharzen behandelten Geweben. Forschungsberichte des Landes Nordrhein-Westfalen, Nr 1298*, Westdeutscher Verlag, Cologne und Opladen, 1963.
12. Kellum, G. E., Barger, J. D., *Anal. Chem.*, **42**, 1428 (1970).
13. Averell, P. R., in *High Polymers, Vol. XII, Analytical Chemistry of Polymers*, Kline, G. M. (Ed), *Part I, Analysis of Monomers and Polymeric Materials*, Interscience, New York, 1959.
14. Bertz, T., Neundorf, C., Köhler, G., *Plaste u. Kautschuk*, **10**, 84 (1963).
15. Vašta, M., Seidl, J., *Chem. Listy*, **50**, 2034 (1956).
16. Cornish, G. R., *Plastics*, **10**, 99 (1946).
17. Haslam, J., Clasper, M., *Analyst*, **77**, 413 (1952).
18. Reid, V. W., Turner, L., *Analyst*, **86**, 36 (1961).
19. Muroi, K., *Japan Analyst*, **11**, 351 (1962).
20. Maltese, P., Clementini, L., Panizzi, S., *Mat. Plastiche*, **35**, 1669 (1969).
21. Muroi, K., Ogawa, K., *Bull. Chem. Soc. Japan*, **36**, 1278 (1963).
22. Zimmermann, H., Hoyme, H., *Faserforsch. Textiltechn.*, **18**, 393 (1967).
23. Gmelin, P., *Chem. Fabrik*, **3**, 449 (1930).
24. Vieweg, R., Müller, A., *Kunststoff Handbuch*, Band VI, *Polyamide*, Carl Hanser Verlag, Munich, 1966.
25. David, D. J., Baumann, G. F., Steingiser, S., *SPE Transactions*, **2**, 231 (1962).
26. Illing, G. V., Hobe, D., *Z. Anal. Chem.*, **230**, 418 (1967).
27. Schmalz, E. O., *Faserforsch. Textiltechn.*, **21**, 209 (1970).
28. Okubo, N., Mashimo, S., Watanabe, T., Jono, W., *Japan Analyst*, **15**, 949 (1966).
29. Nadeau, H. G., Oaks, D. M., *Anal. Chem.*, **32**, 1760 (1960).
30. Keister, D. C., Harrington, R. C., Jr., *Tappi*, **50**, 81A (1967).
31. Forbes, J. W., *Anal. Chem.*, **34**, 1125 (1962).
32. Forbes, J. W., Schissler, D. O., *J. Polymer Sci.*, *Pt C*, **8**, 61 (1965).
33. Langbein, G., Seufert, W., *Kolloid-Z. Z. Polym.*, **193**, 37 (1963).
34. Slonim, I. Ya., Urman, Ya. G., Konovalov, A. G., *Plast. Massy*, **1963**, [5], 58.
35. Salzer, F., Geldermann, H., *Kunststoffe*, **60**, 668 (1970).
36. Barabanov, N. N., Aleksandrov, Yu. A., Seliverstova, V. M., *Plast. Massy*, **1969**, [2], 71.
37. Mandler, H. G. A., *Gummi Asbest Kunststoffe*, **22**, 438 (1969).

Chapter 2

INFRARED AND ULTRAVIOLET ABSORPTION SPECTROSCOPY

2.1 INTRODUCTION

Spectroscopic techniques are increasingly applied in investigations on high molecular weight compounds. Absorption spectra provide a substantial amount of information on the relative arrangement of the structural elements in the system of interest and on the interactions, operating between these elements, which hold them in a definite spatial configuration. Moreover, the spectra reflect the stoichiometric relationships among individual structural elements in the molecule.

The mechanism underlying absorption spectra may be visualised schematically as follows: if electromagnetic radiation falls on an assembly of molecules, as a result of interaction of the radiation with these molecules some of the radiation quanta are absorbed. This attenuation of the original radiation intensity is recorded as the absorption spectrum by a spectrophotometer.

The molecule of a chemical compound is a complex system in a state of dynamic equilibrium. The molecule rotates as a whole with a certain frequency about an axis that passes through the centre of its mass, the nuclei of atoms vibrate with definite frequencies along the bond and at certain angles to the bonds, and finally there is a definite electrical charge distribution in the molecule. Thus, the molecule exists in a definite state of rotational, vibrational and electronic energies which determine the total energy associated with this molecule. Disregarding kinetic energy, the total energy of the molecule may be expressed approximately as

$$E_{tot} = E_{rot}+E_{vib}+E_{el}$$

Transitions of the molecule from one energy state to another are associated with an absorption or emission of energy. The difference in energy between two rotational levels at the same vibrational and electronic levels amounts to 0·005–0·025 eV. The difference in energy between two vibrational levels lies in the range 0·05–0·5 eV. On the other hand, the difference in energy between two electronic levels is of the order of several electron-volts. Hence, the ratios of the changes in these energies are as follows

$$\Delta E_{rot}:\Delta E_{vib}:\Delta E_{el} = 1:10:1000$$

The absorption spectra related to the changes in the rotational energy alone occur in the far infrared region > 100 μm, vibration-rotation spectra fall within the classical infrared region (2–25 μm), whilst the changes in the electronic energy of the molecule give rise to absorption in the visible and ultraviolet parts of the electromagnetic spectrum. Since, however, transitions in electronic energy are invariably accompanied by changes in vibrational energy, electronic-vibration-rotation spectra in the range 200–800 nm (often referred to as vibronic for short) are observed.

In view of the specific character of individual spectral ranges and resulting differences in the design of the instruments, absorption spectra are independently studied in the following ranges:

ultraviolet	200–400 nm
visible	400–800 nm
infrared	0·8–25 μm

For investigations of polymers the most common spectral range is the infrared, together with the near infrared region 0·8–2·5 μm. Studies in the ultraviolet and visible parts of the spectrum are of rather less value. For this reason the question of the analysis of polymers by infrared spectrophotometry has been covered in this book at greater length.

2.2 PRINCIPLES

Vibrations of Molecules

As mentioned before, the infrared spectrum results from changes in the vibrational and rotational energy of a molecule. A diatomic molecule, according to classical mechanics, may be looked upon as a harmonic oscillator with a frequency ω expressed by the formula

$$\omega = \frac{1}{2\pi}\left(\frac{K}{\mu}\right)^{\frac{1}{2}} \tag{1}$$

where:

$\mu = m_1 m_2/(m_1+m_2)$ = the reduced mass of the atoms,
m_1, m_2 = the mass of the atoms,
K = the force constant of the bond.

The energy levels of this harmonic oscillator in terms of quantum theory are given by the equation

$$E = h\omega\left(v+\frac{1}{2}\right) \tag{2}$$

where:

E = the total energy of the level,
h = Planck's constant,
v = the vibrational quantum number (0, 1, 2, 3) which is allowed to change by unity during the energy transitions.

In conformity with the quantum theory the harmonic oscillator may assume only definite values of energy as expressed by equation (2).

The spectrum of a harmonic oscillator should be a single line of frequency

$$\nu = \Delta E/h$$

in view of the restrictions imposed on the changes in the vibrational quantum number and because the energy differences between the adjacent energy levels are the same.

For real diatomic molecules, however, the vibration spectrum is not so simple, and the model of an anharmonic oscillator has to be applied to account for vibrations of the atoms. For an anharmonic oscillator the selection rules for the quantum number are not so rigid as for a harmonic oscillator. The quantum number v may change here not only by 1 but also by 2, 3, etc. The spectrum of an anharmonic oscillator consists of several lines, and the frequency of transition from the level $v = 0$ to $v = 1$ is expressed by the equation

$$\nu_1 = \Delta E_{01}/h = \omega(1-2x)$$

and the second harmonic (1st overtone) for transition from the level $v = 0$ to $v = 2$, as

$$\nu_2 = \Delta E_{02}/h = 2\omega(1-3x)$$

where x denotes the anharmonicity constant. The numerical value of x is usually very small (ca. 0·01), and for this reason the first and second overtones occur at frequencies nearly double and treble the values of the fundamental band frequency (2ν, 3ν). In practice the anharmonicity is neglected in calculations of frequency of the spectral lines, with a resulting error not higher than 3%.

Vibrational modes of atoms in a polyatomic molecule are very complex, but they may be reduced to a number of roughly harmonic vibrational modes, often called normal vibrational modes. In each normal vibrational mode of a definite frequency all the atoms participate. The atoms vibrate in phase but with unequal amplitudes. In view of the considerable differences in the values of the amplitudes of individual atoms which participate in a definite normal vibrational mode, certain groups of atoms can be distinguished which vibrate in an essentially independent way. These groups are called "characteristic" ones, since their presence in different compounds results in similar vibration frequencies. Among such characteristic groups are, for instance, stretching vibrations of C—H, O—H, N—H, C=O and C≡N bonds. Their well-defined frequencies are often used in the qualitative analysis of polymers.

The number of normal vibration modes of a molecule is strictly connected with the number of its constituent atoms. A non-linear molecule of n atoms will give rise to $3n-6$ normal modes, whereas a linear one gives rise to $3n-5$ normal vibrational modes. These relationships also hold good for polymers if some simplifications are assumed. Specifically, if the polymer chain is assumed to be infinitely long, the recurring structural units being of the same spatial arrangement, and if the adjacent chain vibrations do not affect each other, then the number of normal vibrational modes may be found from the number of atoms contained in a single structural unit. For example, in a polystyrene chain the following 16-atom molecule may be assumed as a structural unit

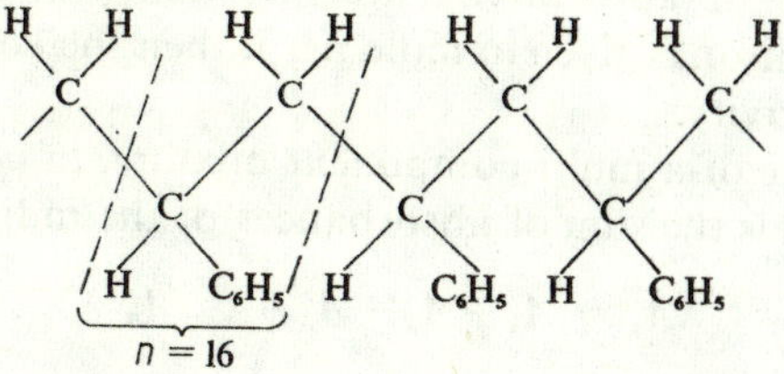

$n = 16$

which gives rise to $3 \times 16 - 6 = 42$ normal vibrations.

The changes in vibrational energy effected by electromagnetic radiation may occur only if the dipole moment of a definite bond or of the molecule as a whole varies with vibration. Thus not all vibrational modes of a molecule are active in the infrared spectrum. For example, no infrared absorption spectra can be observed for such homopolar molecules as H_2, O_2 or N_2. When a molecule is asymmetric it may happen that all of its vibrations are infrared active. This happens in the case of the vibrational modes of the group of atoms in the polystyrene structural unit.

The normal vibrational modes of molecular anharmonic oscillators are not mutually independent, and in the spectrum bands of other frequencies may appear which are the sums or differences of frequencies of various fundamental modes. These are referred to as combination bands.

Despite the complexity of infrared absorption spectra and problems with their interpretation, a considerable body of information on the structure of polymers can be obtained from the observed characteristic frequencies.

Absorption Laws

In all three spectral ranges mentioned the fundamental absorption law is

$$A = \log I_0/I = Klc$$

where:

A = absorbance,
K = absorption coefficient,
l = the thickness of the absorbing layer,
c = concentration of the solution,
I_0, I = the radiation intensities of the incident and transmitted beams respectively.

This relation is of interest in analytical determinations by absorption spectral analysis and is known as the Bouguer–Lambert–Beer law. It states that the absorption of a monochromatic radiation beam which passes through a one-component solution is directly proportional to the concentration of the solution and to the thickness of the absorbing layer. The absorption coefficient K is numerically equal to the absorbance at unit concentration and unit layer thickness. If the concentration is expressed in g per l. and the layer thickness in cm, then the absorption coefficient is termed specific absorbance or absorptivity, and its dimensions are l. cm^{-1} g^{-1}. When the concentration is given in mole per l. then the coefficient is referred to as molar absorptivity.

The absorbance of a multi-component mixture, in conformity with the additivity law, equals the sum of absorbances of the individual components

$$A_s = A_1 + A_2 + A_3 + \ldots A_n$$

where $A_1, A_2, A_3, \ldots A_n$ denote the absorbances of individual components at a definite wavelength. The fact that absorbance obeys the additivity law enables quantitative analysis of multi-component systems to be made.

In quantitative analysis the integral absorbance of an analytical band is increasingly used. This quantity B is expressed by the equation

$$B = \frac{K}{cl} \int_{-\infty}^{+\infty} \ln(I_0/I)\mathrm{d}\nu$$

Integral absorbance is found by measuring the area contained under the band contour. This area approximately equals the product of the band half-width and the band absorption value at its maximum. Hence we have

$$B = \frac{K}{cl} \ln[(I_0/I)]\, d_{1/2}$$

where $d_1/_2$ is the band half-width. In this way absorbance is characterised by three points instead of a single point on the band, thereby reflecting more truly the intra- and intermolecular interactions.

2.3 SAMPLE PREPARATION

According to the nature of the investigations, and the properties of the compound under examination, a suitable sample preparation technique is adopted. In the case of high molecular weight compounds analysis can be conducted with capillary films, solutions, films, Nujol suspensions (mulls), and in potassium bromide pellets.

Capillary Films

Liquids are most conveniently studied as capillary films. A drop of a liquid is placed between two potassium bromide or sodium chloride plates, which are fixed in the mounting and inserted into the spectrophotometer. In the reference beam of the instrument is placed a potassium bromide or sodium chloride plate twice as thick as that used in the sample cell.

The thickness of a capillary film depends primarily on the viscosity of the sample. Thin enough capillary films can rarely be prepared by this method with highly viscous liquids (ca. 70,000 cP) which exhibit strong absorption bands (such as silicone fluids). In such instances satisfactory results are achieved by applying a thin film on one of the plates. The high viscosity of the liquid ensures a constant film thickness during the measurement of the spectrum.

For highly volatile liquids of very low viscosity the capillary film technique is unsuitable and cells should be used instead.

Solutions

Spectral examinations of solutions are primarily used in quantitative analysis and are largely concerned not with the polymers themselves but with secondary compounds, additives and monomers present in polymers. The solvents are the same as those commonly employed in infrared and ultraviolet spectral analyses, the choice being controlled by the solubility of the polymer concerned. In single spectral ranges uncommon solvents may be used, for which compensation should be made.

The solution concentrations used depend on the absorption of the unknown and range from 10 to 100 g per l. at a cell thickness of 0·1 mm.

An advantage of solution analysis is that concentration can be controlled exactly, and absorbance can be measured under optimum conditions by a suitable selection of concentration and cell thickness. Moreover, the solution technique eliminates the effect of crystallinity and molecular interactions.

Films

One of the most frequent techniques of sample preparation in the spectral analysis of polymers is the preparation of films which are pressed or cast from solutions. Polymer solutions are spread over sodium chloride or potassium bromide plates, and the spectra measured after removal of the solvent under an infrared lamp or under reduced pressure. Since sodium chloride and potassium bromide plates are expensive, the films may also be cast from solutions onto glass plates or mirror surfaces, from which, on removal of the solvent, the films are stripped and inserted directly into the spectrophotometer beam.

Films of a uniform thickness may be obtained from solutions by the following procedure.

One end of a glass tube is closed with a damp cellophane sheet which becomes taut when dried. The tube is placed vertically on the flat smooth surface of a table, a suitable volume of a 5–10% polymer solution is poured in, and the top is partially covered with a watch glass to ensure slow evaporation of the solvent in order to produce a smooth film surface. When the solvent is removed, the film is separated from the cellophane by wetting with water. The thickness of such films can be controlled by the volume and concentration of the solution used. Prior to analysis the films are dried under high vacuum.

Solution-cast films are sometimes advantageous, especially when the sample contains some filler which strongly distorts the spectrum and must be removed. To do this, the polymer is dissolved in a suitable solvent, the solution is centrifuged free of the filler, and the film is prepared from the supernatant solution.

Very thin films of polymers from solutions may also be obtained on a water or mercury surface, but these techniques require considerable skill.

In many instances the pressed film technique is preferable. A hydraulic press with heating and cooling systems is required along with a metal mould with a well polished surface, and equipped with a metal spacer to make films of the required thickness.

The precision achieved in the determination of the composition of ethylene-propylene copolymers on samples prepared by either the pressed film technique or solution-cast technique was compared by Drushel and Iddings [1]. They concluded that the pressed film technique gives measurements of higher precision compared with the solution-cast technique. Non-uniformity of the films cast from solutions results in some error even if the ratio of the analytical band to the reference band is taken into consideration.

Occasionally polymer films show strong interference due to double reflection of the light beam from the two interfaces of the thin film [2] (Fig. 2.1).

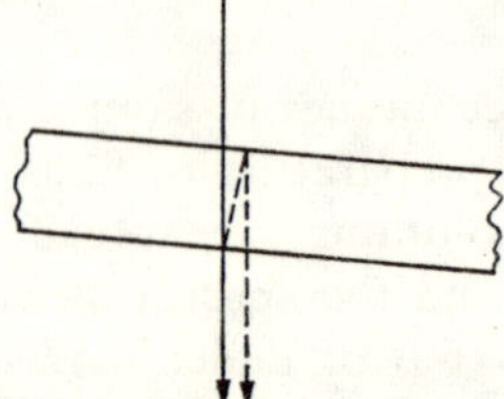

Fig. 2.1 The interference mechanism; broken line—the doubly reflected beam at the film-air interface; solid line—the direct beam

If the wavelength is 10 μm for instance, the index of refraction 1·5, and the film thickness 20 μm, then the doubly reflected beam is delayed by $2 \times 20 \times 1{\cdot}5 = 60$ μm. Since the difference in the path length is a multiple of the wavelength (60:10 = 6), the reflected beam will enhance the intensity of the unreflected beam. However, a beam of wavelength 10·8 μm will be attenuated by the doubly reflected beam, as the delay of 60 μm in the path length constitutes 5·5 wavelengths, which gives rise to a shift of a half wave. Owing to the alternately enhanced and attenuated main light beam, evenly spaced maxima and minima of absorption appear in the spectrum. The interference effect (Fig. 2.2) is observed especially in these spectral ranges where the absorption due to the polymer is low. The effect may be easily eliminated by careful rubbing of the film surface with sandpaper.

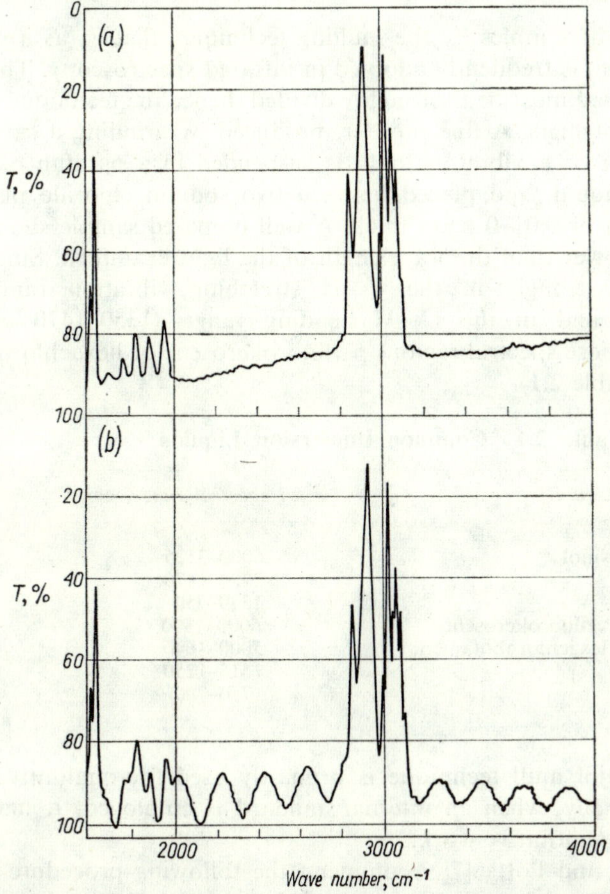

Fig. 2.2 Spectra of polystyrene: *a*—after elimination of interference by roughening one of the film sides, *b*—with interference fringes [2] (by permission of the copyright holders, Carl Hanser Verlag, München)

The thickness of the films to be measured may be varied over a wide range from 2 to 100 μm according to the absorption intensity of the band of the polymer under analysis. For polyolefins the film thickness should be about 20 μm and for polycarbonates or polysiloxanes about 4 μm.

For the examination of low intensity bands, or when minor quantities of functional groups in polymers are analysed, as for instance in the testing of the aging and weathering processes in polymers [3–6], a greater thickness of film must be used.

Among the major advantages of the film technique is easy storage of the samples and the possibility of repeated examination of the same sample. The latter is of particular value for examining aging processes in polymers.

Mulls

Preparation of samples by the mulling technique, that is, as a suspension in paraffin oil, is frequently adopted in infrared spectroscopy. The material to be examined must be thoroughly divided, hence the technique is suitable for brittle polymers. A fine powder, produced by grinding a sample in an agate mortar or a vibratory mill, is suspended in a paraffin oil or other immersion liquid, and placed between two sodium chloride plates so as to make a layer 0·01–0·1 μm thick. A well prepared sample should exhibit dispersion lower than the wavelength of the light examined. Since paraffin oil absorbs strongly in the C—H stretching vibration range (2800–3000 cm^{-1}) and in the C—H bending range (1350–1470 cm^{-1}), for analysis in those spectral regions perfluorokerosene or hexachlorobutadiene are used (Table 2.1).

Table 2.1 Common Immersion Liquids

Liquid	*Useful spectral range*, cm^{-1}
Nujol	5000–3125
	2780–1470
	1370–750
Perfluorokerosene	5000–1390
Hexachlorobutadiene	5000–1640
	1515–1250
	750–700

The Nujol mull technique is primarily used for qualitative analysis, but occasionally, when an internal standard is employed, it may also be adopted in quantitative work.

Bradley and Potts [7] recommend the following procedure in similar cases. A few mg of a suitable compound to serve as internal standard and about the same quantity of the unknown compound are ground in an agate mortar. Nujol oil is added and the grinding is continued to the desired dispersion. The mull is placed between two sodium chloride plates, inserted into the spectrophotometer, and the analytical lines of both the reference compound and the internal standard are recorded.

The ratio of the unknown to the internal standard should be 2:1 or 5:1 on a weight basis, depending on the ratio of the analytical lines chosen. Compounds usually taken as internal standards include potassium periodate, potassium bromate and lead thiocyanate.

Considerable difficulties are encountered in obtaining the spectra of polyurethane foams [8,9]. Their poor solubility makes the preparation of films cast from solution impossible, while owing to their high flexibility they cannot be ground in a mortar. To overcome these problems a procedure is recommended, in which the material is frozen to low temperatures and then ground in a vibratory mill or cut with a microtome into 10–20 μm slices which are next placed between two sodium chloride plates and immersed in Nujol. This latter operation is required to eliminate the interference effect due to double reflection at the air-polymer interface.

Pellets

Examination of compounds in the form of transparent pellets made of potassium bromide or chloride is used in both infrared and ultraviolet spectroscopy. In this technique the unknown is intimately ground with potassium bromide, then the powder is placed in a suitable mould. The sample is evacuated and pressed under a pressure of about 10 ton per cm^2. Too high a pressure may affect the crystalline structure of certain polymers. The chance of reactions such as substitution or association with potassium bromide in polymers is unlikely. For potassium bromide of high purity the transmittance of pellets 1 mm thick is about 95%.

Commonly the pellets contain 1–3% of the unknown. Rigid polyurethane foams may also be examined by this technique, but repeated grinding and pressing operations are required [8].

Adequate drying of potassium bromide and the sample is essential, as trace amounts of water result in strong bands at 1650 and 3450 cm^{-1}. Likewise, if air is not completely removed from the sample the tablets produced are of low transparency.

The potassium bromide pellet technique may be used occasionally in quantitative analysis. Here, however, the conditions of the preparation of pellets must be rigorously defined and observed, especially the way the sample is ground. The values of absorptivity of the compound examined in pellets are usually lower than in solutions. To eliminate the effect of non-uniform thickness internal standards such as potassium thiocyanate are added.

2.4 ULTRAVIOLET SPECTROPHOTOMETRY

Spectra in the ultraviolet range (200–400 nm) result from the changes in the electronic absorption of a molecule and are related to the transitions of π electrons in conjugated systems. This imposes a limitation on the usefulness of this spectral range for the analysis of polymers; only polymers with a considerable proportion of conjugated systems can be analysed by ultraviolet spectroscopy. On the other hand, even in polymers which are transparent in the ultraviolet region, a range of additives such as stabilisers and plasticisers, or impurities such as residual monomers or catalysts, can be identified and determined by this technique. Owing to high absorptivity in the ultraviolet region (an order of magnitude larger than in the infrared region) and because thicker layers can be used, minor quantities of the compound under analysis may be identified and determined. Frequently the order of detectability of a component under analysis may be increased by extraction or steam distillation.

Applications of ultraviolet absorption spectroscopy in the analysis of polymers are listed in Table 2.2.

Table 2.2 List of References to Ultraviolet Studies of Polymers

Polymer	*Determination*	*Reference*
Siloxane copolymers	copolymer composition	10, 58
Polyacrylonitrile	hydrolysis product	11
copolymer with butadiene	inhibitors	12
copolymer with methyl isopropenyl ketone	methyl isopropenyl ketone	13
copolymer with vinylmethylpyridine	vinylmethylpyridine	14
Poly(vinyl alcohol)	crosslinking	15
Polyamides	aging, monomer	16,17
Polybutadiene	styrene	18,19
styrene copolymer	inhibitors	12,20
Poly(vinyl chloride)	degradation products	21,22,23,24
vinyl acetate copolymer	phthalates	25,26
Polyethylene	polymerisation kinetics	27
	antioxidants	28,29,30
	aging	31
Polyisoprene		
natural rubber	colour	32
natural rubber/butadiene-styrene rubber composition	butadiene-styrene rubber	18
isobutylene copolymer	inhibitors	12
phenolformaldehyde copolymer	phenolic hydroxyl group	33
Poly(methyl methacrylate)	degradation products, plasticisers	11,34,35
acenaphthylene copolymer	copolymer composition	36
N-vinylcarbazole copolymer	copolymer composition	37
ethyl acrylate copolymer	additives, by-products	38
Polypropylene	antioxidants	39,40,41
Polystyrene	antioxidants, degradation	25,42,43,44
	monomer	45,46,47
	polymerisation kinetics	48
Polyurethanes	degradation in UV light	49
Polyvinylpyrrolidone	monomer	50
Polyvinyltoluene	monomer	51
Alkyds modified with styrene	combined styrene	52
Polyesters	styrene, phthalate	26,53,54
Phenolics	phenol	55
Urea and melamine resins	melamine	56
	urea, thiourea	57

2.5 INFRARED ABSORPTION SPECTROPHOTOMETRY

Qualitative Analysis

Infrared spectrophotometry is a versatile technique by which a wide range of investigations on polymers may be pursued. These include the identification and arrangement of functional groups in macromolecules, the formation of hydrogen bonds, and the progressive changes which occur during copolymerisation, oxidation, aging and other processes.

Identification is an essential objective in qualitative analysis. Using infrared spectrophotometry there are two distinct approaches to this objective. In the first, the spectrum of the unknown is compared with spectra of known compounds until a coincident known spectrum is found, and in the second the structure of the unknown is deduced by identifying the functional groups it contains. In the former case catalogues of spectra are needed [59–63] and in the latter correlation tables such as those drawn up by Bellamy [64].

Considerable care should be exercised in the interpretation of the spectrum since the nature and position of the absorption bands are affected by a number of factors such as intra- and intermolecular interactions, steric hindrance effects, high electronegativity of adjacent atoms, tacticity and crystallinity.

These factors make the differentiation of compounds of similar structure possible by the infrared absorption technique. To give a specific example, poly(ethyl acrylate) and poly(vinyl propionate) having the following structures may be considered:

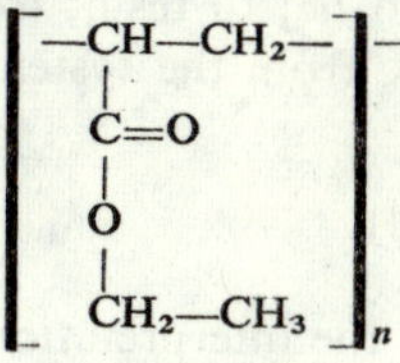

$$\left[\begin{array}{c}-CH-CH_2-\\ |\\ C{=}O\\ |\\ O\\ |\\ CH_2-CH_3\end{array}\right]_n\qquad\left[\begin{array}{c}-CH-CH_2-\\ |\\ O\\ |\\ C{=}O\\ |\\ CH_2-CH_3\end{array}\right]_n$$

Poly(ethyl acrylate) 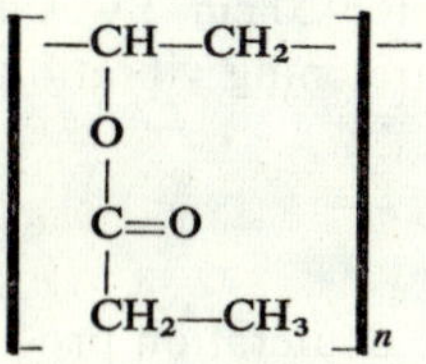 Poly(vinyl propionate)

These polymers differ only in the position of an oxygen atom in the molecule, but their spectra are clearly different (Fig. 2.3) so that the polymers can easily be distinguished.

Not infrequently model compounds have to be synthesised to interpret the spectrum. For the interpretation of the spectra of polymers it is advantageous to use model compounds incorporating a hydrogen or carbon isotope. The isotope shift of a fundamental frequency is found from the formula

$$\nu'/\nu = (\mu/\mu')^{1/2}$$

where:

ν, ν' = the fundamental frequency prior to and after the isotope substitution,

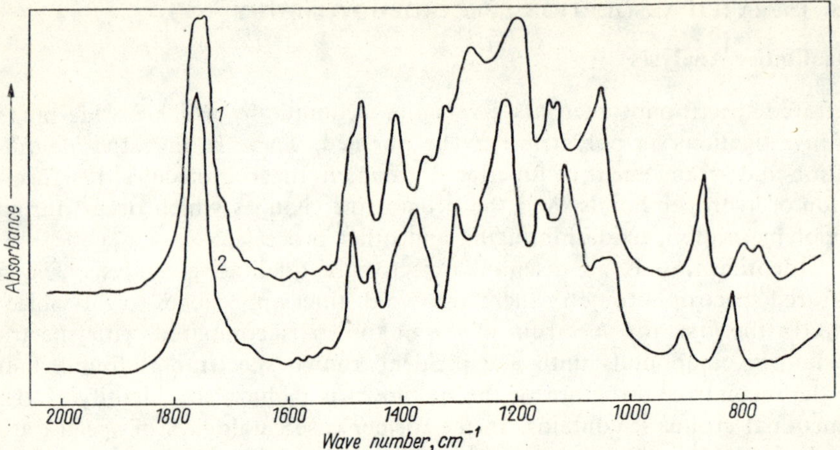

Fig. 2.3 Infrared spectra of poly(ethyl acrylate) (*1*) and poly(vinyl propionate) (*2*) [64a] (by permission of the copyright holders, Academic Press, London)

μ, μ' = the reduced masses of the atoms prior to and after the isotope substitution.

By replacing hydrogen by deuterium the fundamental frequency is decreased by almost 30%. By comparing the spectra of a molecule prior to and after the deuteration it can be found which absorption bands are shifted. This procedure was found helpful in establishing which of the 1100–1400 cm^{-1} polystyrene bands are due to C—H groups on the chain and which originate from the ring C—H vibrations [65]. Also in polypropylene the band at 2920 cm^{-1} was proved [66–68] to result from the tertiary carbon C—H stretching vibrations, that is, from the system:

$$\begin{array}{c} \text{D} \\ | \\ \text{—CH}_2\text{—C—CH}_3 \\ | \end{array}$$

Examples of the deuteration procedure for the interpretation of the spectra of polymers are listed in Table 2.3.

It should be mentioned that polymers cannot always be deuterated directly and that often hydrogen must be replaced by deuterium at the monomer stage.

Table 2.3 Infrared Studies Using Deuterated Polymers

Polymer	*Reference*
Polyacrylonitrile	69
Polybutadiene	70
Poly(vinyl chloride)	71
Poly(methyl methacrylate)	72
Polypropylene	66,67,68
Polystyrene	65

If the polymer is to be identified in a finished product, any plasticisers and fillers present should first be removed, as these additives materially distort the spectrum of the polymer and make its interpretation much more difficult.

Infrared spectroscopy is frequently helpful in differentiating homopolymer compositions from the corresponding copolymers, because the spectra of mixtures of homopolymers exhibit distinct differences from the spectra of their copolymer counterparts, owing to the diverse arrangement of the structural monomer units in the copolymer chains.

Frequently even minor quantities of individual monomer units can be identified by infrared technique in homopolymer compositions and copolymers, provided these monomer units exhibit bands of high absorptivity in a range entirely free of absorption by other components. Examples of this situation are styrene-acrylonitrile and styrene-butadiene copolymers (Figs. 2.4 and 2.5). The bands due to acrylonitrile (2250 cm^{-1}) and butadiene (973 cm^{-1}) are easy to identify, as they occur in the region free of polystyrene absorption.

Fairly clear differences are likewise observed in the spectra of polymers comprising *trans* and *cis* systems, as may be observed by comparing the spectra of *cis*- and *trans*-1,4-polybutadiene in Figs. 2.6 and 2.7.

The near infrared region (0·8–2·5 μm) is particularly useful for the analysis of polymers because spectrophotometers of simple design may be used (quartz prism, tungsten lamp as the light source and glass cells). Also many instruments designed for the ultraviolet and visible spectral range may be used in the near infrared region as well. Furthermore, in this spectral range thick sample layers can be employed (1–10 cm). This makes the use of higher overtones possible (2ν, 3ν) and these are of assistance in providing additional evidence on the structure [73–76].

Absorption bands in the near infrared region are usually due to the harmonics of the fundamental vibration modes of the C—H, N—H and O—H bonds, as well as to combination bands. Since the intensity of overtones is appreciably lower than that of the fundamentals, the only overtones

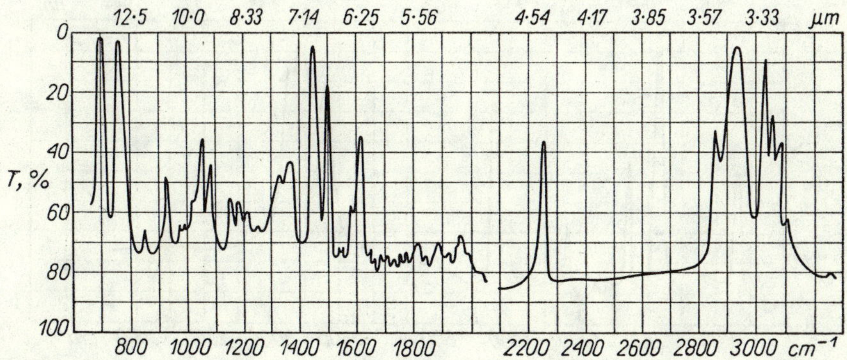

Fig. 2.4 Infrared spectrum of the (70:30) styrene-acrylonitrile copolymer; pressed film

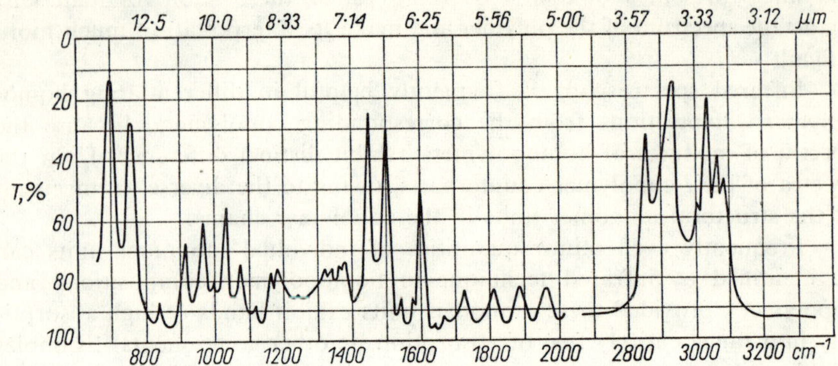

Fig. 2.5 Infrared spectrum of the (95:5) styrene-butadiene copolymer; pressed film

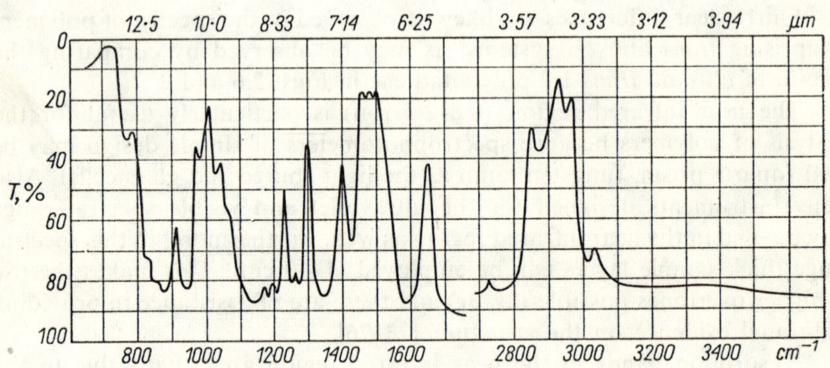

Fig. 2.6 Infrared spectrum of *cis*-1,4-polybutadiene; pressed film

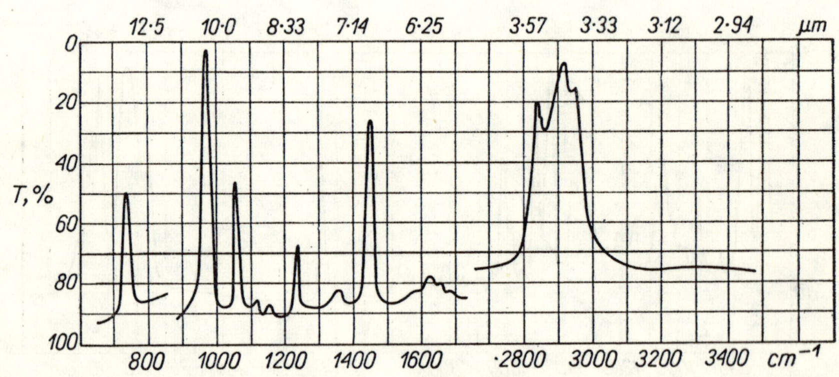

Fig. 2.7 Infrared spectrum of *trans*-1,4-polybutadiene; pressed film

which can be detected are those resulting from groups possessing strong absorption bands.

In overtones in the near infrared region a better separation of bands is accomplished than in the fundamental band region. If for instance two bands in the fundamental frequency region are separated by 5 cm^{-1}, for the first overtone this distance will be 10 cm^{-1}, and for the third overtone, 15 cm^{-1}. Thus, not only can CH_2 groups be identified in the presence of CH_3 groups, but also the position of the CH_2 groups in the molecule can be determined.

The positions of the characteristic bands of typical groups are listed in Table 2.4.

Table 2.4 The Near Infrared Absorption Ranges of Functional Groups

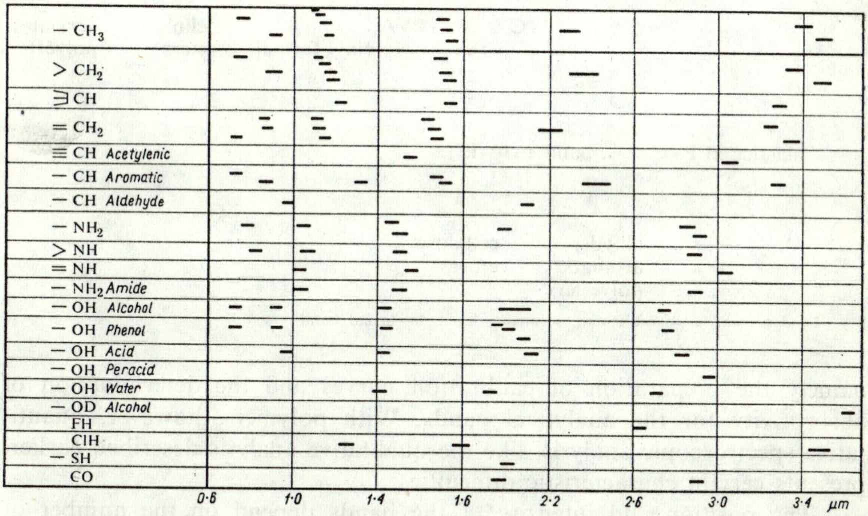

Group	0·6	1·0	1·4	1·8	2·2	2·6	3·0	3·4 μm
$-CH_3$								
$>CH_2$								
$\exists CH$								
$=CH_2$								
$\equiv CH$ *Acetylenic*								
$-CH$ *Aromatic*								
$-CH$ *Aldehyde*								
$-NH_2$								
$>NH$								
$=NH$								
$-NH_2$ *Amide*								
$-OH$ *Alcohol*								
$-OH$ *Phenol*								
$-OH$ *Acid*								
$-OH$ *Peracid*								
$-OH$ *Water*								
$-OD$ *Alcohol*								
FH								
ClH								
$-SH$								
CO								

Correlation of the bands of the CH_2, NH, OH and CO groups for several polymers was made by Foster [74].

A diagram for systematic identification of polymers in terms of their infrared absorption spectra is given in Tables 2.5 and 2.6.

Major collections of the spectra of polymers are described in the monographs by Hummel [59], and by Haslam and Willis [60]. The spectra of a number of plasticisers are given in the work by Meise and Ostromow [62] and the spectra of hardeners and catalysts for epoxy resins are gathered in the paper by Serboli [63].

Quantitative Analysis

INTRODUCTION

Quantitative analysis is used to determine the polymer content and the quantities of additives or other compounds present in the polymer. The principles and procedures common to all spectrophotometric methods are used,

Table 2.5 Polymer Identification Diagram from Infrared Spectra
(after Tryon and Horowitz [78], by permission of the copyright holders, John Wiley & Sons, Inc.)

Spectrum of unknown†
The C=O band at ca. 1740 cm^{-1} is present
Bands of aromatic compounds at 1605, 1590, 1490 cm^{-1}

- p: s. band at 830
  - p: modified epoxy resins
  - a: sharp band at 1430
    - p: plasticised PVC
    - a: s. band 1330–1212
      - p: alkyds, aromatic polyesters
      - a: cellulose ethers
- a: sharp band at 1430
  - p: b. band at 690
    - p: PVC copolymer
    - a: PVA, poly(vinyl formal)
  - a: b. s. band at 1050
    - p: cellulose esters
    - a: acrylates, polyesters

† p—band present, a—band absent, s—strong band, b—broad band.

namely the preparation of calibration curves and the determination of absorptivity for the analytical bands. With polymers, however, quantitative spectroscopic analysis, like the qualitative analysis described earlier, presents certain characteristic difficulties.

The position and intensity of the bands depend on the number of atoms combined in a group, on the strength and nature of the bonds, and also on the geometry of the molecule as a whole. The last factor is of major importance in high molecular weight compounds. Absorption bands of the same functional groups differ in position and intensity depending on the environment or chain length. Thus, their determination requires the use of some standard polymers in which the content of functional groups is known.

The use of polymer standards is not always possible, as the functional groups of interest have to be determined in the standard by another method and this is sometimes difficult. One has often to resort to oligomers or compounds of similar structure as standards. Specifically, for the determination of the CH_3 groups in polyethylene, higher alkanes such as octane, nonane or decane were frequently taken as standards [79,80], whereas for the analysis of C=C groups, higher alkenes [81] were used.

It should be emphasised that the values of absorbance reported in the literature cannot be used directly in studies made with other instruments,

Table 2.6 Polymer Identification Diagram from Infrared Spectra
(after Tryon and Horowitz [78], by permission of the copyright holders, John Wiley & Sons, Inc.)

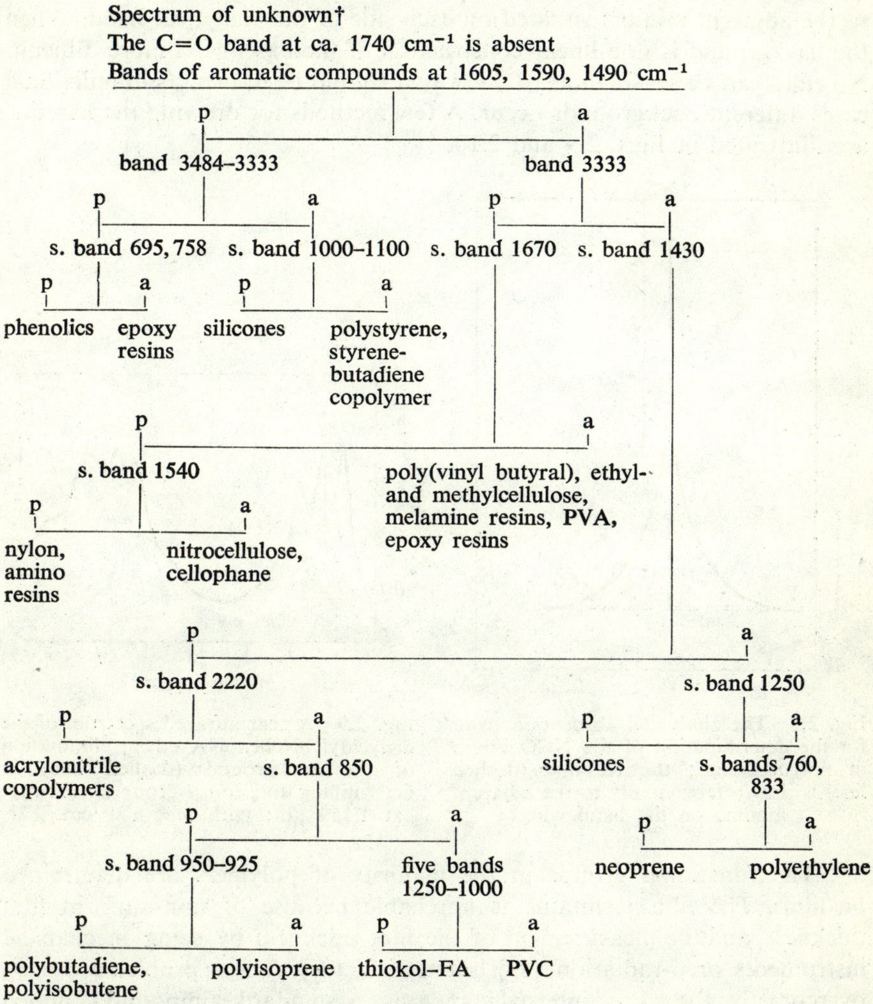

Spectrum of unknown†
The C=O band at ca. 1740 cm^{-1} is absent
Bands of aromatic compounds at 1605, 1590, 1490 cm^{-1}

- p: band 3484–3333
  - p: s. band 695, 758
    - p: phenolics
    - a: epoxy resins
  - a: s. band 1000–1100
    - p: silicones
    - a: polystyrene, styrene-butadiene copolymer
- a: band 3333
  - p: s. band 1670
    - p: s. band 1540
      - p: nylon, amino resins
      - a: nitrocellulose, cellophane
    - a: poly(vinyl butyral), ethyl- and methylcellulose, melamine resins, PVA, epoxy resins
  - a: s. band 1430
    - p: s. band 2220
      - p: acrylonitrile copolymers
      - a: s. band 850
        - p: s. band 950–925
          - p: polybutadiene, polyisobutene
          - a: polyisoprene
        - a: five bands 1250–1000
          - p: thiokol–FA
          - a: PVC
    - a: s. band 1250
      - p: silicones
      - a: s. bands 760, 833
        - p: neoprene
        - a: polyethylene

† p—band present, a—band absent, s—strong band.

as these values depend on the slit width, resolving power, and various aberrations (such as light scattering). Thus the mechanical and electronic precision of a spectrophotometer as well as the way the baseline is drawn, determine to some extent the apparent absorbance [64a].

The recording spectrophotometers in widespread use today are commonly designed to give the percentage transmittance. The position of the band maximum on the transmittance scale (T) in the spectrum depends not only on concentration but also on the properties of the sample, and on many other factors as well. For this reason the absorbance is read off by drawing a baseline as illustrated in Fig. 2.8.

When the background is linear the baseline is drawn as a tangent to the adjacent minima situated on each side of the analytical band. When the background is non-linear construction of the baseline is more difficult. No rules can be set forth as to the way it should be drawn, as in individual cases different backgrounds occur. A few methods for drawing the baseline are illustrated in Figs. 2.9 and 2.10.

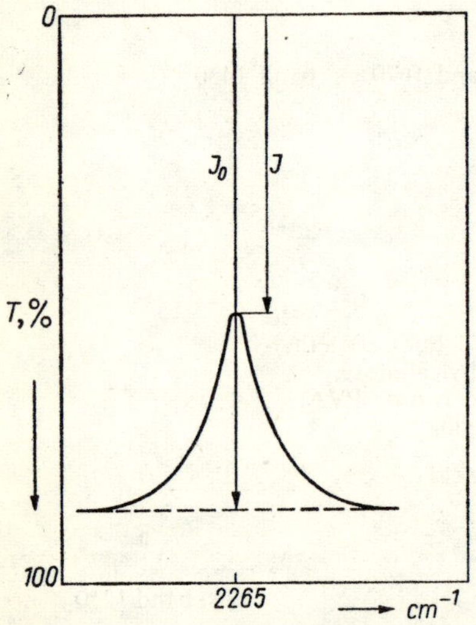

Fig. 2.8 The analytical absorption band for the determination of the NCO group in polyurethanes; the baseline (dashed line) is drawn tangentially to the adjacent minima on the band wings

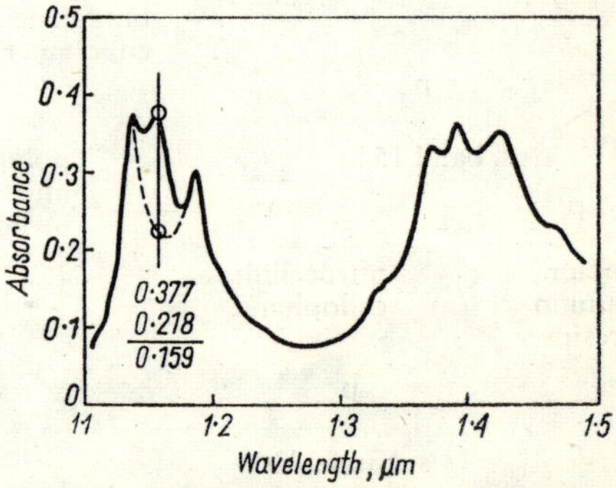

Fig. 2.9 A near infrared spectrum of the diglycidyl bisphenol A ether; elimination of the background (dashed line) in determining the epoxy group absorbance at 1·159 μm; path length 1 cm [82]

The functional groups in the majority of polymers are determined on films. The film technique is unreliable because of non-uniform film thickness, and the measurement of the film thickness by using mechanical instruments or β-radiation [83] is unsatisfactory. These problems can be overcome by the use of internal standards. A standard compound is added in known quantity to the sample under analysis. The standard should exhibit absorption in the region free of absorption from the polymer. The ratio of absorbance of two bands is determined. If the absorbance of the band due to the functional group of interest is A_1, and that of a band from the standard of known concentration is A_2, then in conformity with the Bouguer–Lambert–Beer law one can write:

$$A_1/A_2 = K_1 l c_1 / K_2 l c_2 = K_1 c_1 / K_2 c_2$$

The equation may be simplified by dividing the numerator and denominator by l since the values of the thickness of the absorption layers are equal.

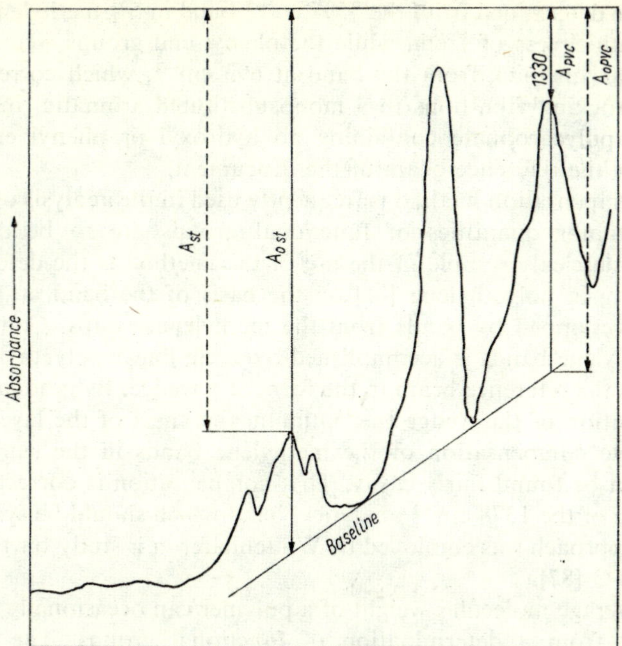

Fig. 2.10 Infrared spectrum of poly(vinyl chloride) stabilised with calcium laurate; the baseline drawn for the determination of the stabiliser content in PVC

To eliminate the effect of film thickness a suitable absorption band of the polymer itself is frequently taken. This serves as an internal standard and is referred to as the reference band (see Fig. 2.10). For instance, in the determination of the composition of a vinyl chloride-vinyl acetate copolymer the band at 1435 cm^{-1} (due to the CH_2 deformation vibrations), which is common to both monomer units, is used as an internal standard. The ratio of absorbance of the carbonyl band at 1730 cm^{-1} to that of the reference band at 1435 cm^{-1} determines the composition of the copolymer [84].

Similarly, in the determination of the CH_3 group content in polyethylene the following bands were used as the reference bands: 1470, 4270 and 4354 cm^{-1} [85].

The reference band and the internal standard techniques are extensively used in quantitative analysis of polymers.

Determination of the end groups is of great help in studies on the structure of polymers and in investigations on polymerisation reaction mechanisms. The phenyl and hydroxyl end groups in polycarbonates, present in quantities of 0·05%, were determined by Horbach *et al.* [86]. Analysis of such low quantities requires operating with thick layers and is possible only when a compensation technique is adopted. The hydroxyl groups were determined from the 3597 cm^{-1} band in 5% methylene chloride with a cell thickness of 1 cm, while the phenyl end groups were measured on films 150 μm thick from the band at 693 cm^{-1}, which corresponds to the C—H rocking vibrations in a monosubstituted aromatic ring. In both analyses a polycarbonate containing no hydroxyl or phenyl end groups was used in the reference beam of the instrument.

The compensation method is frequently used in the analysis of polymers whenever minor quantities of functional groups are to be determined [85,86]. A classical example of the use of this method is the determination of branching in polyethylene [85] on the basis of the band at 1378 cm^{-1} which is overlapped by bands from the methylene groups. Compensation of the methylene bands is accomplished by using linear polyethylene which is placed in the reference beam in the form of a wedge. By gradually adjusting the position of the wedge the optimum thickness of the layer required for complete compensation of the methylene bands in the sample under analysis can be found fairly easily. That compensation is correct is judged by the form of the 1378 cm^{-1} analytical band which should be symmetrical. A similar approach was employed by Whitenhafer in a study on the crystallinity of PVC [87].

The average molecular weight of a polymer can occasionally be evaluated [88,89] from a determination of functional groups. The molecular weight of polyoxymethylene diacetate was thus estimated by Majer [90], who measured the absorbance of the carbonyl band at 1755 cm^{-1}.

Likewise, a method for the determination of the average molecular weight of poly(ethylene oxide) and poly(propylene oxide), on the basis of the OH and =CH— bands, was worked out by Rossi and Maganasco [89], while in low molecular weight polyesters (mol. wt. 420–1260) this parameter was determined from the bands at 3420 and 2632 cm^{-1} by Shabalin and Kiva [91].

In numerous instances polymerisation, isomerisation and cyclisation reactions can be followed with a high degree of accuracy by determining the variations in absorbance of the bands related to the C=C system [88,81]. In analytical practice advantage is taken primarily of bands in the range of 800–1000 cm^{-1}, which are due to the deformation modes of the C—H bonds attached to the double bond. In this range both *trans* and *cis* isomers, and vinyl in the presence of vinylidene groups, can be determined [81,92,93].

Sometimes the ratio of absorbance of some functional groups is sufficient. Thus Fishl and Young [94] and other investigators [95,96], characterised silicone resins by means of the ratio of absorbance of the methylene and phenyl groups, A_{1260}/A_{1430}.

The analysis of the composition of copolymers resolves itself essentially into the determination of the functional groups characteristic of the individual components [97]. Frequently the polymer composition may be established by using homopolymers as standards [98,99]. Thus the composition of a propylene oxide-tetrahydrofuran copolymer was determined by Kurengina *et al.* [100]. The bands at wavenumbers 2800 and 1340 cm^{-1} were selected as the contributions of tetrahydrofuran and propylene oxide respectively. A calibration curve was constructed, using the absorbance at these wavelengths for homopolymer mixtures of known composition, for subsequent copolymer analyses. Similarly the composition of styrene-isooctyl methacrylate copolymer was determined by means of homopolymer standards [98].

This procedure is reliable, however, only if the absorbances, of the analytical band in the homo- and copolymer are the same and if homopolymers are not formed in the copolymerisation process.

Alternatively, the copolymer composition may be found indirectly by evaluation of the unreacted monomers remaining in the post-reaction mixture.

Quite often in the quantitative analysis of polymers use is made of the near infrared region (0·8–2·5 μm). Mention may be made here of the determination of end groups in epoxy resins on the basis of the first overtone of the epoxide ring C—H stretching vibrations near 6070 cm^{-1} and the combination band at 4550 cm^{-1} [82].

Considerable experimental problems are met in the determination of hydroxyl groups in polymers because of association of various kinds. The OH hydrogen may form bonds with another OH oxygen or with another polar group. Thus intra- or intermolecular hydrogen bonding may be produced. In non-polar solvents the intermolecular association can be eliminated and the determination is concerned merely with free hydroxyl groups. However, the investigation of curing and crosslinking processes requires determination of the hydroxyl groups in materials that transform from liquids to solids; the task then becomes more involved.

Sometimes in similar systems the hydroxyl group content can be successfully determined from the isosbestic point. In order to follow the curing process of epoxy resins Dannenberg evolved a procedure for the determination of hydroxyl groups, making use of the isosbestic point at 1·474 μm (Fig. 2.11). He showed the existence of this point by using model compounds and proved its usefulness and obedience to the Bouguer–Lambert–Beer law.

Primary and secondary amino groups can be determined in a variety of polymers from the combination band at 4962 cm^{-1}, the first overtone of the N—H stretching vibrations at 6580 cm^{-1} and 6329 cm^{-1}, and the second overtone at 9615 and 9434 cm^{-1} [82,76].

The composition of the vinyl chloride-vinyl acetate copolymer was estimated by means of the band at 4651 cm^{-1} (C=O), whereas the unreacted vinyl acetate monomer was estimated by the band at 6135 cm^{-1}, which is characteristic of the vinyl group [76].

Ethylene-propylene copolymers were also analysed in the near infrared region [101].

The moisture content in polymers is determined from the band at

5200 cm^{-1} [75,102]. Hydrogen bonding between water and the polymer is likely to occur, so that the intensity and the position of the band depend on the polymer type. Therefore a calibration curve should be made independently for every polymer [102].

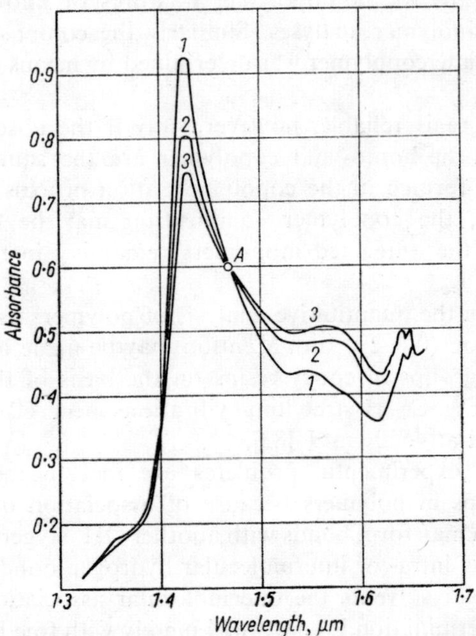

Fig. 2.11 **Infrared spectra of α-phenylglycidyl ether gradually cooled (*1, 2, 3,*) down from a temp. of 115°C [82]; path length 3 mm; *A* —the isosbestic point at 1·474 μm (6784 cm^{-1})**

Investigation of Aging Processes

Nearly all plastic materials undergo aging which manifests itself in altered mechanical and dielectric properties, changed molecular weight, colour, and changes in the infrared spectrum.

The aging process as observed in the infrared spectrum is reflected by decreased intensity of some of the bands and increased intensity of others. On the other hand, some of the bands may increase in intensity in the initial stages of the aging process and decrease later.

Based on a detailed analysis of the infrared spectra of polymers from aged and non-aged materials it was demonstrated that the aging process may be followed in the stretching vibrations of the groups C=O, C—O, O—H, and in the C—H deformation vibrations for C—H groups attached to unsaturated bonds.

The C=O *Stretching Vibration Range* (1600–1900 cm^{-1})

In the aging process of polymers oxidation reactions are of prime significance and result in the formation of carbonyl groups. Identification of these groups by infrared absorption spectrophotometry is fairly simple for a number of polymers, since the C=O stretching vibration gives rise to a band with a high absorptivity, and a great many polymers show no absorption in this range.

The appearance of carbonyl groups is often the first sign of aging in polymers. For example, in aged polyethylene bands appear in the range 1700–1800 cm^{-1} which, as the aging process proceeds, become increasingly complex (Fig. 2.12). A more detailed analysis of this band under various aging conditions and with the help of model compounds, rendered possible the identification of the band components [80,103,104,105].

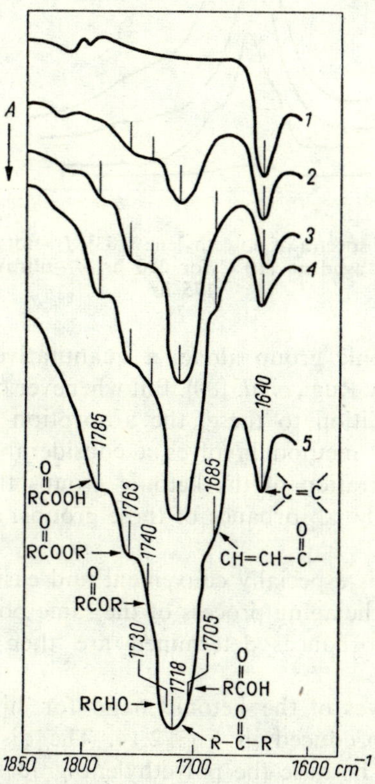

Fig. 2.12 Infrared spectra of thermally aged polyethylene (at 145°C) with time [107]: *1*—not aged, after: *2*—2 hr, *3*—5 hr, *4*—6 hr, *5*—24 hr aging (by permission of the copyright holders, John Wiley & Sons, Inc.)

As seen from Fig. 2.12, bands at 1705, 1718, 1730 and 1740 cm^{-1} may be assigned to acidic, ketonic, aldehydic, and ester carbonyl groups respectively [106]. The bands at 1763 and 1785 cm^{-1} are ascribed to peracids and peresters [103].

The amount of carbonyl group in any individual type in the polymer depends on the conditions and the time of aging. Thermal aging of polyethylene results in the first instance in the formation of ketonic groups, while exposure to ultraviolet light produces mainly aldehydic groups, and γ-irradiation yields a prevalence of ester groups [103,107]. (Fig. 2.13).

A quantitative estimation of the carbonyl content creates considerable difficulties because the use of model compounds is required. For thermally aged polyethylene, for which the band contour is the simplest and can be

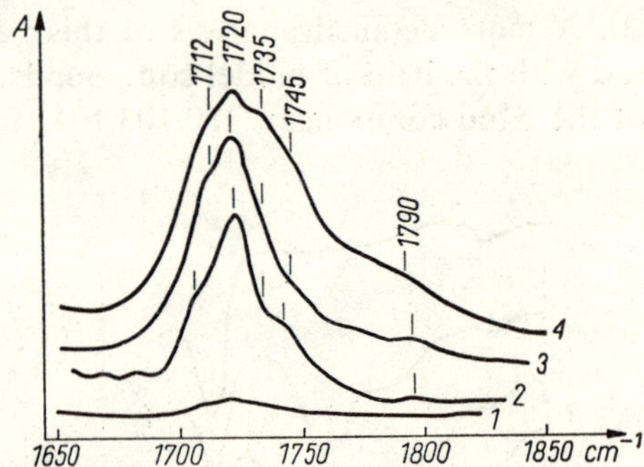

Fig. 2.13 Infrared spectra of polyethylene [103]; *1*—not aged, *2*—γ-radiation aged, *3*—thermally aged at 110°C for 250 hr, *4*—ultraviolet light aged for 175 hr

assigned to the ketonic group alone, a quantitative procedure was successfully developed by Rugg *et al.* [80]. But whenever other types of carbonyl groups occur in addition to these, the absorption band becomes broad and the quantitative method involves a considerable error. In this case it is not the concentration of the ketonic groups that is determined but only the increase in the absorbance of these groups, and the kinetic curves are plotted.

This procedure is especially convenient and easily used in comparing individual stages of the aging process of the same polymer under the same conditions. The absorbances determined are then proportional to the C=O group content.

The kinetic curves of the ketonic band for high- and low-pressure polyethylene are reproduced in Fig. 2.14. The dissimilar nature of the kinetic curves arises because the polyethylene grades have different densities, which directly affect the diffusion of oxygen into the sample bulk.

Investigations on the aging process in polypropylene [108], PVC [103], polystyrene [109–112], and poly(vinyl acetals) in this spectral range lead to the same results as those obtained with polyethylene [4,103,105–107].

Initially carbonyl groups of the ketonic and aldehydic types in the polymer grow in number, but as aging proceeds carbonyl groups of carboxylic and ester origin accumulate.

The C—O *Stretching Vibration Range*

Of lesser diagnostic value is the range of 1000–1300 cm^{-1}, connected with

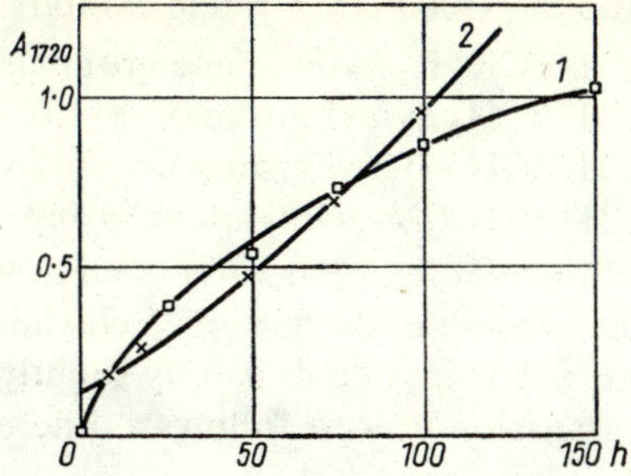

Fig. 2.14 The kinetic curves of the carbonyl bands of low-pressure *1*, and high-pressure *2*, polyethylene obtained during aging with ultraviolet light [103]

the C—O stretching vibrations. In this range nearly all polymers exhibit strong bands, making it difficult to follow changes in the C—O band.

Polyethylene has no absorbance in this range and the variations in the spectrum as the aging process advances can be analysed easily (Fig. 2.15). A fairly strong band at 1175 cm^{-1} and a number of bands of

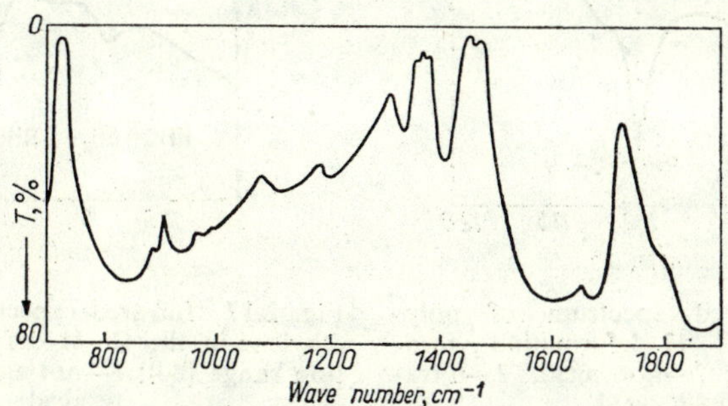

Fig. 2.15 Infrared spectrum of high-pressure polyethylene weathered for 6 months; film 0·2 μm thick

lower intensity at about 1000, 1050, 1100 and 1240 cm were found in aged polyethylene. The bands at 1175 and 1240 cm^{-1} are assigned to ester systems [103], whereas the bands in the region of 1000 cm^{-1} and 1100 cm^{-1} are related to hydroxyl and ether groups. Groups of this type are very likely to appear in other aged polymers.

The Vibration Range of Unsaturated Systems

It has been found that in aged polymers [80] a diversity of unsaturated bonds may be formed, including isolated, conjugated and aromatic types. All these structural features may be observed in the following three spectral ranges:

700–1000 cm^{-1}, deformation vibrations of C—H at C=C,
1500–1700 cm^{-1}, the C=C stretching vibrations,
3000–3100 cm^{-1}, stretching vibrations of C—H at C=C.

Polyethylene in the 850–1000 cm^{-1} range exhibits absorption bands at
888 cm^{-1} from $CH{=}CR_1R_2$ (vinylidene groups),
910 cm^{-1} from $CH{=}CH_2$ (vinyl groups),
990 cm^{-1} from $CH{=}CH_2$ (vinyl groups),
965 cm^{-1} from CH=CH (*trans* vinylidene groups).

On exposure to ultraviolet light the vinyl band at 910–990 cm^{-1} clearly grows in intensity at the expense of the 888 cm^{-1} vinylidene band (Fig. 2.16), whereas in thermal aging the vinyl band is only slightly affected (Fig. 2.17), indicating that the degradation process follows a different path in each case.

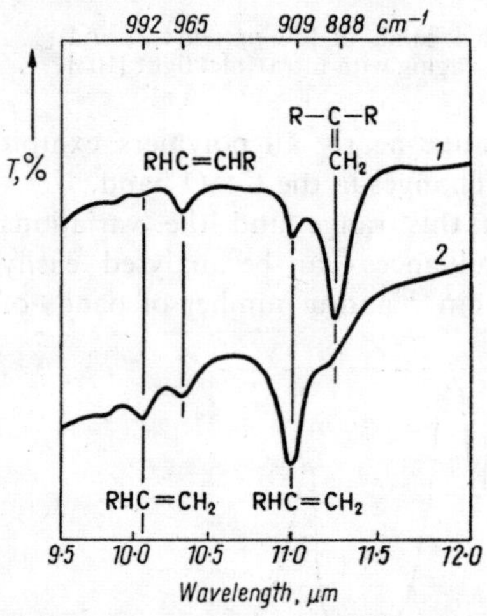

Fig. 2.16 Infrared spectrum of polyethylene in the C—H deformation vibration range [80]; *1*—not aged, *2*—ultraviolet aged
(by permission of the copyright holders, John Wiley & Sons, Inc.)

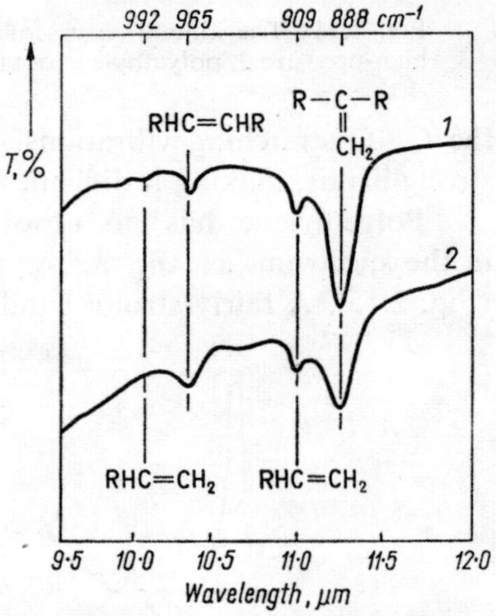

Fig. 2.17 Infrared spectrum of polyethylene in the C—H deformation vibration range [80]; *1*—not aged, *2*—thermally aged
(by permission of the copyright holders, John Wiley & Sons, Inc.)

In the C=C stretching vibration range in the spectrum of thermally aged polyethylene two bands occur at 1640 and 1685 cm^{-1}, as seen from Fig. 2.12. The former is characteristic of the stretching vibrations of unsaturated systems, whereas the latter is observed when these systems are conjugated to a carbonyl group, such as

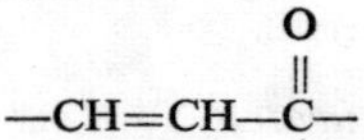

$$-CH{=}CH-\overset{\overset{\displaystyle O}{\|}}{C}-$$

These changes may be explained on the assumption that the oxidation of polyethylene proceeds according to the reaction

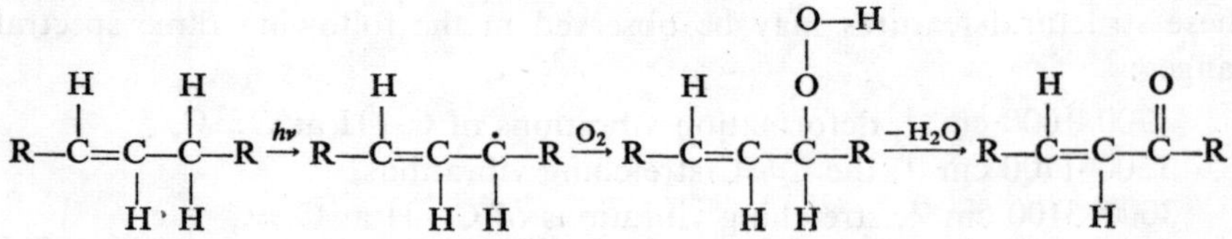

$$R-\overset{H}{\overset{|}{C}}{=}\overset{H}{\overset{|}{C}}-\underset{H}{\underset{|}{\overset{}{C}}}(H)-R \xrightarrow{h\nu} R-\overset{H}{\overset{|}{C}}{=}\underset{H}{\underset{|}{C}}-\underset{H}{\underset{|}{\dot{C}}}-R \xrightarrow{O_2} R-\overset{H}{\overset{|}{C}}{=}\underset{H}{\underset{|}{C}}-\underset{H}{\underset{|}{\overset{O-O-H}{\overset{|}{C}}}}-R \xrightarrow{-H_2O} R-\overset{H}{\overset{|}{C}}{=}\underset{H}{\underset{|}{C}}-\overset{O}{\overset{\|}{C}}-R$$

as a result of which the quantity of CH_2 groups in the polymer decreases (the 2810–2870 cm^{-1} band decreases in intensity) with the simultaneous formation of hydroperoxides and unsaturated ketones (new bands at 3555 and 1685 cm^{-1} respectively).

The validity of this concept was substantiated by the studies made by Luongo [107].

The OH *Stretching Vibration Range*

The presence of hydroxyl groups may be observed in the range 3000–3700 cm^{-1}. In polyethylene and poly(vinyl chloride) (Fig. 2.18), during aging, broad absorption bands appear in this range. The nature of these bands indicates the presence of associated OH groups which may be assigned to the formation of alcoholic systems [80,103,107].

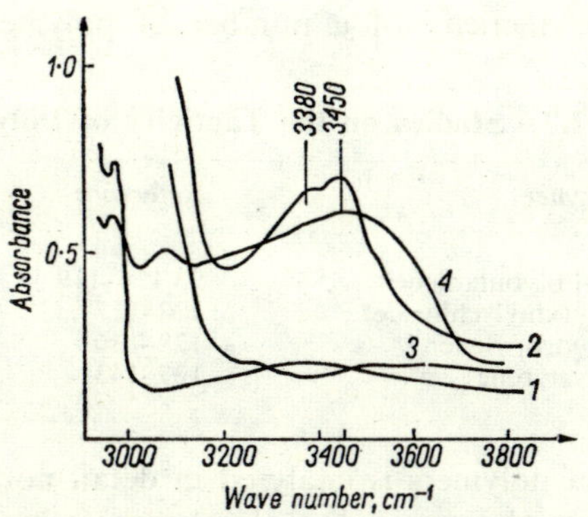

Fig. 2.18 Infrared spectra of polyethylene and PVC in the OH stretching vibration range [103]; *1*—polyethylene not aged, *2*—polyethylene aged at 175°C for 54 hr, *3*—PVC not aged, *4*—PVC ultraviolet aged for 75 hr

In studies of polymer aging it should be borne in mind, however, that the drop in intensity of a band does not always imply that a given group disappears. The variation in intensity of the band of a group under analysis may be due to some new structural systems that arise as a result of aging. For instance in polyethylene the formation of the carbonyl group results in a shift of the 1465 cm^{-1} band, due to methylene groups, to a value of 1410 cm^{-1} when the methylene groups are directly linked to the carbonyl group. As a consequence, the fall in intensity of the 1465 cm^{-1} band is much greater than would be expected solely on the grounds of the degradation reaction of the methylene group [103].

Spectrophotometric studies of the effects of decay of some of the groups and the formation of others enable some interesting conclusions to be drawn concerning the mechanism of the polymer aging process [49,113].

Tacticity and Crystallinity

As distinct from low molecular weight compounds the structure of synthetic polymers is non-uniform in respect of chemical composition, length of macromolecules and tacticity.

For the description of the tacticity of a polymer chain Bovey and Tiers [114], and Miller and Nielsen [115] were the first to advance a concept of triads of the polymer structural units. These triads may be defined as isotactic, syndiotactic or atactic depending on whether the triad central unit is attached to two units of the same configuration, to two units of the opposite configuration, or to one unit of the same and another of the opposite configuration, respectively.

Differences in tacticity are reflected in the infrared spectra of polymers. Hence, the type and degree of tacticity of polymers can be established on the basis of spectrophotometric measurements. Some references to studies concerned with the tacticity of a number of polymers are collected in Table 2.7.

Table 2.7 Studies on the Tacticity of Polymers

Polymer	*Reference*
1,4-Polybutadiene	93,116–119
Poly(vinyl chloride)	120–127
Polypropylene	128–136
Polystyrene	137–143

The tacticity of polymers is analysed in detail not only in view of its considerable theoretical interest, but also because this property is closely related to crystallinity and thus with specific useful properties of polymers. Investigation of the crystallinity of polymers by infrared absorption is based on the bands that characteristically appear in polymer spectra, and which are due to the presence of crystalline or amorphous phases. For convenience the bands are termed here "crystalline" and "amorphous" [144,145]. To determine the content of the crystalline phase the experimental spectrum of the polymer under analysis and a knowledge of the bands characteristic of the crystalline state are required.

If the absorbance in the crystalline (or amorphous) band in question is known together with the density of the sample, the following formula may be used

$$A = Kxdl$$

where:

A = absorbance,
K = absorptivity,
x = the fraction by weight of the crystalline material,
d = sample density,
l = sample thickness.

The degree of crystallinity of polymers may also be determined by the method advanced by Nikitin and Pokrovskii [145]. In this method standard samples of known crystallinity are not required, and the polymer in the molten state is used as a completely amorphous standard. On the basis of two measurements of absorbance of the amorphous band, one taken at room temperature and the other at any temperature t above the melting point, the percentage of the crystalline form in the polymer is calculated from the formula:

$$c_{\text{cryst}} = [1-(A_{\text{room}}/A_t)]\,100$$

This relationship was derived by Nikitin and Pokrovskii on the basis that the absorbance of the amorphous band increases with temperature to attain a constant value at and beyond the polymer melting point, since at the m.p. there is only amorphous polymer present.

The use of the amorphous bands for the determination of the quantity of the amorphous material in the polymer has a number of advantages, such as the following.

1. The amorphous regions do not become oriented as do the crystalline regions of polymer, and the amorphous band intensity is anisotropic (independent of the orientation).
2. The amorphous bands are convenient because most polymers can be obtained in the amorphous state, hence these may be taken as standards for quantitative calibration.

Determination of crystallinity of polymers is usually performed by several parallel methods such as X-ray diffraction, electron microscopy, DTA—differential thermal analysis using a derivatograph, broad-line NMR, and infrared spectroscopy. These other methods of direct determination of crystallinity usually serve for calibrating the results obtained by the infrared technique.

Analysis of Polymers in Polarised Light

Examination of polymers in polarised light represents a separate and fairly vast line of research [146]. The discussion in this section will be restricted to the presentation of methods of measuring dichroism and the advantages to be gained from this technique. The use of polarised light is feasible only when the compound under study has a certain degree of order, as in single crystals or in polymers subjected to orientation treatment.

Orientation of polymer films is achieved by cold stretching. The stretching of polyethylene up to 400–500% elongation results in a high ordering of the chains along the direction of stretching [147,148]. Now, if a beam of polarised light falls on a polymer film thus oriented and the direction of the electrical vector of the incident beam is parallel to the direction of stretching, a sharp increase in absorbance ($A_{\parallel}$) of the bands parallel to the chain will be observed. Conversely, if the direction of the electrical vector is normal to the direction of the orientation, the absorbance ($A_\perp$) of the bonds situated normal to the chain axis will significantly increase.

The ratio of the absorbances of a given band measured perpendicular and parallel to the direction of the electric vector is referred to as the dichroic ratio and is expressed as:

$$R = A_\perp / A_\parallel$$

The quantity R may assume values from zero (no absorption in the perpendicular direction) to infinity (no absorption in the parallel direction). If $R < 1{\cdot}0$ the band is termed the parallel band, if $R > 1{\cdot}0$ it is referred to as the perpendicular band. In most practical cases the dichroic ratio ranges from 0·1 to 10.

The spectra of poly(ethylene oxide) taken in polarised light when the direction of the electrical vector is parallel or normal to the direction in which the foil was stretched are illustrated in Fig. 2.19. Appreciable differences in the absorbances are noticed primarily for the methylene deformation vibrations (such as 842, 960, 1345 cm^{-1}). A more detailed analysis of these vibrations was made by Liang [149] who advanced some concepts on the conformation of the polymer chain.

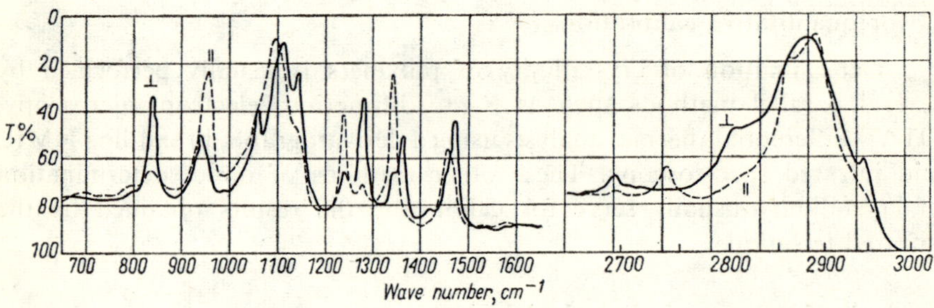

Fig. 2.19 Infrared spectra of poly(ethylene oxide) taken in polarised light [149]; the direction of the electrical vector perpendicular (solid line) or parallel (broken line) to the direction of orientation of the sample; film thickness 3 μm (by permission of the copyright holders, John Wiley & Sons, Inc.)

Also examined in polarised light were polycarbonate [150], poly(vinyl chloride) [151], polyurethanes [152] and polyacrylonitrile [153].

A knowledge of the dichroic ratio may also be of assistance in the interpretation of the infrared spectrum, as the absorption band observed in the spectrum may be assigned in terms of perpendicular or parallel vibrations with respect to the polymer backbone. On the basis of this assignment conclusions can be drawn as to the geometrical structure of molecules.

Attenuated Total Reflection (ATR) Technique

The infrared techniques discussed in previous sections are of no help in solving a variety of problems encountered in the investigation of polymers such as analysis of surface phenomena, filled plastics, laminates, multilayer plastics and others. The development of the ATR technique now enables a number of such problems to be handled easily.

The principle of ATR spectra is based on the phenomenon of total internal reflection. This phenomenon was exploited by Fahrenfort [154] and Harrick [155] in the design of their ATR attachments to be coupled with commercial spectrophotometers of any manufacture. The essential part of the attachment is a prism on which the sample to be analysed is placed directly. The angle of incidence of the light beam, θ, is arranged so that the beam is totally reflected at the prism–sample interface. The beam leaving the prism is attenuated by the sample because of selective absorption at certain wavelengths. The prism is designed to ensure multiple reflection of the beam from the sample as depicted in Fig. 2.20. Thus, even minor changes occurring on the surface of polymer (such as those in the aging process) may be readily detected and analysed. The method of multiple internal reflection (MIR) is sometimes referred to as the frustrated multiple internal reflection (FMIR) technique [156].

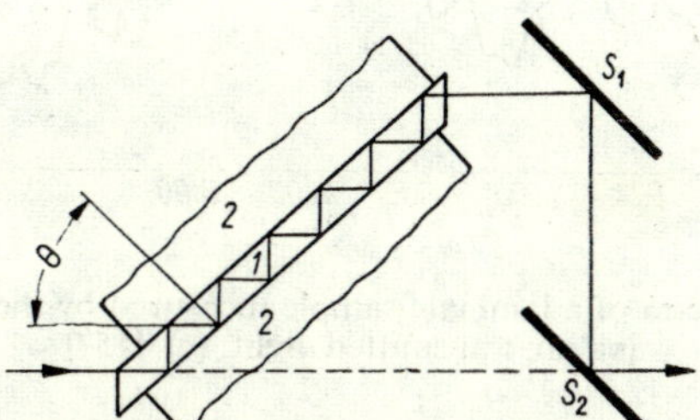

Fig. 2.20 The multiple internal reflection prism used in the ATR technique [158]: *1*—prism, *2*—sample, θ—angle of incidence, S_1, S_2—mirrors (by permission of the copyright holders, Verlag für Grundstoffindustrie, Leipzig)

The spectra obtained by the ATR technique are generally in agreement with spectra obtained in transmitted light [156].

An application of the ATR technique for identification of a laminate plastic is reported by Harris [157] (Fig. 2.21). The spectra of two laminate layers are obtained successively by the ATR technique (Fig. 2.21 *a* and *b*) from which it may easily be shown that the sample is composed of polyethylene and poly(ethylene terephthalate). The spectrum measured in transmitted light, as can be seen from inspection of Fig. 2.21*c*, is much more difficult to identify.

The ATR technique has found a range of practical uses [158,159] and appears to be a valuable tool of great future potential. A number of papers have been published recently on the theoretical aspects of the technique [160] and its practical applications [156,161].

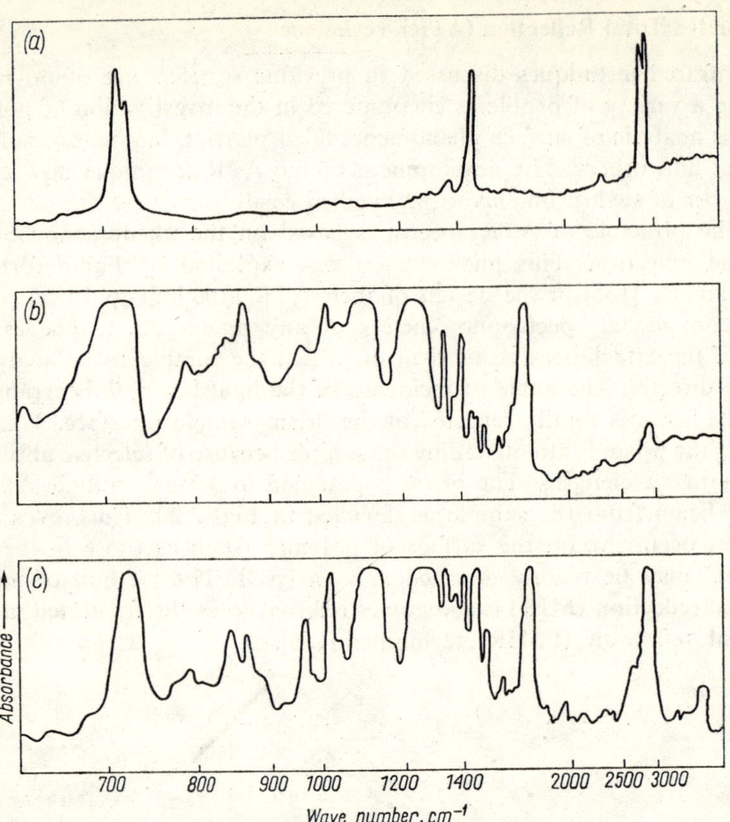

Fig. 2.21 Infrared spectra of a laminate sample measured by the ATR technique (*a*, *b*) and in transmitted light (*c*) [157]

References

1. Drushel, H. V., Iddings, F. A., *Anal. Chem.*, **35**, 28 (1963).
2. Lünebach, E., Hummel, D., *Kunststoffe*, **54**, 210 (1964).
3. Cotten, G. R., Sacks, W., *J. Polymer Sci.*, *Pt. A*, **1**, 1345 (1963).
4. Fratkina, G. P., Kirillova, E. I., Glagoleva, Yu. A., Leitman, K. A., *Plast. Massy*, **1965** [9], 55.
5. Giesen, M., *Plastica*, **16**, 492 (1963).
6. Kirillova, E. J., Matveeva, E. N., Zavitaeva, L. D., Glagoleva, Yu. A., Leitman, K. A., Fratkina, G. P., *Plast. Massy*, **1966** [2], 43.
7. Bradley, K. B., Potts, W. J., *Appl. Spectroscopy*, **12**, 77 (1958).
8. Merten, R., Braun, G., Lauerer, D., *Kunststoffe*, **55**, 249 (1965).
9. Kituchina, G. S., Zarkov, V. V., *Plast. Massy*, **1968** [6], 60.
10. Dolgoplosk, S. B., Klebanskii, A. L., Fomina, L. P., Fikhtengolts, V. S., Shvarts, E. Yu., *Dokl. Akad. Nauk. SSSR*, **150**, 813 (1963).
11. Haslam, J., Grossman, S., *Analyst*, **79**, 238 (1954).
12. Banes, F. W., Eby, L. T., *Ind. Eng. Chem.*, *Anal. Ed.*, **18**, 535 (1946).
13. Pepe, J. J., Kniel, I., Czuha, M., *Anal. Chem.*, **27**, 755 (1955).
14. Stafford, C., Toren, P. E., *Anal. Chem.*, **31**, 1687 (1959).

15. Duncalf, B., Dunn, A. S., *J. Appl. Polymer Sci.*, **8**, 1763 (1964).
16. Turska, E., Kroh, J., Kalinowska, A., *Polimery*, **8**, 272 (1963).
17. Okuhashi, T., *Kobunshi Kagaku*, **27**, 562 (1970).
18. Braun, D., Keppler, H. G., *Makromol. Chem.*, **78**, 100 (1964).
19. Meehan, E. J., *J. Polymer Sci.*, **1**, 175 (1946).
20. Nowakowski, A. C., *Anal. Chem.*, **30**, 1868 (1958).
21. Minsker, K. S., Krats, E. O., Pakhomova, I. K., *Vysokomol. Soedin., Ser. A*, **12**, 483 (1970).
22. Bengough, W. I., Varma, I. K., *Eur. Polymer J.*, **2**, 61 (1966).
23. Golub, M. A., Parker, J. A., *Makromol. Chem.*, **85**, 6 (1965).
24. Salovey, R., Albarino, R. V., Luongo, J. P., *Macromolecules*, **3**, 314 (1970).
25. Balandina, V. A., Korsakov, V. G., Malkina, N. I., Davydova, Z. F., Noskova, M. P., Shevradina, Z. I., *Plast. Massy*, **1967** [10], 53.
26. Shreve, O. D., Heether, M. R., *Anal. Chem.*, **23**, 441 (1951).
27. Hagivara, M., Okamoto, H., Kagiya, Tsk., Kagiya, Tst., *J. Polymer Sci., Pt. A-1*, **8**, 3295 (1970).
28. Yuskevitsyute, S., Shlyapnikov, Yu. A., *Plast. Massy*, **1967** [6], 65.
29. Schröder, E., Rudolph, G., *Plaste u. Kautschuk*, **10**, 22 (1963).
30. Stafford, C., *Anal. Chem.*, **34**, 794 (1962).
31. Böhm, G. G. A., *J. Polymer Sci., Pt. A-2*, **5**, 639 (1967).
32. Kidder, G. A., *Anal. Chem.*, **26**, 311 (1954).
33. Goldschmid, O., *Anal. Chem.*, **26**, 1421 (1954).
34. Haslam, J., Grossman, S., Squirrell, D. C. M., Loveday, S. F., *Analyst*, **78**, 92 (1953).
35. Ohara, K., *Makromol. Chem.*, **142**, 69 (1971).
36. Chernobai, A. V., Shepeleva, A. I., *Plast. Massy*, **1964**, [6], 56.
37. Chernobai, A. V., Shepeleva, A. I., Zubkova, V. S., *Vysokomol. Soedin.*, **7**, 1080 (1965).
38. Turska, E., Kroh, J., Czerwik, Z., *Polimery — tworzywa*, **8**, 222 (1963).
39. Maltese, P., Clementini, L., Frittelli, G., *Materie Plastische Elastomeri*, **31**, 226 (1965).
40. Yamane, T., Suzuki, T., Mukoyama, T., *Kogyo Kagaku Zasshi*, **72**, 2231 (1969).
41. Guillory, J. P., Cook, C. F., *J. Polymer Sci., Pt. A-1*, **9**, 1529 (1971).
42. Selivanov, P. I., Maksimov, V. L., Kirillova, E. I., *Vysokomol. Soedin.*, **8**, 1418 (1966).
43. Newell, J. E., *Anal. Chem.*, **23**, 445 (1951).
44. Klacel, Z., Stauf, M., *Plastické Hmoty Kaučuk*, **1963**, [1], 19.
45. Herzfeld, S. H., Roginsky, A., Corrin, M. L., Harkins, W. D., *J. Polymer Sci.*, **5**, 207 (1950).
46. Reiney, M. J., Tryon, M., Achhammer, B. G., *J. Res. Nat. Bur. Stand.*, **51**, 155 (1953).
47. Tobolsky, A. V., Eisenberg, A., O'Driscoll, K. F., *Anal. Chem.*, **31**, 203 (1959).
48. Dainton, F. S., Ivin, K. J., La Flair, R. T., *Eur. Polymer J.*, **5**, 379 (1969).
49. Hippe, Z., Jabłoński, H., *Polimery*, **12**, 203, 261 (1967).
50. Sidelkovskaya, F. P., Ogibina, T. Ya., Arakelyan, V. G., *Zh. Prikl. Khim.*, **37**, 182 (1964).
51. Grachev, N. M., Petrova, I. B., *Plast. Massy*, **1965** [2], 65.
52. Hirt, R. C., Schmitt, R. G., Stafford, R. W., King, F. T., *Anal. Chem.*, **27**, 226 (1955).
53. Swann, M. M., Adams, M. L., Weil, D. I., *Anal. Chem.*, **27**, 1604 (1955).
54. Shreve, O. D., *Synthetic Organic Coating Resins*, in *Organic Analysis, Vol. III*, Mitchell, J., Jr., Kolthoff, I. M., Proskauer, E. S., Weissberger, A., (Eds.), Interscience, New York, 1956.
55. Roczniak, K., *Polimery*, **10**, 454 (1965).

56. Hirt, R. C., Stafford, E. W., King, F. P., Schmitt, R. G., *Anal. Chem.*, **26**, 1273 (1954).
57. Morath, J. C., Woods, J. T., *Anal. Chem.*, **30**, 1437 (1958).
58. Walędziak, H., *Chem. Anal. Warsaw*, **10**, 579 (1965).
59. Hummel, D. O., *Atlas der Kunststoff-Analyse, Band I. Hochpolymere und Harze*, Carl Hanser Verlag, München, 1968.
60. Haslam, J., Willis, H. A., *Identification and Analysis of Plastics*, 1st Ed., Iliffe Books, London, 1965.
61. Nyquist, R. A., *Infrared Spectra of Plastics and Resins*, Dow Chemical Co., Midland, Michigan U.S.A. 1961.
62. Meise, W., Ostromow, H., *Kunststoffe*, **54**, 213 (1964).
63. Serboli, G., *Infrarotspektren von Härtern und Katalysatoren für Epoxyharze*, Brown, Boveri & Cie., Baden, Schweiz, 1964.
64. Bellamy, L. J., *The Infrared Spectra of Complex Molecules*, Methuen, London, 1962.

64a. Henniker, J. C., *Infrared Spectroscopy of Industrial Polymers*, Academic Press, London, 1967.

65. Cihla, Z., Plíva, J., *Coll. Czech. Chem. Comm.*, **26**, 1903 (1961).
66. Liang, C. Y., Lytton, M. R., *J. Polymer Sci.*, **61**, S45 (1962).
67. Liang, C. Y., Lytton, M. R., Boone, C. J., *J. Polymer Sci.*, **44**, 549 (1960).
68. Liang, C. Y., Pearson, F. G., *J. Mol. Spectroscopy*, **5**, 290 (1960).
69. Tadokoro, H., Murahashi, S., Yamadera, R., Kamei, T. I., *J. Polymer Sci., Pt. A*, **1**, 3029 (1963).
70. Golub, M. A., Shipman, J. J., *Spectrochim. Acta*, **20**, 701 (1964).
71. Shimanouchi, T., Tasumi, M., *Bull. Chem. Soc. Japan*, **34**, 359 (1961).
72. Nagai, H., *J. Appl. Polymer Sci.*, **7**, 1697 (1963).
73. Durbetaki, A. J., Miles, C. M., *Anal. Chem.*, **37**, 1231 (1965).
74. Foster, G. N., Row, S. B., Griskey, R. G., *J. Appl. Polymer Sci.*, **8**, 1357 (1964).
75. Karyakin, A. V., Petrov, A. V., *Zh. Analit. Khim.*, **19**, 1234 (1964).
76. Viller, O. G., *Uspekhi Khim.*, **30**, 777 (1961).
77. Kaye, W., *Spectrochim. Acta*, **6**, 257 (1954).
78. Tryon, M., Horowitz, E., in *High Polymers* Vol. XII, *Analytical Chemistry of Polymers*, Part II, Kline, G. M. (Ed), Interscience, New York–London, 1961.
79. Biernacka, T., Kontnik, B., *Tworzywa-Guma-Lakiery*, **5**, 90, 108 (1960).
80. Rugg, F. M., Smith, J. J., Bacon, R. C., *J. Polymer Sci.*, **13**, 535 (1954).
81. Lomonte, J. N., *Anal. Chem.*, **34**, 129 (1962).
82. Dannenberg, H., *SPE Transactions*, **3**, 78 (1963).
83. Veerkamp, T. A., de Kock, R. J., Veermans, A., Lardinnoye, M. H., *Anal. Chem.*, **36**, 2277 (1964).
84. Takeuchi, T., Mori, S., *Anal. Chem.*, **37**, 589 (1965).
85. Janicka, K., Wiecheć, L., *Polimery*, **11**, 153 (1966).
86. Horbach, A., Veiel, U., Wunderlich, H., *Makromol. Chem*, **88**, 215 (1965).
87. Witenhafer, D. E., *J. Macromol. Sci.* (*Physics*), **B4**, 915 (1970).
88. Magnasco, V., Rossi, M., *J. Polymer Sci.*, **62**, 172 (1963).
89. Korshak, V. V., Sidorov, T. A., Vinogradova, S. V., Komarova, L. I., Valetskii, P. M., Lebedeva, A. S., *Izv. Akad. Nauk SSSR, S. Khim.*, **2**, 261 (1965).
90. Majer, J., *Makromol. Chem.*, **82**, 169 (1965).
91. Shabalin, I. I., Kiva, E. A., *Zh. Fiz. Khim.*, **40**, 2774 (1966).
92. Medalia, A. I., Freedman, H. H., *J. Am. Chem., Soc.* **75**, 4790 (1953).
93. Nikitin, V. N., Mikhailova, N. V., *Vysokomol. Soedin., Ser. A*, **9**, 784 (1967).
94. Fishl, W., Young, I. G., *Appl. Spectroscopy*, **10**, 213 (1956).
95. Grant, G. D., Smith, A. L., *Anal. Chem.*, **30**, 1017 (1958).
96. Lady, J. H., Bower, G. M., Adams, R. E., Byrne, F. P., *Anal. Chem.*, **31**, 1100 (1959).
97. Majer, J., *Chem. Průmysl*, **21**, 237 (1971).

98. Janik, A., Maciejowski, F., Mitus, J., *Chem. Anal. Warsaw*, **11**, 595 (1966).
99. Makarevich, N. J., *Vysokomol. Soedin.*, **8**, 1428 (1966).
100. Kurengina, T. N., Alferova, L. V., Kropachev, V. A., *Vysokomol. Soedin.*, **8**, 293 (1966).
101. Bucci, G., Simonazzi, T., *Chim. Ind. Milan*, **44**, 262 (1962).
102. Langbein, G., Seufert, W., *Kolloid-Z.*, **193**, 37 (1963).
103. Gol'denberg, A. L., Pirozhnaya, L. N., Popova, G. S., Tarutina, L. I., *Molek. Spektroskopya*, Izdatel'stvo Leningradskogo Universiteta, 1960.
104. Chan, M. G., Hawkins, W. L., *Polymer Eng. Sci.*, **7**, 264, (1967).
105. Titratu-Moldovan, L., *Plast. Mod. Elast.*, **21**, [5], 145 (1969).
106. Lomonte, J. N., *Anal. Chem.*, **36**, 192 (1964).
107. Luongo, J. P., *J. Polymer Sci.*, **42**, 139 (1960).
108. Sirota, A. G., Gol'denberg, A. L., Il'chenko, P. A., Ryabikov, E. P., Fedotov, B. G., Karaseva, M. G., Zyuzina, L. I., Kharitonova, O. K., *Plast. Massy*, **1966** [8], 58.
109. Karyakin, A. V., Funtikov, A. I., *Vysokomol. Soedin.*, **7**, 1171 (1965).
110. Bagirov, M. A., Malin, V. P., Gazaryan, Yu. N., Volchenkov, E. Ya., *Vysokomol. Soedin., Ser. A*, **1**, 2323 (1969).
111. Shimada, J., Kabuki, K., *J. Appl. Polymer Sci.*, **12**, 655 (1968).
112. Seymour, R. B., Tsang, H. E., Warren, D., *Polymer Eng. Sci.*, **7**, 55 (1967).
113. Sobiczewski, Z., Wielgosz, Z., *Plaste u. Kautschuk*, **15**, 176 (1968).
114. Bovey, F. A., Tiers, G. V. D., *J. Polymer Sci.*, **44**, 173 (1960).
115. Miller, R. L., Nielsen, L. E., *J. Polymer Sci.*, **46**, 303 (1960).
116. Corish, P. J., *Spectrochim. Acta*, **15**, 598 (1959).
117. Gotlib, Yu. Ya., Boitsev, V. B., *Opt. Spectroscopy*, **15**, 216 (1963).
118. Takeda, M., Iimura, K., Endo, R., *Kogyo Kagaku Zasshi*, **73**, 1509 (1970).
119. Shimanouchi, T., Abe ,Y., Alaki, Y., *Polymer J. Japan*, **2**, 199 (1971).
120. Burleigh, P. H., *J. Am. Chem. Soc.*, **82**, 749 (1960).
121. Germar, H., Hellwege, K.-H., Johnsen, U., *Makromol. Chem.*, **60**, 106 (1963).
122. Krimm, S., Folt, V. L., Shipman, J. J., Berens, A. R., *J. Polymer Sci., Pt. A*, **1** 2621 (1963).
123. Razuvaev, G. A., Zilberman, E. N., Zegelman, V. I., Svetozarskii, S. V., Pomerantseva, E. G., *Dokl. Akad. Nauk SSSR*, **170**, 1092 (1966).
124. Glazkovskii, Yu. V., Zgaevskii, V. E., Ruchin'skii, S. P., Bakardzhev, N. M., *Vysokomol. Soedin.*, **8**, 1472 (1966).
125. Pohl, H. U., Hummel, D. O., *Makromol. Chem.*, **113**, 190 (1968).
126. Millan, J., Nino, E., *Rev. Plast. Mod.*, **21**, 87 (1970).
127. Glazkovskii, Yu. V., Zavyalov, A. N., Bakardzhev, N. M., Novak, I. L., *Vysokomol. Soedin., Ser. A*, **12**, 2697 (1970).
128. Heinen, W., *J. Polymer Sci.*, **38**, 545 (1959).
129. Brader, J. J., *J. Appl. Polymer Sci.*, **3**, 370 (1960).
130. Koenig, J. L., Van Roggen, A., *J. Appl. Polymer Sci.*, **9**, 359 (1965).
131. Woodbrey, J. C., Trementozzi, Q. A., *J. Polymer Sci., Pt. C*, **8**, 113 (1965).
132. Natta, G., Peraldo, M., Allegra, G., *Makromol. Chem.*, **75**, 215 (1964).
133. Kissin, Yu. V., Tsvetkova, V. I., Chirkov, N. M., *Vysokomol. Soedin., Ser. A*, **9**, 1104 (1967).
134. Kissin, Yu. V., Tsvetkova, V. I., Chirkov, N. M., *Vysokomol. Soedin., Ser. A*, **10**, 1092 (1968).
135. Hughes, R. H., *J. Appl. Polymer Sci.*, **13**, 417 (1969).
136. Kissin, Yu. V., Tsvetkova, V. I., Chirkov, N. M., *Eur. Polymer J.*, **8**, 529 (1972).
137. Liang, C. Y., Krimn, S., *J. Polymer Sci.*, **27**, 214 (1958).
138. Tadokoro, H., *Kobunshi Kagaku*, **17**, 231 (1960).
139. Onishi, T., Krimm, S., *J. Appl. Phys.*, **32**, 2320 (1961).
140. Tsuji, W., Kitamuru, R., Sakaguchi, F., *Kobunshi Kagaku*, **23**, 836 (1966).
141. Kobayashi, M., Akita, K., Tadokoro, H., *Makromol. Chem.*, **118**, 324 (1968).
142. Dankovich, A., Kissin, Yu. V., *Vysokomol. Soedin., Ser. A*, **12**, 802 (1970).

143. Helms, J. B., Challa, G., *J. Polymer Sci., Pt. A–2*, **10**, 761, (1972).
144. Ciampelli, F., Cambini, M., Lachi, M. P., *Polymer Sci., Pt. C*, **7**, 213 (1964).
145. Nikitin, V. N., Pokrovskii, E. N., *Dokl. Akad. Nauk SSSR*, **95**, 109 (1954).
146. Elliott, A., *J. Polymer Sci., Pt. C*, **7**, 37 (1964).
147. Glenz, W., Peterlin, A., *J. Macromol. Sci., B*, **4**, 473 (1970).
148. Glenz, W., Peterlin, A., *J. Polymer Sci., Pt. A–2*, **9**, 1191 (1971).
149. Liang, C. Y., *Infrared Polymer Spectroscopy*, in *Newer Methods of Polymer Characterisation*, Ke, B. (Ed.), Wiley, New York, 1964.
150. Yannas, I. V., Lunn, A. C., *J. Polymer Sci., Pt. B*, **9**, 611 (1971).
151. Shindo, Y., Read, B. E., Stein, R. S., *Makromol. Chem*, **118**, 272 (1968).
152. Estes, G. M., Seymour, R. W., Cooper, S. L., *Macromolecules*, **4**, 452 (1971).
153. Koenig, J. L., Wolfram, L. E., Grasselli, J. G., *J. Macromol. Sci., B*, **4**, 491 (1970).
154. Fahrenfort, J., *Spectrochim. Acta*, **17**, 698 (1961).
155. Harrick, N. J., *Ann. N. Y. Academy of Science*, **101**, 928 (1963).
156. Jayme, G., Traser, G., *Angew. Makromol. Chem.*, **21**, 87 (1972).
157. Harris, R. L., *22nd Annual Technical Conference, Technical Papers, Vol. XII, X–3, Soc. of Plastics Engineers*, 1966.
158. Wagner, H., Wuckel, L., *Plaste u. Kautschuk*, **18**, 426 (1971).
159. Bădilescu, S., Toader, M., Oprea, H. O., *Materiale Plastice*, **8**, 422 (1971).
160. Roloff, H., Schwind, A. E., Zilinski, E., Wagner, H., *Plaste u. Kautschuk*, **19**, 14 (1972).
161. Schmitz, D., *Plaste u. Kautschuk*, **16**, 936 (1969).

Chapter 3

CHROMATOGRAPHY

3.1 PRINCIPLES

Any chromatographic separation is associated with a dynamic equilibrium state in a system of stationary and mobile phases. The equilibrium is defined by the distribution ratio K

$$K = c_s/c_m$$

where:

c_s = the concentration of the substance in the stationary phase,
c_m = the concentration of the substance in the mobile phase.

The distribution ratio expressed as the function $c_s = f(c_m)$ is graphically represented as a distribution isotherm (when two liquid phases are involved) or an adsorption isotherm (when a solid adsorbent is used). Since this distribution process occurs at different rates, the substances under analysis undergo separation. Chromatographic separation is usually compared to fractional distillation, over which it has a substantially higher resolving power.

In chromatography the stationary phase may be a solid or a liquid and the mobile phase a liquid or a gas. The type of stationary phase determines the distribution mechanism and two principal types of chromatography are distinguished, namely adsorption chromatography and partition chromatography. In adsorption chromatography the stationary phase is a solid, and in partition chromatography it is a liquid (see Table 3.1).

Table 3.1 Basic Classification of Chromatographic Methods

Type of chromatography	*Stationary phase*	*Mobile phase*	*Method*
Adsorption chromatography	solid solid	liquid gas	column, thin layer gas
Partition chromatography	liquid liquid	liquid gas	column, paper gas

Irrespective of the above division, chromatography may also be classified as column, paper, thin layer, and gas chromatography. Each of these techniques may be either adsorption chromatography or partition chromatography depending on the separation mechanism involved. However, paper chromatography is generally referred to as partition chromatography and

thin layer chromatography as adsorption chromatography. Some other variants of adsorption chromatography are ion-exchange chromatography, in which the stationary phase is an ion exchange resin, and precipitation chromatography, when the stationary phase reacts with the components of the separated mixture.

A measure of the migration rate of individual components of a mixture under separation is the R or R_f value. This represents the ratio of the migration rate of the solute to the migration rate of the mobile phase or the ratio of the distance of the front of the solute from the starting line to the distance of the front of the mobile phase from the starting line. The migration rate of a solute is affected by a variety of factors, of which the distribution ratio is of prime importance.

If a compound under analysis is easily soluble in the mobile phase, then it travels quickly and attains a large R value. If, on the other hand, its solubility in the mobile phase is lower than in the stationary phase (for instance in water) then it migrates slowly resulting in a low R value. Thus, as seen in paper chromatography, the migration rate in partition chromatography is controlled by the solubility of the solute in both phases.

In partition chromatography the equilibrium state is expressed by the equation

$$c_1/c_2 = K$$

where:

c_1 and c_2 = the respective concentrations of the solute in phases 1 and 2,
K = the distribution ratio [1].

In practice the value of the distribution ratio of the solute is directly affected by its solubility in both liquid phases, and depends also on the temperature and on the pH of the column.

Presented in Fig. 3.1 is the plot of the distribution ratio in the form of the distribution isotherm.

In practice pure partition chromatography is rarely observed, and is accompanied by adsorption to a variable extent, which manifests itself

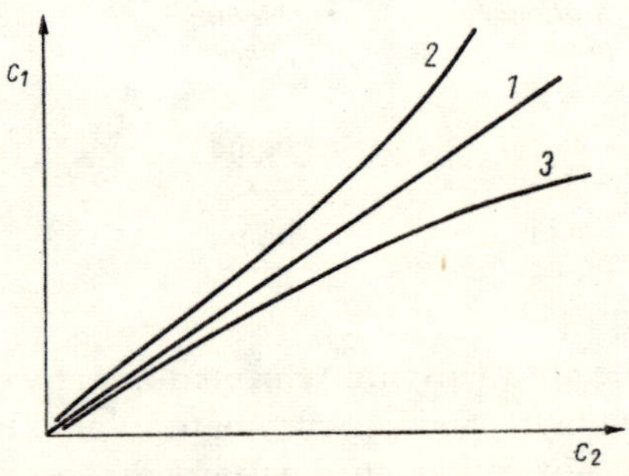

Fig. 3.1 Distribution isotherms for partition of a compound between two liquid phases: *1*—obeying the Nernst law, *2* and *3*—not obeying the Nernst law

as tailing (or trailing) known also as "comets". Occasionally the effect of adsorption is favourable for separation in a liquid–liquid system.

Another factor responsible for deviations from the linearity of the distribution isotherm is the dissociation of weak acids and bases. An acid or an alkali is added to the mobile phase to counteract this effect.

The systems in common use consist of two phases of limited miscibility and a hydrophilic support of low activity. Compounds with a distribution ratio lower than 0·1 can be more efficiently separated by the reversed phase technique.

The separation mechanism in adsorption chromatography is more complex than in partition chromatography, since distribution occurs between a mobile phase and an adsorbent. Among the major factors here are the values of the dipole moments and of the dielectric constant, and the ability to form hydrogen bonds [2]. The types of adsorption forces operative in the system are of equal significance.

The degree of adsorption is also affected by such factors as the choice of adsorbent, the kind of the compounds to be separated, their concentration in solution and the nature of the solvent used.

Since the adsorption process is reversible, the adsorbed molecules leave the adsorbent surface when their energy of interaction with the solvent molecules overcomes the attraction forces of the adsorbent surface.

At a constant temperature the adsorption process is usually presented as the adsorption isotherm, which expresses the quantity of adsorbate adsorbed by 1 g of adsorbent, a, as a function of the concentration of the chromatographed solution, c, under the conditions of adsorption equilibrium. The most commonly used form of the adsorption isotherm is that formulated by Langmuir (Fig. 3.2).

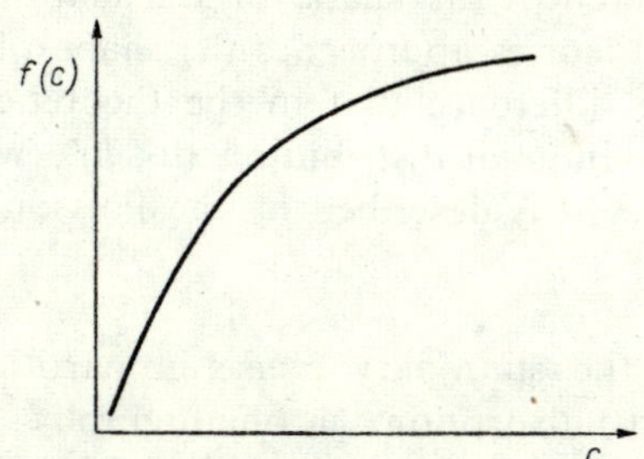

Fig. 3.2 The Langmuir adsorption isotherm

For dilute solutions the adsorption curve is linear, and is described by the equation

$$a = kc$$

where k is a constant.

At higher concentrations the adsorption isotherm becomes curved and runs asymptotically to a certain limiting value. In this case the curve fits the Freundlich equation

$$a = kc^n$$

where k and n are constants.

Recent chromatographic theories describe the chromatographic separation process in an approximate manner. One of these views the chromatographic process as a countercurrent exchange, and uses the concept of theoretical plates [3], in which the effect of diffusion within a single plate is neglected. In other words, a chromatographic column is treated as a system of layers of finite thickness, in which at any time a certain state of partition equilibrium is immediately set up. In each of the layers that correspond to the height equivalent to a theoretical plate (so called HETP units) the concentration of the solute in the mobile phase which leaves the bottom level of the plate, remains in equilibrium with the mean concentration in the stationary phase that stays in that layer [1].

Underlying another theory is the concept of immediate establishment of the partition equilibrium on a column of infinite cross section of an HETP equal to zero. This approach is known as the method of continuous variables [4,5] in which the law of conservation of mass is valid for the solute distributing between the two phases.

Although both theories are aimed at finding optimum conditions for the separation of an unknown compound, the objective is achieved in a different manner. While the theoretical plate theory, being purely empirical, is limited to consideration of the effect of individual parameters on the symmetrical broadening of the concentration bands, the continuous variables theory, which takes into account the diffusion effect, makes it possible to anticipate the diffusion effect on band broadening and on the resulting asymmetry of the chromatographic peaks [6].

Despite these differences the shape of the elution curves derived in terms of either of the theories approaches the shape of a Gaussian distribution curve with the difference that in the theoretical plate theory the shape is given by the binomial distribution $(a+b)^n$, whereas in the continuous variables theory it is described by the Poisson distribution [7].

Supports

The solid support of the stationary phase in partition chromatography and the adsorbent in adsorption chromatography are materials with a large specific surface. A good support differs from an adsorbent in that it lacks the adsorptive properties of the latter. Among the most suitable supports are diatomaceous earth, silica gel, cellulose, and less commonly, other silicates [8,9]. All supports are capable of fixing the mobile phase, especially the strongly polar ones, in a quantity equal to their own weight. The amount of mobile phase used often reaches 250–300 ml for each 100 g of support (stationary phase). The usual grain size of solid supports lies in the range 100–300 mesh.

In reversed phase chromatography it is recommended to use diatomaceous earth made hydrophobic by treatment with dimethyldichlorosilane, or acetylcellulose, or esterified ion-exchange resins, or else rubber powder. Diatomaceous earth with a wide range of grain size is marketed under the name of Celite. Before use the solid supports are usually subjected to suitable treatment. Also used in this chromatographic technique are cellulose, powdered vulcanised rubber [10], microporous granular polyethylene and the cation-exchanger Amberlite IRC × 50 in its partly esterified form [11].

Adsorbents

The stationary phase (adsorbent) in adsorption chromatography is a fine grain porous solid of definite properties, the most significant being the adsorption activity which is determined by the degree of hydration, the pH, the size and shape of the grains, and the size and number of the pores present.

In column adsorption chromatography adsorbents commonly used are alumina, magnesium silicate, silica gel or silicic acid alone or in mixtures with diatomaceous earth. Of lesser importance are magnesium carbonate, aluminium silicate, or hydrophobic charcoal. An adsorbent with changed properties resulting in improved separation may be achieved by mixing adsorbents of different activity or by gradual activation or deactivation of one of these adsorbents. A partial deactivation treatment is frequently of much value in the separation of compounds of strong polarity. The treatment consists in addition of definite quantities of water to deactivate an adsorbent, whereas activation is effected by dehydration of the material by heating, which is particularly useful in the case of hydrophilic adsorbents. With decreasing grain size the adsorptive power of an adsorbent increases on account of an increased specific surface of the material. Too low a grain size, however, is accompanied by an undesirable fall in the migration rate, but this can be counteracted by altering the nature of the adsorbent by mixing it with another adsorbent of low polarity such as diatomaceous earth.

In selecting a suitable adsorbent its polarity and the polarity of the compound to be separated are the two major factors to be considered. The more polar the groups present in a compound or the more groups there are of unsaturated character, the more strongly they adsorb on a hydrophilic adsorbent. Hence, for the chromatographic separation of weakly polar compounds strongly polar adsorbents are used, and conversely, for strongly polar compounds adsorbents of low polarity should be used. Of major importance for ionised substances is the pH value of the adsorbent. Weakly polar compounds are more strongly adsorbed on hydrophobic adsorbents.

Solvents

Solvents for partition column chromatography should be rather volatile. The solvents commonly used to form the phase system are: mobile phase—

a relatively weakly polar organic solvent, stationary phase—a relatively strongly polar organic solvent (both phases are of limited miscibility); either or both phases may contain an aqueous component. Under certain conditions water may be used alone as the stationary phase.

A major factor in the selection of solvents is the solubility in them of the compounds to be chromatographed. Unless there is some solubility no migration will be possible and no separation will occur. Furthermore, the mutual miscibility of both phases should not exceed a certain limit, which is usually ca. 10%. To facilitate the selection of suitable solvents special mixotropic series have been developed [12]. The tailing effect resulting from adsorption in partition chromatography may be suppressed by increasing the amount of water in the mobile phase, by increasing the mutual miscibility of both phases, or by increasing the temperature of the system. In adsorption chromatography the column efficiency (separating power) is also controlled by the properties of the solvent. The more strongly the solvent is bound by the adsorbent, the more it competes with the solute in occupying the active adsorbent centres, and hence the less strongly the solute is fixed. The eluotropic series formulated by Trappe [13] comprises a list (Table 3.2) of a number of solvents of increasing adsorptive power relative to the hydrophilic adsorbent, causing a decrease in the adsorption of the solute. For hydrophobic adsorbents such as charcoal the reported eluotropic series is valid but in the.

The chromatographic column is usually preconditioned with one of the solvents from the eluotropic series. For this purpose the solvent used is the one in the series preceding the solvent used for preparing the solution of the compound under analysis.

Table 3.2 Eluotropic series according to Trappe
(from [13] by permission of the copyright holders, Springer Verlag)

	Solvent	*Boiling point*, °C
Increasing polarity ↓	n-Hexane	68·7
	n-Heptane	98·4
	Cyclohexane	81·4
	Carbon tetrachloride	76·8
	Benzene	80·1
	Chloroform	61·3
	Ether	34·6
	Ethyl acetate	77·1
	Pyridine	115·3
	Acetone	56·5
	Ethanol	78·5
	Methanol	64·4
	Water	100·0

Apparatus

The apparatus used in column chromatography consists of a glass tube (column), equipped with a ground-glass joint at the top to connect with a reservoir holding the solvent, and the lower end is provided with a stopcock. To introduce the mobile phase onto the column pipettes and separating funnels are used and also special assemblies of vessels for continuous elution techniques.

The use of fraction collectors provided some advance in partial automation of the chromatographic separation process. Column efficiency depends on the ratio of its diameter to its height. The sample size taken is matched to the weight of the adsorbent or solid support used. The ratio of the weight of the solute to the weight of the adsorbent for alumina ranges from 1:20 to 1:2000, usually being 1:30.

Packing the Column. There are two ways of packing the chromatographic column, namely the dry technique and the wet technique. In wet packing, the column tube is first filled with a solvent, followed by addition of the adsorbent. In the dry packing procedure the dry adsorbent is first introduced, followed by the solvent.

The Chromatographic Separation. The general principles of applying this procedure are given for the case of an alumina-packed column. First a 1–5% solution is prepared of the mixture to be separated in a solvent of lowest possible polarity (e.g. 1 g of the mixture is dissolved in 100 g of petroleum ether). This solution is introduced into the column packed with 30 g of alumina and allowed to percolate. The first filtrate is collected as a single preliminary fraction. Then a portion of 100 ml of the pure solvent is passed through the column and collected in a number of small separate fractions. The solvent is distilled off from each fraction and the residue weighed. As soon as the weight of the residue from a single fraction decreases below 5% of the weight of the mixture loaded onto the column, a petroleum ether-benzene mixture is introduced into the column. The proportion of benzene in the mixture is gradually increased so that finally a pure benzene eluent is used.

A somewhat different procedure is followed when separating compounds that are insoluble in petroleum ether and in benzene.

Usually it is possible to separate a mixture into some fifteen 2–5 ml fractions at optimum flow rates of 2–10 ml per min.

Instead of stepwise variation of the composition of the solvent used for development of the chromatogram, it is advantageous to employ the continuous gradient elution technique using only two solvents.

In the stepwise elution technique a series of solvents, or their mixtures, of increasing eluting power are successively introduced into the column. In the gradient elution technique the eluent composition poured onto the column is continuously varied. This technique provides better separation than the stepwise elution method.

One variant of the elution method is the displacement technique, in which the solution of the most strongly adsorbed component of the mixture under separation is used as developer. Still another method of chromatographic separation of compounds on a column is that of frontal analysis. This method consists in the continuous passing and percolation of the solution of an unknown. Only part of the least strongly adsorbed component is obtained in a pure state.

3.2 PAPER CHROMATOGRAPHY

Principles

Paper chromatography is described in terms of partition chromatography theory.

The important parameter is the R_f value, defined as

$$R_f = \frac{\text{Distance travelled by centre of spot}}{\text{Distance travelled by developer front}}$$

where both distances are relative to the starting line of the chromatogram. The R_f value is a quantity characteristic of a given compound, and is determined mainly by the distribution coefficient between the stationary and the mobile phases.

Paper and Solvents

A prerequisite for separation of a mixture of compounds is the proper type of paper and a suitable solvent. The paper brands most widely used are those made by such manufacturers as Whatman, Schleicher and Schuell, and Durieux and d'Arches.

A mixture can be successfully separated only if it is partly soluble in the stationary phase (water) and partly in the mobile organic phase. The mobile phase should be miscible with water to a certain extent and hence it is usually a water-saturated organic solvent.

Reversed Phase Chromatography

In the separation of hydrophobic substances use is frequently made of a technique referred to as reversed phase chromatography. In this technique the polar solvent becomes the mobile phase, while the non-polar solvent is held stationary on the paper. The mobile phase is always saturated with the hydrophobic solvent of the stationary phase used to impregnate the paper. However, on many occasions, for the saturation of the mobile phase, instead of the solvent used for impregnation a related solvent of lower volatility is taken, for instance dimethylformamide in place of the more volatile formamide.

Other Procedures in Paper Chromatography

Of the three kinds of paper chromatography in common use, namely ascending, descending and circular techniques, descending chromatography is the most widespread.

Horizontal development, of which circular development is a particular case, is of little significance in analysis of polymers. The centrifugal technique is also of minor importance, because of poor reproducibility.

Much more valuable are two-dimensional and multiple developments. Two-dimensional development is occasionally used to accomplish better separation of a mixture, some components of which remain unseparated on development in one direction. In the multiple development technique the chromatogram is first fully developed in a weakly polar solvent, then redeveloped, after a short drying period, in a more polar solvent but to a shorter distance, for example 2/3 to 3/4 of the distance of the first development.

Quantitative Chromatography

Quantitative determinations by paper chromatography are usually effected by direct measurement of the surface of the spots on the chromatogram or indirectly by measuring the concentration of the extracts prepared from the spots. Measurement of the area of the spots is only informative in the quantitative sense, when made against a standard. Quantitative determination may be made by chromatographing simultaneously on the same paper strip the unknown and a standard reference. After development and detection of the spots, the areas of the spots from the unknown and the reference are compared. The error in visual estimation can reach 30%.

More accurate determinations can be made by marking the contours of the spots with a pencil and measuring the spot areas or the length of the spots. In the former the error is ±5%. In the spot length measurement technique a relationship is used [14], according to which the spot length is proportional to the logarithm of the quantity of the substance present in the spot.

Another quantitative procedure consists in colorimetric determination of the extracts of the spots, or ultraviolet or infrared spectrophotometric measurements of these extracts.

Preparative Chromatography

In preparative paper chromatography paper of a special type in the form of a single sheet or a suitably formed pack is used. Up to several grams of a pure individual compound may be isolated by paper chromatography. The amount of substance applied to a single sheet is usually not larger than 0·01 g, and is usually applied to the starting line as narrow streaks rather than single spots, as is done in analytical chromatography.

Apparatus

The essential component of the apparatus for paper chromatography is the tank. The best tanks are all-glass, in various sizes and shapes, usually about 50 cm high. The solvent trough accommodated inside the tank is likewise usually of glass. Other necessary equipment includes capillaries to apply the solutions to the paper and atomisers for spraying reagent to detect the chromatographic spots. High-vacuum silicone grease or a paste made of a 1:1 mixture of glycerol and starch is used to seal the chromatographic tanks.

Detection

Making the resolved spots on a paper chromatogram visible (detection of the spots) is conducted by a number of procedures, depending on the nature of the compounds analysed. Usually the compounds are converted into coloured derivatives, and sometimes ultraviolet irradiation is employed to make the spots visible by a fluorescence effect. The spots are coloured on the chromatogram by immersing the latter in a suitable reagent or by spraying.

Of the reagents used for detection of chromatographically separated substances some are versatile whereas others are specific group reagents. Ranking in the first group are potassium permanganate, iodine and sodium dichromate, while the second group includes diazonium salts such as that of sulphanilic acid, a reagent for phenols and aromatic amines, ammoniacal silver nitrate solution for detection of compounds with reducing properties, and indicators for detection of acids.

This list of detection reagents is far from complete. Of considerable interest is the method of autoradiography used for the detection of radioisotopes and labelled compounds.

In view of the considerable importance of iodine as a detection reagent, a few examples of its use are given below. It is commonly used as a vapour or in petroleum ether solution, and is also employed in combination with periodic acid. This reagent constitutes a non-destructive means of detection with respect to the compounds identified, as it produces unstable addition compounds with many substances. The higher the degree of unsaturation of the compounds, the more easily they can be detected with iodine. Alternatively, the chromatograms may be sprayed with iodine solution in potassium iodide (0·3% in 5% aqueous KI solution); subsequent spraying with ether or another organic solvent intensifies the colour reaction.

Interpretation of Chromatograms

Depending on the concentration of the main compound and its accompanying impurities, certain variations are observed in the nature and intensity of the colours produced. The R_f values are constant only under identical conditions of temperature, saturation, amount of the phases used, developing time, etc. [15]. Thus it is better to compare the migration rate of unknown compounds with the corresponding values for standards instead of using R_f values. The R_s value thus obtained is defined as

$$R_s = \frac{\text{Distance of the spot centre from the starting line for the unknown}}{\text{Distance of the spot centre from the starting line for the standard}}$$

Starting from the relation between the distribution ratio K and R_f value,

$$K = \frac{A_m}{A_s}\left(\frac{1}{R_f} - 1\right)$$

where A_m and A_s are the cross-sectional areas of the mobile and stationary phases respectively. Martin [16] has demonstrated that the logarithm of this function is to a first approximation directly proportional to the sum of the free energies of the groups constituting the compound under separation. Hence, on the basis of the value of the distribution ratio conclusions may be drawn as to the chemical structure of the compound under analysis.

A measure of the above relationship is the R_m value:

$$R_m = \log(1/R_f - 1)$$

The R_m value is an additive property. Every new group introduced into the molecule of a compound of a given distribution ratio affects the value of this ratio, just as it alters the R_f value connected with this distribution ratio. The change in R_f value is denoted as ΔR_m and is calculated from the difference in the R_m value for the parent compound and for its derivative.

3.3 THIN LAYER CHROMATOGRAPHY (TLC)

The essential advantages of TLC are the following:

(a) the very small quantities of compounds, of the order of 0·01 μg, that can be separated,
(b) its rapidity,
(c) the wide range of solvents and adsorbents that can be used,
(d) the possibility of using very strong developing agents.

The TLC technique is essentially adsorption chromatography but under certain conditions it may become partition chromatography. Lipophilic substances are chromatographed by adsorption chromatography, and hydrophilic compounds by partition chromatography.

Thin layer chromatography, despite many similarities, differs substantially from classical column chromatography. First, the degree of granulation of the adsorbent is higher in TLC; second, the ratio of the adsorbent to the chromatographed compound is different; third, the solvent transport is effected by capillary forces and not by hydrostatic pressure.

Fundamental Procedure

TLC analysis is carried out on adsorbent-coated glass plates, the thickness of the layer being 200–1000 μm. The unknown, in the form of a 5–10% solution, is applied onto the plate at a distance of 2 cm from one of its shorter edges. The plate thus prepared is developed with a solvent in an air-tight chamber. After development the plates are removed from the chambers, dried, and sprayed with a suitable reagent to detect the spots. For the sake of reproducibility it is advisable to develop the chromatograms in the saturated atmosphere of the chambers.

Thin layer chromatography is classed into analytical and preparative, qualitative or quantitative, one-dimensional or two-dimensional, single-run or multiple.

According to Brenner *et al.* [17] the migration rate of the solvent in TLC is affected by such factors as: (1) type and activity of adsorbent, (2) layer thickness, (3) type of solvent and the degree of saturation of solvent in the atmosphere of the chamber, (4) the development technique (depth of immersion, chromatogram height), (5) the quantity of the chromatographed substance.

Adsorbents and Supports

Adsorbents and supports used in coating the chromatoplates should have a large surface area (grain diameter below 60 μm), adequate capillary properties, and a fair chemical and mechanical resistance. Among the most popular adsorbents are silica gel, alumina, diatomaceous earths, and cellulose, in order of decreasing activity. To ensure reproducible results the gel activity is ascertained by chromatographing a mixture of three known dyes, namely indophenol blue, Sudan red G, and *p*-dimethylaminoazobenzene. Three varieties of alumina are in use depending on whether the chromatographed substance is an alkaline, neutral or acidic material. Silica gel G or magnesium silicate is more readily used for partition chromatography than for adsorption chromatography. Cellulose adsorbent does not require the addition of any binding agent. Polyamide made by Woelm, like Swedish Sephadex, is used in special chromatographic techniques.

Quantitative Chromatography

The TLC quantitative technique differs insignificantly from its paper chromatography counterpart. It is also based on direct and indirect determinations. Quantitative analysis is made by measuring either the area or the length of the spots on the chromatogram, or else the concentrations of the compounds in the eluates are determined by spectrophotometric methods. Elution of substances from inorganic adsorbents yields individual compounds of appreciably higher purity than when organic adsorbents are used. For the purpose of mass spectrometric determination compounds adsorbed on silica do not have to be eluted [18]. For documentation purposes the chromatograms are photographed or fixed with collodion or Neo-Neatan (manufactured by Merck).

Apparatus

The apparatus assembly used in TLC technique consists of glass plates of dimensions 10 × 20 or 20 × 20 cm and tightly covered glass chambers.

For application of solutions the same micropipettes are used as in paper chromatography, and the same atomisers for spraying.

The most expensive and complicated device is the applicator for coating the glass plates with the adsorbent slurry.

PRECOATED PLATES AND SHEETS

A recent development in the TLC technique are commercially available precoated glass plates or more often plastic sheets, manufactured by a number of companies. The plastic sheets are made of polymeric materials insoluble in common organic solvents. The sheets are coated with such adsorbents as silica gel of very fine structure, polycarbonate, and polyamide. A substantial advantage of these sheets over conventional glass plates is the shortened developing time and improved results. Typical size of the sheets is 47 × 88 mm.

Mobile Phase

The mobile phase in TLC is selected after consideration of the polarity of the substance to be chromatographed. The preliminary test to recognise the behaviour of a compound during development with a variety of solvents is conducted by a technique known as microcirculation [19]. The mixture of compounds under analysis is applied to a chromatoplate, and the resulting spots are developed with a number of solvents, starting with the least polar. After development the selection of the most suitable solvent is made. The best solvent is the one that gives rise to the largest number of rings round the applied spot.

In the TLC partition technique the same one- or two-phase systems are used as in paper chromatography.

In reversed phase TLC the stationary phase is typically paraffin oil, silicone oil, or undecane, and the mobile phase may be acetic acid or mixtures of ketones with acetonitrile.

Analytical and Preparative Procedures

From 0·1 to 100 μg of a compound may be applied in solution volumes of 0·5–1·5 μl as a single spot on a thin layer chromatogram. A variant of the TLC technique is multiple development, although the two-dimensional method is also used. Horizontal development is rarely adopted.

As the technique is efficient for qualitative identification it can be used successfully for preparative purposes. Up to 0·5 mg of a substance may be applied to a single chromatoplate. Increasing the thickness of the adsorbent layer to 1 mm enables larger quantities of the substance to be analysed (up to 50 mg). After developing and detection of the chromatogram edges, individual zones are collected and extracted.

A novel technique [19a] was worked out in Czechoslovakia for the preparation of preparative layers of dry silica gel and other adsorbents.

In order to pass from TLC conditions to those for a column technique the Duncan equation is used [20]

$$R = \frac{a}{b + 0{\cdot}1a}$$

where:

R = resolving power of the column,
a = the R_f value of a substance travelling rapidly on the TLC chromatogram,
b = the R_f value of a substance travelling slowly on the TLC chromatogram.

Compounds a and b were found to be separable on a column only if $R > 1$.

Detection and Interpretation of Chromatograms

The fact that chromatoplates are commonly prepared from an inorganic adsorbent (silica, alumina) permits the use of aggressive detection reagents such as sulphuric acid, which chars the organic compound. A particularly sensitive detection method is fluorescence when used on plates detected with (1:1) alcoholic sulphuric acid. The detection limit is of the order of ca. 10^{-2} μg.

A certain modification of the fluorescence detection method requires the use of bromine [21,22]. The water added to the gel is replaced by a 0·04% aqueous solution of the sodium salt of fluorescein, and after developing, the chromatogram is detected by treatment with bromine vapour.

The most versatile of the reagents used for detection is iodine, which forms unstable addition compounds with the organic substance separated on the chromatogram. Another method in which the identified compound does not react with the detecting reagent, consists in spraying the chromatograms of lipophilic compounds with water. An additional advantage of the detection method employing iodine or water is that the chromatograms can be redetected with other reagents.

The TLC technique offers the possibility of carrying out a number of reactions directly on the plate, such as dehydration with sulphuric acid, oxidation with chromium trioxide in acetic acid, bromination, hydrogenation, hydrolysis or esterification.

3.4 GAS CHROMATOGRAPHY

Gas chromatography is usually restricted to the separation of gases and compounds with boiling points which do not exceed 400°C.

A compound injected into the column immediately vapourises and undergoes partition between the stationary phase (adsorbent or a liquid) and the gas flowing through the column, referred to as the carrier gas, which constitutes the mobile phase. The components separated on the column reach the detector which indicates either their instantaneous concentration (differential detector) or their total content in the carrier gas (integral detector).

Gas chromatography relies on two principal parameters, namely retention volume, V_R, which is the volume of carrier gas required for the elution of a given component from the column, and retention time, t_R, which represents the time that elapses from the moment the sample is introduced till the appearance of the chromatographic peak. These quantities are related in the following manner,

$$V_R = t_R F_c$$

where:

t_R = the retention time,

F_c = the flow rate of the carrier gas.

A quantity of major importance in gas chromatography is the adjusted retention volume, V'_R. This quantity is characteristic of every chromatographed compound at a specified temperature and phase. The adjusted retention volume equals the retention volume minus the gas hold-up. Its dependence on the partition coefficient is given by the equation [23]

$$V'_R = KV_1$$

where:

V_1 = the volume of the stationary phase (a constant independent of temperature),

K = the partition coefficient.

A parameter frequently used in practice is the retention index suggested by Kováts [24]. This quantity is slightly temperature dependent and provides valuable information on the chemical structure of the compounds under examination.

Simultaneous determination of the retention indices of a given substance on two phases of different polarity enables the determination of the increment of the retention index to be made. This quantity is characteristic of any individual class of chemical compounds. The retention index increment is found from the equation,

$$I = I_p - I_n$$

where:

I_p = retention index on a polar column,

I_n = retention index on a non-polar column.

Of considerable significance for chromatographic practice in gas chromatography is the van Deemter–Zuiderweg–Klinkenberg equation [24]. This equation describes the effect of a number of factors on chromatographic peak broadening. The efficiency of a column depends on the height of a theoretical plate (HETP). The lower the HETP, the more plates are contained in the column and the better is the separation.

Thus, the van Deemter equation says that although a decreased grain size in the support results in a lower HETP value, it is accompanied

8*

by a simultaneous increase in the tortuosity coefficient, which is indicative of non-uniform packing of the column.

A somewhat simplified form of the van Deemter equation (see Fig 3.3) was given by Keulemans and Kwantes [25], namely

$$H = A + B/u + Cu$$

This equation in the graphical form represents a hyperbola with a minimum for $H = A + 2\sqrt{BC}$ at $u = \sqrt{B/C}$.

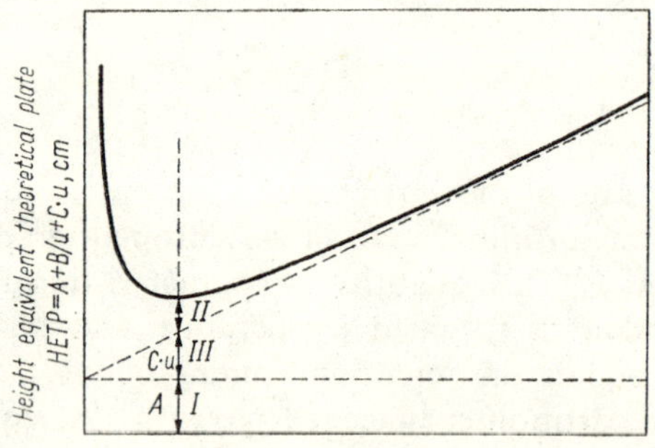

Fig. 3.3 Plot of the van Deemter equation, $H = A + B/u + Cu$, where: H = the height equivalent to a theoretical plate, A = constant dependent on column packing, B = constant dependent on diffusion of molecules in the gas phase, C = constant dependent on diffusion of molecules in the stationary phase, u = flow velocity of the carrier gas

In other words, the best resolution on a chromatographic column occurs when the carrier gas flow rate is equal to $\sqrt{B/C}$. It also follows from the van Deemter equation that with decreasing flow rate of carrier gas longitudinal diffusion arises, since the residence time of vapours in the column then increases. An excessive flow rate of the carrier gas may, however, be attended by impaired resolution since there may be insufficient time to allow equilibrium in the column to be established. Hence, at large flow rates separation is controlled by the term C, whereas at small flow rates it is controlled by B.

Supports

The supports in gas chromatography in widespread use include firebrick (Firebrick C 22, Sterchamol No. 22), and diatomaceous earth (Celite, Kieselguhr). Since every support, despite its general inactivity, has some residual adsorption centres, these centres have to be deactivated by careful impregnation of the material with the stationary phase, otherwise the retention time of certain components may increase, with additional distortion of the corresponding peaks. Furthermore, the presence of such active adsorption centres in the support is often responsible for certain undesirable chemical reactions. For this reason supports should be purified and conditioned.

The conditioning procedure involves washing in concentrated hydrochloric and nitric acids, followed by water, acetone, alkalis (1–3% methanolic potassium hydroxide) and finally with alcohol. Often the support is further treated with dimethyldichlorosilane or polymethylsiloxanes.

Stationary Phase

The stationary phase is composed of high boiling liquids of considerable thermal and chemical stability, which are capable of selectively dissolving the compounds to be separated.

Non-polar phases separate substances not so much through the presence of functional groups in the compounds, as happens in polar phases, but by differences in vapour pressure and molecular weight.

The amount of stationary phase used to impregnate the support is normally in the range 3–30%. A solution of the stationary phase in a suitable solvent is used for impregnation. Among the most popular solvents are aromatic hydrocarbons, chlorinated hydrocarbons and acetone. The impregnation procedure involves pouring the solid support into the solution of a stationary phase and passing an inert gas through the mixture to agitate it and to evaporate the solvent. The solvent residues are removed by heating in a water bath, followed by drying in a vacuum desiccator. The support impregnated in this way is used to pack the column.

Columns

Glass, copper, or stainless steel tubing is used for columns. Their typical length is 1–20 m. Longer columns and finer supports result in an increased number of theoretical plates and, consequently, in improved resolution. Since resolution is proportional to the square root of the number of theoretical plates, n, any increase in the column length might appear to be attended by improved resolution. In practice, however, this is not the case, as there exist interfering factors such as the pressure drop. The column length also affects the sample volume to be injected. This volume should be proportional to the square root of the length of the column used.

Usually the column diameter ranges from 2 to 8 mm, 6 mm being the most typical. The tubes used to make a column are either straight, U-shaped or helical. It is of the utmost importance for satisfactory performance that the column be carefully packed without any small gaps or discontinuities of any kind.

The columns are usually operated at temperatures ranging from 50 to 240°C, but occasionally higher temperatures are available on a specific instrument. Chromatographs are available with an operating temperature up to 500 or even 700°C. It should, however, be borne in mind that at higher temperatures the solubility of the compounds under separation in the stationary phase increases with a resulting drop in resolution. The column temperature may or may not be equal to the detector temperature.

The columns are usually operated under isothermal conditions although occasionally in separating mixtures of compounds of substantially different boiling points the programmed-temperature mode is adopted.

Mobile Phase

Although a heavier carrier gas performs better than a lighter one because of smaller diffusion of vapours within it [26], gases lighter than nitrogen, namely hydrogen and helium, are largely used in gas chromatography with a thermal conductivity detector. On account of its higher thermal conductivity compared to nitrogen, the helium carrier gas improves the sensitivity and allows simpler quantitative interpretation of the chromatograms.

In 6 mm columns the flow rate of carrier gas ranges from 30 to 300 ml per min, being typically less than 100 ml per min.

Apparatus

Essential components of every gas chromatograph include columns, a detector, and a recorder. Often chromatographs are equipped with ancillary units such as an integrator and/or a pyrolysis unit.

A recent development are chromatographs linked with a mass spectrometer.

Detectors

The detectors used may be broadly subdivided into differential and integral types. The former include the thermal conductivity detector (katharometer) and ionisation detectors, the flame ionisation type being the most important. The thermal conductivity detector responds to the difference in thermal conductivity between the carrier gas and the vapours of the compounds under analysis. Ionisation detectors operate on the principle of varied current intensity through the detector chamber when vapours of the compounds chromatographed appear in the interelectrode space. Ionisation detectors are the most sensitive, enabling the detection of compounds down to ca. 10^{-2} ppm.

Recorders and Integrators

The recorders used are standard commercially available electronic compensation instruments with 1–10 mV span and a 1 sec response.

Integrators are used to integrate the results of analysis and to provide them in the form of separate totals corresponding to the quantity of the individual separated component. Among the best integrators are electronic types, some of which provide digital display, others a curve on a recorder chart.

PYROLYSIS UNIT

An attachment used for thermal decomposition of a sample, referred to as a pyrolysis unit, is invaluable in the analysis of polymers. The products, of variable volatility, produced from the sample pyrolysed in the unit, are separated on the column, detected and registered on the recorder chart.

A recent development in this field is a pyrolysis unit based on the Curie point [27,28]. The use of ferromagnetic metals in a high frequency electric field not only shortens the pyrolysis time, but also enables the pyrolysis temperature to be determined [29].

A laser-based pyrolysis unit of special design [30–32] allows the pyrolysis temperature to be raised to 3000–3700°C, and shortens the pyrolysis time to a fraction of a second.

To introduce a sample into the column in quantities from hundredths of a microlitre to ten or even fifty microlitres (depending on the column size) microsyringes are used.

Identification

The components separated by gas chromatography are usually identified either from the chromatograms or by separate examination of collected fractions.

There are several methods by which a compound may be identified on the basis of its chromatogram. The most common is that of comparison of the retention volumes or retention times of the compounds under analysis with the corresponding quantities of standard substances analysed under the same conditions. The standards may be chromatographed alone or together with the sample being examined. Compounds may also be identified in terms of their relative retention volumes [33] or from the plot of the logarithm of the retention volume against the number of carbon atoms in the molecule of the compound, that is, using the relationship $\log V_R = f(N)$ [34]. For identification of unknown substances determination of the retention index and its increment [35] is used.

In addition to the methods described above there are also others, in which the chromatographic column is directly coupled to an infrared spectrophotometer, a far-ultraviolet spectrophotometer [36], or a mass spectrometer.

Quantitative Analysis

The essential relationship in quantitative gas chromatography, referring to differential detectors [10], is the equation

$$w = \int c(v)\,\mathrm{d}v$$

where:

w = the weight of the vapour of the component to be determined,
$c(v)$ = the concentration of the vapour of the component in unit volume of carrier gas flowing through the detector.

A practical measure of the amount of vapour emerging from the column is usually the area under the chromatographic peak from the differential detector.

If the detector response is linear, the quantities of individual components are proportional to the areas under the peaks or, when the peaks are symmetrical, to their heights.

The area under peaks may be found by one of the following methods: (a) using a planimeter, (b) with an integrator, (c) by cutting out and weighing the separate peaks in the chromatogram, (d) by multiplying the peak height by its width at half height. The most accurate results are obtained by measuring the area, using an electronic integrator.

Irrespective of the detector type used, every quantitative analysis requires calibrating, since the response of differential detectors is merely proportional to the concentration of the components leaving the column along with the carrier gas. Among the most commonly used quantitative methods are those of standardisation and the internal standard. The latter provides higher accuracy and does not require previous identification of all the components of the unknown.

Internal Standard Method

In the internal standard method the chromatographic analysis of the unknown mixture is conducted with the addition of a known quantity of a pure compound taken as the standard [25].

The so-called correction factor for areas, F, is calculated from chromatograms of known mixtures of the pure compound with the standard, using the formula

$$F = (S/A)(a/s)$$

where:

A = the weight of the standard, g,
S = the peak area corresponding to the standard, mm^2,
s = the peak area corresponding to the compound to be determined, mm^2,
a = the weight of the compound to be determined, g.

The sample under analysis is mixed with a quantity of the standard similar to that of the compound being determined. The areas under the peaks due to the standard and to the compound are determined from the chromatogram. These peak areas give the percentage Z (as wt. per cent) of the compound being determined in the sample,

$$Z = (s'WF)/S'$$

where:

Z = the quantity of component under analysis, wt. per cent,
s' = the peak area of component being determined, mm^2,
S' = the peak area of the internal standard, mm^2,

W = the quantity, wt. per cent, of the internal standard added to the original sample,

F = correction factor.

The internal normalisation method is based on the assumption that the same quantities of all compounds, irrespective of their structure, yield peaks of equal areas when analysed chromatographically. The percentage Z of the individual components is then found from the formula

$$Z = 100A/(A+B+C)$$

where A, B, and C are the peak areas of the compounds of the mixture under analysis.

This method provides satisfactory results if the components of the mixture have similar properties, and the carrier gas used is helium or hydrogen. Otherwise so-called corrected areas must be used.

When the flame-ionisation detector is used the area of each peak should also be divided by the number of the carbon atoms present in the molecule of the individual compound.

If a component of the mixture is present in very low concentration a known amount of that compound is added to the mixture under analysis [27]. The sample is chromatographed twice under identical conditions, first without addition and the next time after addition of a known quantity of the individual compound which is to be determined in the mixture. From the peak heights measured on both chromatograms the quantity Z, as wt. per cent, of the component under analysis is found according to equation

$$Z = AW/(B-A)$$

where:

A = the peak height for the component determined in the sample,

B = the peak height for the same component being determined after addition of this component as standard to the mixture,

W = the amount of the component added, wt. per cent.

The calibration method consists in independent chromatographic analysis of variable quantities of each component of a mixture. Calibration curves are plotted for each component on the basis of the peak heights or peak areas, and from these curves the absolute quantities of the components are determined in the sample.

Another calibration procedure involves chromatographic analysis of standard mixtures composed of the substance to be determined and a standard compound mixed in several known ratios. The area of the peaks is measured from the chromatograms. A calibration curve of the ratio of peak areas versus ratio of weights of standard and component under analysis is constructed. This should be a straight line, and from it the concentration of the component may be obtained after the sample, containing a known weight of standard, has been chromatographed. Of the above methods the last is the most accurate.

3.5 APPLICATION OF CHROMATOGRAPHY TO ANALYSIS OF POLYMERS

Application of gas chromatography to the analysis of polymers is a fairly recent development. This began with the advent of pyrolysis gas chromatography and considerable improvements have been made in recent years. This technique enables polymer samples to be directly pyrolysed on the column in a carrier gas stream. However, the use of gas chromatography in the field of plastic materials is not confined to the examination of the polymers alone. Monomers and additives such as plasticisers, antioxidants, stabilisers, and solvents used in certain polymer finished products (adhesives, varnishes) can also be analysed by this method.

Fractionation

A major application of chromatography in polymer analysis is in the study of molecular weight distributions by gel-permeation chromatography (also known as gel filtration). Initial studies of synthetic elastomers by this technique revealed [37] that adsorption of a polymer depends on the concentration of the polymer solution. It was found that from solutions of low concentration the molecules of the shortest chains are most strongly adsorbed and the longest chain molecules are most weakly adsorbed. This situation is reversed for adsorption from concentrated solutions.

Claesson [37] was one of the first to investigate polymer fractionation and molecular weight distribution, using gel-permeation chromatography. In his frontal analysis studies of certain synthetic polymers such as poly (methyl methacrylate) and poly(vinyl acetate) he found that for high polymers the quantity of adsorbed molecules decreases with the increasing molecular weight. The molecular weight distribution curves obtained from chromatographic fractionation had the same form as the corresponding curves obtained by other methods such as fractional precipitation. Bannister *et al.* [35] examined this effect, using linear silicones adsorbed on activated carbon. By gradually varying the strength of the elution solvent these authors obtained fractions, and determined their molecular weight and viscosity. Comparison of the plots of both quantities plotted against the volume of eluate gives a picture of the molecular weight distribution in a given polymer.

Among the most studied polymers are polypropylene [38], polychloroprene [39], polybutadiene [40], acetylated poly(propylene glycols), styrene-butadiene copolymers, polyisobutylene [41] and polybutadiene [42]. For a series of phenyl oligomers and urethanes of known molecular weight the effect of molecular weight of the polymer and of the size of its molecules on the course of the chromatographic fractionation on gel were investigated. The gels used were silica gel, styrene-divinylbenzene copolymer, PVC crosslinked with butanediol divinyl ether, poly(methyl methacrylate) and poly (butyl methacrylate) crosslinked with diol dimethacrylates.

Braun [43] reports a computer-based determination of the differential and the integral molecular weight distributions on the basis of chromatograms recorded on a perforated tape and with the use of a versatile calibration method. A digital computer was also used for the determination (by gel-permeation chromatography) of molecular weight and its distribution, and for establishing the nature and extent of branching in polyethylene [44]. Ram and Miltz [45] worked out a novel method for determination of the molecular weight distribution of branched polymers, using gel-permeation chromatography. A recent development is the use of gel-permeation chromatography coupled with other techniques for the study of the synthesis mechanism and the properties of some silicones [46]. Fractionation of polymers by thin layer chromatography was the subject of investigations by Otocka and Hellman [47]. Otocka [48] not only fractionated polymers by thin layer chromatography but also determined their molecular weight distribution by means of a densitometer.

Identification

For identification of polymers gas-liquid chromatography (pyrolysis gas chromatography) and thin layer chromatography are generally used. Until recently only pyrolysis gas chromatography was popular, and the thin layer technique, like paper chromatography, served for identification of polymer additives, such as plasticisers, stabilisers, antioxidants etc. Only polymers that could be split into their monomers and simple components were analysed by these methods. Not until a decade ago did several investigators [48] succeed in identifying a number of polymers by thin layer chromatography after prior degradation of the material, following the procedure reported by Stahl [49].

Originally the interpretation of gas-chromatographic pyrograms was based solely on fingerprint comparison of pyrograms of standards degraded and chromatographed under identical conditions with the chromatograms of the unknowns. Later it was found that for identification of a polymer the identification of one or a certain group of characteristic degradation products is usually sufficient.

Recent successes in polymer identification by means of pyrolysis gas chromatography include thermoplasts such as PVC, poly(vinyl acetate), polyacetals, polystyrene, and also polyolefins, mainly polyethylene and polypropylene [50].

Nelson *et al.* [51], who identified a number of polymers (sample size 0·2–0·5 mg) from the pyrolysis products at 650–750°C, used a column packed with 5% silicone oil supported on Chromosorb P. They found that for polymers of similar structure the pyrograms differed only in the "higher" products which emerge later from the column.

Identification of acrylic, methacrylic, and styrene polymers and copolymers was conducted by pyrolysis gas chromatography by Braun [52], McCormic [53] and others. The major degradation products of methacrylate and styrene polymers were found to be their respective monomers. Phenolic resins were analysed by this technique by a number of authors, and the best results were accomplished with a capillary column packed with tri(2,4-xylenyl) phosphate on Chromosorb P [54]. A slightly poorer resolution was observed in the separation of the degradation products of phenolic resins analysed on the same mobile phase but with conventional 4 mm columns [55]. Resins of varying degree of cure were identified by Biliński [56]. Braun and Arndt developed two identification procedures for phenolics: one based on thin layer chromatography, the other on gas chromatography, both for the pyrolysis products of these polymers.

The most popular methods for identification of polyesters, polyurethanes, and polyamides are paper chromatography [57,58], ion-exchange chromatography [59] and thin layer chromatography. Gas chromatography is of minor importance in these cases. Usually these procedures are based on chromatography of the hydrolysis products of these resins. An identification procedure using TLC was developed by Schöllner and Löhnert [60], and also Kobayashi [61]. Heidemann [62], Braun [63] and Hilt [64] and their co-workers were concerned with identification of polyamides and polyurethanes. Hilt and co-workers succeeded in separating polyesters with hydroxyl end groups from analogous resins with acidic end groups. Identification of polyesters and polyurethanes by paper chromatography, thin layer chromatography and gas chromatography was the object of studies by Zowall and Świątecka [65,66]. A method for identification of polyester resins by gas chromatography of their degradation products was evolved by Luce *et al.* [67]. Gas chromatography of the degradation products of polyurethanes based on xylylene di-isocyanate was the subject of research made by Rumao and Frisch [68]. An identification procedure using gas chromatography for the pyrolysis products of hydrolysates was applied by Zowall and Świątecka [69].

Pyrolysis of polymers followed by thin layer chromatography was investigated by Pastuska [70] and Braun *et al.* Based on polymer degradation procedures Braun and co-workers worked out a thin layer chromatographic identification method for polyamides [71], polyurethanes [72], amino resins [73], and polystyrene containing α-methylstyrene [74].

Another aspect of the field of polymer identification is the analysis of additives, which include plasticisers, stabilisers and antioxidants.

Identification of plasticisers has been dealt with by a number of investigators and suitable analytical procedures using paper chromatography [75], thin layer chromatography [76–78] and gas chromatography [79–81] have been evolved. According to some evidence thin layer chromatography may be better than gas chromatography for identification of plasticisers [82]. An additional advantage of the thin layer analysis is the detection of chromatograms with aggressive reagents.

Identification of stabilisers in PVC was carried out by Bürger [83], Heide [84], Neubert [85], and Türler and Högl [86] using thin layer chromatography.

Antioxidants for polyethylene were analysed after extraction from the polymer [90] or directly in the material by Bancher [87], Braun [88], and Bürger [89]. Kapisinska and Kosljar [91] worked out a method of rapid identification of accelerators, antioxidants and stabilisers by thin layer chromatography.

Related to identification are investigations on the constitution and structure of polymers. Examples in this field are the studies by Gnauck [92] of the distribution in molecular weight of poly(alkylene glycols) by gas chromatography, the determination of stereoblock copolymers of formaldehyde from the gas chromatographs of their degradation products [93], studies of the sequence of monomer units in acrylate-acrylonitrile copolymers by gas chromatography of their degradation products [94].

Quantitative Determinations

Although chromatography is primarily designed for qualitative analysis, certain of its variations are suitable for quantitative work. Chromatography is frequently used for quantitative analysis of impurities in monomers and in polymers, and for residual monomers in the latter.

Of the many available gas chromatographic determinations, some worthy of mention are styrene and other hydrocarbon impurities in polystyrene [95,96], trace acrylonitrile monomer in its polymer [97], phenol in alkyl resols [98], acrolein and furfural in phenolic resins [99], residual solvents in polymer sheets [100] and moisture in poly(vinyl butyral) [101]. Thin layer chromatographic determinations of urea in urea-formaldehyde resins [102], and of furfural and furfuryl alcohol in furan resins [103] have also been recorded.

Direct examination was made of the thermal degradation of poly-(methyl methacrylate) [104] and fractionated polystyrenes [105] and polycarbonate [106], utilising pyrolysis gas chromatography. The yields of each component of the degradation products of polystyrene were examined as a function of molecular weight and the pyrolysis temperature. The monomer yield was found to increase with growing molecular weight of polystyrene and to decrease when the pyrolysis temperature increases. The dependence of yields of pyrolysis products on the molecular weight indicates the possibility of estimating the molecular weight of the original polymer. One method for determining the monomer ratio in ethylene-vinyl acetate copolymers involves measuring the amount of acetic acid produced on pyrolysis of the polymer [107]. Similarly the composition of polyethylene-polyisobutylene copolymers can be evaluated from the pyrograms [108]. Recently Martens and Glas [109] developed a method of numerical evaluation of complex pyrograms based on two fundamental statistical procedures.

References

1. Opieńska-Blauth, J., Waksmundzki, A., Kański, M., *Chromatografia* (*Chromatography*), PWN, Warsaw, 1957.
2. Neher, R., *Steroid Chromatography*, Elsevier, Amsterdam, 1964.
3. Martin, A. J. P., Synge, R. L. M., *Biochem. J.*, **35**, 1358 (1941).
4. de Vault, D., *J. Am. Chem. Soc.*, **65**, 532 (1943).
5. Wilson, W., *J. Am. Chem. Soc.*, **62**, 1583 (1940).
6. Littlewood, A. B., *Gas Chromatography*, Academic Press, New York, 1962.
7. Bayer, E., *Gas-Chromatographie*, Springer-Verlag, Berlin, 1962.
8. Eger, C., *Experientia*, **12**, 37 (1956).
9. Teuent, D. M., Whitla, J. B., Florey, K., *Anal. Chem.*, **23**, 1748 (1951).
10. Nye, J. F., Marton, D. M., Garst, J. B., Friedgood, H. B., *Proc. Soc. Exptl. Biol. Med.*, **77**, 446 (1951).
11. Seki, T., *J. Chromatogr.*, **2**, 667 (1959).
12. Hecker, E., *Verteilungsverfahren im Laboratorium, Suppl. Angew. Chem.*, Verlag Chemie, Weinheim, 1954.
13. Trappe, W., *Biochem. Z.*, **305**, 150 (1940).
14. Fischer, R. B., Parsons, D. S., Morrison, G. A., *Nature*, **161**, 764 (1948).
15. Kowhabani, G. N., Cassidy, H. G., *Anal. Chem.*, **24**, 643 (1952).
16. Martin, A. J. P., *Biochem. Soc. Symposia* (*Cambridge*), 3, 4 (1949).
17. Brenner, M., Niederwieser, A., Pataki, G., Fahmy, A. R., *Experientia*, **18**, 101 (1962).
18. Heyna, H. F., Grutzmacher H. F., *Angew. Chem.*, **74**, 387 (1962).
19. Barret, C. B., Dal, M. S., Padley, F. B., *J. Am. Oil Chem. Soc.*, **40**, 507 (1963).

19a. Černy, V., Joska, G., Labler, *Coll. Czech. Chem. Comm.*, **26**, 1658 (1961).

20. Duncan, G. R., *J. Chromatogr.*, **8**, 37 (1962).
21. Smith, L. L., Foell, T., *J. Chromatogr.*, **9**, 339 (1962).
22. Stahl, E., *Pharmazie*, **11**, 633 (1956); *Chemiker-Ztg.*, **82**, 323 (1958).
23. Matthews, J. S., Pereda, A. L., Aquilas, A., *J. Chromatogr.*, **9**, 331 (1962).
24. Kováts, E., *Helv. Chim. Acta*, **42**, 1915 (1956).
25. Keulemans, A. J. M., Kwantes, A., *Vapour Phase Chromatography, Proceedings of the First Symposium*, London, June 1956, Academic Press, New York, 1957.
26. Keulemans A. J. M., *Gas-Chromatographie*, Verlag-Chemie, Weinheim, 1959.
27. Simon, W., Giacobbo, H., *Chem. Ing. Techn.*, **37**, 709 (1965).
28. Giacobbo, H., Simon, W., *Pharmac. Acta Helv.*, **39**, 162 (1964).
29. Wilmott, F. W., *J. Chromatogr. Sci.*, **7**, 101 (1969).
30. Levy, R. L., Fanter, D. L., *Anal. Chem.*, **41**, 1465 (1969).
31. Kojima, T., Morishita, F., *J. Chromatogr. Sci.*, **8**, 471 (1970).
32. Folmer, O. F., Azaraa, L. V., *J. Chromatogr. Sci.*, **7**, 665 (1969).
33. Ellis, W. H., Le Tourneau R. L., *Anal. Chem.*, **25**, 1269 (1953).
34. Ray, W. H., *J. Appl. Chem.*, **4**, 2181 (1951).
35. Bannister, D. W., Phillips, C. S., Williams, R. J. P., *Anal. Chem.*, **26**, 1451 (1954).
36. Kaye, W. J., *Pittsburg Conf. on Anal. Chem. and Appl. Spectroscopy*, 1965.
37. Claesson, S., *Disc. Farad. Soc.*, **7**, 321 (1949).
38. Poláček, J., Matyska, B., *Coll. Czech. Chem. Comm.*, **27**, 816 (1962).
39. Hulwe, J. M., McLeod, L. A., *J. Polymer Sci.*, **3**, 153 (1962).
40. Brewer, P. I., *Polymer*, **6**, 603 (1965).
41. Yamada, S., Imasi, S., Kitahara, S., *Kobunshi Kagaku*, **23**, 400 (1966); *Chem. Abstr.*, **66**, 47062 (1967).
42. Heitz, W., Ulhrer, H., Hacker, H., *Makromol. Chem.*, **98**, 42 (1966).
43. Braun, G., *J. Appl. Polymer Sci.*, **15**, 2321 (1971).
44. Wild, L., Ranganath, R., Ryle, T., *J. Polymer Sci., Pt. A-2*, **9**, 2137 (1971).
45. Ram, A., Miltz, J., *J. Appl. Polymer Sci.*, **15**, 2639 (1971).

46. Pittman, C. U., Jr., Patterson, W. J., McManus, S. P., *J. Polymer Sci., Pt. A-1*, **9**, 3187 (1971).
47. Otocka, E. P., Hellman, M. Y., *Macromolecules*, **3**, 362 (1970).
48. Otocka, E. P., *Macromolecules*, **3**, 691 (1970).
49. Stahl, E., *J. Chromatogr.*, **37**, 99 (1968).
50. Groten, B., *Anal. Chem.*, **36**, 1206 (1964).
51. Nelson, D. F., Yee, Y. L., Kirk, P. L., *Microchem. J.*, **6**, 225 (1962).
52. Braun, D., Vorendohre G., *Farbe u. Lack*, **69**, 820 (1963).
53. McCormic, H., *J. Chromatogr.*, **40**, 1 (1969).
54. Zulaica, J., Guiochon, G., *J. Polymer Sci., Pt. B*, **4**, 567 (1966); Zulaica, J., Guiochon, G., *Chim. Ind. (Paris)*, **100**, 1108 (1968).
55. Zowall, H., Świątecka, M., *Polimery*, **16**, 127 (1971).
56. Biliński, C., *Rech. Aerospatiale*, **103**, 27 (1964).
57. Fijolka, P., Radowitz, W., *Plaste u. Kautschuk*, **12**, 207 (1965); Fijolka, P., Kayler, R., Lenz, J., *Kunststoffe*, **49**, 222 (1959).
58. Arendt, I., Schenck, H. J., *Kunststoffe*, **49**, 321 (1959).
59. Schröder, E., *Plaste u. Kautschuk*, **5**, 103 (1958).
60. Schöllner, R., Löhnert, P., *Plaste u. Kautschuk*, **15**, 436 (1968).
61. Kobayashi, Y., *J. Chromatogr.*, **24**, 447 (1966).
62. Heidemann, G., Kusch, P., Nettelbeck H.-J., *Z. Anal. Chem.*, **212**, 401 (1965).
63. Braun, D., Vorendohre, G., *Kunststoffe*, **57**, 821 (1967).
64. Hilt, A., Funke, W., Hamann, A., *Dtsch. Farben-Zeitschr.*, **20**, 567 (1966).
65. Zowall, H., Świątecka M., *in preparation.*
66. Zowall, H., Świątecka, M., *in preparation.*
67. Luce, C. C., Humphrey, F. F., Guild, L. V., Norrish, H. H., Coull, J., Castor, W. W., *Anal. Chem.*, **36**, 482 (1964).
68. Rumao, L. P., Frisch, K. C., *J. Polymer Sci., Pt. A-1*, **10**, 1499 (1972).
69. Zowall, H., Świątecka, M., *Polimery*, **18**, 325 (1973).
70. Pastuska, G., *Gummi, Asbest, Kunststoffe*, **22**, 718 (1969).
71. Braun, D., Vorendohre, G., *Kunststoffe*, **57**, 821 (1967).
72. Braun, D., Mai, E., *Kunststoffe*, **58**, 637 (1968).
73. Braun, D., Jung Jin Chul, *Gummi, Asbest, Kunststoffe*, **23**, 618 (1970).
74. Braun, D., Nixdorf, G., *Gummi, Asbest, Kunststoffe*, **22**, 183 (1969).
75. Gude, A., *Kunststoffe*, **52**, 679 (1962).
76. Braun, D., *Kunststoffe*, **52**, 2 (1962).
77. Nagy, Z., *Lebensm. Untersuch. Forsch.*, **126**, 282 (1965).
78. Dotreppe-Grisard, N., Dumoulin, J., *Rev. Belge Mat. Plast.*, **9**, 529 (1968).
79. Gillio-Tos, M., *Kunststoffe*, **56**, 409 (1966).
80. Wandel, M., Tengler, H., *Plastverarbeiter*, **16**, 711 (1965).
81. Fischer, W., Jachn, L., *Plastverarbeiter*, **17**, 117 (1966).
82. Fischer, W., Lenkroth G., *Plastverarbeiter*, **20**, 107 (1969).
83. Bürger, K., *Z. Anal. Chem.*, **192**, 280 (1963).
84. Heide, R. F., *Z. Lebensmitt.-Untersuch.*, **124**, 348 (1964).
85. Neubert, G., *Z. Anal. Chem.*, **203**, 265 (1964).
86. Türler, M., Högl, O., *Mitt. Lebensmitt.-Hyg.*, **52**, 123 (1961).
87. Bancher, E., Scherz, H., *J. Chromatogr.*, **16**, 157 (1964).
88. Braun, D., Vorendohre, G., *Z. Anal. Chem.*, **199**, 37 (1964).
89. Bürger, K., *Z. Anal. Chem.*, **196**, 259 (1963).
90. Heide, R. F., Wonters, O., *Z. Lebensmitt.-Untersuch.*, **15**, 129 (1962).
91. Kapišinska, V., Kosljar, V., *Plasticke Hmoty Kaučuk*, **6**, 338 (1969).
92. Gnauck, R., Fijolka, P., *Plaste u. Kautschuk*, **19**, 28 (1972).
93. Minin, W. A., Berlin, A. A., *Vysokomol. Soedin., Ser. A*, **14**, 9 (1972).
94. Yamamoto, Y., Tsuge, S., Takeuchi, T., *Macromolecules*, **5**, 325 (1972).
95. Crompton, T. R., Meyers, L. W., Blair, D., *Brit. Plastics*, **38**, 740 (1965).
96. Polish Standard, PN-71/C-89061.

97. Borodulina, R. I., Revel'skii, I. A., Shtylenko, A. D., *Plast. Massy*, **1964** [7], 49.
98. Mlejenek, O., *Plasticke Hmoty Kaučuk*, **5**, 366 (1968).
99. Stevens, M. P., *Anal. Chem.*, **37**, 167 (1965).
100. Scott, R. R., *Chem. Ind. (London)*, **52**, 2130 (1964).
101. Shiryaev, B. V., *Plast. Massy*, **1972**, 69.
102. Zowall, H., Szczepanowska, W., *in preparation.*
103. Zowall, H., Szczepanowska, W., *in preparation.*
104. Belenskii, B. G., Turkova, L. D., Andreeva, G. A., *Vysokomol. Soedin., Ser. B*, **14**, 349 (1972).
105. Tsuge, S., Okumoto, T., Takeuchi, T., *J. Chromatogr. Sci.*, **7**, 253 (1969).
106. Tsuge, S., Okumoto, T., Sugimura, Y., Takeuchi, T., *J. Chromatogr. Sci.*, **7**, 253 (1969).
107. Okumoto, T., Takeuchi, T., Tsuge, S., *Kogyo Kagaku Zasshi*, **73**, 702 (1970).
108. Hagen, E., Hazkoto, G., *Plaste u. Kautschuk*, **16**, 21 (1969).
109. Martens, A. J., Glas, J., *Chromatographia*, **5**, 508 (1972).

Chapter 4

POLAROGRAPHY

4.1 INTRODUCTION

Polarography [1,2] is based on the processes which occur during electrolysis of solutions of electrolytes on the mercury dropping electrode. Owing to the specific properties of this electrode the curves of current intensity plotted against the potential applied across the electrodes, which is continuously varied during the measurement, exhibit a characteristic step-like form (Fig. 4.1). The curves are referred to as polarograms. The significance

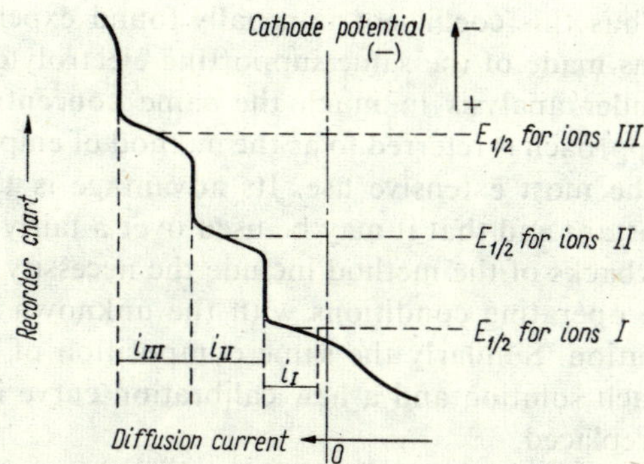

Fig. 4.1 A polarogram of a solution containing various ionic species

of polarography for qualitative and quantitative analyses stems from the fact that under suitable experimental conditions the height of these steps (commonly referred to as polarographic waves) is proportional to the concentration of the depolariser under analysis, and their position relative to the potential axis is characteristic of a specific depolariser. The wave height is proportional to the intensity of the current referred to as the limiting (diffusion) current i_d, which is defined by the Ilkovič equation

$$i_d = 605nD^{1/2}cm^{2/3}t^{1/6}$$

where:

i_d = the mean value of the current intensity, μA,

c = the mean concentration of the depolariser, mmole per litre,

n = the number of electrons participating in the electrode reaction of one molecule of the depolariser,

D = the diffusion coefficient of the reducible or oxidisable substance, cm^2 per sec,

m = the capillary yield, or the weight of mercury flowing through the capillary of the mercury dropping electrode, mg per sec,

t = the drop time, sec.

For a particular depolariser and at specified working conditions of the capillary, including a constant temperature, all the coefficients in the Ilković equation are constant, and the equation may be written in a simplified form

$$i_d = kc$$

where k is a constant for a given solution and capillary.

This equation underlies the quantitative analysis. The concentration value sought could be found by calculations in terms of this equation. In practice, however, the calculative approach is of minor significance, since values of the diffusion coefficient, which must be known, are available for only a few ions, and furthermore, the coefficient k in the equation depends not only on the kind of substance under analysis, but also on its concentration. Thus this coefficient is usually found experimentally, using standard solutions made of the same supporting electrolyte and containing the substance under analysis in much the same concentration as in the unknown. This approach is referred to as the method of empirical calibration and has found the most extensive use. Its advantage is that k does not require to be constant and that it may be used over a fairly wide concentration range. Drawbacks of the method include the necessity of strictly maintaining the same operating conditions with the unknown solution as with the standard solution. Similarly the same composition of the medium has to be used for each solution and a new calibration curve is required when the capillary is replaced.

In polarographic analysis another graphical method is also used, namely that of standard addition. In this approach the polarographic wave height is first determined for the unknown, then a standard solution in which the unknown substance is present in a suitable and known concentration is introduced into the solution. The polarogram is recorded again. Two polarograms have to be recorded here and accurate results are achieved only if the proportionality between the concentration and the diffusion current intensity is maintained; this is not always the case. Since this method also involves extra calculations, it is less common than the method of calibration curves.

It should be mentioned here, that in analytical practice it is not the diffusion current intensity that is employed but the recorded wave height in mm, which is a quantity proportional to i_d.

The fundamental quantity used for qualitative interpretation of polarograms is the potential of the middle of the polarographic wave, referred to as the half-wave potential, $E_{1/2}$. Under strictly specified experimental conditions this is a characteristic property of a given substance, and is independent of the concentration of that substance. The half-wave potential is not, however, a typical physical constant, as it is affected by the type of electrolyte in which the substance is dissolved. This fact has its advantages and disadvantages. Among the latter is the fact that the value of this quantity cannot be strictly established, and the literature data are not always reliable. Occasionally the literature gives no adequate information on the composition of the supporting electrolyte, which makes the $E_{1/2}$ data of little value for identification purposes. On the other hand an advantage is that the interfering effect of other polarographically active substances present in the solution under analysis can be eliminated by changing the polarographic medium.

The value of the half-wave potential is most readily found by a graphical method as illustrated in Fig. 4.2. The portions of the curve corresponding to the residual current *AB* and to the diffusion current *DF* are extended,

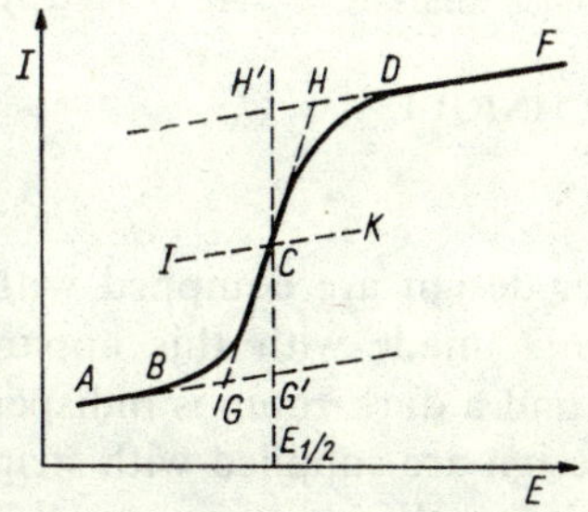

Fig. 4.2 Graphical determination of the half-wave potential

and a tangent is drawn at the inflection point. The section *GH* of the tangent, contained between the points of its intersection with the extended, almost parallel portions of the curve, is divided in half. The line *IK* is then drawn through the mid-point of the tangent, approximately parallel to the extensions of the lines *AB* and *DF* (these portions of the curve are usually not strictly parallel to each other). The abscissa of the point of intersection of this straight line with the polarographic curve yields the value of $E_{1/2}$.

A major advantage of the polarographic method is that, irrespective of the substance analysed, one kind of electrode, namely the mercury dropping electrode, is used (Fig. 4.3). Another advantage is that even rather complex mixtures can be analysed both qualitatively and quantitatively. It should be borne in mind, however, that the polarographic method is very sensitive to the measuring conditions such as temperature, solution pH, the type of solvent, the variety and concentration of ions present in the solution, and other factors. The measuring conditions should be exactly specified for every individual substance to be analysed, and should be strictly maintained throughout the measurement.

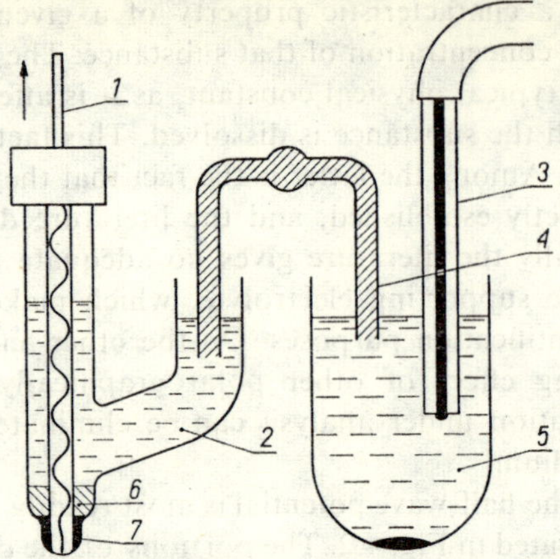

Fig. 4.3 The polarographic assembly: *1*—anode, *2*—saturated calomel electrode (SCE), *3*—the capillary of the mercury dropping electrode (cathode), *4*—salt bridge, *5*—solution under analysis, *6*—Hg + Hg_2Cl_2 paste, *7*—mercury

4.2 EXPERIMENTAL TECHNIQUE

Apparatus

The polarographs of older design are equipped with photographic recording. However, polarograms made with this apparatus require laborious photographic processing and a dark-room is indispensable. For this reason, instruments of current design are supplied with strip-chart recorders which draw polarograms directly on the chart paper.

Solvents

With few exceptions polymers are insoluble in water, which is undoubtedly the most convenient medium for polarographic analysis. Therefore, in the polarographic analysis of polymers [3–6] organic solvents are used extensively. The analyses are carried out either in entirely anhydrous solutions or in mixtures containing variable quantities of water. In the latter case the organic components of such mixtures are usually lower alcohols, dioxan, dimethylformamide, or monoalkyl ethers of ethylene glycol. Of the anhydrous media the following solvents or solvent systems are usually employed: lower monohydric alcohols, glycol, glycerol, dimethylformamide, acetic acid, acetonitrile, dimethylformamide in mixtures with ethanol, methanol, or acetic anhydride, acetonitrile in mixtures with acetic anhydride, mixtures of ethanol or methanol with benzene. Other similar solvents may also be used, provided these do not react with the substances under analysis nor undergo reactions on the mercury dropping electrode and provided they give solutions of the conductance required for polarography.

The considerable effect of non-aqueous solvent on the values of the half-wave potential should be taken into account.

Supporting Electrolytes

In addition to a suitable solvent the solution of a substance to be analysed polarographically should contain certain other compounds, which are indispensable for obtaining a correct polarogram. Of these compounds the first is a conducting or supporting electrolyte, as it has come to be called. A criterion for the selection of this electrolyte is the value of the potential at the start of its cationic wave; this potential should be more negative than the reduction potential of the ions being determined.

The supporting electrolytes commonly used in organic analysis [5,6], including the analysis of polymers, comprise tetraalkylammonium salts and hydroxides, lithium salts and lithium hydroxide. This is because these compounds are soluble in organic solvents and undergo reduction only at very highly negative potentials. This is particularly advantageous when determining substances such as styrene with high negative values of the half-wave potential.

Since hydrogen ions take part in most of the electrode reactions of organic compounds it is necessary to stabilise the pH of polarographic solutions. This is usually done with buffer solutions of weak acids or weak bases. Acetate, citrate, ammonium and other buffer solutions are commonly used in organic polarographic analysis.

4.3 POLAROGRAPHY IN THE STUDY OF POLYMERS

Introduction

Compounds that are reduced or oxidised at the dropping mercury electrode can be directly determined by polarography. In the first category are compounds that contain the following reducible functional groups or groups of atoms:

1. olefinic double bonds conjugated with other double bonds, aromatic rings or with other unsaturated groups,
2. carbonyls in aldehydes and quinones, or in acids and ketones when located in the α position to a double bond, or in the α or β positions relative to another carbonyl group,
3. nitro, nitroso, azo, and amine oxide groups,
4. phenylhydroxylammonium, diazonium, and quaternary ammonium groups,
5. halogen-containing groups,
6. peroxide and epoxy groups,
7. disulphide groups,
8. heterocyclic C=C double bonds.

The group of oxidisable compounds includes:

1. hydroquinones,
2. ene-diols,
3. substances such as mercaptans (–SH groups) which depolarise the

mercury surface owing to the formation of slightly dissociated or insoluble compounds.

Numerous compounds that are unreactive at the dropping mercury electrode can be indirectly determined by converting them into polarographically reactive compounds, for example:

1. aromatic hydrocarbons which can be converted into nitro compounds by nitration,
2. 1,2-glycols which can be converted into aldehydes by oxidation with such oxidants as periodic acid,
3. phenols which can be transformed into nitro or nitroso compounds by treatment with nitric or nitrous acid, respectively,
4. secondary amines which can be converted into nitroso amines by the reaction with nitrous acid,
5. ketones which are converted into (a) imines, by the reaction with primary amines or ammonia, (b) semicarbazones, by the reaction with semicarbazide, (c) oximes, by treatment with hydroxylamine, (d) hydrazones, by the reaction with hydrazine derivatives (notably 2,4-dinitrophenylhydrazine),
6. unsaturated compounds—by bromination to dibromo derivatives or by conversion into pseudonitrosites (in the reaction with nitrous acid).

In the polarographic analysis of polymers both direct and indirect methods are used [3,4].

Investigation of the Physical Properties of Polymers

Under specific conditions, instead of conventional well-defined polarograms in the form of steps of variable slope, the polarographic curves may exhibit maxima. The maxima produced are of two types. Maxima of the first type appear in a narrow potential range as sharp peaks. These are produced in solutions of low concentrations of the supporting electrolyte. The maxima of the second type occur over a wide potential range and are smoother in shape. These are produced in solutions of high concentration of supporting electrolyte when fast dropping electrodes are used.

It has long been known that surface active compounds suppress both types of maxima. These compounds are therefore indispensable ingredients of the supporting electrolytes. In practice, gelatin and agar–agar are most often employed, but occasionally other substances such as poly(vinyl alcohol) are also used.

Some studies have recently been published on the surface activity of certain polymers and its effect in the polarographic investigations of some of the physical properties of these polymers. Rusznák *et al.* [7] found that the size of polymer molecules is of prime significance in suppressing the maxima. In another study these authors [8] established the relation between the molecular weight of cellulose monophthalate diacetate and the suppressing power of solutions of this polymer with respect to the oxygen maximum. Bolewski and Lubina [9] observed the effect of molecular weight of poly(methacrylic acid) on the extent of lowering of the oxygen maximum, and demonstrated that this polymer can be determined by polarography. Finally Bezuglyi and Saliichuk [10] developed analytical procedures for the determination of the molecular weight of polymers such as polystyrene, poly(methyl methacrylate), and poly(vinyl toluene) on the basis of the first type of the oxygen maximum and of poly(vinyl alcohol) and cellulose monophthalate diacetate on the basis of the copper maximum of the second type. These authors believe that the maxima of the second type are more suitable for such analyses. The method was used in the routine determinations of molecular weight of a wide variety of polymers [3], including novolacs.

The polarographic method was also resorted to in tests of the resistance of polymers to certain solvents [11], and of changes in solubility [12] of polymers caused by such polymer-decomposing factors as elevated temperature, light and radiation.

Qualitative Analysis

A method of identification for a range of polymers by polarographic analysis of their pyrolysis products was developed by Bezuglyi *et al.* [3,13,14]. The reason for the use of polarography here is that the pyrolysis products of numerous polymers, either initially formed or after suitable chemical conversion, are reducible at the dropping mercury electrode. The products are mainly vinyl monomers and certain other compounds containing C═C double bonds. Many of these products, though not reducible themselves, easily undergo bromination or nitration and are thereby converted into electroactive substances.

Tests

Polymer (1–1·5 g) is pyrolysed under controlled conditions, and the volatile decomposition products are absorbed in 10 ml of methanol. The solution is made neutral towards neutral red indicator, and a 0·1–1·0 ml aliquot of the solution is transferred to the polarographic vessel filled with a 0·05*N* solution of tetramethylammonium iodide in 50% methanol (or 0·05*M* tetraethylammonium iodide in dimethylformamide [14]). Acidic solutions are first neutralised with tetraalkylammonium hydroxide.

To another aliquot of the methanolic solution is added 2–3 ml of 1*N* methanolic bromine solution to a permanent yellow colour. The excess bromine in a coloured solution is detected with starch-iodide paper. The solution is now neutralised with 5% ammonia solution in 80% methanol, and the polarogram of the bromination products is recorded in 0·1*N* lithium chloride solution in 50% methanol.

Since bromide ions interfere with the measurement of half-wave potentials for positive waves, the bromination products are isolated when required, by ether extraction. The ether is removed by distillation, the residue is dissolved and the polarogram is run in 0·1*N* lithium sulphate solution in 50% methanol.

Yet another aliquot (5 ml) of the methanolic solution is nitrated with 5 ml of nitrating acid (5 volumes of conc. sulphuric acid with 4 volumes of conc. nitric acid). After the nitration is complete the reaction mixture is carefully diluted with 25 ml of distilled water. The nitration products are extracted with ether; the ether layer is separated and washed with 5% potassium hydroxide solution, followed by water in three portions. The ether is removed by distillation, the residue is dissolved in methanol, and the polarogram, in 0·1*N* lithium chloride in a (1:1) methanol–water mixture, is recorded.

Results of the polarographic analysis of the pyrolysis products of a number of polymers are summarized in Table 4.1.

The polarographic method was successfully used for identification of various polyamide grades [3,15]. As both the polymers themselves and their hydrolysis products are polarographically inactive, an indirect approach was adopted, in which the polyamides were first hydrolysed to amino acids which were converted into polarographically active Schiff bases by condensation with formaldehyde.

Since the products of thermal decomposition of acyl derivatives of cellulose contain the aldehydes characteristic of individual derivatives [16], the polarographic method is also suitable for identification of this group of polymers.

Quantitative Analysis

In the plastics industry a wide variety of compounds, which rank among the electroactive compounds mentioned above, are used both as raw materials and auxiliary substances. The polyreaction products contain some of these substances either as unreacted monomers or residual polymerisation initiators, both in minor quantities, or else as structural units of the macromolecules. Other electroactive additives such as stabilisers, antioxidants and plasticisers also constitute essential ingredients of plastic materials.

Unreacted Monomers

Monomers with Olefinic Double Bonds. Usually polarography is used for the determination of minor quantities of monomers with olefinic unsaturation such as styrene and its derivatives, derivatives of acrylic and methacrylic acids.

Styrene is determined in polystyrene by direct methods in non-aqueous solution in the presence of quaternary ammonium salts as the supporting electrolyte (see p. 375). Because of the highly negative value of the half-wave potential, this monomer is best analysed indirectly as its pseudonitrosite (see p. 375) or its mercuri-acetate derivative (see p. 376).

Table 4.1 The values of half-wave potentials of polymer pyrolysis products and their bromo and nitro derivatives [3,13]
(by permission from Izdatel'stvo Metallurgiya)

Polymer type	*Expected pyrolysis product*	$E_{1/2}$ *of the pyrolysis products* *in* 0·05*N* $[N(CH_3)_4]I$	*the bromo derivative in* 0·1*N* LiCl	*the nitro derivative in* 0·1*N* LiCl
Ethylcellulose	—	−1·90[a]	−1·78	—
Natural rubber	$CH_2{=}CH{-}C(CH_3){=}CH_2$	—	−0·6; −1·18; −1·42	−1·20
Styrene-methyl methacrylate copolymer	$CH_2{=}C(CH_3)COOCH_3$ $CH_2{=}CHC_6H_5$	−1·81[c] −2·34[c]	+0·14[b]	−1·0
Poly(vinyl alcohol)	—	−1·46[a] −2·04[a]	−1·18; −1·70	−1·28
Poly(vinyl butyral)	—	−1·45[a] −2·04[a]	−1·18; −1·70	−1·32
PVC	HCl, chloro derivatives	+0·30[a]	—	—
PVC plasticised	HCl, chloro derivatives	+0·30[a]	—	−0·88
Polyethylene	—	—	−1·06	−1·11
Polyisobutylene	$CH_2{=}C(CH_3)_2$	—	−1·08	−1·15
Polycaprolactam	$CH_2CH_2CH_2{-}CH_2{-}$ / $NHCOCH_2$ (ring)	—	—	−1·0
Poly(butyl methacrylate)	$CH_2{=}C(CH_3)COOC_4H_9$	−2·00[a]	+0·02[b]	−1·10
Poly(methyl methacrylate)	$CH_2{=}C(CH_3)COOCH_3$	−1·91[a]	−0·02[b]	−1·05[b]
Polystyrene	$CH_2{=}CHC_6H_5$	−2·34[c]	+0·14[b]	−1·2
Amino resin	—	−2·0[a]	—	−1·0
Phenolic resin	—	—	−1·78	−0·90

[a] 0·05*N* $N(CH_3)_4Cl$
[b] 0·1*N* Li_2SO_4
[c] $N(C_4H_9)_4I$

Styrene in its copolymers is determined as in polystyrene [17, 18]. Likewise in polyester resins, styrene can be determined by the polarographic method [19] (see p. 267).

Ring-substituted styrene derivatives (methylstyrene, divinylbenzene), which have even more negative values of the half-wave potential compared with styrene, are most conveniently determined by indirect methods, as either pseudonitrosites or as mercuri-acetate derivatives.

Methyl methacrylate in poly(methyl methacrylate) is determined directly under essentially the same conditions as for styrene (see p. 409). Methacrylates and acrylates, as well as the corresponding free acids, may be analysed indirectly as their bromo derivatives [20].

Acrylonitrile and styrene in acrylonitrile-styrene copolymers are determined directly in dimethylformamide solution in the presence of quaternary ammonium salts or hydroxides [17,18] (see p. 385).

Vinyl acetate in poly(vinyl acetate) is determined as acetaldehyde after hydrolysis with lithium hydroxide (see p. 391).

Aldehydes. Determination of aldehydes is an important analytical task in the field of polymers, since these compounds are components of phenolic and amino resins which are commercially important.

Formaldehyde in phenolic resins was determined by polarography by Domanský and Berger [21]. This method is also used in the determination of formaldehyde in the process control during the manufacture of urea-formaldehyde resins (see p. 210) and thiourea-formaldehyde resins (see p. 211). It is also applied to the cured resins (see p. 221). Furfural in the presence of formaldehyde in furfural-formaldehyde resins is determined in the alcoholic extracts [22].

Monomers of Other Kinds. Among other monomers which can be determined by polarography are isocyanates in polyurethanes both as unreacted monomers (phenyl isocyanate, 2,4-tolylene di-isocyanate), and as end groups incorporated in the polymer. This method is the only one available, by which these components can be determined in mixtures one with the other.

Caprolactam is determined polarographically in nylon-6 after conversion into the Schiff base by reaction with formaldehyde (see p. 289).

POLYMERISATION INITIATORS

Polarographic analysis is suitable for the determination of incompletely removed residual initiators either of the free-radical or ionic types. The former group includes such compounds as peroxides, hydroperoxides, and diazo compounds. With the latter group the analysis is concerned with traces of metals such as aluminium, titanium, iron, vanadium, and others, which originate from organometallic initiators or inorganic compounds.

Peroxides and hydroperoxides. Benzoyl peroxide is determined in a wide range of resins such as acrylics (see p. 410) and polyesters (see p. 267), and α,α-dimethylbenzyl hydroperoxide in styrene-butadiene latexes (see p. 383).

Diazo Compounds. Minor quantities of azoisobutyronitrile in polymers (including polystyrene) were determined polarographically by Dmitrieva and Bezuglyi [23].

Trace Metals. Trace quantities of metals of various kinds, including potassium and sodium, can be determined in polymers by polarography after dry or wet ashing of the samples under analysis. This method was used for estimation of trace amounts of metals in polyolefins (see p. 351) and of organometallic compounds in PVC [24].

The polarographic technique is also employed in analysis of the catalysts for the esterification reactions which yield polyester resins. These catalysts include oxides or acetates of such metals as zinc, lead, cobalt, manganese and germanium (see. p. 268).

Stabilisers and Polymerisation Inhibitors

Many of these additives in polymers, which are aimed at improving resistance to chemical agents, elevated temperature or light, are polarographically determinable, as are polymerisation inhibitors. These compounds belong to a variety of classes.

The metals incorporated in PVC stabilisers can be analysed polarographically or oscillopolarographically in an ashed polymer sample (see p. 361). Tetraphenyltin was estimated thus in polystyrene and poly(vinyl chloride) by Bezuglyi *et al.* [25].

4,4′-Dihydroxydiphenyl sulphide and sodium dimethyldithiocarbamate, used as polymerisation inhibitors, were determined polarographically by Paściak in styrene-butadiene latexes (see p. 382).

Plasticisers

The most extensively used plasticisers, esters of phthalic acids, are electroactive substances. The polarographic analysis of such plasticisers is a subject of considerable interest.

A method of polarographic determination of a number of phthalates in acetyl- and ethylcellulose-based plastics was worked out by Whitnack *et al.* [26,27]. Octyl phthalate was analysed in high impact polystyrene by Balandina *et al.* [28], while 2-ethylhexyl phthalate was determined in poly(vinyl chloride) (see p. 362).

Macromolecular Constituents

Polarography was also used in studies on the chemical constitution of macromolecules, and permits certain monomer units or functional groups to be determined.

The method is especially useful in the quantitative analysis of unsaturated dicarboxylic acids and phthalic acid in polyester resins (see p. 249).

Likewise isophthalic acid can be determined after conversion into the 5-nitro derivative (see p. 251).

Acetal groups in poly(vinyl acetals) can be polarographically evaluated after acid hydrolysis of these polymers [29] (see p. 400).

Chlorine was determined in vinyl chloride copolymers by Pražak *et al.* [30] after combustion of the sample in gaseous oxygen. This analysis is based on the anodic oxidation of the chloride ions.

Residual acetyl groups and 1,2-glycol groups in poly(vinyl alcohol) were analysed polarographically by Imoto *et al.* [31]. The 1,2-glycol groups were first oxidised with periodate and the unreacted excess of reagent was determined polarographically.

The end groups in poly(methyl methacrylate) were determined by Goode *et al.* [32] in the polymer made by anionic polymerisation in liquid ammonia.

The SiH groups in organisilicon polymers were determined indirectly by a procedure involving the reaction with mercuric chloride (see p. 445).

The polarographic method can also be used in the determination of peroxide and hydroperoxide groups in polystyrene and in other readily soluble polymers.

References

1. Lingane, J. J., *Electroanalytical Chemistry*, Interscience, New York, 1958.
2. Heyrovský, J., Kuta, J., *Principles of Polarography*, ČSAV, Prague, 1965.
3. Bezuglyi, V. D., *Polarografiya v khimii i tekhnologii polimerov* (*Polarography in Chemistry and Technology of Polymers*), Izd. "Khimiya", Leningrad, 1968.
4. Brauer, G. M., in *High Polymers, Vol. XII, Analytical Chemistry of Polymers*, Kline G. M., (Ed.), *Part II, Analysis of Molecular Structure and Chemical Groups*, Interscience, New York, 1962.
5. Zuman, P., *Organic Polarographic Analysis*, Pergamon Press, Oxford, 1964.
6. Zuman, P., Perrin, L., *Organic Polarography*, Wiley, London, 1969.
7. Rusznák, J., Fukker, K., Králik, J., *Naturwissenschaften*, **42**, 643 (1955); *Acta Chim. Acad. Sci. Hung.*, **9**, 49 (1956).
8. Rusznák, J., Fukker, K., Králik, J., *Z. Phys. Chem.*, **17**, 61 (1958).
9. Bolewski, K., Lubina, M., *Polimery*, **12**, 265 (1967).
10. Bezuglyi, V. D., Saliichuk, E. K., *Vysokomol. Soedin.*, **6**, 605 (1964).
11. Bezuglyi, V. D., Saliichuk, E. K., *Plast. Massy*, **1964**, [8], 47.
12. Saliichuk, E. K., Nagornaya, L. L., Bezuglyi, V. D., *Radiokhimiya*, **9**, 221 (1967).
13. Bezuglyi, V. D., Dmitrieva, V. N., *Zavodsk. Lab.*, **25**, 1180 (1959).
14. Shtal, S. S., Dmitrieva, V. N., Bezuglyi, V. D., *Zavodsk. Lab.*, **36**, 1191 (1970).
15. Shtal, S. S., Dmitrieva, V. N., Bezuglyi, V. D., *Plast. Massy*, **1968**, [9], 61.
16. Bezuglyi, V. D., Shtal, S. S., Dmitrieva, V. N., Kononenko, L. V., *Zavodsk. Lab.*, **38**, 1067 (1972).
17. Simpson, D., *Brit. Plastics*, **41**, [5], 78 (1968).
18. Uhde, W. J., Köhler, U., *Z. Lebensm.-Untersuch. u. -Forsch.*, **135**, 135 (1967).
19. Chromy, L., Filipska, M., *Chem. Anal. (Warsaw)*, **9**, 981 (1964).
20. Ryabov, A. V., Panova, G. D., *Dokl. Akad. Nauk SSSR*, **99**, 547 (1954).
21. Domanský, R., Berger, V., *Chem. Zvesti*, **5**, 441 (1951).
22. Malyugina, N. I., Korshunov, I. A., *Zh. Analit. Khim.*, **2**, 341 (1947).
23. Dmitrieva, V. N., Bezuglyi, V. D., *Vysokomol. Soedin*,. **4**, 1672 (1962).

24. Schröder, E., Malz, S., *Plaste u. Kautschuk*, **5**, 416 (1958).
25. Bezuglyi, V. D., Preobrazhenskaya, E. A., Dmitrieva, V. N., *Zh. Analit. Khim.*, **19**, 1033 (1964).
26. Whitnack, G. C., Gantz, E. S., *Anal. Chem.*, **25**, 553 (1953).
27. Whitnack, G. C., Reinhart, J., Gantz, E. S., *Anal. Chem.*, **27**, 359 (1955).
28. Balandina, V. A., Korsakov, V. G., Malkina N. I., Davydova, Z. F., Noskova, M. P., Shevardina, Z. I., *Plast. Massy*, **1967**, [10], 53.
29. Gurvich, D. B., Shevardina, Z. I., *Plast. Massy*, **1960**, [12], 55.
30. Pražák, M., Benc, J., Bartušek, Z., *Chem. Průmysl*, **3**, 297 (1953).
31. Imoto, S., Ukida, J., Kominami, T., *Kobunshi Kagaku*, **14**, 127, 214 (1957).
32. Goode, W. E., Snyder, W. H., Fettes, R. C., *J. Polymer Sci.*, **42**, 367 (1960).

Chapter 5

NUCLEAR MAGNETIC RESONANCE (NMR) SPECTROSCOPY

5.1 PRINCIPLES

In the past two decades nuclear magnetic resonance spectroscopy has been widely applied to the analysis of polymers. The phenomenon of nuclear magnetic resonance occurs in the case of atomic nuclei possessing spin and a nuclear magnetic moment μ. Such nuclei, when placed in a constant magnetic field H and irradiated with radiofrequency ν such that

$$h\nu = \frac{\mu}{I} H$$

where I is the spin quantum number, absorb energy from the applied radiofrequency field, $E = h\nu$, and transitions occur from one energy level to another. This condition is referred to as resonance between the RF field and the nucleus, and the absorption is recorded on an NMR spectrometer in the form of a characteristic spectrum.

In practice in studies of materials and individual compounds by the NMR technique, we have to deal not with a single nucleus isolated from the environment, but with a vast number of mutually interacting nuclei present in a diversity of groups of atoms and molecules. Without going into details of the mechanism of these complex interactions it is simply stated that in solids (that is, in the majority of polymers) the magnetic nuclei placed in a magnetic field of intensity H are also influenced by a local magnetic field produced by the surrounding magnetic nuclei. A certain definite distribution of this local magnetic field exists owing to a definite spatial arrangement of the nuclei in the crystal lattice (or the matrix) of the material under examination, and the resonance condition is fulfilled over a relatively wide frequency range. This is why NMR spectra of solids have the appearance of a single line or, depending on the material and the experimental conditions, several wide bell-shaped lines (Fig. 5.1). From an examination of the width or shape of the NMR line of polymers much valuable information may be gained concerning their structure.

In the case of liquids and low-viscosity solutions, the random thermal motions of molecules results in an averaging of local field intensities to zero, and the resonance condition is realised over a very narrow frequency range. As a consequence, hyperfine interactions between the magnetic nuclei and their surrounding electrons may be observed. The applied external magnetic field affects the motion of these electrons, inducing an opposing magnetic field. Thus the electrons exert a shielding effect on the nucleus. The extent of this screening is not too great but has a significant effect on the NMR experiment. The extent of this effect depends on the electron density round

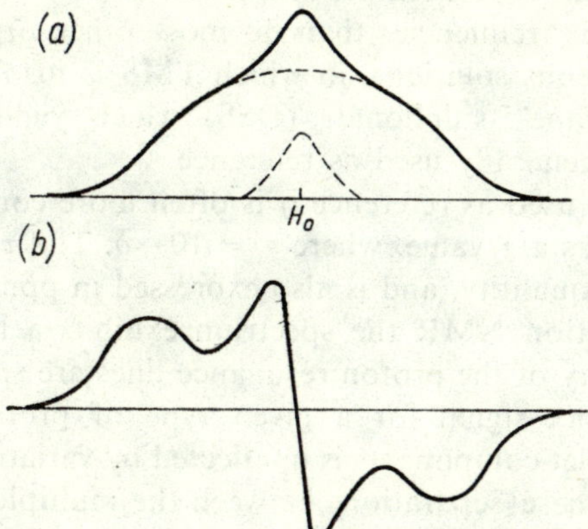

Fig. 5.1 Broad-line NMR signals of polymers: (*a*) absorption mode, (*b*) dispersion mode

the magnetic nucleus, and hence on both the value of the electric charge and its spatial distribution round that nucleus. These latter are in turn controlled by the chemical nature of the resonating atom, that is to say by the type and spatial arrangement of other atoms around the magnetic nucleus, and by the nature of the chemical bonds involved.

The screening effect is responsible for a shift in the resonance signal of a nucleus from the position corresponding to the isolated nucleus of a given kind. This shift is very small. Most of the protons in a variety of organic compounds exhibit a shift within a range of about 700 Hz at a magnetic field intensity of $H_0 = 14{,}000$ Oe, whereas the resonance frequency of the hydrogen nucleus under these conditions is 60 MHz.

The shifts of the resonance signals are usually measured in relation to a signal of a suitable reference compound. In order to eliminate the effect of variations in the intensity of the applied magnetic field, the shift is expressed as the chemical shift δ which is defined as

$$\delta = (H_s - H_x)/H_s$$

where H_x and H_s denote the values of the magnetic field intensity for the resonance line of a given nucleus in the unknown and in the reference respectively.

For convenience the δ value is usually expressed in ppm by multiplying the above formula by 10^6. For calculations the following formula is usually employed

$$\delta = \frac{10^6 \Delta}{\text{RF generator frequency}}$$

where Δ is the difference (in Hz) between the resonance line for the unknown and that for the reference in the spectrum.

The reference compound now in common use is tetramethylsilane (TMS). This compound is miscible with the majority of common organic solvents employed in NMR studies and gives rise to a single strong signal. It absorbs at higher frequencies than do most other organic compounds. For studies in aqueous solutions, in which TMS is insoluble, sodium 2,2-dimethyl-2-silapentane-5-sulphonate (DSS) which yields a strong sharp resonance line, is generally used as reference.

When TMS is used as reference it is often more convenient to express the chemical shift as a τ value, where $\tau = 10 - \delta$. The τ parameter, like δ, is a dimensionless quantity, and is also expressed in ppm.

In high-resolution NMR the spectrum exhibits a hyperfine structure because the majority of the proton resonance lines are split into multiplets. Unlike the resonance signal for a given type of proton the separation between the multiplet components is unaffected by variations in the magnetic field intensity. These separations between the multiplet components are known as the spin-spin coupling constants; they are denoted by J and reported in Hz, not in δ or τ, since they are independent of the field intensity.

The number and relative intensities of multiplet components arising from a given group of equivalent protons depend on the number of protons in the neighbouring groups of atoms, as well as on a variety of other factors. The relation is usually complex in nature.

The chemical shift values, as well as the multiplet splittings, are used in establishing molecular structure on the basis of specially compiled correlation tables or by comparison with the spectra of suitable reference compounds. In cases difficult to interpret it may be necessary to simulate the spectrum from theory and compare the simulated and observed spectra.

NMR spectroscopy is generally suitable for all materials containing atomic nuclei possessing a magnetic moment. Isotopes of that kind are known to represent about one third of the total number of stable isotopes. Despite this only a few of these have found practical application in the NMR technique, namely ^{1}H, ^{19}F, ^{13}C, ^{11}B, ^{31}P, ^{29}Si, ^{14}N, and ^{17}O. Proton magnetic resonance (PMR) spectroscopy has been the most widely used to date. The reason for this is primarily because of the great number and importance of proton-containing compounds, and also because of experimental difficulties with NMR spectroscopy of other nuclei. Of other isotopes suitable for the NMR technique only fluorine exhibits a resonance line intensity comparable to that of hydrogen. This is of great value for NMR spectroscopy of fluorine-containing compounds, including polymers containing fluorine. There is, however, an increasing interest in applications of ^{13}C NMR spectroscopy to studies of polymers.

Owing to the exceptional position of hydrogen, the PMR method has found widespread use in structural studies of organic compounds, including polymers. An additional favourable circumstance is that the ^{12}C and the

^{16}O isotopes, also fundamental constituents of organic compounds, have no magnetic moment and thus are NMR inactive.

5.2 EXPERIMENTAL TECHNIQUE

Solvents

High-resolution NMR spectroscopy is restricted to liquids and solutions, thus raising the question of a suitable solvent. The solvent should be chemically inert towards the compound under analysis, and should not give rise to a resonance signal in the spectral range of interest. For these reasons, the most suitable solvents for NMR experiments are those not containing hydrogen atoms, most frequently carbon tetrachloride, carbon disulphide and deuterochloroform. The range of solvents of value for NMR spectroscopy is increasing, and even includes such substances as arsenic or antimony trichloride. In view of the low intensity of the signals, solutions of fairly high concentrations must be studied. Since the solubility of a wide number of compounds in the above solvents is often insufficient, other more unusual, deuterated solvents may have to be used; these include deuterated acetonitrile, dimethylsulphoxide (DMSO), benzene, acetic acid, or water.

When a restricted range of the spectrum is examined, non-deuterated solvents may also be used, such as water, acetone, trifluoroacetic acid, pyridine, or DMSO.

Every solvent should be free of any traces of dissolved oxygen, which broadens NMR signals and affects the chemical shift values. For nuclei such as ^{13}C with large relaxation times paramagnetic compounds are added to the sample to be examined.

Sample Preparation

The sample for the NMR study is usually placed in a tube ca. 15 cm long and ca. 5 mm in diameter. Larger diameters are usually unsuitable because of field inhomogeneity. The sample volume in the tube amounts typically to 0·50–0·75 ml. Since the PMR signal is sufficiently strong at a concentration of at least 0·1 mole per litre, for a compound of $M = 100$ the quantity used should equal ca. 10 mg. Lower concentrations may be studied when the compound possesses larger number of atoms of the same chemical shift. Higher concentrations are often undesirable on account of association effects. If one is compelled to work with solutions of lower than optimum concentration because of restricted quantities or limited solubility, the line may be increased in intensity by multiple scanning and the use of time-averaging devices.

The chemical shift of a compound, as mentioned previously, is measured relative to the characteristic signal of a reference compound. The reference may be introduced into the sample in two ways. The simplest, most widespread method is dissolution of the reference directly in the solution under examination; it is then referred to as an internal standard. When the reference compound is either insoluble in the solvent or reacts with the solvent, another reference compound is taken or else the reference is placed in the solution in a sealed capillary; this is the external reference method.

Results

Results of studies by high-resolution NMR spectroscopy are expressed as τ or δ values, or else in Hz. In the latter case the reference used and the oscillator frequency must be specified for unambiguous interpretation of the position of the resonance signals. When the δ value is adopted the reference must be known, whereas the value is defined precisely by the signal position.

A typical high-resolution NMR spectrum is illustrated in Fig. 5.2. The signal intensity is on the y-axis and the τ or δ value, or else the intensity of the applied magnetic field, is on the x-axis. The abscissa scale may be expressed in the RF oscillator frequency, in Hz, or in the constant

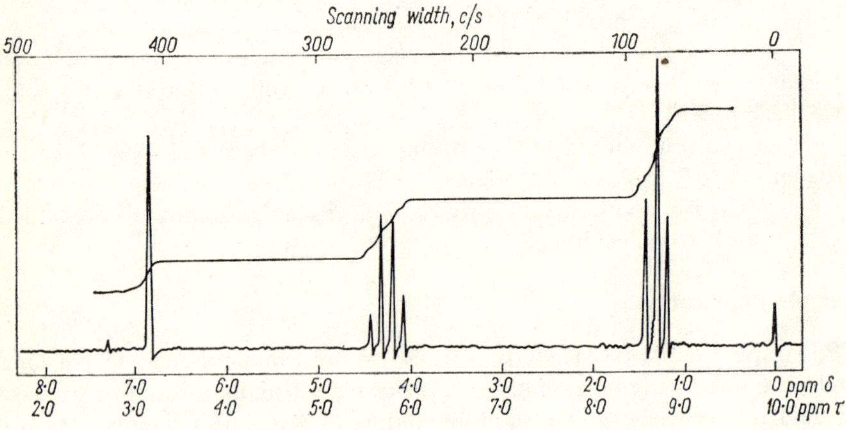

Fig. 5.2 NMR spectrum of ethyl fumarate

magnetic field units, Oe, in broad-line NMR spectroscopy. Furthermore, in studies by the latter technique the spectrum is usually recorded as the dispersion mode of the resonance signal (see Fig. 5.1).

The signal intensity is generally of interest, because it is proportional to the number of nuclei responsible for the signal observed. In view of the complex shape of the signals produced in high-resolution NMR spectra, accurate results are achieved only in terms of the total area contained under the contours of individual signals rather than on the basis of the signal heights. Up to date NMR spectrometers are equipped for that purpose with electronic integrators which record the results of integration of the signal area as an integral curve drawn directly on the same spectrum (Fig. 5.2).

5.3 NMR SPECTROSCOPY IN POLYMER STUDIES

Introduction

Numerous properties of polymers such as mechanical strength, elasticity or thermal resistance, important from the viewpoint of their practical uses as plastic materials, depend on both the chemical structure of macromolecules and the kinetic characteristics of the polymer chains or their segments. NMR spectroscopy allows both aspects of polymer structure to be studied. The high-resolution technique is suitable for the determination of chemical structure of polymeric substances, whereas wide-line NMR spectroscopy is a valuable tool for the study of supramolecular structure, mobility and flexibility of the macromolecule as a whole, or of particular fragments.

The experimental technique in NMR studies of polymers is generally the same as the one adopted for low molecular weight compounds. It should be emphasised, however, that in polymer studies the method is frequently attended by experimental difficulties. Specifically, the lines obtained by the high-resolution technique in polymer solutions are appreciably wider than the spectral lines observed in small molecule compounds. To improve resolution, measures are taken such as diluting the solutions, using the most suitable solvent, or raising the temperature while the spectrum is recorded.

It should be stressed that the use of NMR spectroscopy in polymer studies is also coupled with certain difficulties concerning interpretation of the spectra, these difficulties arise from the peculiar structure of polymeric materials. The main sources of such problems are the complexity of the long macromolecular chains, the diversity of the structural elements of polymers (monomer units, segments, molecules, crystallites) and the low mobility of macromolecules in solution compared with small molecules. The low mobility of macromolecules is due to their size and to polymer-polymer and polymer-solvent interactions which together give a wide range of frequencies to macromolecular motions.

Applications of NMR spectroscopy to polymer studies are now covered by numerous and detailed monographs [1–4].

Broad-line NMR Spectroscopy

From the examination of the line width or line shape of NMR spectra of polymers much valuable information can be obtained, including chain structure, stereoregularity, the presence of characteristic groups, and the like. A characteristic quantity of exceptional value is the second moment $\Delta \overline{H}_2^2$ of the resonance lines.

Of prime significance for the form (width, shape, and second moment) of the polymer NMR line is molecular motion, either of the whole macromolecule or of segments or groups. Therefore, broad-line NMR spectroscopy is a valuable tool in studies of mechanical strength and physicochemical properties of polymers, as well as in the investigation of effects such as phase transitions, degree of crystallinity, variations in polymer orientation, and of chemical processes such as vulcanisation and other crosslinking procedures. Some valuable information on the chemical structure of polymers may occasionally be gained by this technique [1,4,5].

Effects Related to the Supramolecular Structure of Polymers

Degree of Crystallinity. Wilson and Pake [6] were the first to observe that within a definite temperature range the NMR signals of polyethylene and polytetrafluoroethylene consist of two components, one of which is broad and the other narrow. These investigators advanced a suggestion that the broad component corresponds to the molecules occurring in crystalline regions with a limited degree of freedom, whereas the narrow one is due to the more mobile molecules in amorphous regions. The two components are more clearly visible in the dispersion lines. From the ratio of the area of both components the degree of crystallinity of the polymers can be inferred. Used for the estimation of the degree of crystallinity of other polyolefins and of poly(vinyl chloride), the method gave good results coincident with those furnished by X-ray diffraction, infrared spectroscopy and by other techniques [7–9].

The method of determining the degree of crystallinity on the basis of the dispersion mode of the resonance line is not, however, a versatile one and cannot be used for any polymer and at every temperature [10]. An instance of the non-appearance of the complex structure of the resonance line is the NMR spectrum of natural rubber.

Glass Transition. The temperature effect of the second moment was studied in natural rubber, and in atactic polypropylene and polyisobutylene [1]. In all these polymers a fall in the value of the second moment with increasing temperature was observed, attended by narrowing of the resonance lines in the region of the glass transition temperature (T_g), which is related to the transition of the glass to the rubbery state. T_g is conventionally determined by dilatometry.

Good agreement between the T_g value found by NMR and by other conventional methods was also achieved for poly(methyl methacrylate) [11], polystyrene [11,12], poly(vinyl chloride) [11] and styrene-butadiene [12,13] and styrene-isoprene [13] copolymers. It seems that broad-line NMR spectroscopy is as useful as dilatometric, optical and mechanical methods for the determination of T_g. It should be mentioned, however, that the method fails with some polymers such as poly(vinyl acetate) [14].

Polymer Orientation [1]. It was observed that the second moment of the NMR line increased in poly(ethylene terephthalate) and in a number of other polyesters with increased orientation. This suggests impaired molecular motion in oriented polymer films. A rise in the line width is also observed in cold-stretched linear polyethylene. On the other hand no apparent difference is seen between the NMR lines of high-pressure polyethylene oriented by stretching and unoriented. During the stretching of poly(methyl methacrylate) the second moment initially increases until the elongation attains a value of 135%; this may be accounted for by reduced molecular motion. Subsequently, when the elongation exceeds 300% the second moment of the line drops as a consequence of breakdown of the structure.

The stretching of nylon-6,6 is accompanied by a rise in the width of the broad component of the NMR line and an abrupt decline in intensity of the narrow component. The effect may be explained by assuming that the polymer lattice is more rigid in the oriented than in the unoriented polymer, despite the fact that the degree of crystallinity, as determined by X-ray measurement, decreases during stretching of the fibre.

This method was also used in studying the orientation of such polymers as poly(vinyl alcohol), polytetrafluoroethylene, PVC fibres [15], and polyoxymethylene [16,17].

Molecular Motions. As the motion of macromolecules is the major factor that controls the form of the line in the broad-line NMR spectra of polymers, this method is suitable for studying the nature of this motion, and also for investigating molecular interactions in polymer solutions. The majority of works devoted to polymer studies by this method are eventually concerned with these questions [1,3,18,19,20].

Study of Chemical Reactions

Polymerisation Processes. By following the variations in the NMR line width in the spectrum of a reaction mixture, resulting from the changes occurring in the mobility of the growing polymer chains during the polymerisation process, some information may be derived on the kinetics of the process. The method was used for this purpose by Shibata *et al.* [21] (in studies on ethyl acrylate polymerisation), Bonera *et al.* [22] (for the polymerisation of vinyl compounds: styrene, methyl methacrylate, acrylic acid, and vinyl acetate, induced by heat or UV irradiation), and many other authors.

The method was successfully adopted in studies of solid state polymerisation processes. Specifically, the process of direct formation of the polyoxymethylene chain in the monomer crystal lattice was followed in the radiation-induced polymerisation of trioxan [23,24].

Crosslinking Processes. Covalent bonds formed between polymer chains cause a reduction in mobility, which is reflected in the resonance line width. Hence broad-line NMR spectroscopy is suitable for the study of processes such as vulcanisation of rubbers or curing of resins.

The sulphur-vulcanisation of natural rubber was examined by this method by a number of investigators including Gutowsky *et al.* [25], and

Oshima and Kusumoto [26]. The latter recommend this method for the process control of the degree of vulcanisation.

The vulcanisation process in fluorocopolymers was studied by the broad-line NMR technique by Lyubimov *et al.* [27], while Tanaka [28], Matsushita [29] and Chuvaev *et al.* [30] investigated the curing processes in unsaturated polyester resins. Slonim *et al.* [31] and Lösche [32] were concerned with the curing process of epoxy resins, while Waldrop and Kraus [33] used this technique to study the reactions of butadiene-styrene copolymers with carbon black.

Decomposition Processes and Other Reactions. Broad-line NMR spectroscopy also allows investigation of the mechanism of transformations which occur by the action of elevated temperature, atmospheric oxygen, or ultraviolet radiation. The method was employed to follow the degradation processes effected by the above factors in a wide diversity of polymers.

Numerous studies are concerned with broad-line NMR spectroscopy applied to the reactions induced in polymers by the influence of ionising radiation. These may belong to both the detrimental processes, regarded as aging (at high irradiation doses), and the beneficial processes which improve the polymer properties and which occur at strictly controlled irradiation doses [1]. The method was also used in studies of the interaction between a polymer and a plasticiser [34].

Chemical Structure of Macromolecules

Certain compact groups of atoms composed of 2 or 3 magnetic nuclei (such as CH_2, CH_3, CF_2, CF_3), when the distance between these and other groups considerably exceeds the distances between the magnetic nuclei in the group, give rise to resonance lines of a characteristic complex shape that can be theoretically simulated. The NMR lines of polymers containing methylene groups, together with those of methylchloroform recorded at ca. − 200°C, are shown in Fig. 5.3 [35].

It is worth mentioning that the complex structure of the NMR line, which is due to the presence of weak interactions between the groups of atoms, appears as a rule only in spectra measured at low temperatures, when molecular motions are hindered.

From the difference between the second moments of the wide NMR lines it may sometimes be estimated how the asymmetric monomer units are linked in a macromolecule (head-to-tail or head-to-head). Thus by comparing the $\Delta \bar{H}_2^2$ values calculated theoretically for structures I and II

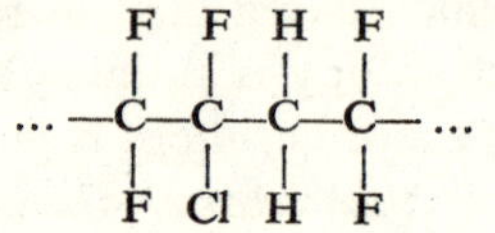

$$\cdots-\underset{\displaystyle F}{\overset{\displaystyle F}{\underset{|}{\overset{|}{C}}}}-\underset{\displaystyle Cl}{\overset{\displaystyle F}{\underset{|}{\overset{|}{C}}}}-\underset{\displaystyle H}{\overset{\displaystyle H}{\underset{|}{\overset{|}{C}}}}-\underset{\displaystyle F}{\overset{\displaystyle F}{\underset{|}{\overset{|}{C}}}}-\cdots \qquad \cdots-\underset{\displaystyle F}{\overset{\displaystyle F}{\underset{|}{\overset{|}{C}}}}-\underset{\displaystyle Cl}{\overset{\displaystyle F}{\underset{|}{\overset{|}{C}}}}-\underset{\displaystyle F}{\overset{\displaystyle F}{\underset{|}{\overset{|}{C}}}}-\underset{\displaystyle H}{\overset{\displaystyle H}{\underset{|}{\overset{|}{C}}}}-\cdots$$

I 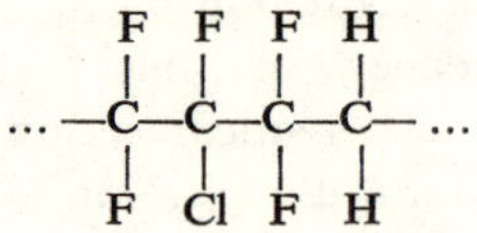 **II**

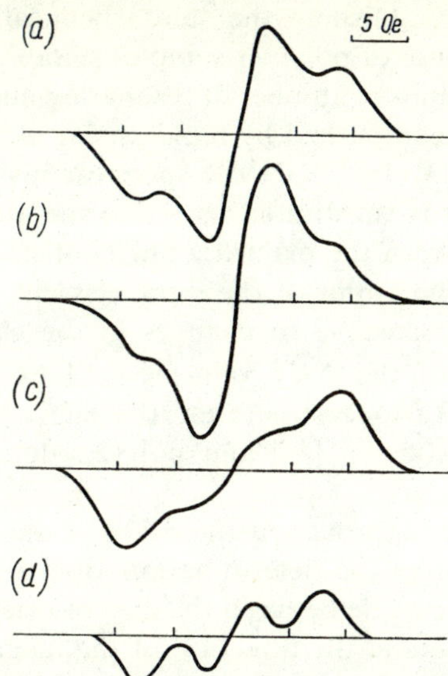

Fig. 5.3 Broad-line NMR spectra of polymers containing CH_3 groups: (*a*) poly(methyl methacrylate) (at −196°C), (*b*) polycarbonate based on bisphenol A (at −196°C), (*c*) butyl rubber (at −189°C), (*d*) methylchloroform (at −196°C)

with the values obtained experimentally, Lyubimov *et al.* [27] suggested that the more likely is structure II.

Branching in linear polymers may sometimes be studied successfully by broad-line NMR spectroscopy. Branching affects the density of packing of molecules in solid polymers, and thus also the mobility of the groups on the molecules. As expected, the NMR line in highly branched polyethylene was found [9,36] to be narrower than in the linear polymer. Branching also affects the form of the temperature dependence of $\Delta \bar{H}_2^2$ [37].

The method is also useful in distinguishing the configuration of *cis* and *trans* isomeric polymers with regular double bond structures in the chain. In particular, *cis* and *trans* polyisoprene and polybutadiene varieties were studied by Takeda *et al.* [38], and by Gupta [39].

The broad-line NMR technique may also prove of assistance in determining the stereochemical structure of polymers, since the chain configuration and conformation considerably affect the shape, width, and second moment of the resonance lines, and also the relaxation times. It should be emphasised that the results may be unreliable, unless special experimental techniques are adopted.

One of these techniques consists in studies of the temperature effect on $\Delta \bar{H}_2^2$ or on the relaxation time. This technique was employed in studies on stereoregular poly(methyl methacrylates) by Nishioka *et al.* [40], whereas Slichter and Mandell [41], and Gupta [42] investigated polypropylenes of different tacticity.

Another way of establishing the stereochemical structure of macromolecules involves the use of polymer samples suitably modified chemically so as to contain the minimum number of resonating nuclei. In PMR spectroscopy this is usually accomplished by replacement of most of the hydrogen atoms by deuterium [43]. In ^{19}F NMR spectroscopy one of the hydrogen atoms in the monomer is substituted by a fluorine atom [44,45]. Since the mutual interaction between the magnetic nuclei of unlike elements is much weaker than between the atoms of the same element, the second moments for such polymers are sensitive to changes in the chemical structure [4]. Thus by comparison of the $\Delta \overline{H_2^2}$ values found experimentally with the values calculated for various structures, the stereochemical structure of poly(2-fluoro-5-methylstyrene) [44], and poly(2,5-difluorostyrene) [45] was established.

Finally, yet another approach to the study of the stereochemical structure of polymers involves the determination of the relation between the second moment and the angle between the axis of orientation of the sample of oriented fibre or film and the direction of the magnetic field. Variations in the spatial orientation of such samples results in variations in that angle, hence also in the value of the second moment [4]. By comparing the course of the experimental and theoretical curves for a few of the most likely conformations, the polymer chains in polyacrylonitrile fibres were demonstrated to have mainly a planar isotactic structure. Also, in poly(vinyl alcohol) the planar arrangement prevails, whereas poly(vinyl chloride) has many non-planar segments of isotactic structure.

Chemical Analysis

Broad-line NMR spectroscopy is equally good for analytical purposes, primarily in the determination of the chemical composition of two-phase systems [1]. When both phases contain different magnetic nuclei (for instance ^{1}H and ^{19}F) the task is fairly simple. The number of resonating nuclei is then determined from the intensities of the signals obtained separately for each of the two phases. If both phases have the same magnetic nuclei, the problem is much more involved, and the analysis can be accomplished only if the NMR line has a complex form, and if this can be separated into two components. This is the case when a solid polymer incorporates as the other phase a liquid or a gas very weakly bonded with the solid phase so that molecular motions in this phase are unaffected.

Because of its advantages, broad-line PMR spectroscopy has found widespread application in the important problem of the determination of water in certain polymers. This has already been discussed (p. 62). This method was also used by Mansfield [46] for the estimation of plasticiser in PVC in the range 20–50%.

The method was successfully used in the determination of the number average molecular weight of certain insoluble and infusible (but non-crosslinked) polymers [47].

High-Resolution NMR Spectroscopy

While broad-line NMR spectroscopy is primarily used for studies of physical structure and properties of polymers, high-resolution NMR is mainly resorted to for chemical studies.

It will be remembered that for the fine structure of the NMR spectrum of a polymer solution to appear, the molecular motion must be sufficiently rapid to average out the local magnetic fields due to neighbouring magnetic nuclei.

High-resolution NMR spectroscopy is, in the case of many polymers, the only method available for the exact elucidation of chemical structure, since by its aid groups of atoms, differing insignificantly in chemical composition, can be quantitatively analysed. The manner in which these groups are linked together can also be established. It is remarkable that the structure of phenolic resins was not fully elucidated until this method became available (see p. 167). The technique is, however, of particular assistance in studying the chain structure of polymers. It has been exploited to solve such problems as the manner in which the unsymmetrical monomers are linked in homopolymers, the content and distribution of the monomers in copolymers and polycondensation resins, structural irregularities (branching and end groups) and, of primary importance, the determination of configuration. In the last mentioned both *cis-trans* and stereoisomerism, and occasionally also the conformation of the polymer chains, have been explored.

Chemical Structure of Macromolecules

Irregularities in the Macromolecule Structure. Brame [48] determined the chlorine distribution in chlorosulphonated polyethylenes. Saito *et al.* [49] studied the arrangement of chlorine atoms along the chain of chlorinated polyethylene and found that products containing less than 40 per cent chlorine have a random distribution, whereas products with higher degrees of substitution contain geminal pairs of chlorine atoms. Heintke and Keller [50] also studied the distribution of chlorine atoms in chlorinated polyethylene. The high-resolution NMR technique was used in investigations on the sequence distribution of CH_2, CHCl and CCl_2 groups in chlorinated PVC [51] and in studies of chemical heterogeneity in vinylidene chloride-vinyl acetate copolymers by Unterforsthuber [52]. The same technique was also used for determination of the end groups in PVC [53], in cellulose triacetate acetolysis products [54], in polybutadiene [55], and the C=C bonds in polyethylene [56,57].

Branching in polyethylene may also be estimated (see p. 151, 341) by NMR spectroscopy although this presents difficulties especially in the case of high density products. Here CAT (Computer for Averaging Transients) devices can help [56]; the measurement is based on the methyl group signals. Branching was also determined in this way in epoxy resins [58].

Arrangement of Unsymmetrical Monomer Units in Homopolymer Chains. In the spectrum of poly(vinylidene fluoride), shown in Fig. 5.4, four fluorine lines can be seen. These result from the following structures:

$$\ldots\text{—}CF_2\text{—}CH_2\text{—}CF_2^{*}\text{—}CH_2\text{—}CF_2\text{—}CH_2\text{—}\ldots \quad \text{(I)}$$

$$\ldots\text{—}CH_2\text{—}CH_2\text{—}CF_2^{*}\text{—}CH_2\text{—}CF_2\text{—}CH_2\text{—}\ldots \quad \text{(II)}$$

$$\ldots\text{—}CF_2\text{—}CH_2\text{—}CF_2^{*}\text{—}CF_2\text{—}CH_2\text{—}CF_2\text{—}\ldots \quad \text{(III)}$$

$$\ldots\text{—}CH_2\text{—}CH_2\text{—}CF_2^{*}\text{—}CF_2\text{—}CH_2\text{—}CH_2\text{—}\ldots \quad \text{(IV)}$$

From the intensity of the signals the content of individual arrangements, such as head-to-head, head-to-tail, and tail-to-head can be found [59,60].

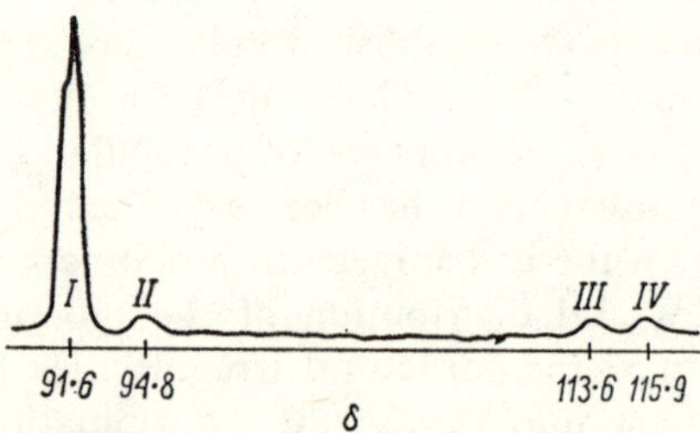

Fig. 5.4 High-resolution fluorine NMR spectrum: 25% solution of poly(vinylidene fluoride) in *N,N*-dimethylacetamide, room temperature, frequency 56·4 MHz, $CFCl_3$ as standard reference (from [59] by permission of the copyright holders, John Wiley & Sons, Inc.)

Based on the high-resolution NMR spectrum the arrangement of unsymmetrical monomers in other polymers was also established, for example in poly(vinyl fluoride) [61].

The method was also applied in determining the structure of polymers prepared from conjugated diene monomers. These include polybutadiene (1,4 and 1,2 additions) [62], polyisoprene (1,4, 1,2, and 3,4 additions) [62], and polychloroprene (1,4, 1,2 and 3,4 additions) [63].

Copolymer Chain Configuration. The chain configuration of a great number of copolymers has been studied. In the NMR spectrum of the solution of vinyl chloride-vinylidene chloride copolymer in *o*-dichlorobenzene at 100°C the —CH_2— protons give a triplet. The two outermost strong signals correspond to segments composed of vinyl chloride and vinylidene chloride units alone, whereas the middle, weak signal is due to the boundary structure —CHCl—CH_2—CCl_2—CH_2—. Based on the intensity ratios of these lines the content of vinyl-vinyl, vinyl-vinylidene, and vinylidene-vinylidene structures may be evaluated [64].

Similarly, the chain microstructure can be established in styrene-methyl methacrylate copolymers. In the spectra of solutions of these copolymers in carbon tetrachloride the methyl methacrylate methoxy protons give two lines [65]. One of these has the same chemical shift value as observed in poly(methyl methacrylate). If in the copolymer chain a styrene unit is neighbour to the methyl methacrylate unit, the proton signal of this methoxy group shifts upfield, resulting in a new line in the spectrum. The shift results from the diamagnetic screening effect of the styrene phenyl group. A similar effect is noticed in the spectra of chloroform solutions of methyl methacrylate-*p*-xylylene copolymers [66].

The chain microstructure was also elucidated in other copolymers by using high-resolution NMR. These copolymers include vinylidene fluoride-hexafluoropropylene copolymer [67], isoprene-butadiene copolymer [62], and also copolymers of vinyl chloride with vinyl acetate [68], ethylene with vinyl acetate and with acrylates [69], acrylonitrile with styrene, isoprene, or butadiene [70], acrylonitrile with methyl methacrylate [71], methyl methacrylate with styrene, and vinylidene chloride with vinyl acetate [72], and trioxan with dioxolan [73].

The monomer distribution in acrylonitrile-styrene copolymers was studied by Schaefer [74], with the aid of high-resolution pulsed ^{13}C NMR spectroscopy. The NMR technique was used to investigate the sequence distribution in poly(ethyleneterephthalate-sebacate) terpolyesters [75].

Geometrical Isomerism. In many instances high-resolution NMR spectroscopy is of help in the differentiation and quantitative determination of the double-bond *cis-trans* configuration in polymers. It was found by this method that the naturally occurring varieties of polyisoprene, hevea and balata rubber, contain 99% of the *cis* and *trans* configurations and that the 1,4 addition accounts for practically 100 per cent of the structural units. The *cis-trans* content was also established in a great number of synthetic polymers, such as polyisoprene [76], isoprene-butadiene copolymers [62], polybutadiene [62], alicyclic epoxy resins [77], and unsaturated polyesters [78], to name but a few.

Stereoregularity. High-resolution NMR spectra of the sequences in macromolecule chains of various configurations and conformations differ appreciably among themselves by both chemical shifts and multiplicity of resonance lines. Because of this, the determination of stereoregularity has become a major application of the technique in the field of polymers. Actually it was not until the advent of this technique that an exact quantitative determination of stereoregularity (tacticity) became possible in a vast number of polymers. However, so far the method has been used in only a few cases for conformational analysis.

In view of the complex appearance of the NMR spectra of polymers, determination of stereoregularity on this evidence alone is not a simple task, as tacticity effects are small and often produce tiny differences in the spectra. Here comparisons of the spectra of the polymers under study with the spectra of polymers of known stereoregularity and of their mixtures are helpful. Spectra of model compounds (dimers, trimers, oligomers), may also prove useful. The techniques of deuteration [79,80] and double resonance may help to simplify comparison spectra.

The high-resolution NMR determination of the stereoregularity of poly(methyl methacrylate) is now a classical piece of work [80]. The spectra of isotactic and syndiotactic poly(methyl methacrylate) are illustrated in Fig. 5.5. The spectra were recorded from a 2 per cent solution at 120°C, as at lower temperatures the signals are unresolved.

For the estimation of tacticity the environment of the main chain methylene groups and the position of the methyl groups in the main chain are of importance. Both methylene protons are magnetically equivalent, giving a single line at $\tau = 8{\cdot}18$ if the unit adjacent to this group is of opposite configuration (syndiotactic) relative to the unit of interest (Fig. 5.5 *b*). If the adjacent unit and the unit of interest are of the same (isotactic) configuration, the two methylene protons become magnetically non-equivalent and, because of a spin-spin coupling, give rise to a quartet with its centre at $\tau = 8{\cdot}18$. The centres of the doublets that form this quartet are positioned at τ 7·83 and 8·46 (Fig. 5.5*a*).

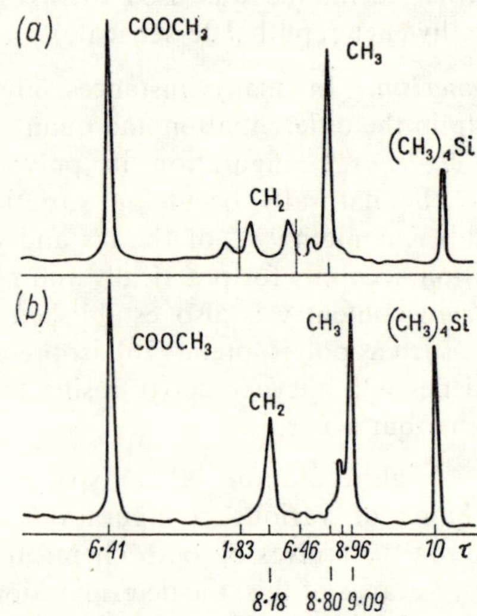

Fig. 5.5 NMR spectra of poly(methyl methacrylate) (2% solution in chloroform at 120°C); (*a*) isotactic content 95%, (*b*) syndiotactic content 85% (from [80] by permisson of Hüthig and Wepf Verlag, Basel)

The α-methyl protons result in three lines depending on how two adjacent units are linked, the assignments being as follows:

(a) line at $\tau = 9{\cdot}09$ is due to the middle α-methyl group of syndiotactic triads,

(b) line at $\tau = 8{\cdot}80$ results from this same methyl group but in isotactic triads,

(c) line at $\tau = 8{\cdot}96$ corresponds to this group situated in the middle of heterotactic triads.

As can be seen from Fig. 5.5, in the spectrum of the isotactic polymer the line at $\tau = 8{\cdot}80$ is the strongest, whereas in the spectrum of the syndiotactic polymer it is the line at $\tau = 9{\cdot}09$. Since the intensities of these lines are proportional to the share of individual triads, the percentage of the iso-, syndio- and heterotactic sequences in the polymer may be estimated. It should be further emphasised that on the basis of the character of the methylene group lines from poly(methyl methacrylate), not only can polymers of different tacticity be distinguished, but the type of stereoregularity may be unambiguously established.

Recently very interesting results have been obtained in studies on the stereoregularity of polymers by using the ^{13}C NMR method. The method was used to study stereochemical configuration in a number of polymers including polystyrene, polypropylene and poly(methyl vinyl ether) [81], polypropylene [82], poly(vinyl chloride) (see p. 357) and polyacrylonitrile (see p. 411).

The literature pertinent to the NMR studies on stereoregularity of polymers covered in this book, is given in the respective chapters devoted to individual polymer groups. The subject is reviewed at greater length by Sewell [2].

Study of Chemical Reactions

In view of the fact that high-resolution NMR spectroscopy is suitable for the determination of both the chemical and stereochemical structures of polymers, it provides a valuable method for following the course of a wide variety of chemical reactions in polymers. Some of these reactions have already been discussed in the section on broad-line NMR spectroscopy. The method has proved invaluable in studies on stereospecificity of polymerisation of such monomers as methyl methacrylate [83,84], chlorotrifluoroethylene [85], propylene [86], α-methylstyrene [87], and many others [2].

Qualitative and Quantitative Analysis

High-resolution NMR spectroscopy is increasingly used in solving purely analytical problems in the field of polymer chemistry. Of the major researches on this subject mention should be made of the following: determination of hydroxyl groups directly linked to silicon, alcoholic hydroxyl groups in silicones (see p. 448), and also methyl and phenyl groups in these polymers [88], identification and semi-quantitative evaluation of the acidic and alcoholic components and monomers in unsaturated polyester resins (see p. 236, 244), detection and quantitative determination of maleic and fumaric acids in resins of the above type (see p. 236, 252), detection and quantitative determination of isophthalic acid in the presence of terephthalic acid as acidic components of these resins (see p. 252), determination of the composition of linear terephthalate-sebacate-ethylene glycol terpolyesters (see p. 226, 243), determination of methylene bridges in phenol novolacs (see p.167, 183), determination of the degree of acetylation of cellulose acetate [89], identification and determination of components of polyurethane resins [90], analysis of oil-modified alkyds and polyurethanes (see p. 329), analysis of urethane prepolymers [91], determination of the composition of polyoxyphenylene-polystyrene mixtures [92], determination of hydroxyl end groups in poly(alkylene oxide) polymers and polyesters by PMR [93,94] and as trifluoroacetates by ^{19}F NMR spectroscopy [95], determination of the composition of nylon-6–nylon-6,6 mixtures and of the copolymer of the components of these polyamides [96].

The method is, in particular, extensively used in the quantitative analysis of copolymers. Analytical procedures for determining the composition of the following series of copolymers were developed: vinyl chloride with vinylidene chloride [64], vinyl chloride with vinyl acetate (see p. 366), tetrafluoroethylene with hexafluoropropylene [97,98], vinylidene fluoride with chlorotrifluoroethylene [99], ethylene with vinyl acetate [100,101] and ethylene with ethyl acrylate [102] (see p. 349), ethylene with propylene [103] (see p. 349), isobutylene with propylene [104], butadiene with isoprene [62,105], isoprene with isobutylene [106], styrene with butadiene [13,107], styrene with methyl methacrylate [108,109], styrene with α-methylstyrene [110], styrene with acrylonitrile [111,112], methyl methacrylate with acrylates [113], ethyl acrylate and methacrylonitrile with acrylonitrile [112], formaldehyde with ethylene oxide [114], ethylene oxide with propylene oxide [115], 1,3-dioxolan with trioxan (see p. 333), and vinyl acetate with unsaturated acid esters [116].

References

1. Slonim, I. Ya., Lyubimov, A. N., *The NMR of Polymers*, Plenum Press, New York, 1970.
2. Sewell, P. R., in *Annual Review of NMR Spectroscopy*, Mooney, E. F., (Ed), *Vol. 1*, Academic Press, London and New York, 1968.
3. *NMR Basic Principles and Progress*, Diehl, P., Fluck, E., Kosfeld, R., (Eds), *Vol. 4, Natural and Synthetic High Polymers* (*Lectures Presented at the Seventh Colloquium on NMR Spectroscopy*), Springer Verlag, Berlin, Heidelberg, New York, 1971.
4. Khachaturov, A., *Polimery*, **13**, 381, 433 (1968).
5. Powles, J. G., *Polymer*, **1**, 219 (1960).
6. Wilson, C. W., III, Pake, G. E., *J. Polymer Sci.*, **10**, 503 (1953).
7. Peterlin, A., *J. Macromol. Sci., Pt. B*, **3**, 19 (1969).
8. Smith, D. C., *Ind. Eng. Chem.*, **48**, 1161 (1956).
9. Fuschillo, N., Sauer, J. A., *Bull. Am. Phys. Soc., Ser. II*, **4**, 187 (1959).
10. Slichter, W. P., McCall, D. W., *J. Polymer Sci.*, **25**, 230 (1957); McCall, D. W., Slichter, W. P., *J. Polymer Sci.*, **26**, 17 (1957).
11. Odajima, A., Sohma, J., Koike, M., *J. Chem. Phys.*, **23**, 1959 (1955).
12. Holroyd, L. V., Codrington, R. S., Mrowca, B. A., Guth, E., *J. Appl. Phys.*, **22**, 696 (1951).

13. Angelo, R. J., Ikeda, R. M., Wallach, M. L., *Polymer*, **6**, 141 (1965).
14. Tanaka, K., Yamagata, K., Kittaka, S., *Bull. Chem. Soc. Japan*, **29**, 843 (1956).
15. Glazkovskii, Yu. V., Kol'tsov, A. I., Pyrkov, L. M., *Vysokomol. Soedin., Ser. B*, **12**, 715 (1970).
16. Iwayanagi, S., Sakurai, I., *J. Macromol. Sci., Pt. B*, **3**, 259 (1969).
17. Sobottka, J., Keller, F., *Plaste u. Kautschuk*, **16**, 92 (1969).
18. Olf, H. G., Peterlin, A., *Kolloid-Z., Z. Polym.*, **215**, 97 (1967).
19. Eichoff, U., Zachmann, G. H., *Ber. Bunsenges. Phys. Chem.*, **74**, 919 (1970).
20. Chierico, A., Del Nero, G., Lanzi, G., Mognaschi, E. R., *Eur. Polymer J.*, **5**, 115 (1969).
21. Shibata, T., Kimura, I., Suita, T., *J. Phys. Soc. Japan*, **13**, 1546 (1958).
22. Bonera, G., De Stefano, P., Rigamonti, A., *J. Chem. Phys.*, **37**, 1226 (1962).
23. Komaki, A., Matsumoto, T., *J. Polymer Sci., Pt. B*, **1**, 671 (1963).
24. Urman, Ya. G., Prut, E. V., Slonim, I. Ya., Enikolopyan, N. S., *Vysokomol. Soedin., Ser. B*, **9**, 770 (1967).
25. Gutowsky, H. S., Saika, A., Takeda, M., Woessner, D. E., *J. Chem. Phys.*, **27**, 534 (1957).
26. Oshima, K., Kusumoto, H., *Kogyo Kagaku Zasshi*, **59**, 806 (1956).
27. Lyubimov, A. N., Novikov, A. S., Galil-Ogly, F. A., Gribacheva, A. V., Varenik, A. F., *Vysokomol. Soedin.*, **5**, 687 (1963).
28. Tanaka, K., *Bull. Chem. Soc. Japan.*, **33**, 1702 (1960).
29. Matsushita, A., *J. Inst. Electr. Eng. Japan.*, **81**, 77 (1961).
30. Chuvaev, V. F., Ivanova, L. V., Zubov, P. I., *Vysokomol. Soedin.*, **6**, 1501 (1964).
31. Slonim, I., Ya., Lyubimov, A. N., Kovarskaja, B. M., *Chem. Průmysl*, **13**, 606 (1963).
32. Lösche, A., *Kolloid-Z., Z. Polymer.*, **165**, 116 (1959).
33. Waldrop, M. A., Kraus, G., *Rubber Chem. Technol.*, **42**, 1155 (1969).
34. Novikov, N. A., Shashkov, A. S., Galil-Ogly, F. A., *Vysokomol. Soedin., Ser. B*, **12**, 323 (1970).
35. Slonim, I. Ya., *Vysokomol. Soedin.*, **6**, 1371 (1964).
36. Fuschillo, N., Sauer, J. A., *Bull. Am. Phys. Soc., Ser. II*, **2**, 125 (1957).
37. Thurn, H., *Kolloid-Z., Z. Polym.*, **179**, 11 (1961).
38. Takeda, M., Tanaka, K., Nagao, R., *J. Polymer Sci.*, **57**, 517 (1962).
39. Gupta, R. P., *J. Phys. Chem.*, **66**, 1 (1962).
40. Nishioka, A., Kato, Y., Uetake, T., Watanabe, H., *J. Polymer Sci.*, **61**, S32 (1962).
41. Slichter, W. P., Mandell, E. R., *J. Appl. Phys.*, **29**, 1438 (1958).
42. Gupta, R. P., *Kolloid-Z., Z. Polymer.*, **174**, 73 (1961).
43. Sobottka, J., Uffrecht, H. H., *Plaste u. Kautschuk*, **18**, 737 (1971).
44. Abdrashitov, R. A., Bazhenov, N. M., Vol'kenshtein, M. V., Kol'tsov, A. I., Khachaturov, A. S., *Vysokomol. Soedin.*, **5**, 405 (1963).
45. Vol'kenshtein, M. V., Kol'tsov, A. I., Khachaturov, A. S., *Vysokomol. Soedin.*, **7**, 296 (1965).
46. Mansfield, P. B., *Chem. Ind. (London)*, **1971**, 792.
47. Liepins, R., Crist, B., Olf, H. G., *J. Polymer Sci., Pt. A–1*, **8**, 2049 (1970).
48. Brame, E. G., Jr., *J. Polymer Sci., Pt. A–1*, **9**, 2051 (1971).
49. Saito, T., Matsumura, Y., Hayashi, S., *Polymer J. Japan*, **1**, 639 (1970).
50. Heintke, W., Keller, F., *Plaste u. Kautschuk*, **18**, 732 (1971).
51. Doskočilová, D., Schneider, B., Drahovádová, E., Štokr, J., Kolinský, M., *J. Polymer Sci., Pt. A–1*, **9**, 2753 (1971).
52. Unterforsthuber, K., *Kolloid-Z., Z. Polym.*, **238**, 430 (1970).
53. Pham, Q. T., Rocaniere, P., Guyot, A., *J. Appl. Polymer Sci.*, **14**, 1291 (1970).
54. Gagnaire, D., Odier, L., Vincendon, M., *J. Polymer Sci., Pt. C*, **28**, 27 (1969).
55. Stubbs, W. H., Gore, C. R., Marvel, C. S., *J. Polymer Sci., Pt. A–1*, **4**, 1898 (1966).
56. Nishioka, A., Kato, Y., *Kogyo Kagaku Zasshi*, **68**, 1452 (1965).
57. Ferguson, R. C., *Kunststoffe–Plastics*, **18**, 723 (1965).

58. Mak, H. D., Rogers, M. G., *Anal. Chem.*, **44**, 837 (1972).
59. Wilson, C. W., III, *J. Polymer Sci., Pt. A-1*, **1**, 1305 (1963).
60. Naylor, R. E., Jr., Lasoski, S. W., Jr., *J. Polymer Sci.*, **44**, 1 (1960).
61. Bovey, F. A., Anderson, E. W., Douglass, D. C., Manson, J. A., *J. Chem. Phys.*, **39**, 1199 (1963).
62. Chen, H. Y., *Anal. Chem.*, **34**, 1134 (1962).
63. Ferguson, R. C., *J. Polymer Sci., Pt. A-1*, **2**, 4735 (1964).
64. Chûjô, R., Satoh, S., Ozeki, T., Nagai, E., *J. Polymer Sci.*, **61**, S. 12 (1962).
65. Nishioka, A., Kato, Y., Ashikari, N., *J. Polymer Sci.*, **62**, S. 10 (1962).
66. Nishioka, A., Kato, Y., Mitsuoka, H., *J. Polymer Sci.*, **62**, S. 9 (1962).
67. Ferguson, R. C., *J. Am. Chem. Soc.*, **82**, 2416 (1960).
68. Takeuchi, T., Yamazaki, M., Mori, S., *J. Polymer Sci., Pt. B*, **4**, 695 (1966).
69. Keller, F., Roth, H., *Plaste u. Kautschuk*, **15**, 800 (1968).
70. Patnaik, B., Takahasi, A., Gaylord, N. G., *J. Macromol. Sci., Pt. A*, **4**, 143 (1970).
71. Guillot, J., Guyot, A., Pham, Q. T., *J. Macromol. Sci., Pt. A*, **2**, 1303 (1968).
72. Ito, K., Yamashita, Y., *J. Polymer Sci., Pt. B*, **6**, 227 (1968).
73. Fleischer, D., Schulz, R. C., *Makromol. Chem.*, **152**, 311 (1972).
74. Schaefer, J., *Macromolecules*, **4**, 107 (1971).
75. Murano, M., *J. Polymer Sci., Pt. A-1*, **9**, 567 (1971).
76. Fujiwara, Y., Fujiwara, S., Fujii, K., *J. Polymer Sci., Pt. A-1*, **4**, 257 (1966).
77. Bacskai, R., *J. Polymer Sci., Pt. A*, **1**, 2777 (1963).
78. Curtis, L. G., Edwards, D. L., Simons, R. M., Trent, P. J., Von Bramer, P. T., *Ind. Eng. Chem., Prod. Res. Develop.*, **3**, 218 (1964).
79. Bovey, F. A., Tiers, G. V. D., *Chem. Ind. (London)*, **1962**, 1826.
80. Braun, D., Herner, M., Johnsen, U., Kern, W., *Makromol. Chem.*, **51**, 15 (1962).
81. Johnson, L. F., Heatley, F., Bovey, F. A., *Macromolecules*, **3**, 175 (1970).
82. Inoue, Y., Nishioka, A., Chûjô, R., *Makromol. Chem.*, **152**, 15 (1972).
83. Kato, Y., Watanabe, H., Nishioka, A., *Bull. Chem. Soc. Japan*, **37**, 1762 (1964).
84. Lipscomb, N. T., Weber, E. C., *J. Polymer Sci., Pt. A-1*, **3**, 55 (1965).
85. Tiers, G. V. D., Bovey, F. A., *J. Polymer Sci., Pt. A-1*, **1**, 833 (1963).
86. Woodbrey, J. C., *J. Polymer Sci., Pt. B*, **2**, 315 (1964).
87. Sakurada, Y., Matsumoto, M., Imai, K., Nishioka, A., Kato, Y., *J. Polymer Sci., Pt. B*, **1**, 633 (1963).
88. Kubota, T., Takamura, T., *Bull. Chem. Soc. Japan*, **33**, 70 (1960).
89. Gagnaire, D., Vincendon, M., *Bull. Soc. Chim. France*, **1968**, 3413.
90. Brame, E. G., Jr., Ferguson, R. C., Thomas, G. J., Jr., *Anal. Chem.*, **39**, 517 (1967).
91. Kobayashi, K., Kurihara, K., Hirose, K., *Kogyo Kagaku Zasshi*, **73**, 1664 (1970).
92. Penczek, I., Biały, J., *Polimery*, **14**, 496 (1969).
93. Page, T. F., Bresler, W. E., *Anal. Chem.*, **36**, 1981 (1964).
94. Liu, K. J., *Makromol. Chem.*, **116**, 146 (1968).
95. Manatt, S. L., Lawson, D. D., Ingham, J. D., Rapp, N. S., Hardy, J. P., *Anal. Chem.*, **38**, 1063 (1966).
96. Yamazaki, N., Takeuchi, T., *Kogyo Kagaku Zasshi*, **70**, 460 (1967).
97. Wilson, C. W. III, *J. Polymer Sci.*, **56**, S. 16 (1962).
98. Brame, E. G., Jr., Sudol, R. S., Vogl. O., *J. Polymer Sci., Pt. A-1*, **2**, 5337 (1964).
99. Sibilia, J. P., Paterson, A. R., *J. Polymer Sci., Pt, C*, **8**, 41 (1965).
100. Chen, H. Y., Lewis, M. E., *Anal. Chem.* **36**, 1394 (1964).
101. Yawaka, Y., Tsuchihara, T., Tanaka, N., Kosaka, K., Hirakida, Y., Ogawa, M., *Kobunshi Kagaku*, **28**, 459 (1971).
102. Porter, R. S., Niksic, S. W., Johnson, J. F., *Anal. Chem.*, **35**, 1948 (1963).
103. Porter, R. S., *J. Polymer Sci., Pt. A-1*, **4**, 189 (1966).
104. Barrall, E. M., Porter, R. S., Johnson, J. F., *J. Chromatogr.*, **11**, 177 (1963).
105. Chen, H. Y., *Anal. Chem.*, **34**, 1793 (1962).
106. Stehling, F. C., Bartz, K. W., *Anal. Chem.*, **38**, 1467 (1966).
107. Senn, W. L., Jr., *Anal. Chim. Acta*, **29**, 505 (1963).

108. Kato, Y., Ashikari, N., Nishioka, A., *Bull. Chem. Soc. Japan*, **37**, 1630 (1964).
109. Gruber, U., Elias, H-G., *Makromol. Chem.*, **86**, 168 (1965).
110. Braun, D., Heufer, G., Johnsen, U., Kolbe, K., *Kolloid-Z., Z. Polymer.*, **195**, 134 (1964).
111. Takeuchi, T., Yamazaki, M., *Kogyo Kagaku Zasshi*, **68**, 1478 (1965).
112. Ritchie, W. M., Ball, L. E., *J. Polymer Sci., Pt. B*, **4**, 557 (1966).
113. Grassie, N., Torrance, B. J. D., Fortune, J. D., Gemmell, J. D., *Polymer*, **6**, 653 (1965).
114. Allen, G., Warren, R., Taylor, K. J., *Chem. Ind. (London)*, **1964**, 623.
115. Mathias, A., Mellor, N., *Anal. Chem.*, **38**, 472 (1966).
116. Dietrich, M. W., Keller, R. E., *Anal. Chem.*, **36**, 2174 (1964).

Part II

ANALYSIS OF POLYMERS

Chapter 6

PHENOLIC RESINS

Phenolic resins, also referred to as phenoplasts or simply phenolics [1,2], are the products of polycondensation reactions of phenols with aldehydes. Phenol is most commonly used for their manufacture, followed by cresols and xylenols, and less often bisphenol A, *p-tert*-butylphenol or other phenols. Formaldehyde is largely used as the aldehydic component, and sometimes also furfural, acetaldehyde, or acrolein. The ratio of polycondensation reactants and the pH of the medium determine the reaction path and a diversity of products may be formed.

Commercially, the condensation of phenols with aldehydes is run either in an acidic medium to make fusible resins (novolacs) or in an alkaline medium to produce thermosetting resins (resols).

The fusible, non-thermosetting novolac resins are invariably produced with the use of an acidic catalyst and excess phenol. The products exhibit a rather high polydispersity and an average melecular weight not higher than 1000, indicating that the resins are rather low polymers. When heated, novolacs melt but fail to undergo further condensation. Up to 250°C the resins do not undergo any significant chemical changes. Novolac resins can only be cured by methods such as heating with hexamethylenetetramine.

Commercial novolacs are brittle transparent solids. They may be clear or coloured light-brown, tan, pinkish or dark brown. In view of their fairly low molecular weight the resins readily dissolve in alcohols. Hydrocarbon-soluble products can be obtained by esterification or etherification of the hydroxyl groups in the novolac resins.

Novolacs are primarily used in the manufacture of moulding powders (with hexamethylenetetramine as a hardener), and these are the most common of the phenolic resins employed for this purpose.

The hardening resins, referred to as resols, are obtained in the presence of an alkaline catalyst. These resins are also formed in an acidic medium when the phenol is present in an insufficient excess. They differ from novolacs in that on heating their viscosity and softening point increase while their solubility in organic solvents becomes lower. With subsequent heating the resins become gummy and finally turn into resite—an infusible and insoluble material that is not softened by heat or solvents. Commercial resols are available either in liquid form or as hard brittle solids.

6.1 STRUCTURE

As to chemical structure, phenolic resins represent a mixture of complex condensation products. Their structure has been elucidated in more detail fairly recently by the use of novel instrumental techniques like infrared spectrophotometry, chromatography, pyrolysis gas chromatography, and NMR spectroscopy.

A complete understanding of the chemical composition of phenolic resins, and quantitative determination of their constituents and structure, is not feasible with currently available techniques; this is also the case with other resins. Numerous constituents of phenolic resins have actually been isolated and identified by paper and thin layer chromatography [3,4,5,6,7]. However, in view of the high complexity of these mixtures, resulting primarily from almost unlimited possibilities for the formation of isomers even at a moderate degree of condensation, the identified constituents represent only a fraction of all the resin components. On the other hand, the content of functional groups can be determined with fairly high accuracy, and occasionally also their mutual position in the constituent phenolic rings.

Chemical methods are suitable for the determination of the content of hydroxyphenyl and hydroxymethyl groups, and of ether bridges. These methods were successfully used in the qualitative and quantitative determination of phenolic hydroxyls in both resols and novolacs, whereas hydroxymethyl groups and ether bridges can only be estimated in resols. Resols were also found to contain appreciably fewer active sites compared with novolacs. Dijkstra and de Jonge [8,9] developed a technique for the determination of the hydroxymethyl group content separately at *ortho* and *para* positions. The content of active *ortho* and *para* positions relative to the phenol can also be determined [10,11,12].

Zulaica and Guiochon [13] have demonstrated that much valuable information on the structure of phenolic resins is obtained from the study of their pyrolysis products by gas chromatography. As a result of breaking of the C—C bond between the ring atoms and the CH_2—O—CH_2 ether bridge, the methylene group may stay on or depart from the ring. These processes result in the formation of phenol or its methyl derivatives, respectively. In the pyrolysis products from the resins made from phenol only, phenol, *o*- and *p*-cresols and 2,4- and 2,6-xylenols were found. More 2,6-xylenol and less phenol is produced from resins made with an acidic catalyst than with an alkaline catalyst. Substantially more information on the structure of these resins can be derived from studies by infrared spectrophotometry [14,15,16,17,18].

The infrared method is good for detecting all the functional groups present in the resins, often indicating at the same time their mutual arrangement, and some of these groups can be determined quantitatively. Infrared spectrophotometry is of special value for studying insoluble resins where it can also be used to follow the resin-hardening process. Available pertinent data were evaluated by Secrest [19] who also recorded a number of the spectra of various phenolic resin types. Resols were found to be easily differentiated from novolacs, as the latter exhibit a stronger and narrower band at 6·65 μm which, according to Bellamy [20], should be assigned to 1,4-, 1,2- and 1,2,4-substituted rings. On the other hand, resols, as distinct from novolacs, exhibit absorption at 9·4 μm, which is due to the presence of ether bridges, at 10 μm due to hydroxymethyl groups, and also a stronger absorption at 11·4 μm on account of a larger quantity of benzene rings substituted at positions 1, 2, 3, and 5. Phenolic resins of all types exhibit absorption at ca. 3 μm resulting from phenolic and alcoholic hydroxyls, and at ca. 8 μm for which phenolic hydroxyls are exclusively responsible.

The NMR technique has proved of exceptional value in studying the structure of soluble phenolic resins. It was used in investigations on both novolacs and resols made from phenol and formaldehyde by Woodbrey, Higginbottom and Culbertson [21]. These authors succeeded in making a quantitative determination of all the major functional groups, including their mutual arrangement with respect to phenolic hydroxyls, and in finding the number average molecular weight. An especially valuable achievement of these authors was to find the position of the methylene bridges with respect to phenolic hydroxyls. The studies by Woodbrey *et al.* (*loc. cit.*) show the following.

1. Novolacs have no hydroxymethyl groups and no dimethylene-ether bridges, and that ca. 50% of the constituent benzene rings are linked with methylene bridges at *p* and *p′* positions, whereas *o,o′*-bondings are rather scarce. The ratio of these two bonding types in a resin depends on the kind of phenol used in the polycondensation reaction [22].

2. In addition to methylene bridges and a large quantity of hydroxymethyl groups, resols occasionally contain a considerable number of dimethylene-ether bridges. Here, the quantity of *p,p′*-methylene bridges is substantially larger than the total number of these bridges at positions *o,o′* and *o,p′*, whereas hydroxymethyl groups present are largely in the position ortho to the phenolic hydroxyls. Thus the results obtained by these authors have actually confirmed generally known facts. A novelty in this investigation has been the discovery that hemiformals of a benzyl type are present in considerable amounts in resol resins. Such groupings have never before been taken into account in considering the constitution of phenolic resins, as they were believed to be unstable under the conditions of the synthesis of resols.

Meanwhile, according to Brockmann *et al.* [23], who investigated the structure of phenol-formaldehyde resols by both NMR and mass-spectroscopy, these resins have no ether bridges, only methylene bridges.

Recently Wagner and Greft [24] separated the components of phenolic resols by gel permeation chromatography, and determined the structure of these resins by NMR spectroscopy. The hydroxymethyl groups and ether bridges of the resolic resins are capable of further condensation reactions at the available positions *ortho* and *para* to the phenolic hydroxyl group. The presence of these groupings is responsible for the thermoreactivity of resols. As a result of subsequent condensation by heating, the originally low polymeric resol molecules combine to form an increasingly bulkier crosslinked structure, until resol is ultimately converted into resite—a product of huge crosslinked macromolecules which is infusible and insoluble.

The novolac molecules have no reactive groups, and therefore do not undergo any chemical changes of importance. These resins can only be cured and converted into resite when heated with formaldehyde-liberating compounds such as paraformaldehyde or hexamethylenetetramine.

6.2 QUALITATIVE ANALYSIS

The chemical composition of present-day phenoplasts is highly diversified. It depends on the phenolic and aldehydic substrates used in their manufacture, and also on the ever-increasing number of techniques employed in their modification. Thus, a detailed identification of the components of these resins, and the way they are combined, can be accomplished only after their isolation from a plastic material, followed by fractionation with the use of modern separation techniques.

In practice, however, such detailed studies are usually not needed and a rough estimation of the resin type is sufficient. For this purpose, the qualitative tests based mostly on the colour reactions of phenols are good. Hence, the presence of a phenol in a resin or its formation as a resin degradation product during the test is a condition for the test to be positive. Occasionally the aldehydic resin component should also be identified. In certain types of highly crosslinked resins, the tests for the resin components may fail.

Direct Chemical Tests

Colour Reactions for Phenols

Phenolphthalein Test. A 1 g resin sample is heated in a porcelain evaporating dish with an equal quantity of phthalic anhydride and 3 drops of conc. sulphuric acid until a dark brown melt is formed. On cooling, the melt is diluted with water and made alkaline with 10% sodium hydroxide solution. A characteristic red colour is indicative of the presence of a phenol-formaldehyde resin. If tarry materials mask the colour, the solution is diluted with water, and 1*N* acid solution is added dropwise. At the neutralisation point a distinct change in the colour occurs.

The Millon Test. The Millon reagent is prepared by dissolving 10 g of mercury in 10 ml of nitric acid (d = 1·42 g per ml) with gentle heating, followed by dilution with 15 ml of distilled water.

PROCEDURE. A small resin sample is heated to boiling for 2 min with 1 ml of clear reagent. A red colour indicates the presence of phenol in the resin. Epoxy, casein, and aniline resins yield a similar reaction in this test. Some of the phenoplasts fail to give this reaction.

Test with 2,4-Dinitrobenzenesulphonyl Chloride. Phenolic hydroxyls react with 2,4-dinitrobenzenesulphonyl chloride in an alkaline medium [25,26] to form 2,4-dinitrobenzenesulphonates. These esters yield a red colour with the excess of hydroxide used in a water-dioxan (15:1) solution. The reaction is characteristic for phenolic compounds and is highly sensitive, as up to 0·002% of phenolic groups can be detected in plastics.

PROCEDURE [26]. A small quantity of a resin (from several to about 50 mg) is dissolved in 1 ml of dioxan, 0·5 ml of 0·15% dioxan solution of 2,4-dinitrobenzenesulphonyl chloride is added, followed by 1·5 ml of 0·15% solution of tetrabutylammonium hydroxide in dioxan containing 7% water. A red colour immediately develops if free phenolic hydroxyls are present in the resin. The test is suitable for the analysis of uncured resins or polymer pyrolysis products.

Colour Reactions for Aldehydes

Chromotropic Acid Test for Formaldehyde (see p. 28)

Aniline Test for Furfural. A resin sample or its methanolic extract is mixed with 1 ml of 5% aniline solution in glacial acetic acid. A red colour develops in the presence of furfural.

Differentiation of Novolacs from Resols

A 15 g resin sample [27] is carefully heated in an open crucible while stirring the melt with a thermometer. If after several minutes of heating at 150°C the resulting melt turns into a resinous material, then the resin sample is a resol. Otherwise the resin ranks among novolac plastics.

Differentiation of Phenolics from Cresolics

A resin sample [28,29] (0·2–0·3 g) is heated under reflux with a solution of 0·5 g of sodium hydroxide in 6 ml of glycol monomethyl ether. Two drops of the reaction mixture are treated with 10 ml of water, 10 ml of 10% sodium hydroxide, and 10 ml of ethanol. Subsequently one drop of aniline is added, and the mixture is shaken. Six drops of 3% hydrogen peroxide are added, the mixture is shaken again, then one drop of 5% hypochlorite solution is added. After 5 min a fairly stable red-brown colour develops in the presence of phenol. Cresols give a blue or turquoise tint.

Phenolic resins can also be distinguished from alkylphenolic resins by chromic acid mixture oxidation. Acetic acid is produced from alkylphenolic resins, whereas only carbon dioxide and water are formed from exclusively phenol-based resins.

Chemical Tests after Pyrolysis

Free phenol and its homologues are produced from pyrolysis of highly cured phenolic resins. The presence of phenols in the volatile pyrolysis products may be detected by using one of the qualitative tests described below, or simply by their smell. Aldehydic components of the resins may also be determined in these pyrolysis products.

Thermal decomposition of the resin is performed as described on p. 21.

Detection of Phenols

Hummel [15] cites the results of very detailed studies by Schouten and Nijveld [30] on the qualitative reactions of phenolic resins specially characteristic of phenol. The tests were made using an aqueous solution of volatile products of the resin pyrolysate. The following tests were found to be the most suitable.

a. The indophenol test [31] after Gibbs (see p. 29). The test is so sensitive that even in the analysis of highly cured resins only an aqueous or methanolic extract of the material examined can generally be used in place of the pyrolysate. For this, 1 g of finely powdered resin or 2 g of varnish is carefully heated in a test tube with 8 ml of methanol or water for several minutes. The resulting extract is decanted or filtered free of the solid into another test tube. When aqueous extraction is used, the test is run directly with 1 ml of the solution obtained. The methanolic extract is evaporated to dryness, and the residue is dissolved in several ml of water. A grain of 2,6-dibromoquinone chloroimide (volume of a half wheat-grain) is added to the aqueous solution followed by vigorous shaking. Subsequently, 5 drops of $0{\cdot}1N$ sodium hydroxide solution are introduced until $pH = 9{\cdot}4$, and the solution is shaken once. The colour developed is observed first after several minutes, then after 30 min. In the presence of phenol a blue colour is produced within a short time. Too large quantities of phenol or the reagent are undesirable.

b. A mixture of 2–3 ml of conc. sulphuric acid with a drop of 3% hydrogen peroxide is prepared, to which a few drops of the solution under analysis are added. The colour of the resulting two-phase mixture is noted, then a few drops of dilute (1:1) nitric acid are added, and the colour is observed again. A green ring is produced if phenol is present. The ring turns orange upon addition of the nitric acid. Other phenols give different colours.

c. The phenol under test is coupled with freshly diazotized sulphanilic acid, *p*-nitroaniline, or *m*-nitrotoluidine in a mildly alkaline solution. A red dye is formed with phenol.

d. The phenol to be identified is coupled in a mildly alkaline solution with the diazonium salt of 2-nitro-4-chloroaniline. A bright red colour develops with phenol. The dye produced settles on the test tube walls.

For resins manufactured from phenol all the tests described give a positive result.

Detection of Aldehydes

The Fuchsin Test. Reagent (fuchsin-sulphurous acid, i.e. the Schiff reagent). Powdered fuchsin (0·5 g) is dissolved in 500 ml of water, 10 ml of 38% sodium bisulphite solution is added, and the mixture is allowed to stand for 10 min. Next 10 ml of conc. hydrochloric acid is added. After 2 hr the solution is ready for use. The reagent should be stored in a dark glass bottle.

PROCEDURE. One ml of the solution obtained from pyrolysis of the resin is treated in a test tube with 2 ml of the Schiff reagent. The test tube is stoppered and shaken carefully. A violet colour appearing within 30 min indicates the presence of an aldehyde.

The test is based on the strong reactivity of formaldehyde to form an addition product with sulphurous acid, but the reaction also occurs with higher aldehydes.

The Chromotropic Acid Test. The test is carried out with one drop of the solution under analysis following the procedure described before (p. 28). The test is very sensitive and characteristic of formaldehyde only.

The Phloroglucinol Test after Jorissen. To 1–2 ml of the solution under analysis is added 0·5 ml of 0·1% phloroglucinol solution followed by 1–2 drops of dilute sodium hydroxide. Appearance of a red colour indicates the presence of formaldehyde (limiting concentration 5×10^{-3}%). With acetaldehyde, the colour formed is orange-yellow, which increases in intensity after some time [15].

Tests after Alkali Fusion

In the analysis of highly hardened resins, fusion with sodium hydroxide may also be used to effect their partial decomposition, which is necessary to liberate phenol.

A sample of 1 g of finely powdered resin is fused with excess sodium hydroxide. On cooling, the melt is diluted in water and the solution is neutralised with hydrochloric acid. The resulting solution can be analysed for the presence of phenols as in the tests described above.

Instrumental Methods

Gas Chromatography

As already mentioned, in the volatile products of pyrolysis of even highly hardened phenolic resins, phenol and its homologues can be found. A more detailed examination of these volatile products, which was accomplished recently by gas chromatography, enabled the characteristic components of a variety of phenolic resin types to be detected. Gas chromatography therefore offers another means of identifying these components.

Parriss and Holland [32] have established that chromatograms of pyrolysis products of phenol-derivative phenoplasts decomposed at 800°C exhibit a large peak due to phenol. For the resins containing 40–42% *m*-cresol as well as phenol, the phenol peak is rather small. Another product characteristic of phenol-derivative phenolics is benzene, which is absent in the pyrolysis products of cresol-derivative resins.

Extensive pyrolysis gas chromatography studies of novolacs and resols based on both phenol and *m*-cresol were carried out by Zulaica and Guiochon [13]. The chromatogram of a *m*-cresol-based novolac resin was found to exhibit peaks originating from *m*-cresol, which are absent in the chromatograms of the phenol-based resins. Chromatograms of the phenol-derivative novolac and resol resins examined also showed certain characteristic differences, as expected. According to these authors, from studies on the phenoplast decomposition products, some information can be gained on the chemical composition of the starting materials used to synthesise the resins, and also on the curing conditions.

Likewise, the results of investigations by Priori and Panetti [33], Cox and Ellis [34], Jackson and Conley [35], Tsuge *et al.* [36], Zowall and Świątecka [37], Mosimann and Weber [38], and Braun and Arndt [39] show great promise for pyrolysis gas chromatography as a means of identifying phenoplasts.

Another variant of the gas chromatography technique, as used for identification of phenoplasts, is the procedure developed by Horiuchi [40], who employed phenolysis of the resin in place of pyrolysis. According to Horiuchi, the resin (for instance a cresolic novolac) is treated with an excess of phenol, *p*-toluenesulphonic acid is added, and the whole is heated for 3–5 hr. The reaction mixture is then steam distilled. The precipitate formed is filtered off and the filtrate is extracted with ether. The ether extract is dried with anhydrous sodium sulphate, the ether is removed by evaporation, and the residue is analysed on a chromatographic column containing lauryl terephthalate as stationary phase. By this procedure *o*- and *p*-cresol components of the resin can be quantitatively determined. The method is also suitable for the analysis of cured resins.

Infrared Spectroscopy

By means of infrared spectroscopy miscellaneous types of phenolic resins can now be conveniently identified owing to the extensive experimental material gathered by a number of authors (covered in the monographs by Hummel [15,18], or by Haslam and Willis [14]). Secrest [19] has summed up the available results in this field and supplemented this material with his own findings. According to Secrest infrared spectroscopy offers the following advantages for identification of phenolic resins.

1. The spectra of phenolic resins differ significantly from the spectra of other thermosetting resins, and also from those of bisphenol A epoxide resins. Consequently, the phenolic resins can be easily identified. The phenolics exhibit absorption at about 3 μm (phenolic and alcoholic hydrox-

yls), absorption bands in the range 6·5–7·0 μm, at about 8 μm (aromatic C—O groups) and at 12·2 μm (see Figs. 6.1 and 6.2). Although melamine resins also exhibit absorption at about 3 and 12·3 μm, an additional very strong characteristic absorption at about 6·5 μm is observed in their spectra. The bisphenol A epoxides exhibit absorption at about 12·0 μm, and a very intense, narrow band at 8·5 μm.

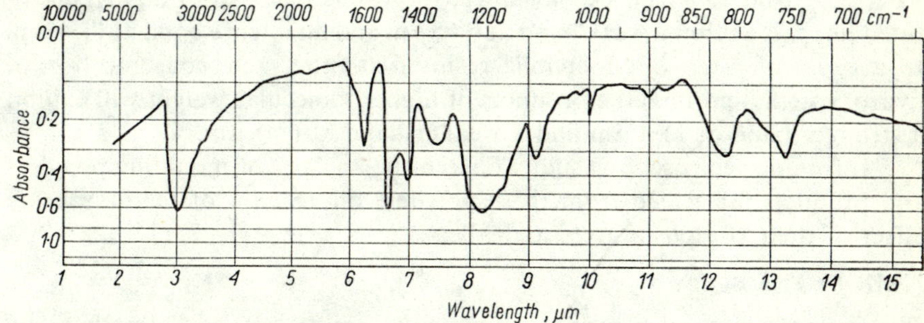

Fig. 6.1 Infrared absorption spectrum of a novolac resin [19]; (KBr disc) (by permission of the copyright holders, Federation of Societies for Paint Technology)

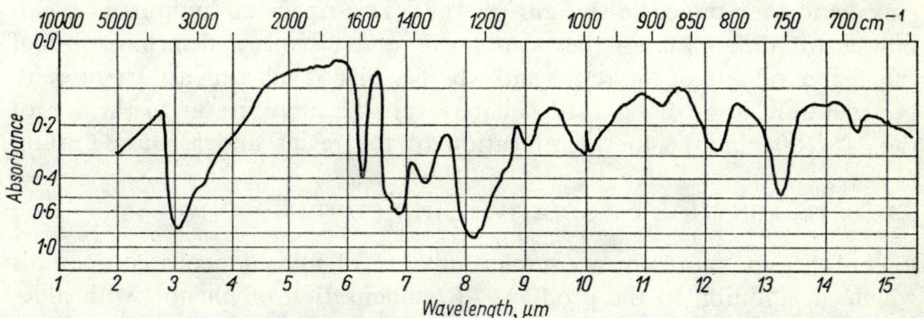

Fig. 6.2 Infrared absorption spectrum of a resol resin [19]; (KBr disc) (by permission of the copyright holders, Federation of Societies for Paint Technology)

2. Resols can be distinguished easily from novolacs. In the spectra of the latter a stronger and narrower band occurs at 6·65 μm (Fig. 6.1). On the other hand, in the spectra of resols, absorption bands are observed at about 9·4 μm from ether groups, at 10 μm (hydroxymethyl groups), and a stronger band at 11·4 μm which is due to a larger quantity of benzene rings substituted at positions 1, 2, 3, and 5 (Fig. 6.2).

3. The presence of alkyd-modified phenolic components may be detected on the basis of the band at 3·4 μm due to the aliphatic C—H bond. The size of the alkyl groups may be determined from the intensity of that band.

4. In novolacs the presence of *para*-substituted phenolic components can be identified on the basis of a relatively strong band at 11·4–11·7 μm. Likewise, the spectra of resols made from *para*-substituted phenols exhibit a stronger band in this range.

5. The resins synthesised from *p*-phenylphenol can be identified easily from very strong bands at 13·2 μm and 14·4 μm, resulting from the mono-substituted benzene rings.

6. The resin type having been established, the relative degree of substitution (degree of conversion of the resin, hydroxymethyl group content, etc.) can be determined by comparing the intensities of bands at 6·65, 11·4, 12·2, 12·8, and 13·3 μm. For instance, the resins made from highly substituted phenols exhibit a weak band at 6·65 μm and an intense band at 11·4 μm. In a series of *p-tert*-butylphenolic resins (with the same concentration of hydroxymethyl groups) the products of higher molecular weight will exhibit a stronger band at 11·4 μm and a weaker band at 6·65 μm.

Infrared spectroscopy usually fails in the analysis of moulding powders and products fabricated from these powders on account of the interfering effect of their composite dyes and fillers.

NMR Spectroscopy

The major structural elements of phenolic resins can be determined by NMR spectroscopy, and similarly the position of these elements with respect to the phenolic hydroxyl, the degree of substitution of the phenolic rings, and even molecular weight (p. 167). This opens up undoubted possibilities for differentiation between resols and novolacs, determination of the degree of cure of the resin, and whether or not substituents are present. No mention, however, is made in the available literature as to the use of the NMR technique for identification of the resins under consideration.

6.3 DETERMINATION OF ADDITIVES AND CONTAMINANTS

Phenolic resins represent a complex mixture of miscellaneous compounds which, in addition to the products of condensation of phenols with aldehydes, also contain unreacted starting materials (phenols and aldehydes), a variable quantity of water, residual catalysts and minor quantities of other additives.

Water

Water, if present in quantities greater than 2%, can be determined in resins by the Dean and Stark procedure so long as the resin is soluble in aromatic hydrocarbons. With lower quantities, better results are achieved by titration with the Karl Fischer reagent (see Section 1.5).

Ash

Inorganic substances contained in the resin usually arise from the residual catalyst, which is not always removed in the manufacturing process.

On heating for several hours at 500–550°, the majority of inorganic compounds used as catalysts remain unchanged in the ash after combustion of organic materials. At a temperature of about 1000°C after ignition for several hours the majority of inorganic catalysts are converted into oxides.

PROCEDURE. A 10 g sample of resin (or moulding powder) is heated in a porcelain crucible over a burner until the volatile components are removed, then the crucible is placed in a muffle furnace at a temperature of 800°C and ignited to constant weight. Prior to each weighing the crucible is allowed to cool in a desiccator.

Unreacted Substrates

Phenolic resins may contain quantities of unreacted phenols although the presence of large amounts of these compounds is undesirable. Determination of free phenols in these resins is also required as an analytical control of the polycondensation process, when unreacted aldehydes are also determined.

Chemical Methods

A fundamental preliminary operation in most techniques for the determination of phenols and aldehydes is their isolation from the resin. This is necessary, as the resin interferes with the determination of free phenol which is usually conducted by bromination as described by Koppeschaar [14,27,41]. The isolation of phenol is generally accomplished by taking advantage of the volatility of phenols with steam.

Steam distillation is, in general, carried out without any preliminary treatment. A resin sample (2–10 g) is either dissolved in 20 ml of methanol and 200 ml of water is added, or it is treated with 100 ml of 10% acetic acid. The mixture is then subjected to steam distillation.

Haslam *et al.* [42] recommend prior dissolution of the resin sample in aqueous sodium hydroxide, precipitation of the resin by addition of acid until pH = 4·5, and separation of the resin by filtration. Steam distillation is carried out with the filtrate. This method is particularly suitable for the analysis of resols which, during steam distillation, may undergo further condensation with any free phenol present in the resin.

Dijkstra and Lammers [43] treat the resin with glycol to soften it before steam distillation, thus substantially shortening the distillation time. In the opinion of these authors, only minor quantities of glycol pass into the distillate, and do not interfere with the determination of phenol.

Dzięgielewski [44] established that unreacted formaldehyde present in the resin reacts with phenols under the conditions of analysis, giving low results. He suggests eliminating this effect by blocking the formaldehyde with hydroxylamine hydrochloride.

Bulygin *et al.* [45] recommend another procedure. Here, phenols are extracted with benzene and subsequently determined in this extract, while formaldehyde is determined on isolation of the resin with hydrochloric acid treatment followed by centrifuging.

Phenols

The distillate contains either only phenol (or only cresols) or a mixture of phenol with cresols (in the case of composite phenol-cresol-formaldehyde resins). In the analysis of a homogeneous phenolic material the total free phenol only is determined, and in the case of composite resins, the ratio of phenols to cresols is also determined.

Analytical procedures for the determination of phenols by Haslam *et al.* [14,42] are given below.

Total Phenols

REAGENTS AND APPARATUS

Potassium Bromide-Bromate, 0·1*N Solution.* Potassium bromate (2·784 g) and potassium bromide (10 g) are dissolved in 1 l. of distilled water.

A steam distillation apparatus.

PROCEDURE. A weighed resin sample (the size of the sample depends on the free phenol content; for resins containing 10–20% phenol it is 2 g, for moulding powders 5 g, or occasionally 10 g) in a 500 ml beaker is treated with 10% aqueous sodium hydroxide of volume 50–200 ml depending on the size of the sample. The contents are heated to boiling for about 10 min until the resin is completely dissolved. In the case of moulding powders only the filler remains undissolved. The solution is allowed to cool and the volume is diluted to about 200 ml. Then it is neutralised, first with 20% sulphuric acid, and finally with a 1*N* solution of the acid until pH 4·5, tested with indicator paper.

The resin precipitate is filtered off and thoroughly washed with cold water. The filtrate and the washings which contain free phenols are transferred to a 500 ml volumetric flask, the volume is made up to the mark, and the contents are thoroughly mixed; 250 ml of this solution is transferred to a steam distillation apparatus, 350–400 ml of the distillate is collected in a 500 ml volumetric flask, then made up to the mark and thoroughly mixed.

Next 250 ml of this solution is transferred to a 500 ml iodination flask and 50 ml of ca. 0·1*N* standard potassium bromide-bromate reagent followed by 10 ml of 20% sulphuric acid are added. The flask is stoppered, and a few drops of 10% aqueous potassium iodide solution are introduced into the wider part of the flask neck around the stopper to prevent bromine losses by evaporation. A blank is run independently. The solutions are allowed to stand for 1 hr in darkness. After that time the stoppers are removed in such a manner that the potassium iodide solution sealant flows down into the flask, and a further 15 ml of the potassium iodide solution is introduced. The liberated iodine is titrated with 0·1*N* sodium thiosulphate solution to a pale yellow colour, then a few ml of chloroform is added to dissolve the bulky tribromophenol precipitate as this holds considerable quantities of iodine, and the titration is completed after addition of 1 ml of 1% starch indicator solution.

The quantity of phenol or cresols in the sample is found from the difference in the titrant volume consumed in the blank and in the sample. One ml of standard 0·1*N* sodium thiosulphate corresponds to 0·00157 g of phenol or 0·00236 g of cresol.

The Ratio of Free Phenol to Cresols

REAGENTS AND APPARATUS

Nitric Acid, Dilute. An air stream is passed through fuming nitric acid until it is decolourised. One volume of the resulting acid is diluted with 4 volumes of water.

The Millon Reagent. Mercury (2 ml) is introduced from a burette into a 100 ml conical flask containing 20 ml of conc. nitric acid. The mixture is occasionally shaken until dissolution of mercury is complete, then 35 ml of water is added. If basic mercuric nitrate precipitates, dilute nitric acid is added dropwise until the solution becomes clear. Then 10% aqueous sodium hydroxide is introduced dropwise with continuous stirring, until the flaky precipitate that forms stops dissolving and a faint but permanent turbidity is produced; 5 ml of dilute nitric acid is added and the contents of the flask are thoroughly mixed. The Millon reagent thus prepared is usable for 2 days.

The Phenol Reference Solution. The stock solution is made by dissolving phenol in water to prepare a 1% wt./vol. solution. The solution is used for the preparation of a reference solution containing 0·025 g of phenol in 100 ml of water. The same day the solution is to be used, 2·5 ml of the stock solution is diluted with water to a volume of 100 ml.

Formaldehyde Solution. Two ml of 40% formaldehyde is diluted with water to 100 ml.

PROCEDURE. An aliquot of the solution under analysis, obtained from steam distillation (see p. 176) and containing a quantity of phenol equivalent to 3·0 ml of 0·1*N* potassium bromide-bromate standard solution, is measured from a burette into a 100 ml volumetric flask, the volume is made up to the mark and the contents are thoroughly mixed.

Five ml of this solution is measured into two test tubes (15 × 2·5 cm), denoted *A* and *B*. At the same time, 5 ml portions of the phenol reference solution is measured into two analogous test tubes, designated *C* and *D*. Five ml of Millon reagent is added from a burette to each test tube, the contents are thoroughly mixed, and heated on a water bath at 100°C for 30 min. The test tubes are then removed from the bath and cooled for 10 min in a cold water bath, 5 ml dilute nitric acid is added to each of the tubes, and the contents are mixed. To test tubes *A* and *C* 3 ml portions of formaldehyde solution are added. The contents of all the test tubes are diluted with water to a volume of 25 ml (the level corresponding to that volume should be previously marked on the tubes), thoroughly shaken, and allowed to stand overnight.

The contents of tubes *A* and *C* are tinted yellow because of the bleaching effect of formaldehyde. The solutions are used as references in the colorimetric determination described below.

Into two 100 ml volumetric flasks marked *C* and *D*, are pipetted 20 ml portions of the solutions contained in the test tubes *C* and *D*, respectively; 5 ml of dilute nitric acid is added to each of the flasks, these are filled to the mark with water and mixed, then two burettes are filled with these solutions (the burettes are also marked *C* and *D*). Thus, burette *C* contains the yellow "phenol reference", whereas burette *D* contains the "phenol reference solution", contaning 0·00001 g phenol per 1 ml.

Next, two 10 ml aliquots from tubes *A* and *B*, that is, "examined solution reference" and the "solution under analysis", are placed in two 50 ml Nessler glasses denoted respectively *A* and *B*. From burette *D* a minute quantity of the "phenol reference solution" is added to glass *A* which contains the "examined solution reference", whereas from burette *C* an equal amount of the "phenol reference" is added to Nessler glass *B*, which contains "solution under analysis". The addition of equal quantities of these solutions to respective Nessler glasses, with shaking after each addition, is continued until the same colour results in both the glasses. The operation should be carried out quickly. The reading is taken of the "phenol reference solution" volume required to match the tint in the Nessler glasses.

CALCULATION

a. Concentration of the "phenol reference solution" in burette *D*

$$(5/100)0{\cdot}025(20/25) = 0{\cdot}001 \text{ g per 100 ml or } 0{\cdot}00001 \text{ g per ml}$$

b. Volume of the "solution under analysis", *V*, introduced into the Nessler glass

$$V = 5(10/25) = 2 \text{ ml}$$

The quantity of phenol, *P*, in g per 100 ml in the "solution under analysis" amounts to

$$P = (v_1\, 0{\cdot}00001)50 = 0{\cdot}0005v_1$$

where v_1 is the volume (ml) of the "phenol reference solution" required to match the tint with that of the "solution under analysis".

The quantity of bromine, *X*, which reacts with phenol, in ml of 0·1*N* potassium bromide-bromate solution, is

$$X = P/0{\cdot}00157$$

The quantity of bromine, *Y*, which reacts with cresol will be

$$Y = Z - X,$$

where *Z* is the total bromine that reacts with 100 ml of the "solution under analysis", in ml of 0·1*N* potassium bromide-bromate solution.

The quantity of cresol, *C*, in g per 100 ml of the solution is

$$C = 0{\cdot}00236\, Y$$

From the *P* and *C* values the percentage composition of free phenols can be easily calculated.

EXAMPLE. The volume of the "phenol reference solution", required to match the colour with that of the "solution under analysis" = 8·0 ml.

1. $P = 8{\cdot}0 \times 0{\cdot}00001 \times 50 = 0{\cdot}0040$ g phenol per 100 ml of the "solution under analysis"
2. $X = 0{\cdot}0040/0{\cdot}00157 = 2{\cdot}55$ ml
   $Y = 3{\cdot}00 - 2{\cdot}55 = 0{\cdot}45$ ml
3. $C = 0{\cdot}45 \times 0{\cdot}00236 = 0{\cdot}0010$ g cresol per 100 ml of the "solution under analysis".

Hence, the composition of the mixture of free phenols amounts to 80% phenol and 20% cresol.

Determination of Formaldehyde

Free formaldehyde can be determined in the same distillate in which phenols are analysed. It should, however, be remembered that in direct steam distillation, resolic resins, in which larger amounts of free formaldehyde usually occur, may undergo further condensation with possible participation of the formaldehyde. For this reason more accurate and more reproducible results are achieved by direct determination of formaldehyde in the resin.

Free formaldehyde in the resin may be determined using hydroxylamine hydrochloride after Haslam and Soppet [46], with the Nessler reagent after Bring [47], and also by the cyanide procedure due to Pfeil and Schroth [48].

PROCEDURE [46]. A resin sample containing not more than 0·4 g of free formaldehyde is weighed into a 250 ml conical flask, and dissolved in 100 ml of 90% ethanol. The solution is neutralised with 1*N* hydrochloric acid, and then with 1*N* sodium hydroxide, using bromothymol blue indicator; 25 ml of 5% ethanolic hydroxylamine hydrochloride solution is added to the neutralised solution, and the mixture allowed to stand for 30 min. The liberated hydrogen chloride is titrated with 1*N* sodium hydroxide solution until a green-yellow colour persists. A blank is run independently.

CALCULATION. The formaldehyde content in the resin, X (%) is found from the formula

$$X = 3{\cdot}003(v_1 - v_2)/m$$

where:

v_2 = the volume (ml) of 1*N* sodium hydroxide solution consumed in titration of the unknown,

v_1 = the volume (ml) of 1*N* sodium hydroxide solution used in titration of the blank,

m = the weight of the resin sample.

Instrumental Methods

Electroanalytical Methods

The hydrogen chloride produced in the reaction of formaldehyde with hydroxylamine hydrochloride according to the equation

$$CH_2O + NH_2OH.HCl \rightarrow CH_2{=}NOH + HCl$$

may also be determined by high-frequency titrimetry [49] with an alkali solution. In this reaction water is formed simultaneously and therefore, in the analysis of non-aqueous solutions of varnishes, a more suitable variant of this method is the aquametric procedure developed by Baibaeva *et al.* [50].

Potentiometric titration by the cyanide method may also be used in the determination of formaldehyde in phenoplasts. The titration is conducted using a silver-calomel electrode system [46].

Chromatography

Isolation of unreacted phenols by steam distillation is laborious and time consuming. Moreover, in the case of resols the resin has to be removed from the distillation residue. Chromatography gives simultaneously rapid isolation and determination of contaminants present in the resin.

A paper chromatography procedure for determination of free phenol was developed by Hudeček and Beranová [51]. As the determination is carried out at room temperature, the method is suitable for both novolacs and resols.

REAGENTS

Eluent. A mixture of cyclohexane (21 ml), chloroform (9 ml) and ethanol (0·6 ml).

Diazotised p-Nitroaniline Solution. To 20 ml of *p*-nitroaniline solution (0·69 g of *p*-nitroaniline and 65 ml of 1*N* hydrochloric acid diluted to 1 l. with water) cooled to 0°C, sodium nitrite solution is quickly added dropwise until the starch-iodide paper shows as excess. The solution is unstable and has to be prepared anew before every analysis.

PROCEDURE. A 10% alcoholic resin solution is applied with a micropipette to the chromatographic paper. Also applied on the same strip are two reference phenol solutions of concentrations somewhat higher and lower respectively, compared with the unknown solution. The chromatogram is developed with the eluent given above, and the spots are made visible with the diazotised *p*-nitroaniline solution. The R_f value of phenol equals 0·35–0·37, and that of *o*-cresol, 0·69–0·71. The phenol content in the resin is found from the plot of the spot area against the logarithm of the concentration of the reference solutions. Accuracy of the method is ±5%.

In gas-liquid chromatography, on account of the rather high operating temperature, the polymer must first be removed from the sample, but an advantage of this procedure is that a variety of volatile impurities can be determined at the same time. Stevens and Percival [52] used this method for the determination of phenol and formaldehyde in phenolic adhesives in plywood.

In order to determine formaldehyde, a resin sample is first dissolved in aqueous sodium hydroxide and subsequently precipitated from the solution with an acid. The filtrate obtained after removal of the resin is introduced on to the column with addition of butanol as internal reference.

A two-column gas chromatograph is employed for the determination. The instrument incorporates a column, for determination of formaldehyde, packed with sucrose acetate (10%) deposited on Teflon, and another column designed for the analysis of phenols, packed with silicone oil deposited on Fluoropak (10 parts to 1 part by wt. respectively). In the determination of phenol, an aqueous resin solution containing *m*-cresol as internal reference, is neutralised and extracted with ether. The ether extract is introduced on to the column.

According to the authors (*loc. cit.*) phenol may be determined with an absolute error not higher than ±0·05% over a concentration range of 0–3%, and formaldehyde with an error of 0·06%.

Mlejnek [53] employed gas-liquid chromatography for the analysis of free phenols in resols based on alkyl derivatives of phenol.

Spectrophotometry

Smith *et al.* [54] demonstrated that free phenol may be determined in the presence of the resin both in novolacs and resols by infrared spectrophotometry from the band at 14·4 μm. This is the only analytical procedure which does not require separation of phenol from the resin. The measurement is made in acetone solution. According to Smith (*loc. cit.*) the presence of *m*-cresol interferes with the analysis. The other two cresol isomers do not interfere. On the other hand, the spectrophotometric procedure for determination of phenol in varnish novolacs, developed by Roczniak [55], is an indirect method and differs principally from conventional techniques only in the last analytical operation. The authoress suggests ultraviolet absorption measurement near 269 nm to replace the bromometric titration. She argues that for low phenol concentrations abnormally low results are obtained with the bromometric method with errors on occasion greater than 50%, whereas the method recommended by her enables phenol to be determined with an error not higher than 2·5%.

6.4 CHEMICAL COMPOSITION OF THE POLYMER

The chemical characteristics of a resin generally include determination of its elemental composition, hydroxyl, phenolic, and hydroxymethyl groups, and occasionally also ether and methylene bridges, active positions and alkoxyl groups.

Elemental Analysis

In the case of phenolic resins elemental analysis is of secondary importance compared with functional group analysis. Elemental analysis may furnish some additional and supplementary data that are of assistance in determining the chemical structure of resins. In such resins only the carbon and hydrogen contents are determined.

Active Sites

The active site content, that is, the number of sites capable of reacting with formaldehyde (or with hydroxymethyl groups), which occupy *ortho* and *para* positions in the phenolic rings, may be determined by the Ingberman procedure [11]. This consists in bromination with a bromine solution in acetic acid in the presence of pyridine. The conventional Koppeschaar method [41] is unsuitable, as the alkyl groups present in the phenolic resin also undergo partial bromination and even oxidation [56].

REAGENTS

Bromine. A 0·15*M* solution in glacial acetic acid.

Pyridine. A 27% solution in glacial acetic acid.

Pyridine, commercial grade (pure), is further purified by bromination by adding 10% bromine. After a time the pyridine is distilled from the mixture, then fractionally distilled over barium oxide.

PROCEDURE. A resin sample (which would take up about three millimoles of bromine) is weighed into a 250 ml iodine flask and treated with 25 ml of 0·15*M* bromine solution in glacial acetic acid. When the resin is completely dissolved the pyridine solution in acetic acid is added with a pipette and the flask is stoppered. It is allowed to stand at room temperature for about 2 min (the minimum reaction time; the reaction should not be allowed to occur for more than 20 min). About 1 ml of 27% pyridine solution in glacial acetic acid is enough to accelerate the bromination reaction efficiently, although Huber and Gilbert [57] recommend the use of 3 ml of neat pyridine for this purpose. Subsequently 5 ml of 50% potassium iodide solution is added and the liberated iodine is titrated with 0·15*N* sodium thiosulphate solution. A blank should be carried out independently.

CALCULATION. The active site content, as mequiv. per g, is calculated from the formula

$$X = n(v_1 - v_2)/m$$

where:

v_2 and v_1 = the volume (ml) of the thiosulphate titrant consumed for titration of the resin solution and the blank respectively,

n = normality of the sodium thiosulphate solution,

m = the weight of the resin sample, g.

Certain phenolic resin grades are sparingly soluble in glacial acetic acid and require the addition of dimethylformamide to effect complete dissolution. The use of dimethylformamide does not affect the results.

The reactive sites in novolacs may be determined by the Ingberman method. Resols were found by Probsthain [12] to give numerous side reactions with bromine.

The quantity of reactive positions may be roughly estimated by determining the average degree of substitution by infrared spectrophotometry on the basis of bands at 6·65, 11·4, 12·2, 12·8, and 13·3 μm [16,19,58].

Ether Bridges

Ether bridges, $—CH_2—O—CH_2$, which occur only in resols, may be detected by the Kretz method [59] which involves reaction with hydrogen chloride. The products formed as a result of scission of the bridges can be detected by the colour reaction with ferric chloride (blue to turquoise colour) or by the reaction with aqueous alkali metal hydroxide solution, when a yellow colour develops.

No conventional methods for the direct determination of ether groups in phenolic resins are presently available. Their content may be found indirectly by subtracting from the total hydroxymethyl group content (as determined by the method based on the reaction with phenol) the difference between the total hydroxyl group content and the phenolic hydroxyls content, since ether bridges react with phenol in the same way as do hydroxymethyl groups. Usually the ether bridge content is insignificant compared with the hydroxymethyl group content, and therefore the accuracy of this method is low [60].

However, ether bridges may be conveniently determined quantitatively by infrared spectrophotometry using the band at 9·3–9·4 μm [19,60]. To do this, a solution of a phenol-free resin is prepared in diacetone alcohol, and the absorption is measured at a cell thickness of about 0·1 mm.

The determination may also be made by the NMR technique after Woodbrey *et al.* [21], when acetylated products are more suitable.

Methylene Bridges

No convenient conventional methods are available for the determination in phenolic resins of $—CH_2—$ bridges, which however, can be analysed by both infrared spectrophotometry [6,18,19] and NMR spectroscopy. In the spectrophotometric determination use is made of strong bands in the region of 12·3 and 13·3 μm. The former band is due to ring substitution at positions 1, 2 and 4 and is suitable for determination of the content of bridges at the *para* position, whereas the latter, resulting from ring substitution at 1, 2 and 3 may be used to determine the content of bridges at the *ortho* position. The total methylene bridge content and the relative content of these bridges in various positions relative to the phenolic hydroxyls can be determined by the NMR method [21,61].

Total methylene and ether bridges (as a ratio of the phenolic hydroxyls to the combined formaldehyde content) in phenoplasts may be determined by infrared spectroscopy on the basis of the work by Yamao *et al.* [62] by measuring the absorbance at 6·29 μm (1590 cm^{-1}).

Hydroxyl Groups

In practice novolacs contain hydroxyl groups attached to a benzene ring, that is, only phenolic hydroxyls. Resols contain in addition alcoholic hydroxyls as part of the hydroxymethyl groups attached to the benzene ring. In phenolic resins the total hydroxyl content is generally determined, and a separate estimation is made of hydroxymethyl and phenolic hydroxyl groups.

Total Hydroxyl Groups

Hydroxyl groups in phenolic resins are usually determined by acetylation in pyridine solution by the Verley and Bolsing method [27,29,60,63,64]. According to Shuter and Berkman [65], in order to prevent fairly ready hydrolysis of the acetyl groups of hydroxymethyls, the acetylated resin is precipitated from an acetone solution with a large excess of water, followed by immediate titration with 0·1*N* sodium hydroxide solution. Hydrolysis of the esterified hydroxymethyl groups may also be avoided by addition of benzene prior to the titration, as the acetylated resin then remains in the benzene layer.

Hydroxymethyl Groups

The hydroxymethyl groups present in the resol molecules react with phenols in the presence of acid catalysts and at elevated temperatures according to the equation:

$$HO{-}C_6H_4{-}CH_2OH + C_6H_5OH \longrightarrow HO{-}C_6H_4{-}CH_2{-}C_6H_4{-}OH + H_2O$$

The method developed by Martin [66] for the determination of hydroxymethyl groups is based on this reaction. This involves separation of the water produced in the reaction by azeotropic distillation using a Dean–Stark apparatus and measurement of the aqueous layer in the receiver. *p*-Toluenesulphonic acid monohydrate is employed as catalyst.

A less accurate method is that worked out by Vansheidt [67] which is based on the same reaction and consists in the determination of the gain in weight of the resin or in the determination of the excess of unreacted phenol which is isolated from the mixture by steam distillation. As catalyst 1*N* hydrochloric acid is used.

The most accurate results in the determination of these groups by the Martin method can be obtained by the modification due to Vorobjov [27,68]. In this method, *m*-cresol is used as the phenolic reagent, and the water formed in the reaction is directly titrated with the Karl Fischer reagent.

REAGENTS

Reagent for the Reaction. In a 1000 ml round bottomed flask equipped with a Dean–Stark trap are refluxed 500 ml of *m*-cresol, 100 ml of benzene and 15 g of *p*-toluenesulphonic acid monohydrate. When the water ceases to collect in the trap, benzene is stripped off by heating to 150°C. The dark-coloured residue is kept in a bottle with a tight glass stopper.

The Karl Fischer Reagent.

APPARATUS. A flat bottomed wide-neck flask of capacity 250–300 ml, equipped with a well fitted glass-joint adapter is connected to the Karl Fischer reagent holder, and a drying tube. Electrodes are provided for titration by the dead-stop method.

Magnetic stirrer.

PROCEDURE. To a thoroughly dried Quickfit flask of capacity 250–300 ml, a finely powdered resin sample (0·3–0·5 g) is introduced and 100 ml of the reagent is added. The flask is tightly stoppered and placed in a thermostat at 90°C. After 60 min the flask is cooled, 2 ml of anhydrous pyridine is added (pyridine condensed on the flask walls should be flushed down). On recooling the stopper is replaced with the titration adapter and the water formed in the reaction is titrated with the Karl Fischer reagent in the usual manner until the potential jump. A blank, not containing resin, is run separately and the water content is determined as for the moisture originally present in the resin.

CALCULATION. The hydroxymethyl group content in the resin, ($X\%$) is calculated from the formula

$$X = \frac{1{\cdot}722\left(v_2 - v_1 - \dfrac{v_3 g}{g_1}\right)T}{g}\,100 - 1{\cdot}033x_1$$

where:

v_1 = the volume (ml) of Karl Fischer reagent used in titrating the blank,
v_2 = the volume (ml) of Karl Fischer reagent used in the titration of the unknown,
v_3 = the volume (ml) of Karl Fischer reagent used in determination of water in the resin,
g = weight of resin sample taken for determination of the hydroxymethyl groups, mg,
g_1 = weight of resin sample taken for the determination of water in the resin, mg,
x_1 = free formaldehyde content, %,
T = water equivalent of the Karl Fischer reagent, mg H_2O per ml reagent,
1·722 = molecular weight ratio of hydroxymethyl group to water,
1·033 = molecular weight ratio of hydroxymethyl group to formaldehyde.

In the above procedure (except with larger quantities of free formaldehyde) dimethylene-ether bridges, which react with phenol similarly to $—CH_2OH$ groups, interfere. Suitable corrections should therefore be made.

Finally, Bulygin *et al.* [69] developed a fast method for determination of hydroxymethyl groups which consists in their oxidation followed by potentiometric titration of the formic acid produced. The hydroxymethyl group content may also be found from the difference between the total hydroxyl and the phenolic group contents.

By the methods discussed above only the total hydroxymethyl group content can be evaluated.

The *ortho* and *para* hydroxymethyl groups can be determined separately by the procedures due to Dijkstra *et al.* One of these [9] consists in colorimetric determination of *o*-hydroxymethylphenol based on the colour reaction with iron ammonium alum solution. The content of *p*-hydroxymethylphenol is found as the difference between the total hydroxymethyl groups and the content of *ortho* hydroxymethyls.

The other method [8] involves potentiometric determination of *o*-hydroxymethylphenol by the reaction with boric acid. *p*-Hydroxymethylphenol, phenol, or formaldehyde do not interfere in this determination. The *para* CH_2OH groups are found in a similar manner as in the first of these procedures.

Phenolic Hydroxyl Groups

Phenolic hydroxyl groups, as opposed to alcoholic hydroxyls, have acidic properties. This distinction forms the basis for the majority of analytical procedures for the determination of phenolic hydroxyl groups in the presence of alcoholic hydroxyls.

Sprengling [70], and Deal and Wyld [71] determine phenolic hydroxyls by acidimetric titration in a visual or a potentiometric variant in ethylenediamine or dimethylformamide. Likewise, the method by Kämmerer *et al.* [72,73] based on the colour reaction with ferric chloride, is possible owing to the ability of phenolic groups to yield phenolate anions which form coloured complexes with ferric ions. Methanol was employed as resin solvent [72], and for sparingly soluble phenolic polycondensates a dimethylformamide–methanol mixture [73].

Shuter and Berkman [65] developed a method for determination of phenolic hydroxyls based on the Schotten–Baumann reaction with *m*-nitrobenzenesulphonyl chloride in acetone–water solution. The method was successfully used for analysis of resols and a satisfactory accuracy was achieved.

The Schotten–Baumann reaction is the basis of recent analytical procedures developed by Trochta and Vorobjov [74] and Salamatova and Zasova [75], for the determination of phenolic hydroxyls in novolacs and resols, but with the use of *p*-toluenesulphonyl chloride. The results obtained by this method are of high accuracy and precision.

6.5 ANALYSIS OF MOULDING POWDERS

In moulding powders of phenolic resins the contents of resin, filler, hexamethylenetetramine, unreacted substrates, and moisture are determined.

Resin

The resin content in phenolic moulding powders can be determined by extraction with such ketones as acetone and cyclohexanone. According to Wallhäusser [76], cyclohexanone is more suited for resolic moulding compositions which fail to dissolve completely in acetone. Certain other materials contained in the sample of moulding powder are first removed by extraction with a suitable solvent such as water for hexamethylenetetramine or benzene for pigments and lubricants.

Unreacted Substrates

To determine free phenols and formaldehyde in moulding powders these compounds are first isolated by steam distillation followed by analysis by one of the methods described previously for uncured crude resins.

Moisture

Water present in a moulding powder can be determined by drying the sample for 24 hr in a vacuum desiccator over phosphorus pentoxide or conc. sulphuric acid.

In the presence of certain fillers good results are also obtained by the Fischer method. This method fails with fibrous fillers, due possibly to a slower penetration of the solvent into the filler particles. The application range and limitations of the Dean–Stark method are the same as in the case of crude resins.

Hexamethylenetetramine

A preliminary operation in the analysis of hexamethylenetetramine in phenolic moulding compounds is its isolation from the remaining components by aqueous extraction. Hexamethylenetetramine can be determined by both direct and indirect methods in an aqueous extract.

The indirect methods are based on the hydrolysis reaction on heating with acids

$$C_6H_{12}N_4 + 2H_2SO_4 \rightarrow 6CH_2O + 2(NH_4)_2SO_4$$

Determination of formaldehyde or free acid in the reaction products gives information on the quantity of hexamethylenetetramine present in the moulding composition.

The direct methods are primarily based on the basic properties of hexamethylenetetramine which can be simply titrated with a standard acid solution.

Indirect Methods

Beranová and Hudeček [77] developed a procedure for the determination of hexamethylenetetramine based on its decomposition to ammonia and formaldehyde. The formaldehyde from the hydrolysis reaction is driven off with steam. Formaldehyde in the distillate is determined by one of the methods discussed previously. The authors [77] recommend the colour reaction with a reagent prepared by treatment of methyl violet with sodium sulphite.

In practice, procedures based on the determination of ammonia produced from the decomposition of hexamine are commonly used. Ammonia combines with some of the acid used in the hydrolysis, and after complete removal of formaldehyde by evaporation, the excess of free acid is determined in the reaction mixture, by visual titration with methyl red as indicator [60,78].

Hexamethylenetetramine can be determined semi-quantitatively by precipitating it from the aqueous extract as its tetrabromide $C_6H_{12}N_4Br_4$. The precipitate formed is heated to boiling with dilute sulphuric acid, after reducing the free bromine with sodium sulphite. The ammonia formed from hydrolytic splitting of hexamethylenetetramine is subsequently determined in the usual manner by driving it off from alkaline solution into a standard acid solution [14].

Direct Chemical Methods

The methods of determination of hexamethylenetetramine based on hydrolysis are inconvenient, as the analysis takes several hours. For this reason rapid, simple techniques of direct titrimetry are more attractive.

According to a procedure given in the Kasterina and Kalinina monograph [27], hexamethylenetetramine can be determined by visual titration with standard sulphuric acid solution. Whitehall and McAdam [79] conducted the titration with 0·1*N* perchloric acid in dioxan and a mixed indicator of bromocresol green and methyl red, whereas Baum and Goodman [80] followed the same procedure but used methyl violet in glacial acetic acid instead. In neither of the methods is hexamine extracted from the moulding powder but is directly titrated in a solution of this powder.

REAGENTS

Perchloric Acid. A 0·1*N* solution in glacial acetic acid.

Methyl Violet. A 2% solution in chlorobenzene.

PROCEDURE [80]. A 0·4 g sample of the moulding powder is dissolved in 50 ml glacial acetic acid and 2–3 drops of the indicator are added. The solution is titrated with 0·1*N* perchloric acid solution until there is a change in colour from violet to blue. The amount of hexamine present is determined on the basis of 1 mole perchloric acid being equivalent to 1 mole hexamine.

Instrumental Methods

The same authors [80] determined hexamine in phenolic moulding powders by potentiometric titration under the same conditions using a glass-calomel electrode system. There they used 0·04 g samples of the powders and dissolved them in 10 ml of glacial acetic acid. Accurate results for the determination of hexamine in phenolic moulding powders were also achieved by Vakhtel and Chernyakina by high-frequency titration [81]. Conductometric titration was successfully employed by Busfield and Chipperfield [82] in determining this amine.

6.6 CURED RESINS

In phenolic moulding materials containing the cured resin the resin, filler, degree of cure, nitrogen and water are determined. In addition, the filler is identified, and for non-laminated products the degree of filler dispersion is determined, whereas for laminates the number of layers is estimated.

Determination of Resin

Until recently resin in cured phenoplast was determined usually by the Esch and Nitsche [83] method which consists in fusion with α- or β-naphthol at 180–190°C.

The residue undissolved in naphthol is washed with a boiling methanol-acetone mixture, then with ether, and finally dried and weighed.

The Esch and Nitsche procedure is not very suitable for the analysis of modern, fully cured products, as the resins fail to dissolve completely in naphthols. Loos [84] remarks that even if he succeeded in achieving correct results, the reason was the loss of resin due to decomposition, which compensated for the error due to incomplete leaching out of the resin from the material. In similar cases better results can be achieved by the benzylamine extraction method at 200°C. The resin dissolves completely on sufficiently prolonged heating [29].

In cellulose-containing products the resin content can be determined as residue after hydrolysis with conc. hydrochloric acid. Glucose produced from hydrolysis of cellulose filler dissolves in hydrochloric acid. The residue is dried and weighed.

The resin content in cured phenoplasts was determined by flash pyrolysis gas chromatography by O'Neil [85] on the basis of the areas under the peaks of the phenolic components.

Determination of Filler

If in the phenolic plastic material under analysis only an inorganic filler is present, its content can be determined by incineration at 550–900°C. An organic filler is determined as the residue from the extraction with naphthols or with benzylamine or, in the case of cellulose, as the difference between the weight of the sample taken for the analysis and the weight of the residue after hydrochloric acid hydrolysis.

Identification of Filler

Filler identification tests are made with the residue after extraction of the resin. In the residue from the naphthol extraction, cellulose is detected by the test with zinc chloride solution and iodine. However, in the case of highly cured products obtained from phenolic moulding powders, as shown by the work by Meyer [86], the phloroglucinol test for wood flour fails, possibly on account of bonds occurring between the resin and the lignin present in the flour. The phloroglucinol test for cellulose also fails in the analysis of the residue from the benzylamine extraction.

The inorganic character of the filler may be established on ignition of a plastic sample.

Wredden [87] examined the fillers by passing polarised light through very thin plates cut out from moulding products. According to Loos [84 adequate information can be obtained on the kind of filler and its dispersion in the bulk of a moulding product from microscopic examination of finely polished sections. By this analysis, the presence of filler grains, or the presence of fibres can be detected along with the orientation of the latter. According to the same author, good results are achieved also by etching the surface of mouldings with boiling hydrogen peroxide (100 vol.) to remove the top layer of the resin. The process is run in a plain beaker covered with a watch glass. After 0·5–1·5 hr, depending on the resin type, the sample is washed for 10 min in water, then with alcohol and dried.

Determination of Low Molecular Weight Components

The bulk of the material of cured phenoplasts comprises large polymeric molecules but also incorporates certain amounts of uncured resin and other low molecular compounds which were introduced along with the substrates or were formed in the curing process. These low molecular weight components include ammonia, free phenol, formaldehyde, water, and others.

Water is determined by the Karl Fischer method. The remaining compounds are analysed in an aqueous extract made by treatment of 10 g of divided polymer with 100 ml of water at 80–90°C. The mixture is allowed to stand for 1 hr without heating and with occasional shaking, then the precipitate is removed by filtration and washed with water. The filtrate is diluted to the required volume.

In the resulting solution phenol and formaldehyde are determined by the methods described previously. In view of the low concentration of these compounds the most suitable methods are colorimetric techniques.

Determination of Ammonia [88]

Finished products made of novolac moulding compositions invariably contain ammonia formed from hexamethylenetetramine during the moulding process. The ammonia may occur both free and combined.

The term free ammonia refers to the ammonia that can be released in gaseous form. On the other hand, combined ammonia is that part of the ammonia that can be leached from the sample with water after degassing. Combined ammonia occurs only in moulding products with wood flour filler. The binding agent in such moulding powders is most likely the acidic groups present in lignin.

Free Ammonia

SAMPLE PREPARATION. Mouldings to be analysed are first broken into small pieces then finely ground in a porcelain-ball mill. The material of a grain size above 1 mm is separated, and the remaining material is further ground for 15 hr. Next, the mill is opened and heated in a drier at 70°C for 5 hr. Finally the mill is repeatedly evacuated to accelerate degassing. The grinding and degassing operations are continued until ammonia can no longer be smelt.

PROCEDURE. Parallel runs are performed for the non-ground, and ground and degassed samples in the analysis for total nitrogen content by the Kjeldahl method in the Parnas–Wagner modification [89] using 0·3 g samples. The difference obtained after recalculation represents free ammonia.

Combined Ammonia

A 0·3 g sample, prepared and degassed as described above, is stirred for 15 min with 20 ml of cold distilled water. The precipitate is then filtered off and washed with water until free of ammonium ions as checked by the Nessler reagent and along with the filter is analysed for total nitrogen content by the Kjeldahl method. The difference between the nitrogen in the degassed material and in the residue from the aqueous extraction, expressed as ammonia content, gives the combined ammonia.

References

1. Pochwalski, J., *Fenoplasty* (*Phenolic Resins*), PWT, Warsaw, 1955.
2. Whitehouse, A. A. K., Pritchett, E. G. K., Barnett, G., *Phenolic Resins*, Iliffe Books, London, 1967.
3. Freeman, J. H., *Anal. Chem.*, **24**, 955, 2001 (1952).
4. Reese, J., *Kunststoffe*, **45**, 137 (1955).
5. Schiemann, G., Hartmann, E., *Z. Anal. Chem.*, **190**, 126 (1962).
6. Šnupárek, J., Beranová, D., *Plaste u. Kautschuk*, **10**, 724 (1963).
7. Yeddanapalli, L. M., Kuriakose, A. K., Gopalakrishna, V. V., *J. Sci. Ind. Research* (*India*), **19B**, 25 (1960).
8. Dijkstra, R., de Jonge, J., *Rec. Trav. Chim.*, **76**, 92 (1957).
9. De Jong, J. I., Dijkstra, R., de Jonge, J., *Rec. Trav. Chim.*, **75**, 1289 (1956).
10. Bender, H. L., *Modern Plastics*, **30** [6], 136 (1953).
11. Ingberman, A. K., *Anal. Chem.*, **30**, 1003 (1958).
12. Probsthain, K., *Z. Anal. Chem.*, **182**, 409 (1961).
13. Zulaica, J., Guiochon, G., *J. Polymer Sci.*, *Pt. B*, **4**, 567 (1966).
14. Haslam, J., Willis, H. A., *Identification and Analysis of Plastics*, 1st ed., Iliffe Books, London, 1965.
15. Hummel, D., *Kunststoff-, Lack- und Gummi-Analyse*, Carl Hanser Verlag, Munich, 1958.
16. Richards, R. E., Thompson, H. W., *J. Chem. Soc.*, **1947**, 1260.
17. Shreve, O. D., *Synthetic Organic Coating Resins* in *Organic Analysis*, *Vol. III*, Mitchell, J., Jr., Kolthoff, I. M., Proskauer, E. S., Weissberger, A. (Eds), Interscience, New York, 1956.
18. Hummel, D. O., Scholl, F., *Atlas der Kunststoff-Analyse*, Carl Hanser Verlag, Munich, Verlag Chemie, Weinheim, 1968.
19. Secrest, P. J., *Official Digest*, **37**, 187 (1965).

20. Bellamy, L. J., *The Infrared Spectra of Complex Molecules*, Wiley, London—New York, 1958.
21. Woodbrey, J. C., Higginbottom, H. P., Culbertson, H. M., *J. Polymer Sci., Pt. A*, **3**, 1079 (1965).
22. Hirst, R. C., Grant, D. M., Hoff, R. E., Burke, W. J., *J. Polymer Sci., Pt. A*, **3**, 2091 (1965).
23. Brockmann, W., Brockmann, H., Jr., Budzikiewicz, H., *Kaut. u. Gummi Kunststoffe*, **21**, 679 (1968).
24. Wagner, E. R., Greff, R. J., *J. Polymer Sci., Pt. A-1*, **9**, 2193 (1971).
25. Urbański, J., *Chem. Anal.* (*Warsaw*), **15**, 615 (1970).
26. Urbański, J., *ibid.*, **15**, 853 (1970).
27. Kasterina, G. N., Kalinina, L. S., *Khimicheskie metody issledovaniya sinteticheskikh smol i plasticheskikh mass* (*Chemical Methods of Analysis of Synthetic Resins and Plastics*), Goskhimizdat, Moscow, 1963.
28. Schönpflug, E., *Textil-Praxis*, **7**, 897, 975 (1952).
29. Krause, A., Lange, A., *Kunststoff-Bestimmungsmöglichkeiten*, Carl Hanser Verlag, Munich, 1970.
30. Schouten, J. M., Nijveld, W. J., *Verfkroniek*, **26**, 58 (1953).
31. Feigl, F., Anger, V., *Modern Plastics*, **37** [9], 151 (1960).
32. Parriss, W. H., Holland, P. D., *Brit. Plastics*, **33**, 372 (1960).
33. Priori, O., Panetti, M., *Poliplasti e Plastici Rinforzati*, **8** [37], 19 (1960).
34. Cox, B. C., Ellis, B., *Anal. Chem.*, **36**, 90 (1964).
35. Jackson, W. M., Conley, R. T., *J. Appl. Polymer Sci.*, **8**, 2163 (1964).
36. Tsuge, M., Tanaka, T., Tanaka, S., *Japan Analyst*, **18**, 47 (1969).
37. Zowall, H., Świątecka, M., *Polimery*, **16**, 127 (1971).
38. Mosimann, H., Weber, W., *Schweizer Arch. Angew. Wiss. Tech.*, **36**, 402 (1970).
39. Braun, D., Arndt, J., *Kunststoffe*, **62**, 41 (1972).
40. Horiuchi, H., *Kogyo Kagaku Zasshi*, **66**, 147 (1963).
41. Koppeschaar, W. F., *Z. Anal. Chem.*, **15**, 233 (1876).
42. Haslam, J., Whettem, S. M. A., Newlands, G., *Analyst*, **78**, 340 (1953).
43. Dijkstra, R., Lammers, M. F., *Rec. Trav. Chim.*, **77**, 933 (1958).
44. Dzięgielewski, J. O., *Chem. Anal.* (*Warsaw*), **6**, 237 (1961).
45. Bulygin, B. M., Mokeeva, R. N., Murashov, Yu, S., *Plast. Massy*, **1968**, [6], 54.
46. Haslam, J., Soppet, W. W., *J. Appl. Chem.*, **7**, 328 (1953).
47. Bring, A., *Chem. Průmysl*, **1**, 272 (1951).
48. Pfeil, E., Schroth, G., *Z. Anal. Chem.*, **134**, 333 (1951/52).
49. Ershov, V. P., Pokrovskaya, V. L., *Plast. Massy*, **1960**, [3], 66.
50. Baibaeva, S. T., Smilga, Kh. V., Tomilova, N. D., *Lakokrasoch. Mater. ikh Primen.*, **1962**, [2], 52.
51. Huđeček, S., Beranová, D., *Chem. Listy*, **50**, 1456 (1956); *Plaste u. Kautschuk*, **4**, 88 (1957).
52. Stevens, M. P., Percival, D. F., *Anal. Chem.*, **36**, 1023 (1964).
53. Mlejnek, O., *Plasticke Hmoty Kaučuk*, **5**, 366 (1968).
54. Smith, J., Rugg, F. M., Bowman, H. M., *Anal. Chem.*, **24**, 497 (1952).
55. Roczniak, K., *Polimery*, **10**, 454 (1965).
56. Kolthoff, I. M., Belcher, R., *Volumetric Analysis, Vol. III*, Interscience, New York, 1957.
57. Huber, C. O., Gilbert, J. M., *Anal. Chem.*, **34**, 247 (1962).
58. Burke, W. J., Ruetman, S. H., Higginbottom, H. P., *J. Polymer Sci.*, **38**, 513 (1959).
59. Kretz, R., *Fette, Seifen, Anstrichmittel*, **57**, 95 (1955).
60. Riley, H. E., in *High Polymers, Vol. XII, Analytical Chemistry of Polymers*, Kline, G. M. (Ed), *Part I, Analysis of Monomers and Polymeric Materials*, Interscience, New York, 1959.
61. Yoshikawa, T., Kumanotani, J., *Makromol. Chem.*, **131**, 273 (1970).

62. Yamao, M., Watanabe, T., Tanaka, S., *Japan Analyst*, **18**, 958 (1969).
63. Bruner, H., Thomas, H. R., *J. Appl. Chem.*, **3**, 49 (1953).
64. Verley, A., Bolsing, F., *Ber.*, **34**, 3354 (1901).
65. Shuter, L. M., Berkman, Ya. P., *Ukr. Khim. Zh.*, **23**, 669 (1957).
66. Martin, H. W., *Anal. Chem.*, **23**, 883 (1951).
67. Vansheidt, A. A., *Prom. Org. Khim.*, **3**, 385 (1937).
68. Vorobjov, V., *Chem. Zvesti*, **9**, 408 (1955).
69. Bulygin, B., M., Foliforova, I. G., Morashov, Yu. S., *Plast. Massy*, **1970**, [7], 62.
70. Sprengling, G. R., *J. Am. Chem. Soc.*, **76**, 1190 (1954).
71. Deal, V. Z., Wyld, G. E. A., *Anal. Chem.*, **27**, 47 (1955).
72. Kämmerer, H., Gölzer, E., Kratz, A., *Makromol. Chem.*, **44/46**, 37 (1961).
73. Kämmerer, H., Gölzer, E., *Makromol. Chem.*, **44/46**, 53 (1961).
74. Trochta, E., Vorobjov, V., *Chem. Průmysl*, **15**, 421 (1965).
75. Salamatova, V. A., Zasova, V. A., *Plast. Massy*, **1970**, [10], 56.
76. Wallhäusser, H., *Kunststoffe*, **49**, 171 (1959).
77. Beranová, D., Hudeček, S., *Chem. Průmysl*, **8**, 218 (1958).
78. Rybnikář, F., Ditrych, Z., Klácel, Z., Ordelt, O., *Analýza a zkoušeni plastických hmot* (*Analysis and Testing of Plastic Materials*), SNTL, Prague, 1965.
79. Whitehall, K. V., McAdam, R. D., *Chem. Ind.* (*London*), **1969**, 412.
80. Baum, D. E., Goodman, K. D., *Brit. Polymer J.*, **2**, 81 (1970).
81. Vakhtel, M. I., Chernyakina, A. F., *Plast. Massy*, **1961**, [2], 65.
82. Busfield, H., Chipperfield, E. H., *Analyst*, **78**, 617 (1953).
83. Esch, W., Nitsche, R., *Kunstharze Andere Plast. Massen*, **8**, 249 (1938).
84. Loos, W., *Kunststoffe*, **56**, 222 (1966).
85. O'Neil, D. J., *Anal. Letters*, **1**, 499 (1968).
86. Meyer, B., *Plaste u. Kautschuk*, **13**, 150 (1966).
87. Wredden, J. H., *Modern Plastics*, **25**, [3], 156 (1948).
88. Hoffmann, H., *Kunststoffe*, **56**, 475 (1966).
89. Parnas, J. K., *Z. Anal. Chem.*, **114**, 261 (1938).

Chapter 7

AMINO RESINS

Amino resins (referred to also as aminoplasts or amino plastics) are the products of polycondensation of aldehydes—commonly of formaldehyde—with urea or related compounds such as melamine, thiourea, and *N*-cyanoguanidine. Included in this group are *p*-toluenesulphonamide-formaldehyde, casein-formaldehyde, and aniline-formaldehyde resins. Urea-formaldehyde and melamine-formaldehyde resins are of most commercial interest [1,2].

Aminoplasts, like phenolics, belong to the group of thermosetting resins, as the soluble and fusible condensation products on heating are converted into insoluble and infusible materials.

The products of the first stage of condensation of formaldehyde with amines take the form of aqueous syrupy solutions, or water-soluble fine-grained powders, or may be amorphous, water-insoluble materials which precipitate from the reaction solution. Their form depends on the reaction conditions, the pH of the reaction medium, the proportion of formaldehyde to amine, and on the temperature.

Amino resins are superior in certain respects to the phenolics; they are colourless, odourless, and more resistant to light. They are widely used for the manufacture of moulding powders, laminates, adhesives, varnishes, and the like. For their use in varnishes the resins are modified by etherification of the hydroxymethyl groups (which are responsible for the hydrophilic properties of the resins) usually with butyl alcohol. This process renders the resins soluble in the organic liquids employed as solvents or diluents for paints and varnishes.

7.1 STRUCTURE

Amino resins, like phenol-aldehyde resins, contain a mixture of products with different degrees of condensation, which usually have rather complex chemical structures. Whereas much progress has recently been made in the study of the chemical structure of phenolic resins, using instrumental analysis, this cannot be said of amino resins. Only the structure of the products of the preliminary stage of condensation of formaldehyde with amine components, primarily with urea or melamine, is known in any detail.

The products were successfully isolated in crystalline form and their structure was established by conventional techniques some time ago [3]. Using paper chromatography Kanto [4] found *N*-hydroxymethyl- and

N, N'-dihydroxymethylurea, and also methylenediurea and its mono- and dihydroxymethyl derivatives in the products of condensation of urea with formaldehyde. In the products of condensation of melamine with formaldehyde, hydroxymethyl derivatives of melamine representing all possible substitutions of amino hydrogen atoms by $—CH_2OH$ groups, were identified. The results are confirmed by the studies of Koeda [5] and Zigeuner and Fitz [6]. From the condensation reaction in butanol Zigeuner identified by means of paper chromatography, not only methyl urea derivatives and their methyl ethers, but products with the repeat unit

$$(—NH—\overset{\overset{\displaystyle O}{\|}}{C}—NH—CH_2—O—CH_2—)$$

that is, linear compounds containing $—CH_2OCH_2—$ bridges. The bridges undergo splitting under the action of $—NH_2$ groups at elevated temperatures to produce methylene bridges between the urea molecules.

Further information about the chemical composition of amino resins, which are the products of further condensation of precondensates, has not as yet been obtained. As with phenolic resins only their constituent functional groups and other characteristic groupings can be identified. Occasionally the way these groups are linked can also be established.

Most of these groups can be determined by chemical methods described in the following part of this chapter. However, on the basis of the results achieved merely by chemical methods nothing can usually be concluded on the way these groups are bonded. Only occasionally, for example in basic research, can the method of splitting, using 2,4-xylenol [7], be of assistance in the structural analysis of amino resins. This reagent cleaves the bonds $—H_2C—O—$ and $—CH_2N—$ present in the resin molecules, producing easily identifiable crystalline hydroxybenzyl carbamides in whose molecules linear groupings characteristic of the parent resin are present.

Much information on the structure of amino resins can be gained from the infrared studies of these materials. In the spectra of the products obtained in an acidic medium and with excess of urea relative to formaldehyde, the presence of the following bands was established in the range 700–1700 cm^{-1} [8,9,10,11]. These are characteristic of the following groups: $C=O$ at 6·1 μm (1640 cm^{-1}), known as amide band I; $C—N$ at 6·4 μm (1560 cm^{-1}), amide band II; $N—H$ at 3·0 μm (3330 cm^{-1}), and at 7·37 and 7·54 μm (1358 and 1326 cm^{-1}) $N—H$ bonds in the $—NH—CO—NH—$ grouping; methylene bridges $—N—CH_2—N—$ at 6·93 and 8·80 μm (1443 and 1136 cm^{-1}); $—NH_2$ groups at 6·25 and 8·8 μm (1600 and 1136 cm^{-1}). Also, in the spectra of products obtained in a weakly acidic medium with an excess of formaldehyde, the methylol bands occur at 6·85, 7·68, and 9·9 μm (1460, 1302, and 1010 cm^{-1}). Bands due to $—CH_2—O—CH_2—$ bridges occur at 8·5, 9·1, and 11·9 μm (1175, 1100, and 840 cm^{-1}), and finally a broadened $N—H$ group band occurs at 3·0 μm.

In the spectra of urea-formaldehyde resins made in butanol, a very strong band of the C—O—C ether bond may be found at 9·2 μm (1087 cm^{-1}) and a band due to methyl groups at 7·23 μm (1382 cm^{-1}).

Spectra of unmodified melamine-formaldehyde resins are very simple. They include a strong triazine ring band at 12·3 μm (813 cm^{-1}), a number of strong bands in the range 6–8 μm (1670–1250 cm^{-1}), also ascribable to that ring [especially the characteristic band at 6·4 μm (1565 cm^{-1})] and also to groups —N—CH_2—N— and —N—CH_2—O, and a very weak band at 9·4 μm (1065 cm^{-1}) which is assigned to —CH_2—O—CH_2— bridges.

Based on the experimental evidence presented above and the considerations on the expected mechanism of the condensation reaction, it is currently believed that the urea-formaldehyde condensates are of linear structure. In a strongly acidic medium (pH < 3) and with excess urea relative to formaldehyde, methyleneureas are produced

$$H(NH—CO—NHCH_2)_n\ NH—CO—NH_2$$

while in a less acidic medium and with excess formaldehyde (the urea to formaldehyde ratio in practice is *ca.* 1:2) the compounds formed contain methylene bridges, and also dimethylene ether bridges and hydroxymethylene groups terminating the molecular chains.

Depending on the conditions of the synthesis employed, the end amino groups of the resin molecules may not be substituted with hydroxymethyl groups. With a considerable excess of formaldehyde some of the imino groups may become substituted with hydroxymethyl groups. Mention should also be made of a possible branching of the type >N—CO—NH— though this is rather infrequent. With a very high ratio of formaldehyde to urea cyclic urones are also likely to occur in the condensates under consideration [6,7].

The condensates produced from thiourea, melamine, *N*-cyanoguanidine, *p*-toluenesulphonamide, like the urea-formaldehyde condensates, are formed through intermediate formation of hydroxymethyl compounds which subsequently react with one another or with the starting amino compounds to make methylene or methylene ether bridges. The structure of these products has been elucidated to a lesser extent than that of the condensates. It can, however, be assumed that all these resins have the same structural pattern.

On the other hand, the behaviour of aniline towards formaldehyde differs substantially from that of urea or melamine. Depending on the pH of the reaction mixture the condensates formed may be of an entirely different structure. The polymers formed in a neutral or weakly acidic medium are of a polystyrene-like structure:

```
~~~~—CH2  CH2  CH2  CH2—~~~~
      \  /  \  /  \  /
       N     N     N
       |     |     |
       Ph    Ph    Ph
```

On the other hand, in the presence of strong acids (in the first instance hydrochloric acid) polymers are formed which contain benzene rings in the main polymer chain:

$$\sim\!\!\sim\!-NH-C_6H_4-CH_2-NH-C_6H_4-CH_2-NH-C_6H_4-CH_2-\!\sim\!\!\sim$$

7.2 QUALITATIVE ANALYSIS

Characteristic of amino resins is the presence of nitrogen which can be detected qualitatively (see p. 27). Equally characteristic is their behaviour on heating. The products of pyrolysis of these resins smell of formaldehyde, ammonia, and amines and exhibit a pronounced alkaline reaction, whereas the pyrolysis products from polyurethanes are acidic with a pungent smell, while polyamides give off a smell of burning hair during thermal decomposition.

Aminoplasts differ from other nitrogen-containing resins, for instance polyamides, by having formaldehyde incorporated in the form of either hydroxymethyl groups (free or etherified) or methylene or dimethylene ether bridges. All these resins with combined formaldehyde decompose on heating with 25% sulphuric acid, and give the reaction with the Schiff reagent (p. 171). The hydroxymethyl groups can be identified separately in their reaction with phenylhydrazine and potassium ferricyanide in alkaline solution. This test is characteristic of free hydroxymethyl groups and of free formaldehyde. Other forms of combined formaldehyde do not react.

Identification of individual amino resin types, in contrast to phenolic resins, is rather simple, as it involves characteristic reactions of the amine components. The amine components of a large variety of aminoplasts differ in chemical properties, as they belong to different classes of organic compounds. Usually these compounds are first liberated by acid hydrolysis of the resins and this, therefore, is the preliminary operation in the majority of qualitative tests on amino resins.

Chemical Methods

UREA RESINS

Test with Potassium Hydroxide [12]

A small quantity of the material is heated with 14% potassium hydroxide solution in ethylene glycol. The resulting urea in the hydrolysis reaction subsequently undergoes decomposition to carbon dioxide and ammonia. The ammonia can be identified either by its characteristic smell and/or by means of a wet red litmus paper. The method is also suitable for quantitative determination of urea.

Since melamine, polyamide, and polyacrylonitrile resins do not interfere with this test, this procedure is followed to distinguish urea resins from others.

Urea resins can also be detected by using other tests such as the urease reaction [13] or the Feigl test [14] (see p. 32). The latter involves the formation of the violet nickel diphenylcarbazide salt. Low molecular weight condensates react directly. Resins with a higher degree of condensation first have to be hydrolytically split with hydrochloric acid, as the colour reaction is produced only with free urea (or thiourea). For thiourea the reaction is about 100 times less sensitive.

Urea resins may be identified by determination of the melting point of *N*-substituted derivatives. The aminolysates produced are diphenylurea (m.p. 288°C) or dibenzylurea (m.p. 167°C) when the resins are reacted with aniline or benzylamine, respectively.

Melamine Resins

Test with Thiosulphate

This test [15] is very characteristic of melamine and is of particular value in differentiating melamine from urea resins.

PROCEDURE. A small sample of resin in a micro test tube is treated with one drop of hydrochloric acid and heated on a glycerol bath at a temperature of 200°C until there is a negative reaction of the evolving vapours with Congo red indicator paper. After cooling, the residue is treated with about 50 mg of sodium thiosulphate. The mouth of the test tube is covered with a disc of Congo red indicator paper moistened with 3% hydrogen peroxide, and the contents are heated to 160°C. In the presence of melamine resin the indicator paper turns blue. Urea resins give a negative result.

Determination as Cyanuric Acid [16]

This test involves conversion of the melamine contained in the resin into cyanuric acid by hydrolysis in 45% phosphoric acid or in 4*N* hydrochloric acid. Cyanuric acid decomposes on heating to produce cyanic acid easily identifiable by its characteristic odour. Cyanuric acid present in the hydrolysis products can also be identified by its reaction with ammoniacal cupric hydroxide solution.

The test is suitable for detection of minor quantities of melamine resins in the presence of large quantities of urea resin.

PROCEDURE. A 3 g resin sample is hydrolysed by heating with 50 ml of 45% phosphoric acid in a distillation apparatus equipped with a condenser. The water evaporating from the flask during the distillation is replaced from a dropping funnel so as to maintain about 50 ml of liquid in the flask. The distillation is continued until formaldehyde is completely removed. On cooling, cyanuric acid crystallises from the reaction solution, and the crystals are isolated by filtration and identified.

If the material under investigation contains as little as 2 parts by weight of urea resin per 1 part of melamine resin cyanuric acid fails to crystallise. When this happens, a new resin sample should be heated with 4*N* hydrochloric acid and formaldehyde should be distilled off as described above.

The contents of the flask are then evaporated to dryness and the residue is dissolved in dilute ammonia solution and filtered. The filtrate is treated with excess ammoniacal cupric hydroxide solution. In the presence of cyanuric acid violet crystals of the copper ammonium salt of this acid are produced in the form of tiny clustered needles or octahedra.

Using this procedure, melamine resin can be detected even in the presence of a ninefold excess of urea resin.

Determination as Melamine

This test [17,18] involves hydrolysis of the resin under examination with 80% acetic acid. The free melamine produced is isolated from the mixture by sublimation and identified after conversion into its picrate or oxalate derivative. Melamine yields a white precipitate when treated with 5% ammonium molybdate solution.

THIOUREA RESINS

Test with Aniline

A small quantity of the resin [9,17] is heated to boiling with a tenfold excess of freshly distilled aniline for 2 hr. Ammonia and hydrogen sulphide are evolved as a result of decomposition of the thiourea in the resin. Ammonia can be detected with wet red litmus paper, while hydrogen sulphide is identified with lead acetate paper. Ammonium sulphide is deposited in the condenser. Its formation is very characteristic of thiourea resins and it can easily be identified by treatment in two parallel tests with a few drops of sodium hydroxide or hydrochloric acid respectively. In the former test ammonia is formed, and in the latter hydrogen sulphide.

On cooling the reaction mixture, crystals of diphenylthiourea precipitate, m.p. 153°C. To remove excess aniline the reaction mixture should be made acidic and extracted with ether. If thiourea resins are present, the residue after evaporation of ether is diphenylthiourea.

The Storfer Test

A small quantity of the resin [19] is boiled for several minutes with water or with some conc. phosphoric acid for 1 hr at 120°C. The aqueous solution is neutralised and filtered. The filtrate is heated to boiling for 2–3 min with several mg of cupric chloride. One drop of the clear solution is spotted on filter paper previously moistened with saturated potassium ferricyanide solution. If the sample under examination contains thiourea resin as its major component, the resulting spot is violet to blue in the centre. If thiourea is present in insignificant quantities the test is positive only when phosphoric acid is employed in the hydrolysis. The test is highly specific. Urea does not interfere.

N-CYANOGUANIDINE RESINS

Test with Silver Picrate

A resin sample [18] is extracted with hot water and to the extract is added aqueous silver picrate solution. In the presence of *N*-cyanoguanidine yellow crystals precipitate. In contrast to melamine, *N*-cyanoguanidine gives a negative reaction with picric acid, but occasionally a yellow turbidity is produced with this reagent. However, the acid hydrolysate of *N*-cyanoguanidine resins gives a positive reaction with picric acid, since the guanylurea resulting from the hydrolysis yields a sparingly soluble picrate.

Alkalimetric Test

Guanylurea resulting from the heating of uncured *N*-cyanoguanidine resins with 2*N* sulphuric acid is such a strong base that the excess sulphuric acid can be back titrated with an alkali with methyl orange as indicator. Cured resins are unaffected by perhydrol, unlike cured urea and melamine resins.

Sulphonamide Resins

The Kappelmeier Test [9]

The resin to be analysed is dissolved in benzene or in a butanol-benzene mixture, and the resulting solution is shaken for several minutes at room temperature with aqueous 4*N* potassium hydroxide solution. *p*-Toluenesulphonamide produced as a resin decomposition product passes into the alkaline layer from which free amide is precipitated by acidification with 4*N* hydrochloric acid. The product, purified by recrystallisation, melts at 137·5°C.

The Feigl Test

The test [14] involves abstraction of the sulphonic group by fusion of sulphonamides with potassium hydroxide. The reaction products are *p*-cresol (or other phenols) and sodium sulphite which is decomposed by acids to yield sulphur dioxide.

PROCEDURE. A small quantity of the well pulverised resin (or a varnish) in a small heat-resistant glass test tube is fused with a potassium hydroxide pellet over a low flame. On cooling, the melt is dissolved in 1–2 drops of water, the solution is made acidic with conc. hydrochloric acid (using the smallest excess of the acid) and the test tube is heated on a water bath. Evolving sulphur dioxide is easily identified by its characteristic odour. Alternatively sulphur dioxide can be detected with a moist paper strip impregnated with nickelous hydroxide. This green paper, if placed in the test tube in fumes containing sulphur dioxide, darkens as a result of the formation of $NiO(OH)_2$.

Phenol and its derivatives present in the mixture after acidification of the melt can be detected by coupling with diazonium salts.

A positive reaction for sulphur dioxide is also produced by sulphonic acids, including their ester and amide derivatives. For this reason plasticisers of these types which are occasionally present in varnishes, interfere with this determination.

Aniline Resins

Coupling with R-Salt or with Phenols

A thoroughly pulverised resin sample is heated to boiling with 20% sulphuric acid. The aniline produced from the hydrolysis reaction is identified by diazotisation with sodium nitrite at 0°C, followed by coupling with an R-salt solution (or with a phenol or α-naphthol solution) previously made alkaline with sodium carbonate. In the presence of aniline-formaldehyde resins a light red colour appears [20].

Aniline in the hydrolysate can also be detected with the furfural test. The hydrolysate is made neutral, then slightly acidic with acetic acid and furfural is added. A red colour appears if aniline is present.

Instrumental Methods

Chromatography

Paper chromatography and thin layer chromatography can be used successfully for identification of free amine components of the resins, which, including those of a high degree of polycondensation, are easily hydrolysed on heating with acids. Paper chromatography was used by Hoc [21] for qualitative analysis of mixtures containing urea, melamine, and ethyleneurea.

PROCEDURE. A paper strip is successively treated with the hot solvents: carbon tetrachloride, ethyl alcohol, and water to remove interfering compounds, and the resin is hydrolysed on the paper with 1% hydrochloric acid for 2 hr. The hydrolysis products are separated by developing the chromatogram with an eluent composed of the mixture (3:1:1) butyl alcohol–ethyl alcohol–water, followed by heating at 170°C. The spots are then detected with chromotropic acid solution. Using this procedure urea, melamine, and ethyleneurea can be well separated.

Plath [22] identified adhesives and other finished products of urea and melamine resins [23], using paper chromatography.

PROCEDURE. Several mg of unknown material is hydrolysed in 2–3 ml of conc. hydrochloric acid. A few drops of hydrogen peroxide are added to oxidise the formaldehyde, and the mixture is heated until a homogeneous solution is produced. The solution is applied to the paper and the chromatogram is developed using a (4:1:5) butyl alcohol–acetic acid–water mixture as eluent. The chromatogram is sprayed with 5% *p*-dimethylaminobenzaldehyde solution in isobutyl alcohol, then it is placed in hydrochloric acid fumes and heated at 50°C. The sensitivity of this method is 0·05 μg for urea and 1 μg for melamine.

Ziener [24] used paper chromatography for analysis of melamine in melamine varnishes. The technique was also employed by Zigeuner and Fitz [6], Hamada [25], Kanto [4], Koeda [5], Dušek [26,27], and Lee [11] in the separation and identification of products of the initial stages of the condensation of urea, melamine and thiourea with formaldehyde (in studies on the condensation reaction mechanism).

Column chromatography was used for the identification of aminoplasts by Tanaka *et al.* [28]. The column was packed with finely divided cellulose and the fractions obtained were identified by infrared spectrophotometry.

Thin layer chromatography was successfully used for identification of aminoplasts [11]. Chene *et al.* [29], using this technique, analysed urea and melamine resins in paper. The resin is hydrolysed in conc. hydrochloric acid with some hydrogen peroxide (30%) added. The thin layer chromatograms, on Silica Gel G, are eluted with a mixture of methanol and ethyl acetate. The dried chromatograms are sprayed with ethanolic *p*-dimethylaminobenzaldehyde. This technique was used for identification of melamine, urea, and benzoguanidine resins in paints by Willems [30]. After removal of the solvent the resins were hydrolysed in 0·1*N* hydrochloric acid and separated on a cellulose packed column using a (3:1:1) butanol–ethanol–water mixture as eluent. Finally Braun and Jung [31] developed an analytical procedure for urea, melamine, thiourea, and aniline resins, as described below.

PROCEDURE [31]. The resin is hydrolysed in 1*N* sulphuric acid, by heating for 2 hr on a water bath. The unhydrolysed residue is removed by filtration. The formaldehyde is removed from the filtrate by steam distillation and the solution is neutralised with barium hydroxide to pH 6·5. The clear solution is applied to a silica-gel coated plate. The chromatogram is developed with a (5:8:6:0·5) pyridine–benzene–acetonitrile–water mixture. The plate is then placed for 10–20 min in a carbon dioxide atmosphere to detect the chromatographic spots.

Infrared Spectroscopy

Amino resins may be fairly easily differentiated from each other on the basis of their infrared spectra [9,10,32]. This method is equally useful in distinguishing amino resins from resins of other types. Urea resins exhibit the bands characteristic of secondary amide groups at 6·1 μm (C=O) and 6·4 μm (C—N), which are also observed in the spectra of polyamides, and in addition strong bands in the range 9–11 μm, which are due to alcohol and ether groups. In the spectrum of butylated urea-formaldehyde varnishes the band at 9·2 μm, which is characteristic of ether bonds, is very strong. This band is absent in the spectra of polyamides, but it does appear in other resins containing ether groups. Melamine resins can be identified from the bands at 6·4 and 12·3 μm, which are characteristic of triazine rings. Typical bands are shown in Figs. 7.1 and 7.2.

The spectra of thiourea-formaldehyde condensates differ from the spectra of their urea counterparts by the absence of a band at 6·1 μm characteristic of C=O groups. Instead, a strong band due to C=S is observed at 6·5 μm, and a band in the region of 7·5 μm. However, the band at 6·4 μm in the spectra of urea resins may partly overlap the 6·5 μm band when a mixture of both types of resins is analysed. Consequently, the presence of thiourea will be masked. On the other hand similar interference is not encountered with the 7·4 μm band, which is probably also due to C=S groups.

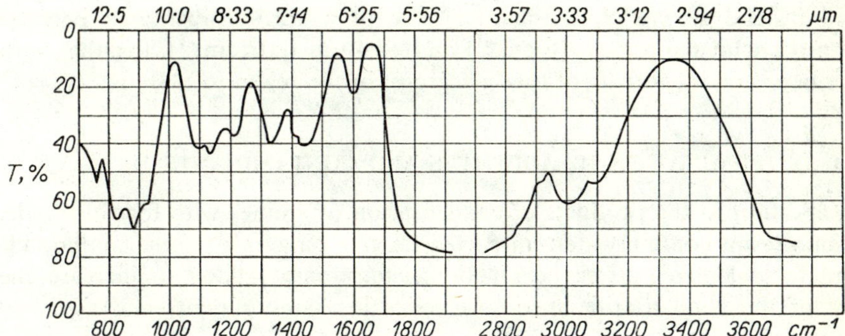

Fig. 7.1 Infrared absorption spectrum of urea-formaldehyde resin; formaldehyde urea ratio 2·2:1; film on As_2S_3 plate from aqueous solution

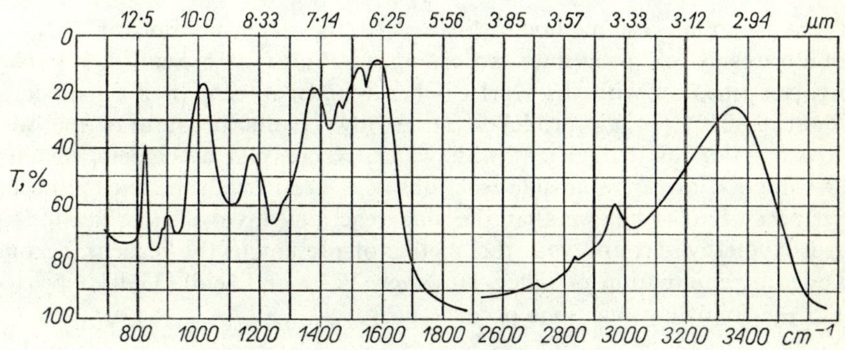

Fig. 7.2 Infrared absorption spectrum of melamine-formaldehyde resin; film on As_2S_3 plate from aqueous solution

In the spectra of sulphonamide resins bands occur at 8·65 and 7·5 μm (1156 and 1333 cm^{-1}) characteristic of —SO_2— groups. Another band at 12·2 μm (820 cm^{-1}) results from the *para*-disubstituted benzene ring. These bands are absent in the spectra of other amino resins.

N-Cyanoguanidine resins can be identified on the basis of the characteristic C≡N band at 4·6 μm (2175 cm^{-1}).

The spectra of both types of aniline resins referred to above, that is those with polystyrene-like structure or those containing benzene rings in their main polymer chains (p. 197), are very characteristic and are substantially different from each other [9]. In the spectrum of the polystyrene-like resin, bands due to the monosubstituted benzene ring occur at 13·2 and 14·5 μm (752 and 690 cm^{-1}), as well as a series of characteristic bands in the range of 5–6 μm. The band characteristic of the NH group at 3·05 μm (3279 cm^{-1}) is missing in the spectrum of this resin.

The spectrum of the aminoplast incorporating benzene rings in the main polymer chain (commercial aniline-formaldehyde resins) is much simpler. The bands due to the monosubstituted benzene ring are absent, and instead strong bands characteristic of a *para*-disubstituted benzene ring occur at 6·6 μm (1516 cm^{-1}) and 12·4 μm (807 cm^{-1}), together with the bands characteristic of the NH group in the region of 3·0 μm.

7.3 DETERMINATION OF ADDITIVES AND CONTAMINANTS

In addition to the products of condensation of amines with formaldehyde, amino resins contain water, unreacted starting materials (amine compounds and formaldehyde), free hexamethylenetetramine (added to increase the pH of the condensation medium) and other contaminants and additives in variable quantities.

Water

There is no precise method available for the determination of water in amino resins. A procedure according to Vašta and Seidl [33] which involves titration with the Karl Fischer reagent in dimethylformamide at about −40°C [34,35] produces rather low results because of the very slow reaction of the reagent with water, whereas at a temperature only 10°C higher, rather rapid side reactions of the resin with the reagent occur. Bertz *et al.* [34] believe that the side reactions involve etherification of hydroxymethyl groups with the methanol present in the reagent, accompanied by elimination of water. However, Vašta and Seidl [33] have proved that the following reactions occur:

$$\text{—NHCH}_2\text{OH} \rightarrow \text{—N=CH}_2 + \text{H}_2\text{O}$$

and

$$\text{—NHCH}_2\text{OCH}_2\text{NH—} \rightarrow 2\ \text{—N=CH}_2 + \text{H}_2\text{O}$$

According to those authors the best results are achieved when water is determined by drying the resin over phosphorus pentoxide at room temperature at reduced pressure (2 mm Hg). The method, however, is not selective, as other volatile matter contained in the resin is trapped by the drying agent along with water.

Hexamethylenetetramine

Tanaka *et al.* [36] used infrared spectrophotometry for the determination of hexamethylenetetramine in the primary products of condensation of urea with formaldehyde. Hexamethylenetetramine is first isolated from the mixture by chloroform extraction in a Soxhlet apparatus, and its concentration in the extract is determined at 8·12 μm. To eliminate absorption due to chloroform carbon disulphide is added to the extract in the ratio of 2:1 by volume. Up to 8 mg of hexamethylenetetramine in 1 ml can be determined by the above procedure, and the standard deviation is 0·015–0·020 mg.

Ash

Ash in urea-formaldehyde resins and moulding compositions is determined by igniting the sample in a muffle furnace at 800°C.

Volatile Matter

The content of volatile matter is determined together with water by drying over phosphorus pentoxide at reduced pressure.

Free Formaldehyde

On account of the peculiar properties of the resins under consideration determination of free formaldehyde is closely connected with the determination of structural units derived from it (except for aniline resins). For this reason the two determinations will be treated together (p. 206). Free formaldehyde in aniline-formaldehyde resins can be determined as in novolacs by its removal from an acidic medium by steam distillation.

Free Aniline

Free aniline in aniline resins is determined in the distillate from an alkaline medium by bromination with the bromide-bromate reagent [37]. Minor quantities of free aniline can be determined in the distillate colorimetrically by the reaction with *p*-phenylenediamine.

Free Alcohols

Amino resins always contain some methanol produced from formalin. Varnish resins may also contain butyl alcohol.

Methyl alcohol can be determined with the highest possible accuracy by evaporating at reduced pressure and trapping the vapours in a specially designed receiver cooled in "dry ice". This is followed by weighing. The procedure is also suitable for the determination of butyl alcohol.

Another method involves acylation with phthalic anhydride at 0°C in pyridine [38]. Under these conditions the acylation reaction fails to go to completion (90·6%), and a compensatory factor is introduced in the calculation of the results.

The first of the two methods is laborious, but preferable especially in the analysis of unmethylated resins containing free hydroxymethyl groups since in the phthalic anhydride method these groups also acylate and produce erroneously high results. The latter method, whose main advantage is speed, can be used only in analysis of etherified resins.

7.4 CHEMICAL COMPOSITION OF THE POLYMER

Analysis of the resins consists of elemental composition, amine content, content of structural units derived from formaldehyde (referred to as total formaldehyde), hydroxymethyl groups, dimethylene ether (or methylene ether) bridges, and alkoxy groups.

Elemental Analysis

The elemental constitution of amino resins is much more varied than that of phenolics, as, in addition to carbon, hydrogen, and oxygen there is nitrogen, and occasionally sulphur, and, in the case of casein-formaldehyde resins, phosphorus as well. Elemental analysis provides much more information on the structure of amino resins than it does in the case of phenolic resins. For instance Staudinger and Wagner [3] based their conclusions regarding the structure of urea and thiourea resins primarily on the evidence of elemental analysis.

Structural Units Derived from Formaldehyde

In the synthesis of amino-formaldehyde resins formaldehyde combines chemically with the amine groups of the other component. Depending on the conditions of the condensation reaction, either hydroxymethyl groups may form, or dimethylene ether bridges $—CH_2OCH_2—$, or else methylene bridges $—CH_2—$ from which loss of oxygen is complete. Furthermore, hydroxymethyl groups in the presence of an alcohol may undergo an etherification reaction to yield an ether group $—CH_2—O—R$. All these groupings are referred to as combined formaldehyde.

In addition to combined formaldehyde amino-formaldehyde resins always contain free formaldehyde in variable amounts. The overall content of combined and free formaldehyde is determined as total formaldehyde.

Determination of combined formaldehyde is of major importance in the analysis of aminoplasts, as on this basis the ratio of amine to formaldehyde in the resin is established.

Hydroxymethyl groups are the least strongly bonded form of formaldehyde but at the same time the most reactive. In a strongly alkaline medium the groups are split to form free amino groups and in an acidic medium they react with one another or with amino groups of other molecules to form dimethylene or methylene ether bridges.

Formaldehyde in the form of etherified hydroxymethyl groups and ether bridges is somewhat more strongly bonded. However, all of these forms of combined formaldehyde undergo cleavage to free formaldehyde when treated with alkalis even at room temperature. The most strongly bonded formaldehyde is in the form of methylene bridges which are split only on heating with strong mineral acids.

In determining free formaldehyde the reactivity of its combined forms should always be remembered. In other words, conditions of high acidity or alkalinity, as well as elevated temperatures, should be avoided.

Total Formaldehyde

Taking into account the above remarks, the determination of total formaldehyde in amino resins consists primarily in liberating the formaldehyde which is present, even in its most unreactive form as methylene bridges. To achieve this Coppa-Zucari [39] used alkaline hydrolysis (18% aqueous sodium hydroxide). It is common practice, however, to use acid hydrolysis at an elevated temperature for the determination of total formaldehyde. Liberated formaldehyde is distilled off and estimated. Thus Nerad [40] subjected amino-formaldehyde resins to hydrolysis by 1*N* sulphuric acid, and Levenson [41] resorted to phosphoric acid (1:1). According to Grad and Dunn [42] phosphoric acid is superior, as its heat of dissolution in water is lower, and it is completely non-volatile under the conditions of distillation.

The methods available in the literature also differ in the choice of solution in which the distilled formaldehyde is absorbed, and in the way formaldehyde is determined in this solution. Thus Levenson [41] distilled formaldehyde into an alkaline hydrogen peroxide solution, to oxidise it to formic acid, which was subsequently determined by alkalimetric titration. Grad and Dunn [42] collected the distillate in alkaline potassium cyanide solution. Formaldehyde reacts with cyanide to yield cyanhydrin:

$$CH_2O + KCN + H_2O \rightarrow CNCH_2OH + KOH$$

The excess of unreacted potassium cyanide was determined gravimetrically as silver cyanide. Cyanide ions can be determined volumetrically in the reaction mixture by the argentometric technique [43] or by direct titration with mercuric nitrate, using diphenylcarbazone as indicator [44]. Described below is the Grad and Dunn procedure as modified by Haslam [43].

Reagents

Alkaline Potassium Cyanide Solution. Potassium cyanide (24·8 g) and potassium hydroxide (10 g) are dissolved in a 1 l. volumetric flask in distilled water and the volume is brought up to the mark with water.

Silver Nitrate. A 0·2*N* solution.

Sodium Chloride. A 0·2*N* solution.

Procedure. A 0·5 g resin sample is weighed into a 500 ml distillation flask. The flask is connected to a condenser and receiver, which is a 500 ml volumetric flask containing 50 ml of the alkaline potassium cyanide. Into the distillation flask 50 ml of (1:1) phosphoric acid is added, and the contents are heated. The formaldehyde distils off with steam, at a rate of 65–70 drops per min; 450 ml of the distillate is collected. On completion of the distillation the receiver is chilled and made up to the mark with water; 100 ml of the resulting solution is pipetted into a beaker containing 25 ml of 0·2*N* silver nitrate solution. The mixture is made just acid with nitric acid, and the excess silver ions are determined by titration with 0·2*N* sodium chloride solution using a calomel electrode–silver electrode system (the calomel electrode is connected with a salt bridge to the solution under examination).

A blank test is run independently by diluting 50 ml of the alkaline potassium cyanide solution with water in a 500 ml volumetric flask, and repeating the same procedure as with the distillate.

CALCULATION. Total formaldehyde content *X*(%) is calculated from the formula

$$X = \frac{30{\cdot}03 \times 5(v_2 - v_1)n \times 100}{1000m} = \frac{15{\cdot}01(v_2 - v_1)n}{m}$$

where:

v_2 = volume (ml) of the 0·2*N* sodium chloride solution used in titration of the sample under examination,

v_1 = volume (ml) of 0·2*N* sodium chloride solution used for titration of the blank,

n = normality of the sodium chloride solution,

m = weight of the resin sample, g.

The method is laborious as are other methods based on hydrolysis and distillation.

Morath and Woods [35] have demonstrated that with certain modifications, consisting mainly of replacing conc. sulphuric acid by a 5*N* solution, the most suitable method for determination of minor quantities of formaldehyde, and for rapid analytical control purposes, is that developed by Bricker and Johnson [45,38]. This method is based on the reaction with chromotropic acid.

REAGENTS

Chromotropic Acid. A 10% solution. Freshly recrystallised chromotropic acid (5·0 g) is dissolved in 40 ml of water and diluted to 50 ml; 2 ml of this solution, diluted with 5*N* sulphuric acid to 250 ml, should exhibit transmission with respect to water of not less than 90%.

PROCEDURE. A 1 g resin sample is weighed into a 1 l. measuring flask, then dissolved in water and diluted to the mark with water (Solution *A*); 10 ml of this solution is pipetted into a 100 ml measuring flask and diluted to the mark (Solution *B*). Solution *B* (5 ml) is pipetted into a large test tube containing 2 ml of the chromotropic acid solution; 25 ml of sulphuric acid is then added with stirring. A blank is run independently in another test tube, taking 5 ml of distilled water in place of Solution *B*. Both test tubes are immersed in a boiling water bath for 30 min, then cooled. The contents are transferred to 250 ml measuring flasks, and diluted to the mark with 5*N* sulphuric acid. The temperature of the mixture should be 30°C. Absorbance of the solution under test is measured relative to the blank at 570 nm using 1 cm cells. The formaldehyde content (in mg) in the solution examined, that is, in 5 ml of Solution *B*, is found from the calibration curve previously prepared using reference solutions which contained 0·05–0·30 mg formaldehyde in 250 ml of the liquid.

CALCULATION. Total formaldehyde content $X(\%)$ is calculated from the formula

$$X = \frac{200a}{m}$$

where:

a = the formaldehyde content (mg) in the coloured solution found from the absorption measurement,

m = weight of resin sample, g.

FREE FORMALDEHYDE

Of the many conventional techniques for the determination of free formaldehyde in the analysis of aminoplasts the hydroxylamine and sulphite methods are commonly used. The latter is perhaps most widely accepted.

The reaction of sodium sulphite with formaldehyde produces sodium hydroxymethanesulphonate and an equivalent amount of sodium hydroxide:

$$CH_2O + Na_2SO_3 + H_2O \rightarrow HOCH_2SO_3Na + NaOH$$

Two variants of the sulphite method are known: the method [37] involving titration of the hydroxide with standard acid solution, and the iodometric method, worked out by de Jong and de Jonge [46], in which the sulphite produced from decomposition of sodium hydroxymethanesulphonate with sodium carbonate is titrated with standard iodine solution:

$$HOCH_2SO_3Na + Na_2CO_3 \rightarrow CO_2 + Na_2SO_3 + NaHCO_3$$

In view of its simplicity and speed the alkalimetric variant is commonly used [37]. The iodometric variant is resorted to whenever higher accuracy is required.

Determination by the sulphite method should preferably be performed with a parallel addition of hydrochloric acid with the sulphite to neutralise the sodium hydroxide produced in the reaction since the presence of sodium hydroxide favours elimination of hydroxymethyl groups. The lowest possible temperature should also be maintained during the determination [35]. Another way to avoid undesirable side reactions, which are likely to occur in the determination of free formaldehyde, is by previous isolation of this compound from the resin by extraction with amyl alcohol, as recommended by Petz and Cherubim [47].

The hydroxylamine method is also used for the determination of free formaldehyde. According to Smythe [48] the acid liberated in the reaction of formaldehyde with hydroxylamine hydrochloride is titrated with sodium hydroxide 30 sec after the hydroxylamine hydrochloride has been added. The time of the analysis is critical, as the longer the resin stays in the acid solution the higher are the results. According to the modified Smythe method as reported by Mobers [49] the analysis is preferably performed in a neutral medium from the outset, as the titration with sodium hydroxide solution is carried out immediately after hydroxylamine hydrochloride has been added.

For thiourea resins Probsthain [50] recommends a method involving combination of free formaldehyde with potassium cyanide in a mildly alkaline solution at a low temperature. The excess of unreacted potassium cyanide is determined iodometrically.

PROCEDURE [50]. Potassium cyanide solution (0·1*N*, 25 ml) is transferred into a 200 ml conical flask with a tight-fitting glass stopper. The flask is stoppered, the contents are cooled to a temperature of 0–4°C, and 2 drops of 1% thymolphthalein are added. The solution is neutralised by dropwise addition of 0·5*N* sulphuric acid until a faint blue colour appears. Next a resin sample (containing about 1·5 mequiv of free CH_2O) is introduced and if required sulphuric acid is added again to maintain the faint blue colour. Five minutes later the solution is made strongly acidic with conc. hydrochloric acid, followed by immediate addition of enough bromine to make the solution yellow. The excess bromine is subsequently removed by adding 1–2 ml 5% aqueous phenol solution, and 10 ml 10% potassium iodide solution is then added. The flask is stoppered and allowed to stand for 30 min. The liberated iodine is titrated with 0·1*N* sodium thiosulphate solution. A blank is run independently.

CALCULATION. Free formaldehyde content X(%) is calculated from the formula

$$X = 1{\cdot}501\,\frac{(v_1 - v_2)n}{m}$$

where:

v_2 and v_1 = the volumes (ml) of the 0·1*N* sodium thiosulphate used in titration of the solution under test and of the blank, respectively,

n = normality of the sodium thiosulphate solution,

m = weight of resin sample, g.

Polarographic Method

The polarographic method for determination of free formaldehyde was used for the first time by Crowe and Lynch [51] in their study of the condensation of urea with formaldehyde. As a supporting electrolyte a mixture of lithium hydroxide and chloride was used, that is, a strongly alkaline medium. Apparently such an alkaline medium adversely affects the accuracy of the determination, as the combined formaldehyde may be liberated as a result of splitting.

For the same reason Smythe [48] maintains a nearly neutral pH 8, so that the results are obtained with much higher accuracy. Likewise Dušek [27], in determining formaldehyde in the products of the reaction of thiourea with formaldehyde in an alkaline and in a neutral solution, uses an almost neutral supporting electrolyte (pH 8·2). In his studies on this reaction in an acidic solution Dušek (*loc. cit.*) employed a method of photometric titration (after previous conversion of formaldehyde into its oxime by treatment with hydroxylamine hydrochloride in a neutral solution).

Hydroxymethyl Groups

Hydroxymethyl groups incorporated in amino resins are bonded to the nitrogen atom $—NH—CH_2OH$, unlike the hydroxymethyl groups in phenolic resins, and undergo a quantitative cleavage in an alkaline solution even at room temperature to yield free formaldehyde:

$$—NH—CH_2OH \rightarrow —NH_2 + CH_2O$$

Thus these groups can be determined using the method employed for free formaldehyde determination in which alkaline reagents are employed. The most suitable is the cyanide method [52,53]. The iodometric method [37,54] may be used only in the absence of reducing compounds.

The Cyanide Method

REAGENTS

Potassium Cyanide. A 0·1*N* solution (0·1*M*).
Mercuric Nitrate. A 0·1*N* solution (0·05*M*).
Diphenylcarbazone. A 0·1% alcoholic solution.

PROCEDURE. A resin sample (containing *ca.* 1 mmole of $-CH_2OH$ groups), is weighed into a 200 ml conical flask and 15 ml of 0·1*N* potassium cyanide is added, followed by 5 ml of 2*N* sodium hydroxide. After 30 min the mixture is neutralised with 2*N* sulphuric or nitric acid in the presence of a few drops of diphenylcarbazone solution (until yellow). Then 10 ml of 10% aqueous pyridine is added, the mixture is neutralised with 0·5*N* sulphuric acid until yellow orange, and titrated with 0·1*N* mercuric nitrate solution until a violet colour persists. A blank is titrated in a parallel run. In the case of sparingly soluble resins the reaction should last for 2–48 hr at room temperature instead of 30 min.

CALCULATION. Hydroxymethyl group content $X(\%)$ is found from the formula

$$X = 3{\cdot}103\,\frac{(v_1 - v_2)n}{m}$$

where:

v_2 and v_1 = the volumes (ml) of mercuric nitrate solution used for titration of the resin and of the blank, respectively,
n = normality of the mercuric nitrate solution,
m = weight of resin sample, g.

In thiourea resins hydroxymethyl groups can be determined by the method devised by Probsthain [50]. This method is a modification of the cyanide procedure described above.

By the methods outlined here hydroxymethyl groups can be determined in aminoplasts only in the absence of dimethylene ether bridges, which are usually present in minor quantities, and when alkoxymethylene groups are absent, since the latter undergo cleavage to yield free formaldehyde [53]. Free formaldehyde is concomitantly determined by the foregoing methods. Hence, the methods actually determine the total —CH_2OH groups and free formaldehyde content. This should be remembered in calculating the content of —CH_2OH groups.

In this connection the —CH_2OH group content alone can be determined, according to Morath and Woods [35], simply by their etherification with anhydrous methanol in the presence of hydrochloric acid. The water formed is titrated with Karl Fischer reagent. The water present in the resin is likewise determined in a parallel run with the same reagent in dimethylformamide at a temperature of $-40°C$. The analytical procedure is described in the paper cited.

Ether Bridges

Ether bridges in amino resins can be determined by infrared spectrophotometry [55] as in phenolic resins, making use of the band at 9·3 μm. There is no other method available for the determination of ether bridges alone. In the cyanide method the bridges are determined together with hydroxymethyl groups, as they react quantitatively with potassium cyanide according to the equation:

$$—NH—CH_2OCH_2—NH— + 2KCN + 3H_2O = 2\ —NH_2 + 2CNCH_2OH + 2KOH$$

The same is true of the iodometric method.

Alkoxyl Groups

Amino resins invariably contain a certain amount of etherified hydroxymethyl groups, due to the presence of methanol in technical grade formalin. For certain uses (varnishes, finishes) amino resins are purposely etherified to render them hydrophobic and readily soluble in organic solvents.

Alkoxyl groups are commonly determined by the Zeisel method. Originally this method was employed only for the determination of methoxyl and ethoxyl groups but Houghton and Wilson [56], and later Shaw [57], modified this method achieving quantitative results with derivatives of higher alcohols as well, including butanol (but with the exception of *sec*-butanol).

Butoxyl Groups

In the flask of the apparatus [57] (Fig. 7.3) 5 ml of freshly distilled hydriodic acid (d = 1·7 g/cm^3) is placed along with 25 ml of 80% aqueous phenol. Some water is poured into the drip tube, the apparatus is assembled, and the contents of the flask are heated under reflux for 1 hr in a nitrogen stream (6 ml per min); 5 ml of the absorbing solution is placed in the absorber. This solution contains 17·5 g of potassium acetate in 227 ml of acetic acid and 2 ml of bromine. The sample is then introduced into the flask, preferably in the form of a film deposited on a glass tube. Nitrogen is passed at a rate of 1 ml per min and the flask is heated with a microburner. As soon as

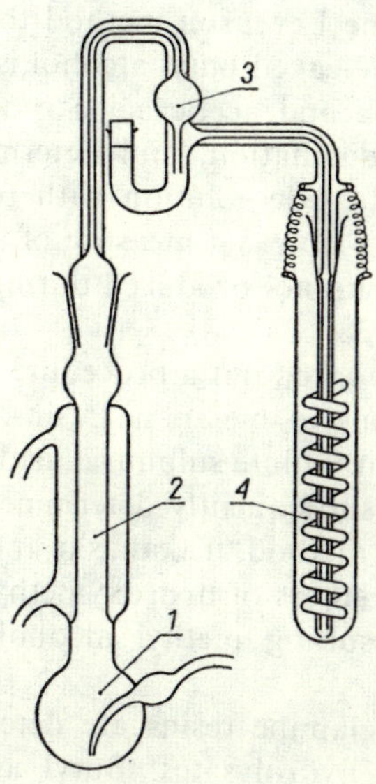

Fig. 7.3 Apparatus for determination of butoxyl groups after Shaw [57]; *1*—reaction flask, *2*—condenser, *3*—drip tube, *4*—absorber

the liquid starts boiling, the gas flow rate is raised to 6 ml per min. Nitrogen is passed for the whole period of the reaction (3 hr). After that time the absorber contents are rinsed with distilled water into a 100 ml conical flask, containing 5 ml of 25% sodium acetate solution. The colour due to the excess bromine is removed by adding 90% formic acid. Subsequently, a few drops of 10% sulphuric acid are added along with several crystals of potassium iodide. Liberated iodine is titrated with 0·1*N* sodium thiosulphate. A satisfactory end point is achieved even without addition of starch. A blank is run independently.

By filling the absorber with fresh bromine solution after every determination ten analyses can be carried out in the apparatus with the same solution in the reaction flask.

CALCULATION. The BuO group content $X(\%)$ is found from the formula

$$X = \frac{1{\cdot}2186(v_2 - v_1)n}{m}$$

where:

v_2 and v_1 = the volumes (ml) of the titrant used in titration of the solution under test and in the blank, respectively,

n = normality of the thiosulphate titrant,

m = weight of resin sample, g.

The simplicity of this method is an advantage but the results are not as accurate as those from the Levenson method [41] using chromic acid oxidation of butanol. In the Levenson method the resin is first hydrolysed with phosphoric acid. The cleaved butyl alcohol is distilled off and oxidised with chromic acid. Butyric and acetic acids produced are removed from the reaction mixture by distillation, and determined in the distillate by titration with sodium hydroxide solution with phenolphthalein indicator. The result of the titration is an exact measure of the butoxyl group content in the resin, as 1 mole of butanol produced during hydrolysis forms 1 mole of butyric or acetic acid [38].

Tanimato *et al.* [58] worked out a procedure for the analysis of methylated hydroxymethyl groups in melamine resins. This involves hydrolytic splitting of these groups with dilute sulphuric acid to produce free hydroxymethyl groups, which are subsequently determined in an alkaline solution by a modified iodometric method. Takahashi and Yamazaki [59] adopted acid hydrolysis for methyl ethers of hydroxymethyl derivatives of melamine and of urea, and the resulting methyl alcohol was determined colorimetrically.

Butoxyl groups in melamine resins are determined by Miyauchi [60] in their phosphoric acid hydrolysates. Butyl alcohol and formaldehyde resulting from the cleaved ether groups are distilled off into water. This solution is treated with silver oxide to remove formaldehyde and the butanol present is determined colorimetrically using a conc. sulphuric acid solution of *p*-dimethylaminobenzaldehyde.

Ether groups in amino resins can also be determined spectrophotometrically by measuring the absorbance at 11·1 μm in the case of MeO groups and at 12·1 μm in the case of BuO groups [61].

Determination of Amine Constituents

The commonest method for determining amine constituents in amino resins is the Kjeldahl method because of its simplicity, rapidity, and accuracy. Unfortunately, one cannot always rely on the results of this method on its own. From the total nitrogen provided by this method the amine content can be found only when a single type of amine constituent is present. If various amine constituents or components are present in the resin the analyses are conducted by the techniques presented below.

Urea

In urea resins, and also in their mixtures with melamine resins, urea is commonly determined by the Kappelmeier method subsequently modified by Grad and Dunn [42]. This method involves aminolytic splitting of the resin by heating to boiling with benzylamine:

$$(H_2N)_2C{=}O + 2H_2NH_2C{-}C_6H_5 \rightarrow O{=}C(NHCH_2{-}C_6H_5)_2$$

The resulting dibenzylurea is precipitated by acidification and determined gravimetrically. Good accuracy is obtained but the procedure is very long as the aminolysis itself requires 8 hr heating. According to Haslam [43], by using the apparatus designed by him, the reaction time can be shortened to 2 hr.

The reaction with aniline [16] used for identification of urea resins is unsuitable for quantitative analysis since the yield of the diphenylurea produced is too low (*ca.* 80%) [17,38].

Urea and its hydroxymethyl derivatives can be determined in the primary condensation products from urea and formaldehyde by the method of Landqvist [62]. This is a colorimetric method based on the yellow reaction product with *p*-dimethylaminobenzaldehyde (λ_{max} at 427 nm). The method was adopted [63] for the determination and detection [64] of urea in urea-formaldehyde resins or in their mixtures with melamine resins.

REAGENTS

The Ehrlich Reagent. *p*-Dimethylaminobenzaldehyde (2 g) is dissolved in 70 ml of ethanol, and 10 ml of conc. hydrochloric acid is added by pipette. The solution is filtered into a 100 ml measuring flask, and the volume is made up to the mark with alcohol. The resulting solution is filtered again and the first 10 ml of the filtrate is discarded.

Methanolic Hydrochloric Acid. Concentrated hydrochloric acid (50 ml) is diluted to volume in a 500 ml volumetric flask with methanol.

PROCEDURE. A 0·1 g sample of urea resin is weighed into a 100 ml round-bottomed flask, 10 ml of methanol is added, followed by 1 ml of conc. hydrochloric acid, and the mixture is refluxed for 2 hr. The walls of the condenser are then washed by pouring down several ml of methanolic hydrochloric acid solution; the contents of the flask are cooled and quantitatively transferred to a 50 ml volumetric flask, and the volume is made up to the mark with methanolic hydrochloric acid solution. A 10 ml aliquot of this stock solution is transferred to a 50 ml volumetric flask, 10 ml of the Ehrlich reagent is added, and the contents are mixed and diluted to the mark with water. The absorbance of the solution is immediately measured at 420 nm relative to the reference solution prepared by combining 20 ml of methanolic hydrochloric acid with 5 ml of the Ehrlich reagent and diluting to 50 ml with water in a volumetric flask. The urea content of the resin under analysis is found by comparing the absorbance of the test solution with the absorbance of a reference solution prepared by mixing

10 ml of a urea solution (0·1 g urea in 200 ml methanolic hydrochloric acid) with 5 ml of the Ehrlich reagent and diluting to 25 ml in a volumetric flask. The results are highly sensitive to changes in temperature and pH.

Melamine

The procedure based on the reaction of cyanuric acid with ammoniacal cupric oxide solution mentioned earlier in this Chapter (p. 198) has also been used successfully in quantitative determinations [38]. In this method the resin is first subjected to a 12 hr hydrolysis with 3*N* hydrochloric acid. In the first reaction stage melamine is formed and is subsequently hydrolysed to cyanuric acid. The hydrolysate is treated with ammoniacal cupric oxide to precipitate cupric ammonium cyanurate which is then dissolved in 2% hydrochloric acid and titrated with a standard iodine solution. However, the procedure is rather involved and laborious.

Prior hydrolysis is likewise required in more recently published analytical procedures for melamine, for example the gravimetric methods of Widmer [65] (in which melamine is converted into a sparingly soluble picrate), and of Swann and Esposito [66] in which the melamine is precipitated as its trihydrochloride. Hydrolysis is also required in the ultraviolet spectrophotometric technique developed by Hirt, King, and Schmitt [67].

Recently Knappe and Peteri [68] have succeeded in evolving a straightforward and rapid method for the determination of melamine in amino resins. The method is based on the fact that melamine residues incorporated in the resin retain the properties of a monoacidic base and do not differ in this respect from the free compound. Thus the combined melamine in the resin can be directly determined by titration with a standard acid solution without having to resort to prior hydrolysis.

Alkalimetric Method. A 0·5 g sample of the resin under analysis, which contains 50–70% of dry material, is dissolved in 20 ml of a (1:4) acetic anhydride–glacial acetic acid mixture, and a few drops of 0·1% crystal violet solution in glacial acetic acid are added. The solution is titrated with 0·1*N* perchloric acid in glacial acetic acid until a yellow colour persists.

CALCULATION. The melamine content in the resin under analysis *X*(%) is found from the formula

$$X = 1261{\cdot}3\,\frac{vn}{mP}$$

where:

v = the volume (ml) of the 0·1*N* perchloric acid used in the titration of the resin sample,

n = normality of the perchloric acid solution,

m = weight of resin sample, g,

P = the dry weight of resin (%) in the sample.

Other bases (including urea) present in the resin interfere with the analysis as they neutralise the acid, although not in proportion to their content.

Thiourea

In the majority of analytical procedures for thiourea in resins containing other amine constituents or other nitrogen-containing substances, advantage is taken of the presence of sulphur. Sulphur can be converted into sulphates by using oxidising agents, and determined as barium sulphate. The oxidation may be performed in a Parr micro bomb for example, a common procedure in elemental analysis of sulphur. Alternatively oxidation may be carried out with nitric acid [17,38].

If other sulphur-containing compounds are also present in thiourea resins, thiourea should be analysed by some other technique. To determine thiourea selectively Morath and Woods [35] took advantage of the fact that it exhibits a strong absorption in the ultraviolet region at 237 nm.

PROCEDURE. A resin sample of 0·200±0.05 g is weighed into a 250 ml flat-bottomed flask containing 10 ml of 1*N* hydrochloric acid and the reagents are refluxed for 1 hr. On cooling the contents of the flask are transferred quantitatively to a 500 ml volumetric flask and diluted to the mark with 0·1*N* hydrochloric acid (Solution *A*). A 10 ml aliquot of this solution is diluted to 200 ml in a volumetric flask with the same 0·1*N* hydrochloric acid (Solution *B*). Absorbance is measured at 237 nm with reference to the 0·1*N* hydrochloric acid.

CALCULATION. Thiourea content in the resin *X*(%) is found from the formula

$$X = \frac{A_{237}100}{a_{237}bc}$$

where:

A_{237} = the absorbance of the solution under analysis measured at 237 nm,

a_{237} = absorption coefficient of thiourea (146·9 at 237 nm determined after 1 hr in 0·1*N* hydrochloric acid),

b = cell thickness, cm,

c = resin concentration of Solution *B*, g per l.

Analysis of Moulding Powders

Aminoplast moulding powders are analysed for amine constituents, total formaldehyde, content and type of filler, hexamethylenetetramine, plasticisers, moisture and ash content.

Amine Components. In the analysis of moulding powders containing nitrogen-free fillers the urea content can be found from the Kjeldahl nitrogen determination in the crude solid separated by filtration from the benzylamine hydrolysate (p. 215). Melamine can be determined by hydrolysis of the resin with phosphoric acid, following the procedures outlined on p. 198.

Total Formaldehyde. In urea and melamine moulding compositions total formaldehyde is determined after phosphoric acid hydrolysis by the procedures described on p. 207.

Fillers. Fillers present in the urea-based moulding powders can be determined as insoluble residues after treatment with 20% aqueous acetic acid. In the case of insoluble resins one has to resort to phosphoric acid hydrolysis, or to aminolysis to break down the resin, as for cured resins (p. 219). Inorganic fillers can also be identified by igniting the moulding powder under analysis (p. 189).

Microscopic examination and application of tests similar to those made for cured resins (p. 220) are of value in establishing the kind of filler used.

Water and Ash. Water and ash in aminoplast moulding powders are determined as in the resins themselves (p. 204, 205).

Hexamethylenetetramine. Hexamethylenetetramine (or hexamine) was determined by Baum and Goodman [69] in urea-formaldehyde moulding powders by a method similar to that used for phenol-formaldehyde moulding compositions, namely by visual or potentiometric titration of the chloroform extract of the composition with standard perchloric acid.

REAGENTS

Perchloric Acid. A 0·1*N* solution in glacial acetic acid.

Methyl Violet. A 2% chlorobenzene solution.

PROCEDURE. The resin compound is finely ground and dried overnight in a vacuum desiccator over phosphorus pentoxide. For the best results the residual moisture should be less than 1% by wt. The dried sample (20 g) is weighed into a 250 ml conical flask, 30 ml of chloroform is added and the flask is stoppered and shaken for 1 hr. The suspension is filtered off on a No. 3 porosity sintered glass filter and the filter cake is washed twice with 20 ml portions of chloroform. The combined washings are evaporated to dryness preferably by means of a rotary evaporator. The residue is dissolved in glacial acetic acid and made up to 50 ml. A suitable aliquot of this solution is titrated, after addition of 2–3 drops of the indicator solution, to the methyl violet end point.

For the potentiometric titration a glass-calomel electrode system is used.

Results consistent with those obtained by both titrimetric procedures were also obtained by these authors by the NMR technique on the basis of a singlet due to twelve equivalent hydrogen atoms in the hexamine molecule at $\delta = 4{\cdot}8$ ppm.

7.5 CURED RESINS

The usual determinations in cured amino resins include the resin content itself, filler, moisture, formaldehyde, degree of cure, and occasionally free hydroxymethyl and alkoxyl groups. Furthermore, as in the case of phenolics, the filler type is identified and in the uncured products the filler dispersion in the bulk of the product or the number of layers in the laminates is established.

Resin

The resin content in cured urea and melamine products can be determined as the loss in weight in a powdered sample of the resin on complete oxidation with a (1:1) mixture of 5% acetic acid and 15% hydrogen peroxide at a temperature of 50°C, or with 30% hydrogen peroxide alone [64]. When following this procedure it should be remembered that if the oxidant is more concentrated than prescribed or if the contact time is too long the oxidant may also strongly react with the filler. The resulting errors may be as high as 12% for wood flour.

Another method for determination of the resin content in aminoplasts was evolved by Widmer [65,70], and involves determination of the amine constituents of the resin. The method, which is especially suitable for the analysis of plastics composed of urea and melamine resins, is given below.

Similarly Wise and Smith [71] determined the urea resin content in paper by urea analysis, using infrared spectrophotometry at 6·05 μm (amide II band).

Urea

A 0·5 g sample of powdered resin is heated in an autoclave for 24 hr at 160°C with an excess (10 ml) of benzylamine. Then 200 ml of anhydrous methanol is added and the mixture is reheated to dissolve the dibenzylurea formed. The filler is removed by filtration. Methanol is driven from the filtrate under reduced pressure, 100 ml of 1*N* hydrochloric acid is added, and the solution is heated for 1 min at 60°C. The precipitated dibenzylurea is collected by filtering off on a sintered glass funnel and the filter cake is washed with three 10 ml portions of water, dried for 2 hr at 105°C and weighed.

Melamine [65, 70]

A 0·5 g sample of powdered resin is heated in an autoclave with 10 ml of conc. (25%) ammonia solution for 24 hr at 160°C. After cooling, 30 ml of water is added, the excess ammonia is driven off and the filler is removed by filtration. A hot picric acid solution (2 g per 150 ml of water) is added to the filtrate and the precipitated melamine picrate, after cooling for 3 hr, is filtered off, washed twice with water (5 ml), dried at 105°C, and weighed.

Usually no problems are encountered in either method. Dissolution of the resin generally exceeds 90%.

On the basis of the urea and melamine concentrations the content of the respective resins can be roughly estimated as follows.

The urea resin content = the urea content × 1·55 to 1·75.

The melamine resin content = the melamine content × 1·55 to 1·65.

The factors appearing in the formulae depend both on the way the condensation was carried out and on the degree of cure, as well as on the amount of resin dissolved, and cannot be derived from a definite stoichiometric ratio. In doubtful cases determination of nitrogen content is advisable to check the analyses of urea and melamine [64]. It should be remembered, however, that the resin may additionally contain ammonia from the synthesis process. Synthetic fibres (such as nylon) used as fillers may also affect the total nitrogen content.

In the case of melamine-phenolic moulded products the content of the components can be determined as follows:

a. total resin content—by treatment with benzylamine or with ammonia,

b. total melamine resin content—from the nitrogen content (after previous evaporation of ammonia).

Thiourea

A sample of powdered resin is heated under reflux with conc. nitric acid. From the quantity of sulphates formed by precipitation as barium sulphate from the filtrate after filtering off the filler, the thiourea content is calculated. Other oxidising agents may also be used in this determination.

Fillers

Organic fillers are identified as residues in the determination of the resin content by the foregoing method. Inorganic fillers can also be analysed by igniting the resin as in the case of phenolic resins (p. 189).

For identification of cellulose the test with a zinc chloride and iodine solution is used. Wood flour is detected by using the phloroglucinol test. Both of these tests can reasonably be employed in examining the residue from the nitric acid–hydrogen peroxide treatment. When β-naphthol is used in the analysis only the test for cellulose can be done successfully, whereas neither test is suitable for the residue from the benzylamine treatment.

The kind of filler used, and its dispersion in the bulk of the finished moulded product, can be established from microscopic examination of polished surfaces of cross sections of these products as in the case of phenoplasts (p. 190).

Water

Water in finished moulded and laminated aminoplastic products can be determined by drying the powdered material under reduced pressure at room temperature over phosphorus pentoxide.

Free Formaldehyde

Free formaldehyde in cured amino resins can be determined by aqueous extraction as for water-insoluble resins. Sensitive methods of analysis, such as colorimetry or polarography, are most suitable for determinations in aqueous solutions in view of the low content in the substance under analysis. Thus, free formaldehyde is determined in chipboards impregnated with urea resin by heating the material for 30 min in boiling methanol in an air stream, and absorbing the formaldehyde in aqueous phenylhydrazine solution [32]. In this solution formaldehyde is subsequently identified or determined, by using a colour reaction with potassium ferricyanide. The colour develops after acidification with hydrochloric acid of a suitable concentration.

References

1. Wirpsza, Z., Brzeziński, J., *Aminoplasty* (Amino Resins), WNT, Warsaw, 1970.
2. Bachmann, A., Bertz, T., *Aminoplaste*, VEB Deutsch. Verlag, Leipzig, 1970.
3. Staudinger, H., Wagner, K., *Makromol. Chem.*, **12**, 168 (1954).
4. Kanto, *Kagaku* (*Kyoto*), **10**, 83 (1955).
5. Koeda, K., *Nippon Kagaku Zasshi*, **75**, 571 (1954).
6. Zigeuner, G., Fitz, H., *Monatsh. Chem.*, **90**, 211 (1959).
7. Zigeuner G., *Fette, Seifen, Anstrichmittel*, **57**, 14, 100 (1955).
8. Becher, H. J., *Ber.*, **89**, 1593, 1951 (1956).
9. Hummel, D., *Kunststoff-, Lack- und Gummi-Analyse*, Carl Hanser Verlag, Munich, 1958.
10. Hummel, D. O., Scholl, F., *Atlas der Kunststoff-Analyse*, Carl Hanser Verlag, Munich; Verlag Chemie, Weinheim, 1968.
11. Lee, W. Y., *Anal. Chem.*, **44**, 1284 (1972).
12. Swann, M. H., Esposito, G. G., *Anal. Chem.*, **28**, 1984 (1956).
13. Saechtling, H., *Kunststoffe*, **42**, P21 (1952); Saechtling H., *Kunststoff-Bestimmungstafel*, Carl Hanser Verlag, Munich, 1963.
14. Feigl, F., *Spot Tests in Organic Analysis*, 7th Ed., Elsevier, Amsterdam, 1966.
15. Feigl, F., Anger, V., *Modern, Plastics*, **37**, [5], 151 (1960).
16. Kappelmeier, C. P. A., Mostert, J., *Verfkroniek*, **29**, 40 (1956).
17. Kappelmeier, C. P. A., *Chemical Analysis of Resin-Based Coating Materials*, Interscience, New York, 1959.
18. Rybnikář, F., Ditrych, Z., Klácel, Z., Ordelt, O., *Analýza a zkoušeni plastických hmot*, SNTL, Prague, 1965.
19. Storfer, E., *Mikrochim. Acta*, **1**, 260 (1937).
20. Brauer, G. M., Newman, S. B., in *High Polymers, Vol. XII, Analytical Chemistry of Polymers*, Kline, G. M. (Ed), *Part III, Identification Procedures and Chemical Analysis*, Interscience, New York, 1962.
21. Hoc, S., *Allg. Pap. Rundschau*, **1962**, 650.
22. Plath, L., *Holz Roh- u. Werkstoff*, **19**, 489 (1961).
23. Plath, L., *Adhäsion*, **14**, 174, 176, 179 (1970).
24. Ziener, H., *Deutsch. Farben-Z.*, **19**, 57 (1965).
25. Hamada, M., *Kogyo Kagaku Zasshi*, **58**, 286 (1955).
26. Dušek, K., *Chem. Listy*, **50**, 1948 (1956).
27. Dušek, K., *J. Polymer Sci.*, **30**, 431 (1958).
28. Tanaka, S., Miyamoto, Y., Yoshimi, N., Matsuda, Y., *Kogyo Kagaku Zasshi*, **62**, 653 (1959).

29. Chene, M., Morel, D., Dubourgeat, M., *Papeterie*, **87**, 264 (1965).
30. Willems, J. H. J., *Verfkroniek*, **42**, 53 (1969).
31. Braun, D., Jin Chul Jung, *Gummi Asbest Kunststoffe*, **23**, 618 (1970).
32. Haslam, J., Willis, H. A., *Identification and Analysis of Plastics*, 1st Ed., Iliffe Books, London, 1965.
33. Vašta, M., Seidl, J., *Chem. Listy*, **50**, 2034 (1956).
34. Bertz, T., Neundorf, Ch., Köhler, G., *Plaste u. Kautschuk*, **10**, 84 (1963).
35. Morath, J. C., Woods, J. T., *Anal. Chem.*, **30**, 1437 (1958).
36. Tanaka, S., Miyamoto, Y., Yoshimi, N., *Bunseki Kagaku*, **5**, 87 (1956).
37. Kasterina, G. N., Kalinina, L. S., *Khimicheskie Metody Issledovaniya Sinteticheskikh Smol i Plasticheskikh Mass* (*Chemical Methods of Analysis of Synthetic Resins and Plastics*), Goskhimizdat, Moscow, 1963.
38. Averell, P. R., in *High Polymers, Vol. XII, Analytical Chemistry of Polymers*, Kline, G. M., (Ed), *Part I, Analysis of Monomers and Polymeric Materials*. Interscience, New York, 1959.
39. Coppa-Zucari, G., *Inds. Plastiques*, **4**, 183 (1948).
40. Nerad, Z., *Chem. Listy*, **44**, 35 (1950).
41. Levenson, H., *Ind. Eng. Chem., Anal. Ed.*, **12**, 332 (1940).
42. Grad, P. P., Dunn, R. J., *Anal. Chem.*, **25**, 1211 (1953).
43. Haslam, J., *Chem. Age*, **71**, 1301 (1954).
44. Ditrych, Z., *Syntetické Pryskyřice*, **5**, 240 (1955).
45. Bricker, C. E., Johnson, H. R., *Ind. Eng. Chem., Anal. Ed.*, **17**, 400 (1945).
46. De Jong, J. I., de Jonge J., *Rec. Trav. Chim.*, **71**, 890 (1952).
47. Petz, A., Cherubim, M., *Holz Roh- u. Werkstoff*, **13**, 70 (1955).
48. Smythe, L. E., *J. Am. Chem. Soc.*, **75**, 574 (1953).
49. Mobers, L. M., *Plastica*, **7**, 598 (1954).
50. Probsthain, K., *Z. Anal. Chem.*, **187**, 104 (1962).
51. Crowe, G. A., Lynch, C. C., *J. Am. Chem. Soc.*, **70**, 3795 (1948); **71**, 3731 (1949); **72**, 3622 (1950).
52. De Jong, J. I., *Rec. Trav. Chim.*, **72**, 653 (1953).
53. Seidl, J., Vašta, M., *Chem. Listy*, **50**, 2031 (1956).
54. De Jong, J. I., de Jonge, J., *Rec. Trav. Chim.*, **71**, 643 (1952).
55. Lady, J. H., Adams, R. E., Kesse, I., *J. Appl. Polymer Sci.*, **3**, 65 (1960).
56. Houghton, A. A., Wilson, H. A. B., *Analyst*, **69**, 363 (1944).
57. Shaw, B. M., *J. Soc. Chem. Ind.*, **66**, 147 (1947).
58. Tanimoto, K., Nemoto, T., Akita, T., *Kabunshi Kagaku*, **13**, 288 (1956).
59. Takahashi, A., Yamazaki, I., *Kobunshi Kagaku*, **15**, 228 (1958).
60. Miyauchi, N., *Kobunshi Kagaku*, **20**, 42 (1963).
61. Yoshimi, N., Yamauchi, T., Tanaka, S., *Kagyo Kogaku Zasshi*, **65**, 1243, 1246 (1962).
62. Landqvist, N., *Acta Chem. Scand.*, **11**, 776 (1957).
63. Adams, M. L., Swann, M. H., *Official Digest*, **31**, 1247 (1959).
64. Loos, W., *Kunststoffe*, **56**, 222 (1966).
65. Widmer, G., *Kunststoffe*, **46**, 359 (1956).
66. Swann, M. H., Esposito, G. G., *Anal. Chem.*, **29**, 1361 (1957).
67. Hirt, R. C., King, F. T., Schmitt, R. G., *Anal. Chem.*, **26**, 1273 (1954).
68. Knappe, F., Peteri, D., *Z. Anal. Chem.*, **194**, 417 (1963).
69. Baum, D. E., Goodman, K. D., *Brit. Polymer J.*, **2**, 81 (1970).
70. Krause, A., Lange, A., *Kunststoff-Bestimmungsmöglichkeiten*, Carl Hanser Verlag, Munich, 1970.
71. Wise, J. K., Smith, C. D., *Anal. Chem.*, **39**, 1702 (1967).

Chapter 8

POLYESTER RESINS

8.1 INTRODUCTION

The term "polyester" commonly refers to high molecular weight compounds containing ester links in the main chain of the molecules [1,2,3]. In other words, these compounds incorporate acidic and alcoholic entities bonded alternately to one other. The polyester molecules may have a linear (from bifunctional alcohols and acids, or hydroxy-acids) or a branched structure (if one of the constituents has a functionality greater than two).

Polyesters may be obtained in a simple way by esterification of compounds containing hydroxyl and carboxyl groups. Polycarboxylic acids and polyalcohols, or hydroxy-acids are therefore common starting materials. Besides the acids and alcohols, their derivatives such as anhydrides, esters, or acyl halides are used as feedstock for the manufacture of these plastics. Polyester resins may also be synthesised by a polyaddition reaction of anhydrides with epoxy compounds, by hydrolytic polymerisation of cyclic esters, or by polymerisation of unsaturated esters.

Several varieties of polyester resins are commonly recognised which differ from each other in their chemical constitution and in their commercial uses. These include saturated linear polyesters such as polyethylene terephthalate, polycarbonates and polyesters for the manufacture of polyurethane resins, alkyd resins of branched structure (both modified and unmodified), and also linear unsaturated polyesters that are capable of forming crosslinked structures by copolymerisation with reactive monomers. Poly(ethylene terephthalate) and polycarbonates, being linear polymers, are thermoplastic.

Poly(ethylene terephthalate) is a transparent material soluble in only a few solvents such as phenol, chlorophenol, cresols, and nitrobenzene, at elevated temperatures. The polymer is used in the manufacture of synthetic fibres (Terylene). This product is manufactured by transesterification of methyl terephthalate with ethylene glycol.

A variety of simple polycarbonates may be prepared either by polycondensation of bisphenol A with phosgene in the presence of alkalis or by ester exchange between diphenyl carbonate and bisphenol A. Polycarbonates of this type are transparent materials with a yellowish tint. They are soluble in a great number of organic solvents even at room temperature. This thermoplastic resin is gaining in popularity as a high melting, transparent plastic with good electrical properties.

Unmodified alkyd resins are prepared from polyfunctional acids and alcohols. For this reason alkyd resins are distinguished from other polyester resins by their branched molecular structure. The polyfunctional constituent is usually the alcohol. Alcohols used as starting materials include glycerol, trihydroxymethylpropane, pentaerythritol, hexanetriol, ethylene or diethylene glycol, 1,2-propanediol, and 1,3-butanediol. Phthalic anhydride or adipic acid, alone or in mixtures with sebacic acid, are common acid components.

Depending on the materials used in the synthesis and their molar ratio, and also on the degree of condensation, the alkyd resins produced may be viscous liquids or solids. Alkyd resins are readily soluble in oxygen-containing solvents, but are insoluble in hydrocarbons.

Owing to their branched structure unmodified alkyd resins are used as components of nitro- and acetylcellulose impregnating compositions, and in mixtures with phenolic, urea-formaldehyde, melamine, epoxy and polyurethane resins.

Modified alkyd resins, unlike unmodified alkyds, contain esterified carboxyl groups of monobasic, as well as dibasic, acids. The acids are constituents of both drying and non-drying vegetable oils added to modify the resin. For this reason modified alkyd resins are water-repellent and film-forming materials soluble in hydrocarbon solvents. The resins have major outlets in the paint and varnish industry.

The principal alcohols used in the manufacture of modified alkyd resins are glycerol and pentaerythritol, and less frequently 3-hydroxymethyl-2,4-pentanediol (hexanetriol) and trihydroxymethylpropane. Phthalic acid or its anhydride, and recently isophthalic acid, are the main acids used. The monocarboxylic acids employed are fatty acids contained in vegetable or fish oils.

Unsaturated polyesters are formed as products of polycondensation reactions of unsaturated dicarboxylic acids with diols. The presence of the —C=C— double bonds distinguishes them from other linear polyesters. These double bonds are capable of copolymerisation reactions with monomers to change the resin into hardened crosslinked products.

Commercial unsaturated polyesters are available in the form of solutions of the resin in the monomer, the quantity of the latter in the mixture being usually 30–40%. At the time of application the resins are treated with polymerisation initiators. As a result of the copolymerisation reaction occurring at room temperature the polyesters change into products with a crosslinked three-dimensional molecular structure.

Polyalcohols used in the manufacture of unsaturated polyesters include 1,2-propanediol, 1,3-butanediol, ethylene and diethylene glycols. Maleic anhydride, and sometimes fumaric or itaconic acids, are used as the acid components. To improve mechanical properties of unsaturated polyesters saturated dicarboxylic acids are also added, mainly in the form of their anhydrides. The most common of these is phthalic anhydride, but for special purposes tetrahydrophthalic or hexachloro-Δ^4-endomethylene-tetrahydrophthalic (HET) anhydrides are sometimes used.

The monomer most frequently involved is styrene, but vinyltoluene, methyl methacrylate, and others are also used.

8.2 STRUCTURE

Determination of the chemical structure of polyester resins is a relatively simple task in the case of polyesters composed of only one kind of glycol and one kind of dicarboxylic acid. Like all polyester resins these readily undergo hydrolysis reactions into the component acids and alcohols in the presence of a number of bases such as inorganic bases, alkali metal alkoxides, amines, and hydrazine. Acid hydrolysis is also effective. These polyester constituents may be identified easily by a number of methods. As a consequence of the polycondensation reaction the acid and glycol units alternate along the polyester chain

$$-A-B-A-B-A-B-$$

Polyester molecules are terminated by acid or alcohol functional groups, most often the latter. Terminal phenolic groups may occur in the case of polycarbonates from bisphenol A. Only in linear polyesters obtained from hydroxy-acids does the hydroxyl necessarily equal the carboxyl content. When polyesters are made from other intermediates the content of terminal groups depends on the ratio of the reagents. However, even for an alcohol to acid ratio of 1:1 the numbers of terminal groups of each kind may not be equal owing to side reactions that occur during synthesis.

The polyester resin molecules may be terminated by other groupings besides hydroxyl and carboxyl groups, depending on the polycondensation reaction conditions and the starting compounds. These other entities may be ester groups, which constitute the majority of terminal groups in polyesters made by the ester exchange method, and which also occur in modified alkyd resins, or acyl chloride groups —COCl, or OCOCl [4].

The theoretical pattern of the linear polyester structure as a "sandwich" of acid and alcohol entities is in practice much more involved owing to the side reactions occurring under the vigorous conditions of the polycondensation process. As a result of these reactions new compounds are formed which affect the reaction path and give rise to the formation of new structural elements in the resultant product. For instance, monocarboxylic acids are produced as a result of decarboxylation of dicarboxylic acids and these terminate polyester molecules. In the case of unsaturated acids used in polyester synthesis the production of dicarboxylic acids of new kinds as a result of addition of glycols to the double bonds should be taken into account [5,6]. Another side reaction worth mention is isomerisation of maleic components of unsaturated polyesters to fumaric acid units. These critically affect the properties of the resulting resins [7,8,9,10]. In the case of polycarbonates prepared by transesterification the Kolbe reaction [11] should be taken into consideration. This reaction enables crosslinking and branching to occur during the synthesis.

For determining the structure of a linear polyester of ethylene glycol and terephthalic and sebacic acids, Khramova *et al.* [12] applied NMR spectroscopy.

In the case of alkyd resins synthesised from di- and trifunctional alcohols and di- and monocarboxylic acids (the latter in modified alkyds) a branched crosslinked structure is expected. The structure of molecules of these polyesters, in contrast to linear polyesters, cannot be deduced simply from the fact that only esterification reactions are involved. Likewise, examination by solvolysis fails to provide enough information on the structure of such macromolecules because this technique enables only the total content of individual resin components to be found. Some information on the structure of alkyd resins may be obtained by classical fractionation and by adsorption chromatography techniques [13,14,15,16]. Fractionation of alkyds with polar solvents (alcohol, acetone) results in fractions differing in polarity, whereas fractionation with non-polar solvents allows separation into fractions of varying molecular weight.

Considerable attention has been given to the structure of cured resins prepared from unsaturated polyesters [17]. The degree of cure of such resins was studied by infrared spectrophotometry by Hayes *et al.* [18], Alekseeva *et al.* [19], Lomakin *et al.* [20] and Rosso [21]. Funke and Hamann [22], and Funke *et al.* [23] investigated the structure of the resins in question by chemical methods. Gilch *et al.* [24] first hydrolysed the cured resins and then determined the composition of the hydrolysates, after esterification with diazomethane, by infrared spectrophotometry, while Bohdanecký *et al.* [25] followed the same chemical procedure but used a conductometric technique. Ghanem [26] studied the composition of the solvolysis products of the cured resins by the labelled atom technique using ^{14}C-labelled maleic anhydride in the synthesis. Finally Mleziva and Vladyka [27] and also Nollen *et al.* [28] determined the molecular weights of the solvolysis products.

8.3 QUALITATIVE ANALYSIS

Qualitative analysis of polyesters most commonly starts from solvolysis of the resins, that is, the splitting of the polyester into its alcoholic and acidic constituents. Sometimes, however, solvolysis is not required. This is the case when only one of the polyester constituents is to be detected and this can be done by pyrolysis. Solvolysis may also be avoided whenever the relevant information can be obtained by non-destructive instrumental methods.

Detection of Dicarboxylic Acids

Chemical Methods

Test for Carbonate [29] *and Bisphenol A*

A sample of the resin (≈0·5 g) is hydrolysed by heating for 4 hr with 40 ml of 0·5*N* ethanolic potassium hydroxide. The precipitate formed is filtered off and treated with dilute hydrochloric acid. The evolved carbon dioxide is identified with barium hydroxide solution. The crystalline precipitate is washed with water and after drying is recrystallised from alcohol. M.p. of the crystals 153–156°C and unchanged m.p. of the mixture with authentic bisphenol A indicates the presence of polycarbonates.

Tests for Terephthalic Acid

Colour Reaction with o-Nitrobenzaldehyde. A sample of the resin is carefully heated in a test tube, at the upper part of which a paper strip is placed, moistened with a freshly prepared solution of *o*-nitrobenzaldehyde in 2*N* sodium hydroxide. A bluish-green colour develops in the presence of polyethylene terephthalate.

In the presence of adipic acid the colour produced is bluish-violet.

Identification of Terephthalic Acid after Solvolysis

1. Hydrolysis of poly(ethylene terephthalate) (PET) with aqueous sodium hydroxide solution produces terephthalic acid which, in contrast to phthalic acid, does not melt but sublimes at 300°C.
2. Hydrolysis of PET with ethanolamine yields terephthalic acid-bis-2-hydroxyethylamide which is identified by its m.p. (230°C) and nitrogen content.
3. The saponification number of PET determined by heating with potassium ethoxide is 588–602.

Tests for Phthalic Acid

Detection as Phthalic Anhydride. A sample of the resin is heated in a test tube. If phthalic acid is present phthalic anhydride deposits on the test tube walls in the form of needles of m.p. 131°C.

Detection as Thymolphthalein. A small sample of the resin (ca. 0·1 g) is heated for 10 min on a glycerol bath at 120–130°C with about 0·3 g of thymol and 1 drop of conc. sulphuric acid. After cooling, the reaction mixture is dissolved in 50% ethanol and made alkaline with dilute sodium or potassium hydroxide solution. A dark-blue colour indicates the presence of phthalic acid. In the presence of nitrocellulose the colour is green.

Detection as Phenolphthalein. A small sample of the resin (ca. 0·1 g) is carefully melted in a test tube with ca. 0·2 g of crystalline phenol and 1 drop of conc. sulphuric acid. After cooling the melt is dissolved in water (10–20 ml) and the solution made alkaline with 5% sodium hydroxide. The red colour of phenolphthalein produced in the reaction indicates the presence of phthalic acid.

Tests for Succinic Acid

Colour Reaction with Hydroquinone [30]. Several mg of the resin is heated in a test tube to 190°C with 1 g of hydroquinone and 2 ml of conc. sulphuric acid. After cooling, the reaction mixture is diluted with 25 ml of water and shaken with 50–75 ml of benzene. A red colour in the benzene layer indicates the presence of succinic acid.

To confirm the presence of succinic acid the benzene layer should be washed with water then shaken with 0·1*N* aqueous sodium or potassium hydroxide. A blue colour should develop when the test is positive. Phthalic acid interferes with this test.

Colour Test with a Pinewood Chip. The solution under test, which contains free succinic acid is neutralised with ammonia and evaporated to dryness. The residue is vigorously heated with a burner, and a pinewood chip is placed in the evolving fumes. The chip turns red when succinic acid is present.

Tests for Adipic Acid

Test with Resorcinol. A small sample of the resin is melted in a test tube with an equal amount of resorcinol and 2 drops of conc. sulphuric acid. When made alkaline the resulting melt turns a dark purple colour in the presence of adipic acid. A fluorescence may be observed in this test if any of the isomeric phthalic acids occurs in the resin.

Test with o-Nitrobenzaldehyde. The procedure is identical to that described for terephthalic acid. In this test adipic acid gives a bluish-black or bluish-violet colour.

Test for Sebacic Acid

Sebacic acid may be detected by the resorcinol colour reaction (for procedure see *Tests for Adipic Acid*, above). The solution turns orange with greenish fluorescence if sebacic acid is present. Since succinic acid gives a dark red colour with a similar fluorescence in this test, sebacic acid may be detected by the resorcinol test only when there is no succinic acid in the unknown.

Tests for Maleic Acid

Colour Test with Acetic Anhydride. A sample of the potassium dicarboxylates produced from hydrolysis of the unknown (see p. 246) is treated with 25 ml of water acidified with nitric acid, and the solution is evaporated to dryness. The dry residue is treated with 2 ml of acetic anhydride, and the whole is heated. Three drops of sulphuric acid are added after cooling the solution. When maleic acid is present, a wine-red colour develops, changing to olive-brown on standing.

Lieberman–Storch–Morawski Reaction. This test gives a wine-red colour changing to olive-brown, with pure maleic resin (see p. 28).

Tests for Fumaric Acid [31]

Reaction with Copper Sulphate and Pyridine. A sample of the potassium salts produced in hydrolysis of the resin (see p. 247) is treated with several ml of a mixture composed of 4 ml of 10% aqueous copper sulphate, 1 ml of pyridine and 5 ml of water. Emerald green crystals precipitate in the presence of fumaric acid.

Colour Test with Acetic Anhydride and Pyridine. A 0·1 g sample of the potassium salts obtained from hydrolysis of the resin (see p. 247) is treated in a test tube with 3 ml of pyridine and 1·5 ml of acetic anhydride. A dark violet colour, which develops in the presence of fumaric acid at room temperature, turns to blue on heating. Maleic acid in this test gives a pink colour in the cold, which changes to brown on heating.

Systematic Tests after Hydrolysis or Aminolysis

The method of identification and quantitative determination of individual dicarboxylic acids in the presence of other acids worked out by Swann [32] is good for the following acids: phthalic, sebacic, fumaric, maleic, adipic, and succinic. It should be recalled here that the method involves hydrolytic splitting of the resin under examination with alcoholic potassium hydroxide followed by separation of the dicarboxylic acid fraction.

The manner in which the acids are isolated and purified is essentially the same as the corresponding procedures employed for quantitative purposes. Similarly, the qualitative detection of phthalic, sebacic, fumaric, and maleic acids is based on the same reactions as the quantitative methods (see pp. 247, 248).

For the purpose of identification of the acid components, polyesters may be subjected to aminolysis with benzylamine. The *N*-benzylamides thus formed are separated by fractional precipitation. The separation is based on the poor solubility of amides of saturated acids and the good solubility of amides of unsaturated acids in 50% ethanol, and also the poor solubility of amides of either class in petroleum ether and benzene.

The isolated and purified *N*-benzylamides of individual acids may be identified by their m.p., nitrogen content and infrared spectra.

Instrumental Methods

Instrumental methods of analysis permit simultaneous detection of several dicarboxylic acids present in a polyester sample. Some of these methods even make a complete qualitative analysis possible, or at least detection of a number of constituents of the resins. The techniques widely used in analysis of polyester resin are treated below.

Paper and Thin Layer Chromatography (TLC)

The paper chromatography technique for detection of dicarboxylic acids was used by Tawn and May [33], Arendt and Schenck [34], Fijolka, Kayler, and Lenz [35], and Fijolka, Radowitz and Runge [36]. Braun and Geenon [37] employed the TLC technique.

According to Arendt and Schenck [34] the resin is hydrolysed with 0·5*N* potassium hydroxide solution in anhydrous ethanol. The separated potassium salts of the acids are filtered off, dried, and converted into free acid with an ion-exchanger. The mixture of free acids is analysed by two-dimensional chromatography using Schleicher-Schuell No. 20 436 paper and a phenol-water-formic acid mixture as eluent. The chromatograms obtained are detected with a 0·04% alcoholic bromocresol green solution.

A good separation was achieved for a mixture of terephthalic, maleic, fumaric, phthalic, glutaric, adipic, and sebacic acids. However, the ethoxysuccinic acid produced from fumaric acid under the hydrolysis reaction conditions [38] overlapped the glutaric acid. The eluent of 100 parts by vol. of benzene, 100 parts by vol. of acetone, 20 parts by vol. of formic acid, and 60 parts by vol. of water initially used to eliminate the interfering effect of ethoxysuccinic acid [35], did not yield a sharp separation. Satisfactory results were achieved by Arendt and Schenck (*loc. cit.*) by developing the chromatogram with a mixture of 100 parts of benzene, 50 parts of acetone, 20 parts of formic acid, and 60 parts of water (all components by vol.). The use of bromocresol purple in place of bromocresol green as reagent improved detectability of the acids to 5–10 μg.

Fijolka, Radowitz and Runge [36] have improved the procedure for hydrolysing polyester resins. They examined several eluents and worked out a novel technique for making the chromatogram spots visible. This proved to be more sensitive than previously published methods. To avoid formation of alkoxysuccinic acid the resins were hydrolysed with a solution of potassium hydroxide in aqueous acetone.

The following mixtures were used as eluents: acetone–benzene–water–formic acid, phenol–water–formic acid, isopropyl alcohol–ammonium carbonate–ammonia.

The best results were obtained with the first mixture. Furthermore, the acid-base indicators of low sensitivity were replaced by a mixture composed of 5% aqueous silver nitrate and 10% alcoholic vanillin solutions, thereby lowering the limit of detectability of the acids to 4 μg.

Described below is the method of identification of dicarboxylic acids after Fijolka *et al.* [36].

REAGENTS AND SOLUTIONS

Eluents

a. Acetone–benzene–water–formic acid. Acetone (400 ml) is mixed with 400 ml of benzene, 240 ml of water, and 80 ml of conc. formic acid. The mixture is shaken and allowed to separate. The two-phase mixture is used, and the paper is immersed in the bottom layer for 2 min.

b. Phenol–water–formic acid. To 750 g of phenol 250 ml of water and 1·25 ml of conc. formic acid are added.

c. Isopropyl alcohol–ammonium carbonate–ammonia. To 750 ml of isopropyl alcohol is added 250 ml of a solution containing 220 ml of 25% ammonia and 35 g of ammonium carbonate in 1 l. of water.

Reagent for the Detection of Chromatograms. A 5% solution of silver nitrate in distilled water is combined immediately before use with an equal volume of a 10% ethanolic vanillin solution.

PROCEDURE

Hydrolysis of the resin. The resin (0·3 g) is dissolved in 25 ml of acetone, an excess of a 0·5*N* aqueous potassium hydroxide solution is added, and the mixture is shaken for 1–2 hr at room temperature. The reaction mixture is neutralised with a 0·5*N* hydrochloric acid, and passed through a cation-exchange column (Wofatite KPS 200, height 10 cm, diameter 2·5 cm) from which it is eluted with acetone. The eluate containing free acids is evaporated under reduced pressure at 40°C. The residue is dissolved in acetone and the solution is transferred to a 25 ml graduated flask.

Chromatographic separation. The acetone solution of free acids is placed on chromatographic paper, and the chromatogram is developed with one of the eluents mentioned above. The eluents are complementary for the separation of a mixture of dicarboxylic acids. The best results are obtained with eluent *a*. The development of the chromatogram takes 18 hr at 20°C.

The developed chromatograms are dried for 3–4 hr in the air in order to remove residual formic acid prior to detection. After drying they are sprayed with the silver nitrate–vanillin solution and dried again in air at room temperature. Care should be taken not to expose the chromatogram to direct sunlight but only to weak ultraviolet radiation. This can be achieved by adequate artificial lighting of the chromatograms and protection against sunlight. After 2 hr the dicarboxylic acids present appear as white spots on a light brown background.

As shown by Braun and Geenon [37] thin layer chromatography can be successfully used in the analysis of the very complex mixtures of dicarboxylic acids commonly encountered in polyester resins and plasticisers. The acids, in the form of ammonium salts, are deposited on plates coated with silica gel, using a mixture of alcohol, water, and 25% ammonia (25:3:4).

Gas-Liquid Chromatography

Chromatographic methods are very valuable techniques for the identification of the components of polyester resins especially when neither the chemical properties nor the infrared and ultraviolet spectra of the acidic and alcoholic constituents of the resin are substantially different. Of these methods gas-liquid chromatography offers particular advantages. It is fast (the time required for a determination is usually less than an hour) and ensures high resolution. Application of this technique to the analysis of alkyd resins is reviewed by Haken [39].

Owing to the low volatility of the majority of components of polyester resins only the lower acids, which are rather seldom encountered, and lower glycols can be directly determined by this method. For this reason the polyester components have to be converted into volatile compounds prior to analysis by this technique. This is done by esterification or by pyrolysis reactions. When the components are rendered volatile by esterification, the alcoholic fraction is analysed separately from the acidic fraction after the initial chemical treatment. When pyrolysis is used all the polyester components can be detected in a single operation.

For identification of the acidic components of alkyd resins a transesterification reaction was used by Esposito and Swann [40], using a methanolic lithium methoxide solution. The esters of both mono- and dicarboxylic acids thus produced were isolated from the reaction mixture by extraction with methylene chloride. When fumaric acid was to be differentiated from maleic acid, methanolic boron trifluoride solution was employed in the transesterification reaction.

Two temperature-programmed columns were used by Esposito and Swann. One contained Carbowax 20 M and the other silicone rubber as stationary phase. By using the second column, isophthalic acid could be separated from phthalic acid. This also gave better resolution for the separation of maleic from fumaric acid and lauric from adipic acid.

REAGENTS

Lithium methoxide. A 0·5*N* solution in methanol.

Boron trifluoride. A methanolic solution. The concentration of this solution should be such that 1 ml is neutralised by 11–12 ml of 0·5*N* potassium hydroxide.

PROCEDURE. A solution containing 0·3 g of the resin is refluxed with 15 ml of lithium methoxide in methanol until completely dissolved and thereafter for 2 min more; 5 ml of 6*N* sulphuric acid is added through the condenser, the contents of the flask are transferred to a separating funnel, and 50 ml of water and 35 ml of methylene chloride are added. After careful shaking the organic layer is separated and washed with water. The solvent is removed and the residue is introduced onto the chromatographic column.

By this method, employing a Carbowax 20 M column, a good separation was obtained for the following acids in order of their retention times: pelargonic, succinic, benzoic, fumaric, maleic, lauric, adipic, itaconic, diglycollic, myristic, azelaic, palmitic, sebacic, phthalic, isophthalic, stearic, oleic, linoleic, and linolenic. The presence of modifying resins of the urea, melamine, phenolic, or nitrocellulose types does not interfere with the analysis.

If the presence of maleic or fumaric acids is suspected, the transesterification with boron trifluoride catalyst should be repeated as before on another sample of approximately 0·3 g, using 5 ml of the reagent. The mixture should be boiled for 5 min, then transferred to a separating funnel with 50 ml of water and 35 ml of methylene chloride and shaken vigorously. Finally the methylene chloride layer is filtered and the solvent removed by evaporation.

Rawlinson and Deeley [41] analysed polyester resins by gas chromatography after transesterification with methyl acetate at 175°C.

Volatilisation of the polyester components by pyrolysis was carried out by Sadowski and Kaufeld [42]. The pyrolysis products were separated at 220°C on a column with a polyester of succinic acid as stationary phase. The assignments of the chromatographic peaks, however, could not be easily made for individual acid components except phthalic acid. The kind of fatty acid in the resin has no effect on the chromatogram of the pyrolysis products. On the basis of the chromatograms alone it is also difficult to establish if the alkyd resin is modified with the pure fatty acid and glycerol or with the corresponding oil only.

The method of pyrolysis gas chromatography proved to be more successful in the analysis of cured laminated polyester resins [43] which had as acid components maleic, phthalic and isophthalic acids. Substantially diffe rent chromatograms were obtained for various acid components, enabling them to be distinguished from one another.

Infrared Spectroscopy

Dibasic acids are the polyester components which are perhaps the easiest to identify by direct infrared spectroscopy [44,45]. This is especially true of aromatic acids. Particularly characteristic are the spectra of poly-*o*-phthalates (Fig. 8.1 and 8.2). There are bands at 5·8 μm (1724 cm^{-1}) from C═O groups, at 7·8 μm (1282 cm^{-1}) from C—O—C bonds, and the bands from the aromatic ring at: 6·25 μm (1600 cm^{-1}) and 6·35 μm (1575 cm^{-1})

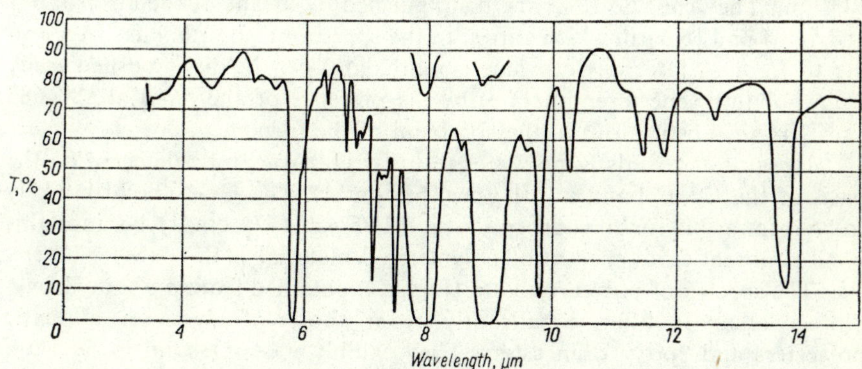

Fig. 8.1 Infrared absorption spectrum of poly(ethylene terephthalate) [44] (by permission, from *Kunststoff-, Lack- und Gummi-Analyse*, Carl Hanser Verlag, München, 1958)

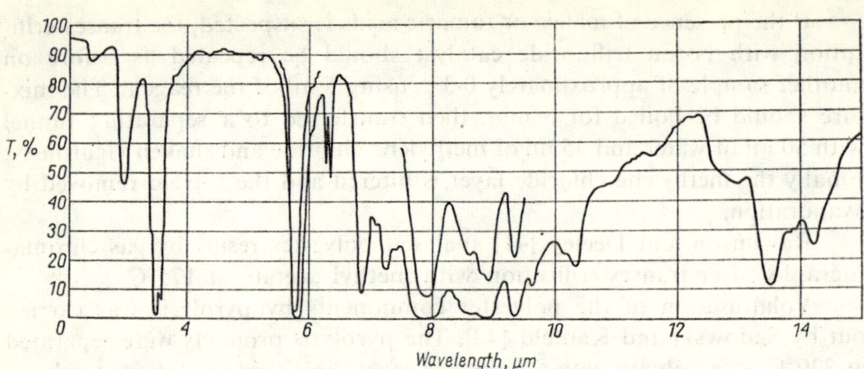

Fig. 8.2 Infrared absorption spectrum of Alphthalat 423D alkyd resin (a polyphthalate resin modified with linseed oil in 65%) [44]; film on NaCl plate (by permission, from *Kunststoff-, Lack- und Gummi-Analyse*, Carl Hanser Verlag, München, 1958)

(doublet), 8·6 μm (1163 cm^{-1}), 9·35 μm (1070 cm^{-1}), 13·5 μm (741 cm^{-1}) and 14·2 μm (704 cm^{-1}) [44].

The spectra of polyisophthalates contain bands at 5·85 μm (1710 cm^{-1}) (C=O), at 7·7 and 8·1 μm (1299 and 1235 cm^{-1}) (C—O—), and also at 6·23 and 6·32 μm (1605 and 1581 cm^{-1}) (doublet). Other bands are observed at 8·6 μm (1163 cm^{-1}), 8·85 μm (1130 cm^{-1}), 9·15 μm (1093 cm^{-1}) 9·3 μm (1076 cm^{-1}), 13·7 μm (730 cm^{-1}), and a weak band at 15·3 μm (653 cm^{-1}).

The spectra of polyterephthalates exhibit characteristic absorption bands at 5·82 μm (1718 cm^{-1}) (C=O), 7·9 and 9·0 μm (1299 and 1111 cm^{-1}) (C—O—) and at 9·8, 11·5, and 13·7 μm (1022, 869, and 730 cm^{-1}).

Isophthalates and terephthalates may be distinguished easily from *o*-phthalates by means of their infrared spectra. On the other hand, both isophthalates and terephthalates exhibit a characteristic absorption at 13·7 μm. They can, however, be distinguished by means of the absorptions at 7·7, 8·1 and 15·3 μm which appear in the spectra of isophthalates. According to Luongo [46] poly(ethylene isophthalate) can be distinguished easily from poly(ethylene terephthalate) by the presence of the band at 680–685 cm^{-1} (*meta*-substitution) in the spectrum of the former.

Unsaturated acids in polyester resins (Fig. 8.3) can be detected on the basis of the C=C band at 6·1 μm (1639 cm^{-1}). The C=C bond is likely to be responsible for the absorption at 12·95 μm (772 cm^{-1}) because this band is absent from spectra of polymerised material [44].

The spectra of polycarbonates (Fig. 8.4) obtained from *p,p'*-dihydroxydiphenylalkanes differ significantly from those of common aliphatic polyesters and epoxy resin esters. They exhibit a C=O band at 5·65 μm (1770 cm^{-1}), characteristically shifted towards shorter wave lengths on account of the neighbouring phenyl ring, and a band at 12·05 μm (830 cm^{-1}) of medium intensity characteristic of *para*-substitution. The bands due to

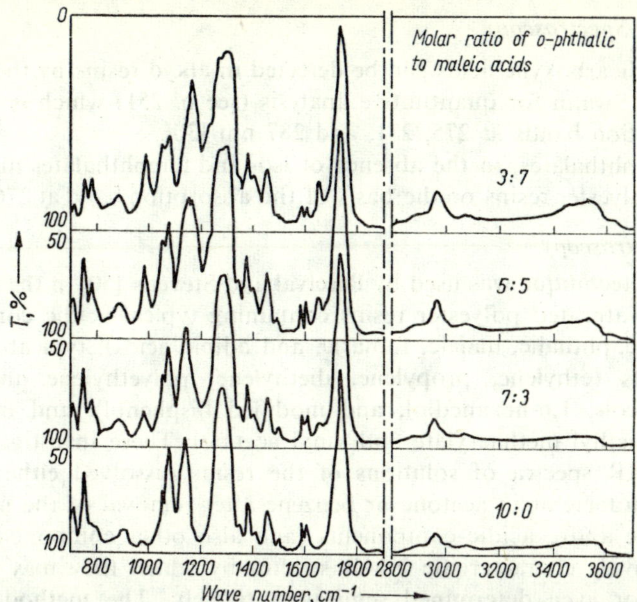

Fig. 8.3 Infrared absorption spectra of unsaturated polyester resins (from ethylene glycol and a mixture of phthalic and maleic anhydrides) [47]; films on NaCl plates from ethyl acetate solution (by permission of the copyright holders)

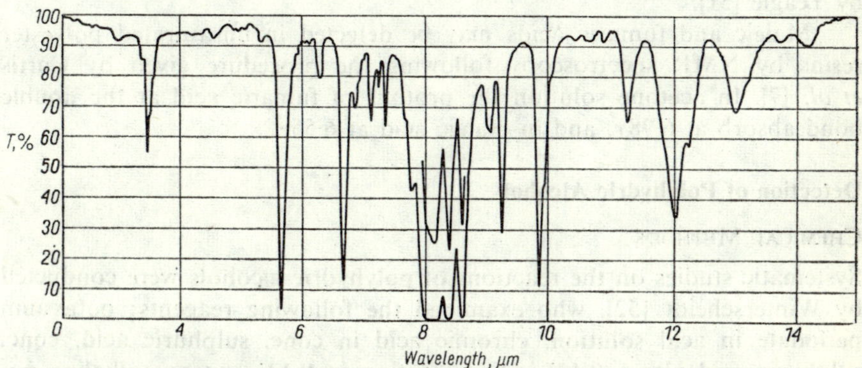

Fig. 8.4 Infrared absorption spectrum of polycarbonate derived from bisphenol A [44]; film from ethylene dichloride solution (by permission, from *Kunststoff-, Lack- und Gummi-Analyse*, Carl Hanser Verlag, München, 1958)

C—O— are clearly split (8·15, 8·4, and 8·6 μm or 1227, 1190 and 1163 cm^{-1}) [44].

Identification of saturated polybasic fatty acids in polyesters of more complex structure is rather difficult. When interpretation of infrared spectra of polyesters is ambiguous the resins should be subjected to solvolytic splitting. The resulting free acids or their salts, amides or esters may then be identified, using, for example, the infrared method as described by Stafford *et al.* [48].

Ultraviolet Spectroscopy

Aromatic dicarboxylic acids can be detected in alkyd resins by the method devised by Swann for quantitative analysis (see p. 251) which is based on the absorption bands at 275, 281, and 287 nm [30].

Orthophthalates, in the absence of iso- and terephthalates may be detected in polyester resins on the basis of the absorption band at 276 nm [49].

NMR Spectroscopy

The NMR technique was used by Percival and Stevens [50] in the examination of unsaturated polyester resins containing typical acidic components (isophthalic, phthalic, maleic, fumaric, and adipic acids), typical alcoholic components (ethylene, propylene, diethylene, polyethylene and dipropylene glycols, 1,6-hexanediol, and modified bisphenol), and monomers (styrene, methyl methacrylate and vinyl acetate). These investigators examined NMR spectra of solutions of the resins dissolved either directly in the monomer, or in acetone or benzene after removal of the monomer.

All the above acidic components (and also other components of the resins) provide characteristic NMR spectra by which they may be easily identified or even determined semi-quantitatively. The method is rapid and simple, as it does not involve previous hydrolysis or degradation. Polyester resins may also be identified by the NMR technique as reported by Yeagle [51].

Maleic and fumaric acids may be detected in unsaturated polyester resins by NMR spectroscopy following the procedure given by Curtis *et al.* [7]. In acetone solution the protons of fumaric acid at the double bond absorb at $6{\cdot}98\tau$, and in maleic acid at $6{\cdot}55\tau$.

Detection of Polyhydric Alcohols

Chemical Methods

Systematic studies on the reactions of polyhydric alcohols were conducted by Winterscheidt [52], who examined the following reagents: potassium periodate in acid solution, chromic acid in conc. sulphuric acid, conc. sulphuric acid alone, $0{\cdot}1N$ iodine solution in $0{\cdot}1N$ aqueous alkaline potassium iodide (the iodoform reaction), and alkaline copper sulphate solution (the Wagner test). Results of the tests are listed in Table 8.1.

The tests were carried out with a polyalcohol fraction isolated as follows. After hydrolysis of the resin under test (see p. 246) potassium salts of dicarboxylic acids are filtered off, the filtrate is acidified, and the monocarboxylic acids present are removed by petroleum extraction. The aqueous solution is then made alkaline, evaporated to dryness, and extracted with methanol.

Pentaerythritol present in the methanol solution precipitates on addition of acetone. To identify pentaerythritol the following tests are carried out:

a. an aqueous solution of the precipitate, when shaken with benzaldehyde in hydrochloric acid, gives a dibenzal derivative in the presence of pentaerythritol,
b. 0·5–1·0 ml of an aqueous solution of the precipitate under test is added carefully to 5 ml of 0·1% carbazole solution in conc. sulphuric acid. In the presence of pentaerythritol a dark bluish-violet colour appears at the interface between the two layers.

In order to ascertain that the polyester under examination does not also contain erythritol, mannitol, dulcitol or similar polyalcohols, the crystalline precipitate from hydrolysis is treated with aqueous potassium periodate. Pentaerythritol fails to react with this reagent, but other alcohols are oxidised.

After precipitation of pentaerythritol with acetone the methanolic extract is evaporated to a small volume. From the residue crystals of trihydroxymethylpropane of m.p. 55°C may precipitate, as this alcohol is the only solid compound of all the remaining polyalcohols occurring in polyester resins.

The Wagner Test. To 3 ml of aqueous solution of the alcohol under examination, 5 drops of 5% copper sulphate solution are added followed by 1 ml of 10% sodium hydroxide solution. The mixture is shaken and then filtered. A blue colour in the filtrate indicates a positive result.

The Iodoform Test. To 3 ml of the solution under test are added 2 pellets of potassium hydroxide and 5 ml of 0·1*N* iodine solution. A yellowish-white turbidity or a yellow crystalline precipitate indicates a positive result.

Iodine Consumption. A sample containing 0·2 g of the alcohol is treated with 50 ml of 0·1*N* iodine solution and 20 ml of 1*N* sodium hydroxide. The solution is allowed to stand for 30 min in the dark with occasional shaking, then 30 ml of 1*N* hydrochloric acid is added, and the excess iodine is titrated with 0·1*N* sodium thiosulphate solution.

Potassium Periodate Oxidation. A 0·2–0·4 g sample of the alcohol to be analysed is treated with 30 ml of water, 10 ml of 0·5*N* sulphuric acid containing potassium periodate (the periodate is added in a 75% excess with respect to the alcohol), and the mixture is shaken for 1–2 hr. A part of the reaction solution is titrated with 0·1*N* sodium hydroxide solution, and another part is treated with potassium iodide and the liberated iodine titrated with sodium thiosulphate solution.

Test with Concentrated Sulphuric Acid. About 1 ml of the solution under test is shaken with 5 ml of conc. sulphuric acid. A brown colour indicates a positive result.

Colour Reactions after Thinius. Polyhydric alcohols occurring in polyester resins may also be detected by using the colour reactions reported by Thinius and Schröder [53].

Table 8.1 Reactions of Polyhydric Alcohols with Miscellaneous Reagents

Alcohol	*Oxidation products* with KIO_4 equiv. per mole: COOH	*Oxidation products* with KIO_4 equiv. per mole: CHO	*Oxidation products* with CrO_3 *in conc.* H_2SO_4, moles of CH_3COOH per mole	*Wagner test*	*Iodoform test*	*Volume of* 0·1*N iodine solution,* ml per 0·2 g	*Colour with conc.* H_2SO_4
1,3-Butanediol	0	0	ca. 1	−	+	ca. 8 ml	brown
1,4-Butanediol	0	0		−	−	< 2 ml	brown
2,3-Butanediol	0	$2CH_3CHO$	ca. 2	+	+	ca. 20 ml	brown
Glycerol	1	$2CH_2O$	0	+	−	< 2 ml	—
Ethylene glycol	0	$2CH_2O$	0	+	−	< 2 ml	—
1,6-Hexanediol	0	0	0	−	−	< 2 ml	brown
2,3-Hexanediol	0	0	ca. 2	−	+	ca. 20 ml	brown
3-Hydroxymethyl-2,4-pentanediol**	0	0	ca. 2	+*	+	ca. 5 ml	brown
1,2-Propanediol	0	$1CH_2O$, $1CH_3CHO$	ca. 1	+	+	ca. 20 ml	—
Trihydroxymethylpropane	0	0	ca. $1C_2H_5COOH$ some CH_3COOH	+	−	< 2 ml	—

* A green colour develops.

** Commonly known as hexanetriol.

Tests for Bisphenol A (*Detection of Polycarbonates*)

Colour Reaction with p-Dimethylaminobenzaldehyde [54]. A 0·1–0·2 g sample of the resin is heated in a test tube. The products of pyrolysis condense on a fibre-glass plug placed inside.

A part of the product absorbed on the fibre-glass plug is treated with a 1% methanolic solution of *p*-dimethylaminobenzaldehyde. One drop of conc. hydrochloric acid is added and if polycarbonates are present a dark blue colour appears, which becomes somewhat less intense upon addition of methanol or acetone. A red colour that does not change to blue is observed in the case of polyamides.

Colour Reaction with Hydrogen Chloride. Some of the pyrolysis product is treated with gaseous hydrogen chloride or with dilute (1:1) hydrochloric acid. An intense red colour which is unaffected by methanol should appear.

The Gibbs Indophenol Reaction. In the pyrolysis products from polycarbonates phenols are present. These can be identified by suitable tests such as the sensitive indophenol tests (see p. 29).

INSTRUMENTAL METHODS

Paper Chromatography

For qualitative analysis of free polyhydric alcohols paper chromatography was used by Bergner and Sperlich [55]. For the detection of alcohol components of polyester resins this method was first adopted by Tawn and May [33]. It has since become universally accepted.

Arendt and Schenck [34] hydrolysed polyesters with 1*N* alcoholic potassium hydroxide. After filtration of dicarboxylate salts and after evaporation of ethanol under reduced pressure, the mixture of polyhydric alcohols was analysed by two-dimensional chromatography without prior dilution. The chromatogram was developed first with a chloroform–ethanol mixture, then with water-saturated ether in the other direction. This procedure gave a good separation of 1,4- from 1,3-butanediol, and diethylene glycol from 1,4-butanediol (in a mixture also containing ethylene and triethylene glycols, 1,2- and 1,3-propanediols, and glycerol). The chromatogram was detected with a reagent of the following composition: 9 parts by vol. of 5% silver nitrate solution with 1 part by vol. of 25% ammonia. After drying at 100–110°C there appeared brown or dark-brown spots which were due to individual polyhydric alcohols.

Polyester resins were analysed chromatographically using *n*-butyl alcohol saturated with water as mobile phase and developing the chromatogram in one direction only. The results were compared with the results achieved by chemical methods (elemental analysis, m.p. determination of the urethane and 3,5-dinitrobenzoate derivatives) [56].

Conversion of the glycols into 3,5-dinitrobenzoates with 3,5-dinitrobenzoyl chloride followed by chromatographic analysis was adopted by

Gasparič and Borecky [57]. Although very small amounts of glycols can be detected by this procedure, its disadvantage is the parallel formation of monoesters of the alcohols examined. In the case of glycerol three compounds are produced, namely mono-, di-, and triglycerides which give rise to three spots on the chromatogram. This creates problems in the interpretation of the results.

Three mobile phases were used for identification and separation of polyalcohols by Fijolka *et al.* [36], namely: acetone–benzene–water, water-saturated butyl acetate, and water-saturated butyl alcohol. These eluents were found to complement one another.

For the separation of glycerol from pentaerythritol the best eluent was composed of pyridine, butanol and water in the ratio of 5:2:2, whereas bisphenol A was best separated from polyalcohols with a (10:3:1) decalin–cyclohexanol–water mixture. The chromatograms were visualised with the common reagent ammoniacal silver nitrate solution. The usefulness of other colour reactions was also examined and isovanillin was shown to produce a strong colour with polyhydric alcohols.

According to Fijolka and Radowitz [58] the best eluent is a (10:10:1) acetone–benzene–water mixture. The procedure devised by them [58] is described below.

REAGENTS AND SOLUTIONS

Mobile Phase. Ten parts by vol. of acetone are shaken for 1 hr in a separating funnel with 10 parts by vol. of benzene and 1 part by vol. of water. The mixture is allowed to stand for about 10 hr, then it is separated and the aqueous layer discarded.

Ammoniacal Silver Nitrate Solution. Nine volumes of a 5% aqueous silver nitrate solution are combined with one volume of 25% ammonia.

Isovanillin Solution. Isovanillin (1·5 g) and 1 g of *p*-toluenesulphonic acid are dissolved in 100 ml of ethanol.

Diazotised p-Nitroaniline. *p*-Nitroaniline (1 g) is dissolved with heating in a mixture of 20 ml of distilled water and 2 ml of conc. hydrochloric acid. The solution is diluted with distilled water to a volume of 180 ml, cooled to room temperature, 20 ml of 5% sodium nitrite solution added, and the mixture is agitated until dissolution is complete. In the case of persistent turbidity the solution can be filtered.

Schleicher–Schuell No. 602 hP paper was used for chromatographic analysis.

PROCEDURE. In order to isolate polyhydric alcohols, a polyester sample (ca. 0·6 g, but not larger than 1 g) is dissolved in 30 ml acetone or benzene, and the solution is heated for 2–3 hr with 0·5*N* potassium hydroxide in ethanol. After removal of the potassium dicarboxylates by filtration through a porosity 4 sintered glass crucible the alkaline solution of polyhydric alcohols in ethanol is directly applied to the paper, as the potassium ions remain at the starting point and do not interfere. Styrene present in cast resins, being a lipophilic compound, migrates in the solvent front, and does not affect the R_f value of the compounds being analysed.

The solution under examination is applied to the paper in the form of spots, taking care that the starting spot is as small as possible. The chromatogram is developed for 4 hr at 20°C, using the mobile phase described above. Three chromatograms are prepared in parallel and detected in the following manner.

1. The air-dried chromatogram is sprayed with the ammoniacal silver nitrate solution, and heated at 110°C in a drier for 30 min. The glycols appear as dark brown spots on a light brown background. With a decreasing number of hydroxyl groups and increasing distance between them, the sensitivity of the polyalcohols towards the reagent becomes lower.

2. The air-dried chromatogram is treated with the isovanillin solution sprayed through a very fine nozzle, and is then dried for 30 min at 90°C. The temperature should not exceed 90°C, otherwise the background is too dark. The chromatograms may be sprayed in addition with 0·1, 0·01, or 0·001% ferric chloride solution to make the background lighter. The resprayed chromatograms are dried in air. The alcohols appear on a greyish-violet background as spots of various colours.

The reagent is especially sensitive to 1,3-diols, which can be detected in quantities down to about 10 μg. The following glycols can be detected with this reagent: 1,3-propanediol, 1,3-butanediol, 1,5-pentanediol, 1,6-hexanediol, and also glycerol, pentaerythritol, and bisphenol A. All the remaining di- and polyhydric alcohols, including glycerol, are most conveniently detected with the ammoniacal silver nitrate solution.

3. The air-dried chromatogram is sprayed with the solution of *p*-nitrobenzenediazonium hydrochloride solution, then immediately with a sodium carbonate solution. Bisphenol A appears as a red azo-colour, but di-, and polyhydric alcohols do not react with this reagent. Optimum detectability is about 50 μg of bisphenol A.

Gas-Liquid Chromatography

In the original studies by gas-liquid chromatography the alcoholic polyester components were analysed by introducing them without chemical modification into the column. Thus, a mixture of ethylene, diethylene, and triethylene glycols was separated [59] on a column packed with Carbowax 1000 at 150–180°C or on an Apiezon column at 200°C [60]. In the same way Nadeau and Oaks using a column packed with tetrahydroxylethylenediamine at 105°C [61], analysed a mixture of ethylene and propylene glycols, while Clifford [62] separated trimethylene glycol from excess glycerol by means of a Reoplex 400 column at 120°C. Much better results, however, are obtained when the polyhydric alcohols are converted into more volatile esters or ethers before separation.

The polyalcohols contained in paint alkyd resins were converted into corresponding acetates with acetic anhydride by Esposito and Swann

[63,64]. Two temperature-programmed chromatographic columns were used, one of which was packed with Carbowax 20M and the other with silicone rubber. The best results were obtained on the Carbowax 20M column. The silicone column, apart from a twofold shorter time for analysis, failed to resolve certain combinations, for example ethylene and propylene glycol, or mannitol and sorbitol. In this study the alcohols were first separated from the resin by aminolysis using *n*-butylamine. According to Ešnerová [65] benzylamine is superior in certain cases to other amines.

In order to eliminate the interfering effects of amines used in the aminolysis, and also that of phthalic acid, as well as to simplify the analytical procedure, the polyalcohols from the polyesters were converted into the corresponding acetates by transesterification by heating them in acetic acid in the presence of sulphuric acid [66]. The esters thus produced were separated on a column containing poly(ethylene glycol) (Polywax 1550) as stationary phase (5% on Embacel support).

According to Mlejnek [66] interfering substances can be eliminated by extraction of the transesterification products with carbon tetrachloride. This method can therefore be successfully used not only for homogeneous polyesters but also for modified ones. This method was used with excellent results in the analysis of a variety of commercial alkyd resins containing as their major constituents ethylene and diethylene glycols and glycerol, with maleic, adipic, phthalic and terephthalic acids, modified with linseed oil and with other vegetable oils and resols.

In recent years the technique of chromatographic analysis of alcohols of low volatility involving their conversion into highly volatile trimethylsilyl ethers, has gained popularity. For identification of polyhydric alcohols occurring in polyesters this method was used by Li Gotti *et al.* [67].

Presented below is the method worked out by Mlejnek.

APPARATUS. The column is packed with 5% Polywax 1550 [poly(ethylene glycol)] supported on Embacel (grain size 0·15–0·25 mm).

REAGENTS

Acetylating Mixture. A mixture of 600 ml of glacial acetic acid, 100 ml of benzene, and 5 ml of conc. sulphuric acid is heated for 3–4 hr under reflux. The condenser is equipped with an azeotrope receiver. As soon as water droplets cease to appear in the receiver, the condenser and the receiver are detached and benzene is distilled off until the vapour temperature reaches 115°C.

PROCEDURE. A resin sample containing about 1 mequiv of polyalcohols, is weighed into a 25 ml acetylation flask, and 1 ml of the acetylating mixture is added. The flask is stoppered with a silicone rubber stopper, and heated in a thermostat at 80°C for 2 hr. The contents are then transferred to a 5 ml test tube, 1 ml of distilled water is added, and the mixture shaken with two 0·5 ml portions of carbon tetrachloride. The upper, aqueous layer is transferred to another test tube, and similarly extracted with chloroform. The combined chloroform extracts are washed with water, then with aqueous sodium bicarbonate solution, and again with water until neutral. The chloroform solution is transferred to a small dry test tube, most of the chloroform is stripped off in a nitrogen stream, then it is made up to the original volume with carbon disulphide. The solution thus obtained is introduced onto the chromatographic column kept at 260–280°C.

The application of gas-liquid chromatography to the analysis of alkyd resins has been reviewed by Haken [39].

The method of pyrolysis gas chromatography as used for qualitative analysis of polyesters is described in the papers by Sadowski and Kaufeld [42], who examined alkyd and unsaturated polyester resins, and by Luce *et al.* [43] who investigated hardened polyester laminates. In both researches the chromatograms obtained differed considerably with varying composition of the alcoholic components.

Infrared Spectroscopy

The alcoholic constituents have an insignificant effect on the spectra of polyester resins. For this reason, a direct identification of these constituents from the infrared spectra of the resins is not feasible [44,45]. Polyhydric alcohols can, however, be identified successfully by this technique after hydrolysis of the resin. Commercial glycols have substantially different infrared spectra, and also individual resins usually contain no more than two polyhydric alcohols [68,58].

The spectra of 14 mixtures of common glycols, taken in a variety of proportions and combinations were studied by Shay *et al.* [68] over the range 650–1550 cm^{-1}. Pairs of glycols with quite different structures can thus be conveniently identified in binary mixtures when their concentration ratios are 1:2, 2:1, or 3:1. For lower concentrations of either component and also in ternary mixtures, identification of polyhydric alcohols can be accomplished only in certain special cases.

Brudkowska, Hippe, and Jabłoński [47] have investigated the infrared spectra of polyesters containing common dicarboxylic acids, such as phthalic, maleic, and fumaric, and the glycols of ethylene, diethylene, and propylene. Contrary to the available data, certain bands characteristic of the diols used in the synthesis were also discovered in the spectra of polyesters containing aromatic groups. According to the above authors, the range most suitable for identification of the diol component lies at 1330–1410 cm^{-1}, and for propylene glycol the additional band near 2984 cm^{-1}.

Dipentaerythritol in pentaerythritol can be detected, after conversion into its acetyl derivative, from the band at 1115 cm^{-1} (carbon tetrachloride solution) [69].

NMR Spectroscopy

Dihydric alcohols such as ethylene, propylene, diethylene, polyethylene, and dipropylene glycols, 1,6-hexanediol, and modified bisphenol, may be detected directly in unsaturated polyester resins, according to the study by Percival and Stevens [50], by means of NMR spectroscopy.

Detection of Fatty Acids

Paper Chromatography

Paper chromatography was used by Lichthardt [70] in the analysis of alkyd resins modified with linseed, castor, coconut, soya-bean, and cod liver oils. The oils can be identified by this technique not only when used alone but also in their mixtures. As a mobile phase 96% acetic acid saturated with undecane was used, and the chromatograms were detected with solutions containing copper sulphate and potassium ferrocyanide.

Alkyd resins modified with an oil of only one kind, and specifically with soyabean or linseed oil, were analysed by Kaufmann and Büscher [71]. To simplify the analytical procedure, the resins were hydrolysed directly on the paper by moistening the spots of the solution of the unknown with 10% potassium hydroxide in glycerol followed by heating for 30 min at 90°C. After neutralisation and drying of the paper the chromatogram was developed with a glacial acetic acid–acetonitrile–undecane mixture, and the spots were detected with solutions of copper acetate and rubeanic acid.

The individual acids, (palmitic, stearic, linoleic and linolenic acids) clearly visible as distinct spots on the chromatogram, were determined photometrically. In the case of unsaturated acids, and especially when present in large quantities, the results obtained by this method are much too low. This is likely to be due to their oxidation during analysis.

Gas-Liquid Chromatography

Gas-liquid chromatography was used by Zieliński, Mosley, and Bricker [72], for analysis of both oils and the oil-modified alkyd resins used in the manufacture of paints and varnishes. The materials examined were hydrolysed with 10% alcoholic sodium hydroxide solution. The fatty acids liberated on acidifying were extracted with ether and converted into corresponding methyl esters by heating with methanolic hydrogen chloride solution in the presence of 2,2-dimethoxypropane [73]. The column was packed with diethylene glycol succinate supported on Chromosorb and the separation was carried out at 210–225°C. The chromatograms obtained had distinct peaks corresponding to individual fatty acids. This allows the identification of a variety of oils, alone or in mixtures, from the characteristic peaks of the constituent acids, such as ricinoleic acid from castor oil, oleostearic acid from tung oil, and linoleic acid from soya-bean oil.

Infrared Spectroscopy

Identification of oils and other modifying agents in alkyd resins by infrared spectroscopy offers considerable problems on account of strong interference from phthalates. The spectra of modified resins compared with those of individual *o*-, iso- and terephthalates show only insignificant new features. These features become more pronounced at large thicknesses when strong bands appear at 3·4 and 6·9 μm (2941 and 1449 cm^{-1}) (aliphatic C—H bonds), a characteristic band at 8·6 μm (1163 cm^{-1}) (aliphatic ester C—O—), a more intense absorption between 10·0 and10·5 μm (1000–952 cm^{-1}), and a stronger absorption at 13·85 μm (722 cm^{-1}).

The presence of a drying oil in aliphatic polyesters may be detected on the basis of a band in the range 10·0–10·5 μm.

The spectra of alkyd resins containing abietic acid rather than fatty acids (colophony-modified resins) differ substantially from the spectra of other alkyd resins, and also from the spectrum of the unmodified resin. Particularly characteristic is a C=C band of medium intensity at 6·09 μm (1642 cm^{-1}).

Ultraviolet Spectroscopy

Fatty acids in the monocarboxylic acids fraction may be identified by ultraviolet spectrophotometry which has wide application in the analysis of fats [74]. According to Shreve the methods used for this purpose are based on the following facts: (*a*) unsaturated acids that contain two or three conjugated double bonds absorb strongly at 234 and 268 nm; (*b*) unsaturated acids not containing conjugated double bonds and which do not absorb in the ultraviolet range, may be converted reproducibly into products which contain ultraviolet absorbing systems, by heating with a solution of potassium hydroxide in ethylene glycol or glycerol.

Benzoic acid was detected in the monocarboxylic acid fraction by Swann *et al.* [75] after prior removal of other acids by extraction with carbon tetrachloride. They based the analysis on the measurement of the absorption band at 273 nm.

Detection of Styrene

Free styrene in polyester resins may be detected by infrared spectrophotometry on the basis of the band at 11 μm (909 cm^{-1}), which does not normally overlap others, and the bands at 14·4 and 13·9 μm (695 and 719 cm^{-1}). In cured (and also uncured) resins the bands due to polystyrene at 6·7, 13·2, and 14·35 μm (1493, 758, and 697 cm^{-1}) [44,45] may also occur.

Monomers such as styrene, methyl methacrylate, and vinyl acetate in polyester resins may be detected by the NMR technique [50].

8.4 QUANTITATIVE ANALYSIS

In polyester resins of any kind the structural units are usually determined as dicarboxylic acid, polyhydric alcohol, OH groups (hydroxyl number) and carboxylic groups (acid number). The amounts of unconverted starting material and of various additives are also determined. Additional analyses are carried out in specific cases. Thus in alkyd resins monocarboxylic acids are determined along with the iodine value and the content of modifying materials such as colophony or synthetic resins. In unsaturated polyester resins styrene is determined and in paints non-volatile matter as well.

Determination of Structural Components

Functional groups present in polyester resins are determined directly in the samples under examination. If required, the solvents or other substances interfering with a given determination, for instance alcohols in the hydroxyl number determination, are removed. The functional groups are determined as acid number, iodine value, saponification number, and hydroxyl number. On the other hand in the determination of the resin-forming components such as polyols, mono- and dicarboxylic acids, and certain modifying materials, the resin is subjected to previous solvolysis to free and to separate the components of various types. This procedure is followed in all chemical methods and in the majority of instrumental techniques. For this reason the methods for splitting polyesters and separating their components have been described first.

Hydrolytic Decomposition of Polyesters and Separation of the Polymer Components

The commonest method for splitting polyesters into their components is hydrolysis with potassium hydroxide in anhydrous ethyl alcohol after Kappelmeier. The method is described at length in the monograph [76].

A polyester sample is dissolved in benzene and subjected to hydrolysis by heating with excess potassium hydroxide solution in anhydrous ethyl alcohol. Potassium dicarboxylates precipitate. After cooling the reaction mixture to room temperature, the precipitated salts are filtered off, washed with alcohol and dried at 110°C. The crystalline precipitate is used for the determination of individual dicarboxylic acids.

The filtrate is made slightly acidic with conc. hydrochloric acid and the potassium chloride precipitate is removed by filtration. The filtrate is concentrated to a small volume, diluted with water and extracted with ether. Free monocarboxylic acids that pass into the ether layer are recovered from the extract by stripping off the solvent. The aqueous layer is distilled under reduced pressure to remove water and to separate the polyol fraction.

In the determination of dicarboxylic acids present in unsaturated polyesters the hydrolysis with alcoholic potassium hydroxide solution is accompanied by side reactions which involve ethoxide addition at the double bonds of maleic and fumaric acids to form ethoxysuccinic acid. To avoid such side reactions, the hydrolysis is carried out with 10% aqueous potassium or sodium hydroxide solution, on a solution of the resin in acetone, chloroform or benzene [38,77,78,79]. Under these conditions only minor quantities of malic acid are produced.

If neutral compounds separate after the hydrolysis, these can be identified directly. Alcohols such as allyl alcohol may be removed from the solution by steam distillation. Careful acidifying of the solution with conc. hydrochloric acid may result in the liberation of only monocarboxylic acids, whereas dicarboxylic acids may remain in the solution as salts. Monocarboxylic acids are separated by extraction with ether, or, in the case of higher fatty acids, as a layer immiscible with water. Dicarboxylic acids are liberated by addition of an excess of hydrochloric acid. The precipitate is filtered off and recrystallised from water. The acids are determined by the methods described below. Glycols and glycerol are isolated from the acidic filtrate by distillation. The salts are filtered off if they separate as a precipitate, and the residue is distilled under reduced pressure to isolate individual alcohols.

The method of hydrolysis with aqueous alkalis suffers the drawback that it does not give quantitative results of high precision. However, it permits the detection of low boiling alcohols, and facilitates detection of certain less common polyester components.

Determination of Dicarboxylic Acids

Chemical Methods

The most detailed and comprehensive analytical procedure for individual dicarboxylic acids such as phthalic, sebacic, fumaric, maleic, succinic, and adipic in mixtures of these acids as they occur in alkyd resins has been developed by Swann [32]. The resin is hydrolysed with potassium hydroxide in non-aqueous solution, and the acids separated as their potassium salts. According to Swann's method phthalic acid is determined gravimetrically as the lead salt.

A method for the determination of phthalic acid as the neutral potassium salt containing alcohol of crystallisation $C_6H_4(COOK)_2.C_2H_5OH$, has been described by Kappelmeier [80]. Since the potassium salts of other dicarboxylic acids that occur in polyesters do not contain alcohol of crystallisation, the phthalic acid content may be determined by first drying the potassium salt mixture at 60°C and subsequently at 130–140°C, at which the phthalic salt loses its alcohol of crystallisation.

Less precisely but more rapidly, phthalic, as well as isophthalic acid, can be determined by titration methods. Here again polyesters are first hydrolysed, and the dicarboxylate fraction is isolated as the potassium salts. Thus alkali titration was used by Esposito and Swann [81] for the determination of isopththalic acid and certain other dicarboxylic acids. Before titration the acids were liberated from their potassium salts by acidifying. Kruysse [82] recommends direct titration of potassium phthalates and isophthalates in a non-aqueous medium with perchloric acid.

All the above titration methods may only be used when one or at most two acids are present in the mixture to be analysed. On the other hand, the method reported by Kreshkov *et al.* [83], which involves potentiometric titration in a differentiating solvent, is suitable for determination of several acids in the same mixture.

On account of their low solubility in water isophthalic and terephthalic acids can be determined gravimetrically as free acids [76,84], whereas sebacic acid is determined by this method as its zinc salt (see Swann [32]).

Total unsaturated acids, in the absence of saturated acids, are determined colorimetrically as their bromo derivatives.

If other acids are absent, maleic and fumaric acids can be determined, when both occur in the same sample, by first estimating their total content (after the alkaline hydrolysis) by back-titration of the excess alkali with 0·5*N* hydrochloric acid. Fumaric acid is then precipitated with cadmium acetate, and the excess reagent is back-titrated with 0·05*M* EDTA solution [85]. The maleic acid content in the resin is found by difference.

If other dicarboxylic acids are absent, adipic acid can be determined gravimetrically as its silver or potassium salt [76]. Grimmer [86] points out, however, that the latter variant of this method leads to considerable error, due to adsorption of the salts of fatty acids on the surface of potassium adipate crystals, and to dissolution of these crystals in water during washing.

Adipic and sebacic acids in mixtures with phthalic acid may be determined by the method of Clasper and Haslam [87]. The mixture of acids is separated on a silica-gel chromatographic column using first a 2% solution and subsequently a 15% solution of butanol in chloroform as eluents. Sebacic acid is eluted first followed by adipic acid. The acid content is determined in the eluates by titration with 0·05*N* alcoholic potassium hydroxide with *m*-cresol red as indicator.

Sebacic acid in mixtures with maleic acid may be determined by a method developed by Zelenina [88]. Total acids are determined after the alkaline hydrolysis by titration of the excess alkali with hydrochloric acid solution with phenolphthalein as indicator. Sebacic acid is determined in the hydrolysate by precipitation with 0·05*M* zinc acetate. Excess reagent is titrated with an EDTA solution.

Succinic acid is determined, like adipic acid, as its silver salt. If another dicarboxylic acid is present, it may also be determined gravimetrically as its potassium salt [80].

Instrumental Methods

Electroanalytical Methods

Two methods of simultaneous determination of dicarboxylic acids in alkyd resins have been worked out by Kreshkov. These involve potentiometric titration in a non-aqueous medium. In one of the methods [89] the potassium salts contained in the hydrolysate are dissolved, after isolation, in a (1:4) ethylene glycol–methyl ethyl ketone mixture, and titrated in a methyl ethyl ketone solution with a 0·1*N* perchloric acid.

In the other method [90] the lithium dicarboxylates, produced by hydrolysis of the resin with lithium hydroxide in anhydrous methanol, are dissolved in some methanol, and are led onto a cation-exchange column. Free acids are eluted from the column with a (3:1) isopropanol–methanol mixture, and are titrated with a tetraethylammonium hydroxide solution in a benzene–methanol mixture, in a medium of differentiating properties [a (4:1:2) isopropanol–methanol–methyl ethyl ketone mixture].

A method for determination of *cis-trans* isomers of unsaturated acids (fumaric–maleic and citraconic–mesaconic acid pairs) when simultaneously present in the same mixture involves the reaction of *cis* isomers with mercury acetate in a non-aqueous medium. Total acids are estimated by potentiometric titration with 0·1*N* methanolic potassium hydroxide [91].

Kreshkov *et al.* [83] describe the determination of unsaturated mono- and dibasic acids in an isopropyl alcohol medium by potentiometric titration with 0·1*N* tetraethylammonium hydroxide solution. The same authors report an analytical procedure for the determination of all three isomeric phthalic acids [92] by combined parallel titration with the same reagent and potassium hydroxide in methyl ethyl ketone.

Maleic, fumaric, citraconic, mesaconic, and also phthalic acids, undergo electrochemical reduction on a dropping mercury electrode and so these acids could be determined by polarography. The following polarographic methods for their codetermination have been developed [93,94].

1. Total maleic and fumaric acids (along with phthalic acid): strongly acidic medium, e.g. 0·04*N* with respect to hydrochloric acid and 0·5*M* with respect to ammonium chloride [95,96,97]).
2. Maleic acid (individually) in the presence of fumaric and phthalic acids: the Britton–Robinson buffer of pH 7 [95,96,98].
3. Maleic and fumaric acids (simultaneously) when both present: ammonium buffer of pH 8·2 [78,97,79,94].
4. Fumaric acid (individually) in the presence of maleic acid: ammonium buffer of pH 9·7 [78,97] or 9·3 [98].
5. Citraconic, mesaconic and phthalic acids in their mixtures: acidic medium as in 1 above [95, 96].

Described below is the determination procedure for unsaturated acids and phthalic acid after Novák [95].

a. Determination of total maleic and fumaric acids and simultaneously also of phthalic acid content.
b. Simultaneous determination of citraconic, mesaconic and phthalic acids when all three are present.

Hydrolysis of the Resin. A resin sample of 0·1–0·5 g is dissolved in 15 ml of acetone; 10 ml of 1*N* potassium hydroxide is added, and the solution is refluxed in a round-bottomed flask fitted with ground-glass joints for 10–20 min (unsaturated polyesters) or for 90 min (alkyd resins). Afterwards the acetone is distilled off, and the residue is transferred to a 100 ml measuring flask, acidified with 10 ml of 1*N* hydrochloric acid, and diluted with water to the mark. The solution is then filtered into a dry beaker and the clear solution is pipetted to perform the analysis.

Polarographic Determination. An aliquot of the solution (1–5 ml) is transferred with a pipette to a 25 ml measuring flask; 12·5 ml of 1*M* ammonium chloride solution and 1 ml of 1*N* hydrochloric acid are added, and the volume is made up to the mark. The polarogram is recorded after removal of oxygen, versus a SCE electrode. The polarographic curve is recorded beginning from −0·4 V. The half-wave potentials with reference to SCE under the experimental conditions are, respectively: maleic and fumaric acids, −0·64 V; citraconic acid, −0·53 V; mesaconic acid, −0·65 V, and phthalic acid, −1·05 V. A similar run is made with a standard.

c. Determination of maleic acid in the presence of fumaric and phthalic acids.

Hydrolysis of the resin is performed as described above.

Polarographic Determination. The solution after hydrolysis is weakly acidic. For this reason the same volume of the solution as is introduced into the 25 ml measuring flask is neutralised in a parallel run with an aqueous 0·1*N* sodium hydroxide solution and phenolphthalein. The volume of the alkali thus found is added to the measuring flask. The solution in the flask is then made up to the mark with the Britton–Robinson buffer of pH 7, and the polarogram is recorded as described above, beginning the curve from −1·2 V. The value of the half-wave potential under the experimental conditions with reference to SCE is −1·46 V.

Composition of the Britton-Robinson buffer (pH 7): To 100 ml of a solution containing 0·004 mole of phosphoric acid, 0·004 mole of boric acid, and 0·004 mole of acetic acid, is added 50 ml of 0·2*N* carbonate-free sodium hydroxide.

CALCULATION. The acid content *X*(%) is calculated from the formula

$$X = \frac{10hbMf_{st}}{h_1 am}$$

where:

h = the wave height for the unknown,
h_1 = the wave height for the standard,
b = the volume (ml) of standard solution,
M = molecular weight of the acid determined,
f_{st} = molarity of the standard solution (usually 0·01*M*),
a = the volume (ml) of the hydrolysate aliquot examined,
m = weight of the resin sample, g.

In the continued studies on the joint determination of maleic and fumaric acids by the polarographic method Novák and Krátký [99] found that better results are obtained for some complex polyesters if the analysis is carried out in a non-aqueous medium (50% ethyl acetate, 30% ethanol, and 20% water).

Unsaturated acids and phthalic acid were also determined by an oscillopolarographic technique [95,96]. By this technique the following acids may be determined simultaneously in an acidic medium (hydrochloric acid-ammonium chloride mixture): maleic, fumaric and phthalic, and also citraconic, mesaconic and phthalic. A major advantage of the oscillopolarographic technique is its simplicity and rapidity in operation.

Phthalic acid in the hydrolysate solution from linear polyesters was determined by Fijolka and Ziebarth [100] by the method developed by Fijolka, Lenz, and Runge [38] in 0·235*N* acetate buffer at pH 3·6. Saturated acids do not interfere with the determination by this method as their waves in acidic media occur at a common position which is clearly apart from the position of phthalic acid (the difference in half-wave potential is 0·4 V). Neither adipic acid, nor tetrahydro- or tetrachlorophthalic acid nor HET acid* give waves within the range from −1 to −2 V in that medium, and do not interfere with this determination. Finally Krátký and Novák [101] analysed isophthalic acid in unsaturated polyesters by direct polarography after conversion of this compound by nitration into its 5-nitroderivative.

Gas-Liquid Chromatography

The qualitative method evolved by Esposito and Swann (see p. 232) has been modified for quantitative determinations by Percival [102] who prolonged the reaction time of methanolysis to 18 hr. By this method the following could be determined with satisfactory accuracy: maleic, fumaric, itaconic, adipic, phthalic, and isophthalic acids. In a similar manner Esposito [103] determined phthalic anhydride in alkyd resins.

Ultraviolet Spectroscopy

A method based on the measurement of absorbance of an aqueous solution of potassium phthalates at 276 nm has been developed by Shreve and Heether [49]. The method is suitable for determination of phthalates in the presence of other dicarboxylic acids, including unsaturated ones (maleic and fumaric). In the presence of unsaturated acids suitable corrections are introduced. This method was improved by Swann *et al.* [104] extending its application range to cover isomeric phthalic acids. According to this improved method potassium dicarboxylate fractions are dissolved in a 1:1 methanol-water mixture acidified with hydrochloric acid, and the absorbance measured at 275 nm (*ortho*-isomer), 281 nm (*meta*-isomer), or at 287 nm (*para*-isomer).

Maleic and fumaric acids were determined on the basis of their ultraviolet absorption spectra at 220 nm after separation on an Amberlite G-450 ion-exchange column by Funasaka, Kojima and Fujimura [105]. The acids

* 1,4,5,6,7,8-hexachloro-2-norbornene-5,6-dicarboxylic acid.

were eluted from the column separately with 4*M* calcium chloride solution at pH 1·5.

Infrared Spectroscopy

The phthalic acid content in polyester resins is usually estimated on the basis of the band at 15·4 μm (649 cm^{-1}) [45]. The ratio of the phthalic acid content to the modifying oil content in modified alkyd resin paints was measured by Salama and Dunn [106] on the basis of the absorbance at 6·32 μm (1581 cm^{-1}) (the phthalic component), and at 3·42 μm (2927 cm^{-1}) (the modifying component).

The quantitative ratio of the fumaric acid units to the maleic acid units in unsaturated polyesters was determined by Chiang and Bobalek [107] who measured the absorbance at 10·3 μm (967 cm^{-1}) (*trans* configuration) and at 13·8 μm (724 cm^{-1}) (*cis* configuration). The same ratio was determined on the basis of the absorbance at 10·2 μm (981 cm^{-1}) and 7·17 μm (1397 cm^{-1}) for the *trans* and *cis* configurations respectively, by Reichert and Nollen [108].

NMR Spectroscopy

Maleic and fumaric acids in unsaturated polyester resins may be determined by NMR spectroscopy according to the study of Curtis *et al.* [7].

Isophthalic acid was determined in the presence of terephthalic in terephthalate-isophthalate copolymers from NMR spectra on the basis of the multiplet at 8·21, 7·90, 7·80, 7·35, 7·22, and 7·10τ (Murano *et al.* [109]). Terephthalic acid gives a single signal at 7·74τ.

The composition of alkyd resins modified with drying oil (Reynolds *et al.* [110]) or with soya-bean oil (Opp [111]) was also roughly estimated by this method.

Determination of Fatty Acids

Chemical Methods

The filtrate, after removal of potassium dicarboxylates in the procedure for determination of these acids by chemical methods (see p. 246), in addition to the fatty acid salts, also contains excess potassium hydroxide, volatile solvents, polyhydric alcohols, and non-hydrolysable matter. The fatty acids incorporated in this mixture are determined [76] as follows.

The filtrate solution (see p. 246) is evaporated on a water-bath keeping its volume at 250 ml by adding the required quantities of water. As soon as the volatile organic solvents are stripped off, the solution is transferred to a separating funnel and 20% sulphuric acid is added until the solution is acid to Congo red paper. The fatty acids thus liberated are extracted with ether by shaking the solution several times with several portions of ether. The combined ether extracts are washed with water to remove residual sulphuric acid and the ether is evaporated on a water-bath in a carbon dioxide stream to prevent oxidation of the acids. The non-volatile residue is first dried for 10 min at 110°C, then over sulphuric acid in a desiccator until constant weight is reached. The monocarboxylic acids thus isolated are characterised by determining their acid number, iodine value and diene number.

Along with monocarboxylic acids a variety of resin additives may pass into the ether extract. Those soluble in ether, such as high boiling solvents, modifying resins, etc. will give rise to high results. More accurate results may be achieved either by making allowance for the content of non-hydrolysable matter, which can be estimated as described below (see p. 267), or by determining the acid content in the non-volatile residue.

Instrumental Methods

Ultraviolet spectrophotometry was used by Swann *et al.* [75] who determined benzoic acid at 273 nm. This acid was present in the residue after carbon tetrachloride extraction of fatty acids. If benzoic acid is accompanied by its *p-tert*-butyl derivative (encountered in novel alkyd resins), both compounds may be determined by gas-liquid chromatography after conversion of monocarboxylic acids contained in the ether extract into the corresponding methyl esters by means of diazomethane [112] or sodium methoxide [113].

Paylor *et al.* [114] determined tall oil fatty acids in alkyd resins by gas-liquid chromatography after hydrolysis and esterification with methanolic boron trifluoride solution.

Determination of Polyhydric Alcohols

Chemical Methods

Polyhydric alcohols can be determined in polyester resins by chemical methods when the resins are first decomposed hydrolytically by one of the methods described above. Polyhydric alcohols are present in the aqueous layer after the fatty acid extraction. This aqueous solution also contains the potassium salt of the mineral acid used to acidify the solution prior to the fatty acid extraction, and is saturated with ether. The solution is neutralised to pH 7 and concentrated in a beaker on a water-bath by evaporation to a volume of 75 ml. The residue is cooled, filtered into a 100 ml measuring flask, and made up to the mark. The solution thus prepared serves as a stock solution for determination of individual polyhydric alcohols.

In the analysis of polyhydric alcohols the Malaprade method [115] is popular. This consists of treatment with periodic acid or its salts. Alcohols containing vicinal hydroxyl groups are oxidised according to the equations:

$$\underset{\text{glycerol}}{CH_2OHCHOHCH_2OH} + 2NaIO_4 \rightarrow 2HCHO + HCOOH + 2NaIO_3 + H_2O$$

$$\underset{\text{ethylene glycol}}{CH_2OHCH_2OH} + NaIO_4 \rightarrow 2HCHO + NaIO_3 + H_2O$$

$$CH_3CHOHCH_2OH + NaIO_4 \rightarrow HCHO + CH_3CHO + NaIO_3 + H_2O$$
propylene glycol

Pentaerythritol fails to react with periodates.

As seen from the above equations, certain other oxidation products are formed in addition to formaldehyde from the reactions of glycerol and propylene glycol with periodates. For this reason these alcohols can be determined together in mixtures by means of this method.

Glycerol

If other vicinal tri- and polyols are absent glycerol may be determined by acidimetric titration of the formic acid produced in the reaction with periodate.

REAGENTS

Sodium Periodate. In a 1 l. measuring flask 60 g of sodium periodate is dissolved in distilled water; 120 ml of 0·1*N* sulphuric acid is added and the volume is made up to the mark with water.

Sodium Hydroxide. A 0·125*N* standard solution.

PROCEDURE. A sample of the solution of polyhydric alcohols, prepared as described above, containing 0·32–0·5 g of glycerol, is diluted with water to a volume of 50 ml, made just alkaline to pH 8·1±0·1, then 50·0 ml of the periodate solution is added. A blank with 50 ml of water is prepared and treated in the same manner. The solution is allowed to stand at room temperature for 30 min, then 10 ml of 50% ethylene glycol solution in water is added. After 20 min the blank is titrated to pH 6·5±0·1, while the solution under examination is brought to pH 8·1±0·1, with the 0·125*N* standard sodium hydroxide.

CALCULATION. The glycerol content $X(g)$ in the aliquot of the solution of the unknown sampled for the periodate reaction, is evaluated from the formula

$$X = 0{\cdot}09209(v_2 - v_1)n$$

where:

v_2 = the volume (ml) of the 0·125*N* sodium hydroxide used for the titration of the solution examined,

v_1 = the volume (ml) of the 0·125*N* sodium hydroxide used for the titration of the blank,

n = normality of the sodium hydroxide,

0·09209 = coefficient allowing for the mol. weight of glycerol.

Total Glycols

Ethylene glycol, and other 1,2-glycols, in the absence of glycerol and other related polyhydric aclohols except pentaerythritol, may be determined by the periodate oxidation following the method worked out by Pohle *et al.* [116] and subsequently modified by Siggia [117].

REAGENTS

Periodic Acid Solution. Periodic acid (5 g) is dissolved in 200 ml of water and 800 ml of glacial acetic acid is added to the resulting solution.

Sodium Thiosulphate. A 0·1*N* standard solution.

PROCEDURE. An aliquot of the polyol solution (see p. 253) containing 0·01–0·09 g of glycols, is pipetted into a 250 ml iodine flask; 100 ml of the periodic acid solution is then added, and the mixture is allowed to stand at room temperature for 30 min. Next, 20 ml of 20% aqueous potassium iodide solution is added, and the liberated iodine is titrated with the 0·1*N* sodium thiosulphate solution.

A blank is run in a similar manner taking also 100 ml of the reagent.

CALCULATION. The glycol content *X*(g) in the aliquot of the solution taken for the periodic reaction is evaluated from the formula

$$X = \frac{(v_1 - v_2)nM}{2000}$$

where:

v_2 = the volume (ml) of the 0·1*N* thiosulphate solution used in the titration of the solution analysed,
v_1 = the volume (ml) of the titrant used for the blank,
n = normality of the thiosulphate solution,
M = molecular weight of the glycol.

Glycol and Glycerol

Vicinal glycols and glycerol, when present in the same sample, may be determined by a combination of alkalimetric and iodometric methods either after Allen *et al.* [118] or after the Pohle and Mehlenbacher procedure [119] reported below.

REAGENTS

Periodic Acid Solution. Periodic acid ($HIO_4.2H_2O$; 20 g) is dissolved in 1 l. of distilled water. If turbid the solution is filtered through a sintered glass crucible.

Sodium Hydroxide. A 0·125*N* standard solution.

Sodium Thiosulphate. A 0·08*N* standard solution.

PROCEDURE

Glycerol. An aliquot of the polyol solution (see p. 253), containing ca. 0·4 g of alcohols, is pipetted into a 600 ml beaker, 50 ml of water is added and the pH of the solution is adjusted to 6·2 using a pH-meter. A blank of 50 ml water is prepared in the same manner. Then 50 ml of the periodic acid solution is added; the beaker is covered with a watch glass and allowed to stand at room temperature. After 1 hr the solution is diluted to 250 ml with distilled water and titrated potentiometrically with a standard sodium hydroxide solution to pH 6·2 (the volume of the titrant used, *A*). The blank is treated in the same manner except that it is titrated to pH 5·4 instead of 6·2. Both solutions are preserved for the determination of glycols.

Glycols. The solution after titration is transferred quantitatively to a 500 ml volumetric flask and diluted with water to the mark. An aliquot of 50 ml of this solution is pipetted into a 300 ml conical flask; 10 ml of glacial acetic acid and 10 of ml 20% aqueous potassium iodide are added and the solution is thoroughly mixed. After 2 min the liberated iodine is titrated with the standard sodium thiosulphate solution with starch as indicator at the end of the titration (the thiosulphate volume used, *D*).

The blank from the glycerol analysis is run in the same way (the volume of the thiosulphate titrant used, *C*).

If the volume of the thiosulphate used in the titration of the solution under test amounts to less that 80% of the corresponding volume in the blank titration, the determination is repeated, using a smaller volume of the test solution.

CALCULATION. The glycerol content X(g) in the aliquot of the solution taken for the periodic acid reaction is calculated from the formula

$$X = 0{\cdot}09209\ (A-B)n$$

where:

A = the volume (ml) of the 0·125*N* sodium hydroxide used for titration of the unknown solution,

B = the volume (ml) of the sodium hydroxide used for titration of the blank,

n = normality of the sodium hydroxide solution,

0·09209 = a coefficient based on the mol. wt of glycerol.

The polyhydric alcohol content expressed in terms of the glycerol content is evaluated from the formula

$$y = 0{\cdot}2302\ (C-D)n$$

where:

D = the volume (ml) of the 0·08*N* thiosulphate solution used in titration of the test solution examined,

C = the volume (ml) of the titrant used for titration of the blank,

n = normality of the sodium thiosulphate solution,

0·2302 = a coefficient based on the molecular weight of glycerol.

Having evaluated x and y, the ethylene and propylene glycol contents, denoted as G and P respectively, are found from the expressions

$$G = 1{\cdot}348(y-x); \qquad P = 1{\cdot}652(y-x)$$

Pentaerythritol can be determined gravimetrically as its dibenzal derivative [76].

Polyhydric alcohols may also be determined by an alkaline potassium permanganate oxidation using the procedure developed by Thinius and Schröder [53].

Diethylene Glycol

Diethylene glycol is an unwanted alcohol component of polyethylene terephthalate. Kirby *et al.* [120] determined this glycol in the alcohol fraction obtained from hydrolysis of this resin, by oxidation with potassium dichromate and titration of the excess reagent with ferrous ammonium sulphate solution. Ethylene glycol interfering with this determination is previously removed by oxidation with potassium periodate and subsequent distillation of the formaldehyde produced.

Diethylene glycol in the presence of ethylene glycol was determined in poly(ethylene terephthalate) resin by Mifune and Ishida [121] basically in the same manner, although the excess of potassium dichromate was determined spectrophotometrically in this case by measuring the absorbance at 448 nm. Ethylene glycol was determined in a parallel run by oxidation with lead tetraacetate in a water–acetic acid–acetone solution. The formaldehyde produced in the reaction was determined iodometrically.

Instrumental Methods

Gas-Liquid Chromatography

The glycols of ethylene, propylene, 1,3-butylene, diethylene, triethylene, and dipropylene were determined in a variety of polyester resins with semiquantitative accuracy by Percival [102], who adopted the qualitative method developed by Esposito and Swann (see p. 241).

Likewise, the polyhydric alcohols in the form of acetates were determined in alkyd resins by de la Court *et al.* [122] who decomposed the resins under examination by aminolysis using phenylethylamine. The identical procedure was followed by Wittendorfer [123] in his analysis of minor quantities of trihydroxymethylpropane in polyesters and polyurethane foams.

On the other hand, Esposito [124], and Esposito and Swann [125] determined polyhydric alcohols in alkyd resins by a method based on aminolysis combined with trimethylsilyl derivative formation; Atkinson and Calouche [126] employed the same technique in the analysis of poly(ethylene terephthalate) resins.

Minor quantities of alcoholic components containing ether links, formed in the synthesis of polyesters as side products, were determined by the method discussed by Gaskill *et al.* [127], Dinse and Tucek [128] (diethylene glycol) and also Schöllner and Kowal [129] (diglycerol).

Thin Layer Chromatography

Minor quantities of diethylene glycol in poly(ethylene terephthalate) were determined by thin layer chromatography by Janssen *et al.* [130]. According to these authors this technique is actually less precise but more rapid than the gas-liquid chromatographic method.

Colorimetry

Propylene glycol in admixture with other polyols may be determined colorimetrically [76] using the colour reaction of acetaldehyde, formed in the periodate oxidation reaction, with hydrazine hydrate and sodium nitroprusside.

REAGENTS

Sodium Periodate. A 6·6% solution.

Hydrazine Hydrate. A saturated solution, freshly prepared.

Sodium Nitroprusside. A 4% solution, freshly prepared.

PROCEDURE. An aqueous solution of polyhydric alcohols obtained as described previously (see p. 253) after the ether extraction is neutralised, evaporated to a volume of 80 ml, filtered into a 100 ml volumetric flask, and made up to the mark. An aliquot of this solution, corresponding to 5–60 mg of formaldehyde, is transferred to a 100 ml distillation flask; 25 ml of 6·6% sodium periodate solution is added, and the contents are diluted to 50 ml. The reaction mixture is allowed to stand at room temperature for 1 hr, then it is distilled; 40 ml of the distillate is collected in an ice-cooled 50 ml volumetric flask, into which the condensate is introduced through a glass tube extending to the bottom of the flask. The contents of the flask are then diluted to the exact volume.

An aliquot of this solution not exceeding 6 ml, which corresponds to 400–1250 mg of propylene glycol, is transferred to a 15 ml test tube. Water to a volume of 6 ml is added followed by 1·5 ml of hydrazine hydrate solution and 0·5 ml of 4% sodium nitroprusside solution. The absorbance at 570 nm is measured immediately in 1 cm cells with respect to a reference solution containing the same quantities of reagents. Readings of the absorbance values are repeated at certain time intervals, and the maximum absorbance value is noted (usually after 1·5 min). Results of the determination are found from a calibration curve prepared with the use of a solution of propylene glycol containing 2–3 g of the glycol in 1 l. of water (the procedure followed is analogous to that used for the unknown). In order to plot a reasonable calibration curve at least four points are needed.

The ethylene glycol content in a mixture with propylene glycol is found from the difference between the total glycols (determined, after periodic acid oxidation to formaldehyde, with the help of a colour reaction with chromotropic acid), and the propylene glycol content (determined as described above [76]).

Infrared Spectroscopy

Infrared spectroscopy was employed in the determination of dipentaerythritol in pentaerythritol by Jaffe and Pinchas [69]. The unknown was first acetylated with acetic anhydride and the reaction product was dissolved in carbon tetrachloride and the absorbance measured at 1115 cm^{-1}.

NMR Spectroscopy

The ratio of diethylene glycol to ethylene glycol in polyterephthalate resin was determined from the intensities of the signals at 3·8τ (OCH_2 group) and 4·5τ (CH_2OCO group) in the NMR spectrum of a solution of the resin in *o*-chlorophenol (Harada and Ueda [131]).

Determination of Functional Groups

Carboxyl Groups in Alkyds and Unsaturated Polyesters

Polyester resins, in addition to unreacted COOH groups, always contain free anhydrides and acids in varying amounts. For this reason, determination of free COOH groups is feasible only when the low molecular weight impurities are completely removed. In commercial grade resins the total content of all acidic components is the acid number. This characteristic quantity is composed of total and partial acid number. The total acid number is determined by titration with alkalis or with alkoxides in aqueous acetone. The partial acid number is determined by titration with anyhydrous ethanolic potassium hydroxide in an anhydrous non-aqueous medium in exactly the same manner as the acid number of polymers (see p. 48).

As a titration medium *tert*-butanol or dimethylformamide is recommended by Schultz and Fijolka [132] who also suggest tetrabutylammonium hydroxide as titrant.

The determination of acidic components is treated in more detail in subsequent sections of this chapter.

Carboxyl Groups in Poly(ethylene terephthalate)

Carboxyl groups in poly(ethylene terephthalate) resins are usually determined by acidimetric titration. The analysis is commonly performed at elevated temperatures on account of the poor solubility of this polymer in the cold in common organic solvents. As a rule, mixtures of phenols and chloroform, or aniline, alone or with chloroform, are used as the polymer solvents and as the titration medium. Specifically, Chu [133] titrated poly(ethylene terephthalate) in a phenol-chloroform mixture at 80–100°C with a solution of potassium hydroxide in benzyl alcohol, and tetrabromophenol blue as indicator. Majewska and Warzywoda [134] dissolved the polymer in aniline or in an aniline-chloroform mixture and carried out the titration at 75°C with 0·05*N* sodium hydroxide, using phenolphthalein as indicator. The same solvent was used in analyses by Ešnerová [65] and Skwarski [135]. The analytical procedure of Skwarski [135] follows.

A finely ground sample of poly(ethylene terephthalate) (ca. 0·5 g) is dissolved in 20 ml of aniline. The solution is cooled to room temperature, then 15 ml of chloroform is added along with 3 drops of aurin indicator solution, and titrated with a 0·1*N* alcoholic sodium hydroxide solution until the colour changes from yellow to pink.

Carboxyl groups in poly(ethylene terephthalate) may also be determined by potentiometric titration of a polymer solution in an *o*-cresol-chloroform mixture at 90°C with 0·1*N* ethanolic potassium hydroxide, using a glass electrode–SCE system [136].

A recognised method of radiochemical determination of carboxyl groups (and also hydroxyl groups) in poly(ethylene terephthalate) involves treatment of the unknown with ^3H_2O. After isotopic exchange of active hydrogens, the radioactivity of the polymer is measured by a scintillation technique [137].

Hydroxyl Groups

In polyesters containing hydroxyl groups exclusively of the alcohol type which occur in nearly all polymers of this kind (with the exception of polycarbonates), the hydroxyl groups can be determined as a hydroxyl value as described on p. 52. However, drawbacks with this method are that pyridine and a catalyst are introduced into the sample, and also that the analytical procedure is time consuming. The methods employing acetylation of hydroxyl groups in the presence of acid catalysts do not suffer from these shortcomings.

The method of acetylation in the presence of acid catalysts as used for analysis of low molecular weight alcohols was employed by Fritz and Schenk [138]. Ethyl acetate was used as the reaction medium, and a number of acids as catalysts, namely hydrochloric, *p*-toluenesulphonic, perchloric and trichloroacetic. The best results were obtained with perchloric acid. Several authors have developed a number of methods of analysis of polyesters based on this study. Thus, Emelin and Tsarfin [139] used sulphuric acid as catalyst and ethyl acetate as the resin solvent. The unreacted excess of acetic anhydride, after hydrolysis with water, was determined by potentiometric titration with alcoholic potassium hydroxide. Stetzler and Smullin [140] also employed ethyl acetate as reaction medium and *p*-toluenesulphonic acid as catalyst. The reaction was carried out at 50°C for 5 min. Similar acetylation reaction conditions are recommended by Heidel *et al.* [141] and by Gawłowski [142], although the latter used butyl acetate as the reaction medium.

Perchloric acid catalysed acetylation was used by Schrötter [143] in the analysis of hydroxyl groups (as hydroxyl value) in alkyd resins.

Finally, Glinka and Majewska [144] used xylene as the reaction medium and sulphuric acid as catalyst. The method worked out by Schrötter [143] is described below.

REAGENTS

The Acetylating Mixture. Into a 250 ml measuring flask 1·2 g of 72% aqueous perchloric acid is introduced, followed by 150 ml of ethyl acetate; 8 ml of acetic anhydride is then added dropwise by pipette while cooling the solution. The mixture is cooled to a temperature of 5°C, 42 ml of acetic anhydride is carefully added and after standing for 30 min the volume is made up to the mark with ethyl acetate.

Potassium Hydroxide. A 0·5*N* alcoholic solution.

PROCEDURE. A resin sample is dissolved in 5 ml of ethyl acetate and 5 ml of the acetylating mixture is added from a burette (the burette with the reagent should be protected against evaporation). After exactly 6 min 2 ml of water is added, the contents are mixed, and 10 ml of a (3:1) pyridine-water mixture is added. Free acetic acid is titrated with 0·5*N* alcoholic potassium hydroxide with phenolphthalein as indicator.

The alcoholic OH groups in polyesters may also be determined aquametrically by means of acetylation followed by determination of the water produced [145].

Other titrimetric methods for determining hydroxyl groups in polyester resins (and also certain other polymers) are also worth mentioning. These include a method which involves reaction with phenyl isocyanate [146], the Zimmermann and Kolbig method [147] employing acylation with sulphobenzoic anhydride, and another by Zimmermann employing acylation with 3,5-dinitrobenzoyl chloride [148]. A method of high accuracy based on the reaction with phosgene was evolved by Bush *et al.* [149]. In this method, the excess of the volatile reagent is removed by evaporation, then the chloroformyl groups formed in the reaction with phosgene are hydrolysed with sodium hydroxide. The chloride ions produced from the hydrolysis are determined argentometrically after acidification of the solution.

A gasometric method, originally developed for epoxy resins and based on the reaction with lithium aluminium hydride [150], may be used for the determination of minor quantities of hydroxyl groups in polyesters such as polycarbonates.

The hydroxyl groups in polyesters and polyethers were determined by infrared spectrophotometry by Hilton [151] who measured the absorbance of the polymers under examination in a (1:9) chloroform-carbon tetrachloride mixture at 2·83 μm (polyesters) and 2·87 μm (polyethers). Similarly Murphy [152] determined the hydroxyl number of alkyd resins by measuring the absorbance of dichloromethane solutions at 2·85 μm.

End groups and the average molecular weight of some polyalkylene glycols and some polyesters were determined by NMR spectroscopy by Page and Bresler [153].

Phenolic Hydroxyl Groups

Phenolic hydroxyl groups may be determined by the lithium aluminium hydride procedure in the case of polycarbonates. However, this method fails to give results accurate enough for high polymers. This method yields the total active hydrogen content of all substances and groups present and therefore larger quantities of water interfere with the analysis. On the other hand, satisfactory results are achieved by the methods reported by Horbach *et al.* [154]: a colorimetric and an infrared method. The colorimetric method consists in the formation of a coloured complex with titanium ions in a methylene chloride–glacial acetic acid system. In the spectroscopic method the absorbance of polymer solutions in methylene chloride is measured at 3597 cm^{-1}. The colorimetric method is several times as sensitive as the infrared spectrophotometric method. However, the use of titanium tetrachloride, a highly moisture-sensitive reagent, requires the other reagents employed to be thoroughly dried. Precautions have to be taken in handling to prevent contamination by moisture from the air.

The author of this chapter has developed a method of colorimetric determination of phenolic hydroxyl groups in polymers including polycarbonates [155]. The method is based on a colour reaction with 2,4-dinitrobenzenesulphonyl chloride and piperazine. Compared with the titanium tetrachloride method [154] this method is more sensitive and less affected by moisture. The detailed analytical procedure is reported below.

REAGENTS

Triethylenediamine. A 1·5% benzene solution.

2,4-*Dinitrobenzenesulphonyl Chloride.* A 0·3% benzene solution.

Benzoic Acid. A 2% benzene solution.

Piperazine Hydrate. A 0·4% dimethylformamide solution.

PROCEDURE. A solution of the resin is made up in benzene† so that 1 ml contains ca. 0·4 μequiv. of phenolic hydroxyls; 1 ml of this solution is treated in a 10 ml volumetric flask with 1 ml of a 0·3% solution of 2,4-dinitrobenzenesulphonyl chloride in benzene, then immediately 0·6 ml of 1·5% triethylenediamine and 0·4 ml of 2% benzoic acid (benzene solutions) are added with vigorous shaking. After 5 min the flask is filled to the exact volume with the 0·4% dimethylformamide solution of piperazine. After 20 min the absorbance is measured at 390 nm with reference to a blank solution prepared from the same reagents mixed in the same order. Results of the determination are found from a calibration curve made with bisphenol A as reference.

Ester and Ether Groups

The total content of ester groups, both in the main polymer chain and elsewhere in the sample, is determined as a saponification number by reaction with alcoholic potassium hydroxide solution (p. 49). The saponification number of ethylene glycol and diethylene glycol polyesters was determined by Karpov [156] who hydrolyzed the polymer with a mixture of 10 ml of propanol and 25 ml of 1*N* aqueous sodium hydroxide. The excess of alkali was determined by titration with hydrochloric acid with phenol red as indicator. Minor amounts of phenoxyl groups occurring at the ends of polycarbonate molecules may be determined spectrophotometrically from the absorbance at 683 cm^{-1} in the infrared spectrum. Films 150 μm thick are employed in these measurements.

---

† In the case of an epoxy resin both the resin to be examined and benzoic acid are dissolved in a benzene-acetone (3:7) mixture.

Carbon-Carbon Double Bonds

Carbon-carbon double bonds in polyester molecules are due to unsaturated dicarboxylic acids. These bonds may be determined either by evaluation of the unsaturated acids content in the hydrolysate by one of the methods described above (polarography is best) or directly by the reaction with dodecyl mercaptan [157,158]. Acrylic and methacrylic esters also react with this reagent, but monomers such as styrene, vinylstyrene and allyl phthalate do not.

A method for the separate determination of vinyl and fumaric double bonds in polyester resins was worked out by Tarasov *et al.* [159]. The total C=C bond content was determined by bromination in the presence of mercuric sulphate (2*N* solution in 6*N* sulphuric acid) as catalyst, while the vinyl C=C bonds were determined by bromination without a catalyst (or else by iodination with Wijs reagent). The fumaric acid content was found by difference.

The content of *cis* and *trans* configurations in unsaturated maleic-fumaric polyesters was determined by infrared spectrophotometry by Reichert and Nollen [108] who measured the absorbance at 7·17 μm (*cis*) 10·2 μm (*trans*).

The C=C bonds in molecules of alkyd resins are due to unsaturated fatty acids. The total content of these bonds may be determined as an iodine number with the Hanus reagent or with Wijs reagent (p. 51).

Determination of Double Bonds by the Mercaptan Method. The method involves the addition reaction of *n*-dodecyl mercaptan to a double bond in alkaline solution according to the reaction:

$$\gt C{=}C\lt \; + \; HSR \rightarrow \; \gt \underset{RS}{\underset{|}{C}}{-}\underset{H}{\underset{|}{C}}\lt$$

The excess of unreacted mercaptan is titrated with a standard iodine solution in acid to a permanent yellow colour.

REAGENTS

Potassium Hydroxide. A solution of 1 part by wt. of potassium hydroxide in 20 parts by wt. of ethanol.

Dodecyl Mercaptan. A 5% solution in ethanol.

Iodine. A 0·1*N* solution.

PROCEDURE. A sample of 1–2 g of a polyester resin (containing or free of styrene) is dissolved in a 200 ml conical flask in 50 ml of a mixture of equal parts by weight of ethanol and toluene, or alternatively in ethyl acetate, chloroform or acetone alone. If the resin fails to dissolve, either another inert solvent is taken, or the analytical procedure is continued anyway, since the resin usually dissolves as a consequence of the mercaptan addition. In the latter case the solution is made alkaline after addition of the mercaptan and agitated until the resin is completely dissolved. The flask is flushed with a jet of nitrogen, then 20 ml of the mercaptan solution is added by pipette, the flask is sealed with a ground-glass stopper and the contents are shaken for some time. Next, 2–4 ml of the potassium hydroxide solution is added until alkaline; the flask is stoppered again and shaken. The mixture is allowed to stand for 3–5 min at room temperature with occasional shaking. After standing, the mixture is made acidic with glacial acetic acid and titrated with a standard iodine solution until a permanent yellow colour develops.

8.5 DETERMINATION OF OTHER CONSTITUENTS

Free Anhydrides

The determination of unreacted anhydrides in polyester resins offers a rather difficult analytical problem. The resins in this group always contain free carboxylic groups on the polyester molecules and also free acids. Both interfere with the common acidimetric method used for analysis of anhydrides.

In older papers, for example the paper by Klausch [160], the determination of free phthalic anhydride in alkyd resins is based on acidimetric visual titrations in two parallel runs in two media: a non-aqueous medium (acetone–alcohol), and aqueous acetone medium. The method is based, on the one hand, on the rapid reaction of anhydrides with alcohols in a non-aqueous medium in the presence of alkalis with the formation of monoesters, and on hydrolysis of anhydrides in a water-containing medium to form dicarboxylic acids on the other.

Fijolka has developed a method [161] which allows simultaneous determination of acid anhydrides, free acids and carboxyl end groups in polyesters. This method differs from the one mentioned above in that it employed a potentiometric or conductometric titration. In titrating resins containing free maleic or phthalic anhydride, three jumps are noticed on the titration curve with sodium ethoxide titrant in non-aqueous media. The first jump corresponds to neutralisation of the first COOH group of free maleic acid. The second jump arises from neutralisation of the monoester COOH group, which is formed from the reaction of the anhydride with alcohol, and also from neutralisation of the polymer COOH end groups. The third jump corresponds to neutralisation of the second COOH group of free maleic acid.

In titrating in a water-acetone medium with the same sodium ethoxide solution of the same type of resin, three potential jumps are also observed. The first one corresponds to neutralisation of one of the COOH groups of free maleic acid (both the acid contained in the resin and produced as a result of hydrolysis of the anhydride). The second jump is due to neutralisation of the polyester COOH group, and the third jump comes from reaction at the same groups as the third jump in the titration in a non-aqueous medium.

From the volume of the sodium ethoxide titrant used to produce the first potential jump in both titrations, both the free anhydride content and the free acid content can be evaluated. Furthermore, the difference in the ethoxide titrant volume used between the first and the second jumps in the acetone-water titration corresponds to the content of end COOH groups in polyester.

Resins containing HET anhydride† as an acid component are analysed in the same manner with the exception that potentiometric titration here is no more advantageous than visual titration. This method, however, makes possible the simultaneous determination of free anhydrides and free acids in resins containing both maleic anhydride and HET anhydride.

A colorimetric method of analysis of free maleic anhydride in unsaturated polyesters has been devised by Mlejnek [162]. The method is based on a colour reaction with dimethylaniline. The procedure, as improved later by Navyazhskaya and Sporykhina [163], is as follows. A resin sample of 0·1–1·5 g is dissolved in 3–3·5 ml of acetone at a temperature lower than 40°C; 5 ml of colourless dimethylaniline is added, the volume is made up to 10 ml with acetone and after 5 min absorbance is measured, using a blue filter. The minimum determinable concentration of maleic anhydride for this method equals 0·05 mg per ml.

Free Monomer

Chemical Methods

The usefulness of available chemical methods for the determination of styrene in unsaturated polyester resins containing units from unsaturated dicarboxylic acids, was examined by Demmler [157]. According to him, the most convenient is the method involving Wijs reagent.

This method is useful not only in analysis of styrene but also for the determination of other monomers like allyl phthalates, methyl methacrylate, vinyltoluene, allyl cyanurate, and also isolated and conjugated double bonds in unsaturated fatty acids. The Wijs reagent fails to react, however, with the double bonds from maleic and fumaric acid residues or with acrylate double bonds.

Procedures which involve addition of halogens to the monomer double bond, namely those employing Wijs reagent, dibromopyridine [31] or free bromine, are too low in selectivity to ensure adequate accuracy. The most suitable method for analysis of styrene was found to be that using mercuric acetate [164]. Mercuric acetate, like ICl, fails to react with the unsaturated acid double bonds present in polyester resins. In alcoholic solutions the following reactions take place:

(a) $$Hg(OOCCH_3)_2 + CH_3OH \rightarrow Hg\begin{cases} OCH_3 \\ OOCCH_3 \end{cases} + CH_3COOH$$

† Hexachloro-Δ^4-endomethylenetetrahydrophthalic anhydride.

(b) $Hg\begin{matrix} OCH_3 \\ OOCCH_3 \end{matrix} + R{-}HC{=}CH_2 \rightleftharpoons \underset{CH_3O}{R{-}HC}{-}\underset{HgOOCCH_3}{CH_2}$

(c) $\underset{OCH_3}{R{-}HC}{-}\underset{HgOOCCH_3}{CH_2} + HCl \rightarrow \underset{OCH_3}{R{-}HC}{-}\underset{HgCl}{CH_2} + CH_3COOH$

REAGENTS. Mercuric acetate (40 g) is dissolved in a 1000 ml volumetric flask in methanol containing 2 ml of acetic acid and made up to the mark with methanol.

PROCEDURE. Into a 200 ml conical flask a 100–400 mg sample of the unknown mixture is weighed (not more than 200 mg styrene, 220 mg vinyltoluene or allyl phthalate), and immediately 2 ml of 2-ethoxyethanol and 20 ml of the mercuric acetate solution are added. The flask is stoppered with a tight fitting stopper, and the contents mixed and cooled to 10–15°C.

After the required time interval (20 min, 4 hr and 20 hr, for styrene, vinyltoluene and allyl phthalate, respectively) 20 ml of glycol and 2–3 drops of thymol blue are added. The resulting mixture is titrated with 0·1*N* methanolic hydrochloric acid. A blank is run separately in the same manner, taking 20 ml of the reagent.

CALCULATION. The monomer content $X(\%)$ is evaluated from formula

$$X = \frac{(v_1 - v_2)nM}{10m}$$

where:

v_1 = the volume (ml) of 0·1*N* hydrochloric acid used to titrate 20 ml of the reagent (the blank),
v_2 = the volume (ml) of 0·1*N* hydrochloric acid used to titrate the test solution,
n = normality of the hydrochloric acid,
m = weight of the sample, g,
M = equivalent weight of the monomer (styrene 104·15, vinyltoluene 118·18, diallyl phthalate 123·13).

The monomer content in polyester resins may also be determined in terms of volatile matter by drying for 2 hr at 110°C.

Instrumental Methods

Styrene in polyester resins may be determined by gas-liquid chromatography after Esposito and Swann [165] after dissolving the resin in methyl ethyl ketone. A variety of other monomers, such as methyl methacrylate, styrene, vinyltoluene and allyl phthalate, can also be determined in liquid unsaturated polyesters [166].

Polymerised styrene in styrenated alkyds can be measured by the spectrophotometric method of Hirt *et al.* [167]. They dissolve the resins in cyclohexane and measure the absorbance at 269, 282, and 291 nm.

Styrene in polyester polymers may be determined by polarography with alcoholic tetraethylammonium chloride as supporting electrolyte [168].

Unhydrolysable Matter

Alkyd resins may contain amounts of unhydrolysable substances originating from unrefined oils used in their manufacture. These may also be modifying additives or mineral oils. The content of these substances can be determined by ether extraction of the alkaline solution after hydrolysis of the resin with aqueous sodium hydroxide (after ASTM D 1397–56T, DIN 53183).

PROCEDURE. A resin sample of 8–10 g containing 0·05–0·2 g of unhydrolysable material is dissolved in 10 ml of benzene and 50 ml of ethanol; 5 ml of 50% aqueous sodium hydroxide and 5 ml of water are then added and the mixture is heated under reflux on a steam bath for 2 hr. The hydrolysate is transferred to a 500 ml separating funnel, along with washings from the flask which is washed once with water and thrice with 25 ml portions of ether. The contents of the funnel are diluted with water to a volume of 300 ml, 10 ml of ethanol is added, the whole is well shaken, then the ether layer is separated. The extraction of the aqueous layer is repeated with three 20 ml portions of ether. The combined ether extracts are washed several times with water until neutral, then the extracts are evaporated on a water bath in a previously weighed beaker. The residue is dried in a vacuum oven at 80°C to constant weight.

Water

The moisture content was determined in poly(ethylene terephthalate) by Dinse and Praeger [169,170] who followed a method reported by Reid and Turner (see p. 58). This method yields results of high accuracy which coincide with the results obtained by the manometric method and by the vacuum drying procedure [170] (see Section 1.5).

The water content in the same polymer was determined in a similar manner by Zimmermann and Hoyme [171], the only difference being that water was removed in a nitrogen stream and not directly from the polymer but from the solution resulting from extraction of the resin with a (1:3) phenol-tetrachloroethane mixture.

Rough estimates of the water content in polycarbonates, as well as in certain other thermoplasts, may be obtained by observing the size of water vapour bubbles produced on melting the polymer [172].

Catalysts

Benzoyl peroxide in polyester resins was determined, using polarography, by Kladkevich *et al.* [173] who extracted this compound with benzene from the resin before the determination. The same method was followed by Edelmann and Wyden [174] in analysis of catalyst compounds of Sb, Pb, Zn, Co, Mn and Ge in polyester resins.

The transesterification or polycondensation catalysts (Sb, Co, Mn, Mg, Zn and P compounds) were determined in poly(ethylene terephthalate) by colorimetric methods by Zimmermann *et al.* [175]. Likewise Hoyme [176] determined Ge, Ti and Ca in this resin by colorimetric techniques.

Dimethylaniline, used as a hardening agent, was determined in unsaturated polyester resins by Uhde and Spranger [177] by means of thin layer chromatography of the chloroform extract of the resin.

Miscellaneous Contaminants and Additives

The titanium oxide filler present in poly(ethylene terephthalate) was determined polarographically by Dinse and Ewert [178].

Magnesium incorporated in the cured resin ("organically" bound) and not as a filler constituent, was analysed by Newell [179] who pyrolysed the resin at 340°C. The pyrolysate was extracted with buffer solution and the extract was titrated with a 0·05*M* EDTA solution.

8.6 DEGREE OF CURE OF UNSATURATED RESINS

The degree of cure of unsaturated polyester resins may be determined by extraction of soluble (unpolymerised) components with acetone or chloroform [180]. Literature relevant to other methods for determining the degree of cure is reported in the earlier section on polyester structure in this chapter (see p. 226). Recently Gnauck and Fijolka [181] determined the degree of crosslinking of cast unsaturated polyester resins by hydrolysing the resins with a mixture of *tert*-butyl alcohol, dioxan and 1*N* potassium hydroxide (7:1:2), followed by chloroform extraction of the hydrolysate.

References

1. Mleziva, J., Hanzlik, V., Kincl, J., Miklas, Z., *Polyestery, jeich vyroba a zpracovani* (*Manufacturing and Processing of Polyester Resins*), SNTL, Prague, 1964.
2. Ogorkiewicz, R. M., in *Polyesters, Vols. I and II*, Ritchie, P. D. (Ed), Iliffe Books, London, 1965.
3. Kłosowska-Wołkowicz, Z., Królikowski, W., Penczek, P., *Żywice i laminaty poliestrowe* (*Polyester Resins and Laminates*), WNT, Warsaw, 1969.
4. Schnell, H., *Plast. Inst. Trans.*, **28**, 143 (1960).
5. Klaban, J., Ordelt, Z., *Plaste u. Kautschuk*, **12**, 210 (1965).
6. Knödler, S., Funke, W., Hamann, K., *Makromol. Chem.*, **53**, 212 (1962).
7. Curtis, L. G., Edwards, D. L., Simons, R. M., Trent, P. J., Bramer, P. T., *Kunststoff-Rundschau*, **12**, 550 (1965).
8. Gulbins, E., Funke, W., Hamann, K., *Kunststoffe*, **55**, 6 (1965).
9. Szmercsänyi, I. V., Maros, L. K., Zahran, A. A., *J. Appl. Polymer Sci.*, **10**, 513 (1966).
10. Gupta, S. K., Thampy, R. T., *Makromol. Chem.*, **139**, 103 (1970).
11. Schnell, H., *Angew. Chem.*, **68**, 633 (1956).
12. Khramova, T. S., Urman, Ya. G., Mochalova, O. A., Medvedeva, F. M., Slonim, I. Ya., *Vysokomol. Soedin., Ser. A*, **10**, 894 (1968).

13. Brett, R. A., *J. Oil and Colour Chemists' Assoc.*, **41**, 428 (1958).
14. Levasseur, L., *Peint. Pigm. Vernis*, **30**, 44 (1954).
15. Tawn, A. R. H., *J. Oil and Colour Chemists' Assoc.*, **39**, 223 (1956).
16. Wright, H. J., Du Puis, R. N., *Ind. Eng. Chem.*, **36**, 1004 (1944).
17. Sedov, L. N., Pugachevskaya, N. F., Zil'berman, E. G., *Plast. Massy*, **1970**, [8], 6.
18. Hayes, B. T., Read, W. J., Vanghan, L. H., *Chem. Ind.* (*London*), **1957**, 1162.
19. Alekseeva, I. A., Spasskii, S. S., *Vysokomol. Soedin.*, **2**, 1645 (1960).
20. Lomakin, A. T., Sanzharovskii, A. T., Zubov, P. I., *Lakokrasoch. Mater. i ikh Primen.*, **1963** [6], 13.
21. Rosso, J. C., *Rev. Gen. Caoutch. Plast.*, **5**, 303 (1968).
22. Funke, W., Hamann, K., *Angew. Chem.*, **69**, 733 (1957); **70**, 53 (1958).
23. Funke, W., Gebhardt, W., Roth, H., Hamann, K., *Makromol. Chem.*, **28**, 17 (1958).
24. Gilch, H., Funke, W., Hamann, K., *Makromol. Chem.*, **31**, 93 (1959).
25. Bohdanecký, M., Mleziva, J., Šternschuss, A., Zvonař, V., *Makromol. Chem.*, **47**, 201 (1961).
26. Ghanem, N. A., *Makromol. Chem.*, **36**, 109 (1960).
27. Mleziva, J., Vladyka, J., *Farbe u. Lack*, **68**, 144 (1962).
28. Nollen, K., Funke, W., Hamann, K., *Makromol. Chem.*, **94**, 248 (1966).
29. Rybnikař, F., Ditrych, Z., Klácel Z., Ordelt, O., *Analýza a zkoušeni plastických hmot* (*Analysis and Testing of Plastic Materials*), SNTL, Prague, 1965.
30. Swann, M. H., *Anal. Chem.*, **29**, 1352 (1957).
31. Kasterina, G. N., Kalinina L. S., *Khimicheskie Metody Issledovaniya Sinteticheskikh Smol i Plasticheskikh Mass*, (*Chemical Methods of Analysis of Synthetic Resins and Plastics*), Goskhimizdat, Moscow, 1963.
32. Swann, M. H., *Anal. Chem.*, **21**, 1448 (1949).
33. Tawn, A. R. H., May, G. J., *J. Oil and Colour Chemists' Assoc.*, **40**, 528 (1957).
34. Arendt, I., Schenck, H. J., *Kunststoffe*, **48**, 111 (1958); **49**, 321 (1959).
35. Fijolka, P., Kayler, R., Lenz, I., *Kunststoffe*, **49**, 222 (1959).
36. Fijolka, P., Radowitz, W., Runge, F., *Plaste u. Kautschuk*, **10**, 521 (1963).
37. Braun, D., Geenon, H., *J. Chromatogr.*, **7**, 56 (1962).
38. Fijolka, P., Lenz, J., Runge, F., *Makromol. Chem.*, **26**, 61 (1958).
39. Haken, J. K., *Aust. Paint J.*, **13**, 11 (1967).
40. Esposito, G. G., Swann, M. H., *Anal. Chem.*, **34**, 1048 (1962).
41. Rawlinson, J., Deeley, E. L., *J. Oil and Colour Chemists' Assoc.*, **50**, 373 (1967).
42. Sadowski F., Kaufeld H., *Farbe u. Lack*. **69**, 597 (1963).
43. Luce, C. C., Humphrey, E. F., Guild, L. V., Norrish, H. H., Coull, J., Castor, W. W., *Anal. Chem.*, **36**, 482 (1964).
44. Hummel, D., *Kunststoff-, Lack- und Gummi-Analyse*, Carl Hanser Verlag, Munich, 1958.
45. Hummel, D. O., Scholl, F., *Atlas der Kunststoff-Analyse*, Carl Hanser Verlag, Munich; Verlag Chemie, Weinheim, 1968.
46. Luongo, J. P., *Anal. Chem.*, **33**, 1816 (1961).
47. Brudkowska, B., Hippe, Z., Jabłoński, H., *Chem. Anal.* (*Warsaw*), **11**, 497 (1966).
48. Stafford, R. W., Francel, R. J., Shay, J. F., *Anal. Chem.*, **21**, 1454 (1949); Stafford. R. W., Shay, J. F., Francel, R. J., *ibid.*, **26**, 656 (1954).
49. Shreve, O. D., Heether, M. R., *Anal. Chem.*, **23**, 441 (1951).
50. Percival, D. F., Stevens, M. P., *Anal. Chem.*, **36**, 1574 (1964).
51. Yeagle, M. L., *J. Paint Technol.*, **42**, 472 (1970).
52. Winterscheidt, H., *Seifen-Öle-Fette-Wachse*, **81**, 16 (1955).
53. Thinius, K., Schröder, E., *Chem. Techn.*, **8**, 395 (1956).
54. Placzek, L., *Kunststoffe*, **50**, 174 (1960).
55. Bergner, K. G., Sperlich, H. Z., *Lebensm.-Untersuch. u. -Forsch.*, **97**, 252 (1953).
56. Rejhová, H., Ulbrich, W., *Plaste u. Kautschuk*, **6**, 539 (1959).
57. Gasparič, J., Borecky, J., *J. Chromatogr.*, **5**, 466 (1961).
58. Fijolka, P., Radowitz, W., *Plaste u. Kautschuk*, **12**, 207 (1965).

59. Coates, V. J., *Gas Chromatography*, Academic Press, New York and London, 1958.
60. Ginsburg, L., *Anal. Chem.*, **31**, 1822 (1959).
61. Nadeau, H. G., Oaks, D. M., *Anal. Chem.*, **32**, 1760 (1960).
62. Clifford, J., *Analyst*, **85**, 475 (1960).
63. Esposito, G. G., Swann, M. H., *Anal. Chem.*, **33**, 1854 (1961).
64. Esposito, G. G., *Anal. Chem.*, **34**, 1173 (1962).
65. Ešnerová, J., *Plasticke Hmoty Kaučuk*, **2**, 237 (1965).
66. Mlejnek, O., *Chem. Průmysl*, **13**, 105 (1963).
67. Li Gotti, I., Piacentini, R., Bonomi, G., *Mat. Plastiche*, **36**, 213 (1970).
68. Shay, J. F., Skilling, S., Stafford, R. W., *Anal. Chem.*, **26**, 652 (1954).
69. Jaffe, J. H., Pinchas, S., *Anal. Chem.*, **23**, 1164 (1951).
70. Lichthardt, U., *Farbe u. Lack*, **63**, 387 (1957).
71. Kaufmann, H. P., Büscher, F. J., *Fette, Seifen, Anstrichmittel*, **62**, 1141 (1960).
72. Zieliński, W. L., Jr., Moseley, W. V., Jr., Bricker, R. C., *Official Digest*, **33**, 622 (1961).
73. Lorette, N. B., Brown, J. H. Jr., *J. Org. Chem.*, **24**, 261 (1959).
74. Shreve, O. D., *Synthetic Organic Coating Resins*, in *Organic Analysis*, *Vol. III*, Mitchell, J. Jr., Kolthoff, I. M., Proskauer, E. S., Weissberger, A. (Eds), Interscience, New York, 1956.
75. Swann, M. H., Adams, M. L., Weil, D. J., *Anal. Chem.*, **28**, 72 (1956).
76. Jones, J. R., Jr., *Alkyds*, Parker, E. E., *Polyesters*, in *High Polymers*, *Vol. XII*, *Analytical Chemistry of Polymers*, Kline, G. M., (Ed), *Part I*, *Analysis of Monomers and Polymeric Materials*, Interscience, New York, 1959, pp. 17 and 295.
77. Garn, P. D., Gilroy, H. M., *Anal. Chem.*, **30**, 1663 (1958).
78. Navyazhskaya, E. A., *Khim. Prom.*, **1960**, [6], 466.
79. Voigt, J., *Plaste u. Kautschuk*, **4**, 3 (1957).
80. Kappelmeier, C. P. A., (Ed), *Chemical Analysis of Resin-Based Coating Materials*, Interscience, New York, 1959.
81. Esposito, G. G., Swann, M. H., *Anal. Chem.*, **32**, 49 (1960).
82. Kruysse, L., *Verfkroniek*, **37**, 448 (1964).
83. Kreshkov, A. P., Bykova, L. N., Smolova, N. T., *Plast. Massy*, **1962**, [6], 51.
84. Swann, M. H., *J. Paint Technol.*, **40**, 468 (1968).
85. Zelenina, E. N., *Zh. Prikl. Khim.*, **43**, 1630 (1970).
86. Grimmer, J., *Chem. Průmysl*, **9**, 444 (1959).
87. Clasper, M., Haslam, J., *J. Appl. Chem.*, **7**, 328 (1957).
88. Zelenina, E. N., *Zavodsk. Lab.*, **34**, 1440 (1968).
89. Kreshkov, A. P., Yarovenko, A. N., Zel'manova, I. Ya., *Lakokrasoch. Mater. i ikh Primen.*, **1964**, [5], 49.
90. Kreshkov, A. P., Yarovenko, A. N., Zelenina, L. N., *ibid.*, **1965**, [4], 58.
91. Kreshkov, A. P., Balyatinskaya, L. N., *ibid.*, **1964**, [6], 43.
92. Kreshkov, A. P., Bykova, L. N., Smolova, N. T., *Plast. Massy*, **1964**, [10], 49.
93. Elving, P. J., Rosenthal, I., *Anal. Chem.*, **26**, 1454 (1954).
94. Warshowsky, B., Elving, P. J., Mandel, J., *Anal. Chem.*, **19**, 161 (1947).
95. Novák, V., *Plaste u. Kautschuk*, **12**, 202 (1965).
96. Novák, V., Foršt, V., *Chem. Průmysl*, **14**, 541 (1964).
97. Traxton, M. E., *Chem. Ind.* (*London*), **1966**, 1613.
98. Penczek, P., Lewandowska, T., *Polimery*, **13**, 59 (1968).
99. Novák, V., Krátký, B., *Plaste u. Kautschuk*, **19**, 493 (1972).
100. Fijolka, P., Ziebarth, G., *Plaste u. Kautschuk*, **11**, 140 (1964).
101. Krátký, B., Novák, V., *Chem. Průmysl*, **21**, 340 (1971).
102. Percival, P. F., *Anal. Chem.*, **35**, 236 (1963).
103. Esposito, G. G., *U. S. Gov. Res. Rep.*, **38**, 56 (1963).
104. Swann, M. H., Adams, M. L., Weil, D. J., *Anal. Chem.*, **27**, 1604 (1955).
105. Funasaka, W., Kojima, T., Fujimura, K., *Japan Analyst*, **13**, 42 (1964).
106. Salama, C., Dunn, R., *Can. Spectrosc.*, **12**, 175, 178 (1967).
107. Chiang, M., Bobalek, E. G., *Official Digest*, **31**, 1287 (1959).

108. Reichert, K. H., Nollen, K., *Farbe u. Lack*, **72**, 947 (1966).
109. Murano, M., Kaneishi, Y., Yamadera, R., *J. Polymer Sci., Pt. A*, **3**, 2698 (1965).
110. Reynolds, R. J. W., Walker, K. R., Kirby, G. W., *Polymer*, **11**, 333 (1970).
111. Opp, C. J., *Official Digest*, **36**, 42 (1964).
112. Haken, J. K., *J. Gas Chromatogr.*, **2**, 263 (1964).
113. Gerasimova, L. I., Zlobina, V. P., Ermolaeva, T. A., *Lakokrasoch. Mater. i ikh Primen.*, **1970**, [2], 58.
114. Paylor, R. A. L., Feinland, R., Conroy, N. H., *Anal. Chem.*, **40**, 1358 (1968).
115. Malaprade, M. L., *Bull. Soc. Chim. France*, **43**, 683 (1928).
116. Pohle, W. D., Mehlenbacher, V. C., Cook J. H., *Oil and Soap*, **22**, 115 (1945).
117. Siggia, S., *Quantitative Organic Analysis via Functional Groups*, Wiley, New York, 1954.
118. Allen, N., Charbonnier, H. Y., Coleman, R. M., *Ind. Eng. Chem., Anal. Ed.*, **12**, 384 (1940).
119. Pohle, W. D., Mehlenbacher, V. C., *J. Am. Oil Chemists' Soc.*, **24**, 155 (1947).
120. Kirby, J. R., Baldwin, A. J., Heidner, R. H., *Anal. Chem.*, **37**, 1306 (1965).
121. Mifune, A., Ishida, S., *Kogyo Kagaku Zasshi*, **65**, 824 (1962)
122. De la Court, F. H., Van Cassel, N. J. P., Van der Valk, J. A. M., *Farbe u. Lack*, **75**, 218 (1969).
123. Wittendorfer, R. E., *Anal. Chem.*, **36**, 930 (1964).
124. Esposito, G. G., *Army Coating and Chem. Lab., Aberdeen Proving Ground Rep. CCL–260* (1969).
125. Esposito G. G., Swann, M. H., *Anal. Chem.*, **41**, 1118 (1969).
126. Atkinson, E. R., Calouche, S. I., *Anal. Chem.*, **43**, 460 (1971).
127. Gaskill, D. R., Chasar, A. G., Lucchesi, C. A., *Anal. Chem.*, **39**, 106 (1967).
128. Dinse, H. D., Tucek, E., *Faserforsch. Textiltechn.*, **21**, 205 (1970).
129. Schöllner, R., Kowal, K., *Plaste u. Kautschuk*, **13**, 620 (1966).
130. Janssen, R., De Deyne, V., Vetters, A., *Makromol. Chem.*, **123**, 119 (1969).
131. Harada, T., Ueda, N., *Kobunshi Kagaku*, **22**, 685 (1965).
132. Schulz, G., Fijolka, P., *Plaste u. Kautschuk*, **15**, 407 (1968).
133. Chu, Ta-Kang, *Ko Fen Tzu T'ung Hsun*, **8**, 74 (1966).
134. Majewska, J., Warzywoda J., *Melliand Textilber.*, **43**, 480 (1962).
135. Skwarski, T., *Polimery*, **9**, 103 (1964).
136. Maurice, M. J., Huizinga, F., *Anal. Chim. Acta*, **22**, 363 (1960).
137. Možišek, M., Klimánek, L., *Plaste u. Kautschuk*, **15**, 99 (1968).
138. Fritz, J. S., Schenk, G. H., *Anal. Chem.*, **31**, 1808 (1959).
139. Emelin, E. A., Tsarfin, Ya. A., *Plast. Massy*, **1961** [3], 75.
140. Stetzler, R. S., Smullin, C. F., *Anal. Chem.*, **34**, 194 (1962).
141. Heidel, K., Dittmann, W., Stürzenhofecker, F., *Makromol. Chem.*, **99**, 27 (1966).
142. Gawłowski, J., *Chem. Anal.* (*Warsaw*), **12**, 527 (1967).
143. Schrötter, E., *Plaste u. Kautschuk*, **12**, 248 (1965).
144. Glinka, Z., Majewska F., *Polimery*, **11**, 167 (1966).
145. Mlejnek, O., *Chem. Zvesti*, **9**, 27 (1955).
146. Dreher, B., *Farbe u. Lack*, **67**, 703 (1961).
147. Zimmermann, H., Kolbig, C., *Faserforsch. Textiltechn.*, **18**, 536 (1967).
148. Zimmermann, H., Tryonadt, A., *Faserforsch. Textiltechn.*, **18**, 487 (1967).
149. Bush, D. G., Kunzelsauer, L. J., Merrill, S. H., *Anal. Chem.*, **35**, 1250 (1963).
150. Stenmark, G. A., Weiss, F. T., *Anal. Chem.*, **28**, 1784 (1956).
151. Hilton, C. L., *Anal. Chem.*, **31**, 1610 (1959).
152. Murphy, J. F., *Appl. Spectroscopy*, **16**, 139 (1962).
153. Page, T. F., Jr., Bresler, W. E., *Anal. Chem.*, **36**, 1981 (1964).
154. Horbach, A., Veiel, U., Wunderlich, H., *Makromol. Chem.*, **88**, 215 (1965).
155. Urbański, J., *Chem. Anal.* (*Warsaw*), **15**, 615 (1970).
156. Karpov, O. N., *Zh. Analit. Khim.*, **23**, 470 (1968).
157. Demmler, K., *Kunststoffe*, **55**, 840 (1965).

158. Kalinina, L. S., Nikitina, N. I., *Plast. Massy*, **1971**, [7], 55.
159. Tarasov, A. I., Volodina, V. I., Spasskii, S. S., *Zh. Analit. Khim.*, **21**, 360 (1966).
160. Klausch, W., *Farbe u. Lack*, **63**, 119, 162 (1957).
161. Fijolka, P., *Plaste u. Kautschuk*, **10**, 582 (1963).
162. Mlejnek, O., *Chem. Zvesti*, **11**, 425 (1957).
163. Navyazhskaya, E. A., Sporykhina, V. S., *Lakokrasoch. Mater. i ikh Primen.*, **1964**, [2], 52.
164. Schumann, L., Kleeberg, W., *Z. Anal. Chem.*, **213**, 421 (1965).
165. Esposito, G. G., Swann, M. H., *J. Gas Chromatogr.*, **3**, 192 (1965).
166. Matsumoto, K., *Kogyo Kagaku Zasshi*, **70**, 305 (1967).
167. Hirt, R. C., Stafford, R. W., King, F. T., Schmitt, H. R. G., *Anal. Chem.*, **27**, 226 (1955).
168. Ayres, W. M., Whitnack, G. C., *Anal. Chem.*, **32**, 358 (1960).
169. Dinse, H. D., Praeger, K., *Faserforsch. Textiltechn.*, **20**, 449 (1969).
170. Praeger, K., Dinse, H. D., *Faserforsch. Textiltechn.*, **21**, 37 (1970).
171. Zimmermann, H., Hoyme, H., *Faserforsch. Textiltechn.*, **18**, 393 (1967).
172. *Plastics Technol.*, **15**, [3], 12 (1969).
173. Kladkevich, G. P., Ivashchenko, L. N., Shapoval, V. I., *Ukr. Khim. Zh.*, **35**, 968 (1969).
174. Edelmann, K., Wyden, H., *Gummi Asbest Kunststoffe*, **23**, 96 (1970).
175. Zimmermann, H., Hoyme, H., Tryonadt, A., *Faserforsch. Textiltechn.*, **21**, 33 (1970).
176. Hoyme, H., *Faserforsch. Textiltechn.*, **22**, 419 (1971).
177. Uhde, W. J., Spranger, D., *Plaste u. Kautschuk*, **13**, 722 (1966).
178. Dinse, H. D., Ewert, K. H., *Faserforsch. Textiltechn.*, **21**, 541 (1970).
179. Newell, J. E., *Anal. Chem.*, **38**, 1205 (1966).
180. Klaban, J., *Plasticke Hmoty Kaučuk*, **8**, 129 (1971).
181. Gnauck, R., Fijolka, P., *Plaste u. Kautschuk*, **14**, 816 (1967).

Chapter 9

POLYAMIDES

9.1 INTRODUCTION

Polyamides represent linear polymers resulting from a polycondensation reaction of amino acids, lactams, or dicarboxylic acids with diamines. Polyamides from linear, unsubstituted aliphatic monomers are often called nylons. Polycondensation of amino acids for instance may be written as follows:

$$x[H_2N{-}(CH_2)_n{-}COOH] \rightarrow H{-}[{-}NH{-}(CH_2)_n{-}CO{-}]_x{-}OH + (x-1)H_2O$$

The amino acids employed are those in which the amino group is separated from the carboxyl group by at least five methylene groups. The acids of commercial interest are ε-aminocaproic, ω-aminoundecanoic, 7-aminoenanthic and 9-aminopelargonic.

The polycondensation reaction of diamines with dicarboxylic acids proceeds analogously:

$$x[H_2N - (CH_2)_n{-}NH_2] + x[HOOC{-}(CH_2)_m{-}COOH] \rightarrow$$
$$H{-}[{-}HN{-}(CH_2)_n{-}NH{-}CO{-}(CH_2)_m{-}CO{-}]_x{-}OH + (2x-1)H_2O$$

Polyamides of this type are synthesised from dicarboxylic acids such as adipic, sebacic, succinic, pimelic, azelaic, lauric, terephthalic and isophthalic. The other component is usually a primary aliphatic amine, in which the amino groups are at the ends of the methylene chain, or a phenylenediamine such as 1,5-naphthalenediamine, or piperazine. Of wide use industrially, are hexamethylenediamine and *m*-phenylenediamine, and the aromatic amine *m*-xylylenediamine.

In addition to simple polyamides mixed polyamides are prepared in a simultaneous condensation of polymeric plant oil acids with polyalkylamines.

The third group of polyamides comprises modified polyamides resulting from the reaction of a polyamide molecule with compounds capable of substituting the NH hydrogen of the polyamide. The commonest are *N*-hydroxymethyl derivatives [1], that is, the products of the reaction of formaldehyde with polyamides as follows:

$$\begin{matrix} | \\ N{-}H \\ | \\ C{=}O \\ | \end{matrix} + CH_2O \rightleftharpoons \begin{matrix} | \\ N{-}CH_2OH \\ | \\ C{=}O \\ | \end{matrix}$$

Owing to the reactivity of the *N*-hydroxymethyl group in these intermediates, *N*-alkoxymethyl (I) and *N*-alkylthiomethyl (II) derivatives are easily formed in the presence of alcohols and mercaptans respectively:

$$\begin{array}{l}|\\ \text{N—H}\\ |\\ \text{C=O}\\ |\end{array} + CH_2O + ROH \rightleftharpoons \begin{array}{l}|\\ \text{N—CH}_2\text{OR}\\ |\\ \text{C=O}\\ |\end{array} + H_2O \qquad \text{(I)}$$

$$\begin{array}{l}|\\ \text{N—H}\\ |\\ \text{C=O}\\ |\end{array} + CH_2O + RSH \rightleftharpoons \begin{array}{l}|\\ \text{N—CH}_2\text{SR}\\ |\\ \text{C=O}\\ |\end{array} + H_2O \qquad \text{(II)}$$

N-Phenoxymethyl derivatives are obtained from the reaction of formaldehyde in the presence of phenols (or cresols). Other examples of modified polymers are the products of their reaction with ethylene oxide:

$$[\text{—HN—(CH}_2)_n\text{—NH—CO—(CH}_2)_m\text{—CO—}]_x + y\ \underset{\text{O}}{\text{CH}_2\text{—CH}_2} \rightarrow$$

$$\left[\begin{array}{l}\text{—NH—(CH}_2)_n\text{—N—CO—(CH}_2)_m\text{—CO—}\\ \qquad\qquad\qquad\quad |\\ \qquad\qquad\quad \text{(CH}_2\text{—CH}_2\text{—O)}_y\text{—H}\end{array}\right]_x$$

Polyoxide chain growth rather than polymer hydroxyethylation occurs in this reaction. The product is an example of a graft copolymer. These products are formed from polyamides activated with ozone [2] or γ-radiation [3–8].

The numerical notation employed in the chemical name of nylons or polyamides consists of two numbers, the first of which refers to the amine and the second to the acid component. The numbers correspond to the number of carbon atoms present in the molecules of the two monomers. Also, the copolymer name is followed by the ratio of the copolymer constituents in parentheses. Although a wide variety of polyamides may be synthesised [9–15], only a few are of commercial interest, notably nylons-6, 11, 12, nylon-6,6 and nylon-6,10 and their copolymers, and modified nylons-6, and 6,6 [16]

Individual polyamides exhibit diverse physicochemical properties such as melting point, density, solubility and water absorption [17–23]. Melting point is considerably affected by the number of methylene groups occurring between the polar substituents in the polyamide chain unit. The longer the polymethylene chain of the diamine and of the acid component, the lower is the melting point of the polyamide. Also, the polyamides with a diamine constituent containing an odd number of methylene groups have lower melting points compared with their counterparts containing an even number of methylene groups. Analogous correlation is observed in polyamides produced from amino acids.

Solvents for simple polyamides are concentrated acids such as sulphuric, phosphoric, formic and acetic, and also 70% chloral hydrate solution, alcoholic solutions of certain salts such as calcium chloride, and ethylene chlorohydrin. At elevated temperatures these polymers dissolve in higher alcohols (for example benzyl alcohol), in anhydrous glycols and in dimethylformamide. At about 150°C and under increased pressure simple polyamides also dissolve in lower aliphatic alcohols.

Modified polyamides exhibit lower melting points, a lower modulus of elasticity, and inferior tensile strength [24]. Polyamide-6,6 with a 50% substitution with *N*-hydroxymethyl groups melts at 100°C [25] and elastic recovery of the material is substantially enhanced. The polyamides in which half of the amide groups are substituted exhibit rubber-like properties.

Of greatest interest technologically is the change in the properties of modified polyamides related to their increased solubility in common organic solvents such as lower alcohols or their mixtures with chloroform. Aqueous alcoholic solutions containing 60–90% alcohol are better solvents than alcohol alone.

Formaldehyde-modified polyamides are easily converted into crosslinked structures (by increased temperature or organic acids) which are insoluble in alcohol, *m*-cresol, or even in formic acid. Such polyamides undergo acid and alkaline hydrolysis but the alkoxymethyl and alkylthiomethyl derivatives are hydrolysed only in acid conditions.

Ethylene oxide-polyamide copolymers combine the individual properties of both plastics, notably the solubility and high m.p. of the polyamide, and the elasticity characteristic of poly(ethylene oxide). The side chains of poly(ethylene oxide) act as an internal plasticiser [26]. Ethylene oxide grafted polyamides containing carboxylic end groups yield insoluble products on account of the crosslinks formed. The polyamides without carboxylic groups provide soluble grafted products [27]. Polyamides containing phenyl, cyclohexyl [28–32] or thiophene rings, or those with short aliphatic chains, exhibit high thermal stability.

Copolymers of two or more polyamides also exhibit a distinct change in properties. They are usually more soluble, softer, with a lower m.p. and lower tensile strength than the homopolymers. The change in properties of a copolymer is roughly commensurate with the mole fraction of its constituents [33].

All polyamide types are decomposed by heat or light or the action of chemical agents.

9.2 QUALITATIVE ANALYSIS

Chemical Methods

Numerous methods are known for identification of polyamides based on their behaviour in a flame [34], thermal decomposition [35], m.p. determination, nitrogen content, density, solubility [36,37] and colour reactions [38–40]. Unmodified polyamides may be identified following the diagram given below, which is based on the solubility of polyamides in formic acid and on characteristic colour reactions [39–42].

A polyamide sample (ca. 0·1 g) is dissolved in 1 ml of conc. formic acid at room temperature (dissolution time about 2 hr) and is treated according to the diagram:

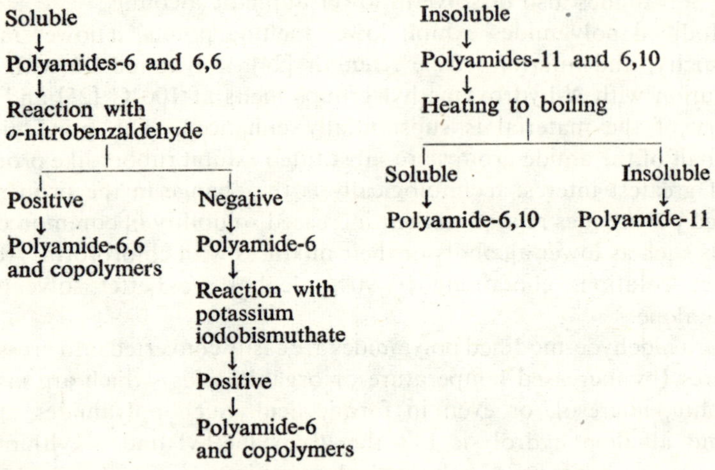

Colour Test with o-Nitrobenzaldehyde for Adipic Acid [39]

A 0·2 g sample of the polyamide is heated with a low flame in a test tube equipped with a glass tube to lead off the pyrolysis products to another test tube containing 2 ml of saturated *o*-nitrobenzaldehyde solution in 2*N* potassium hydroxide. This solution is then heated to boiling and becomes red-brown in the presence of adipic acid due to its pyrolysis to cyclopentanone.

Colour Test with Potassium Iodobismuthate for Caprolactam [39]

A 0·5 g sample is heated with 50 ml of distilled water at constant boiling for 10–15 min. On cooling 2–3 drops of conc. sulphuric acid are added to an aliquot of 0·5 ml of the solution, followed by addition of 2 ml of potassium iodobismuthate (a solution of 5 g of basic bismuth nitrate and 25 g of potassium iodide in 10 ml of 2% sulphuric acid). Precipitation of an orange-red complex $(C_6H_{11}ON)_3 . 2BiI_3 . 6HI . 2H_2O$ indicates the presence of caprolactam. When caprolactam is present, treatment of the aqueous extract of nylon-6 with *m*-dinitrobenzene in 2*N* aqueous alcoholic potassium hydroxide produces a pink-purple colour.

Nylons-6, 6,6 and 6,10 can be distinguished by determining their solubilities in hydrochloric acid solution at various temperatures and concentrations [36].

Another method of polyamide identification [43] involves dissolving the sample in a hot polyol and allowing the solution to cool to precipitate the polymer. The mixture is reheated and the temperature at which the polymer redissolves is noted. This temperature is characteristic of an individual polyamide type.

Ninhydrin Test for Caprolactam

A 5 g sample of finely powdered polymer is refluxed with 30 ml of distilled water for 10–15 min. A 10 ml aliquot of this solution is then treated with 20 drops of conc. hydrochloric acid and reheated for 2–5 min. On cooling 5 drops of 1% aqueous ninhydrin are added; the mixture is neutralised with 1*N* sodium hydroxide until a yellow colour appears, and then heated to boiling. In the presence of caprolactam the solution turns violet-blue. For solutions containing low concentrations of caprolactam the heating with distilled water is continued for more than 15 min.

Differentiation of Nylon-6 from Nylon-6,6

A 0·1 g powdered nylon sample is treated with a mixture of 10 parts by wt. phenol, 10 parts by wt. lactic acid, 20 parts by wt. glycerol and 22 parts by wt. distilled water saturated with the dye cellulose blue. After 30 sec the sample is washed with distilled water. Under such conditions nylon-6 becomes coloured, whereas nylon-6,6 remains unchanged.

Differentiation of Polyamides from Polyurethanes with the Sanger Reagent [44]

A 0·1 g powdered sample is heated for about 10 min with the Sanger reagent (2% *m*-dinitrofluorobenzene solution in 66% ethanol). A yellow colour indicates the presence of polyamide. Polyurethanes give no colour.

Certain properties of major polyamide types listed in Table 9.1 may prove of assistance in identification of these polymers.

Instrumental Methods

Polarography [49]

A 0·5 g polyamide sample is hydrolysed with 30 ml of 6*N* hydrochloric acid for 3 hr. The chilled hydrolysate is made alkaline to pH 10 with solid potassium hydroxide, 10 ml of 36% formalin (technical grade) is carefully added dropwise and the mixture is allowed to stand for 30 min. The Schiff bases produced in the reaction with formaldehyde due to the —CH=N group in the molecule, undergo reduction at a dropping mercury electrode. The precipitate formed is filtered off, and the filtrate is acidified to pH 6·7 with hydrochloric acid, and used in the determination.

Five ml of background electrolyte (4·5 ml 1*N* lithium acetate and 0·5 ml 1*N* acetic acid) is placed in the polarographic cell, followed by 2–4 ml of the solution under examination. Nitrogen is bubbled through for 3–5 min, and a polarographic recording is made. The method is suitable for the identification of all polyamides whose polarographic waves are at a more negative potential than the formaldehyde wave.

Spectrophotometric Methods

Identification of polyamides may also be made by infrared spectrophotometry using the film technique [50]. The spectra of copolymers exhibit

Table 9.1 Certain Properties of Nylons [45, 46, 47, 48]

Nylon type	*M.p.,** °C	*Nitrogen content†,* %	$d^{20°}$,** g per cm³	*Odour*	*Appearance of the hydrolysate§*		*M.p.††* °C
					when hot	*when cold*	
Nylon-6	214–222	12·4	1·21–1·14	burning horn	clear solution	clear solution	123–125 (21·2% Cl)
Nylon-11	190–199	7·6	1·05–1·07	castor oil	clear solution	cryst. 11-aminoundecanoic acid hydrochloride ppte	143–145 (14·9% Cl)
Nylon-6,6	258–264	12·4	12·1–1·14	burning horn	clear solution	precipitate of adipic acid crystals	150–152
Nylon-6,10	208–216	9·9	1·07–1·09	burning horn	sebacic acid crystals precipitate	more sebacic acid crystals precipitate	133–134

* As determined on a Kofler hot stage microscope.
† By the Kjeldahl method.
** Measured in $ZnCl_2$ solution
§ Appearance of the solution is examined after hydrolysis with conc. HCl (1 g sample in 1 ml HCl).
†† Of the acid or the amino acid hydrochloride.

insufficient features to identify the polyamide type. The features in the infrared spectra of polyamides (Fig. 9.1) are the bands due to the peptide link. The major absorptions are the following: the band arising from the —NH— group (3300 cm^{-1}), the amide band related to the carbonyl stretching (1650 cm^{-1}), and the amide band related to —NH— deformations (1550 cm^{-1}). Although these bands fail to differentiate polyamides from

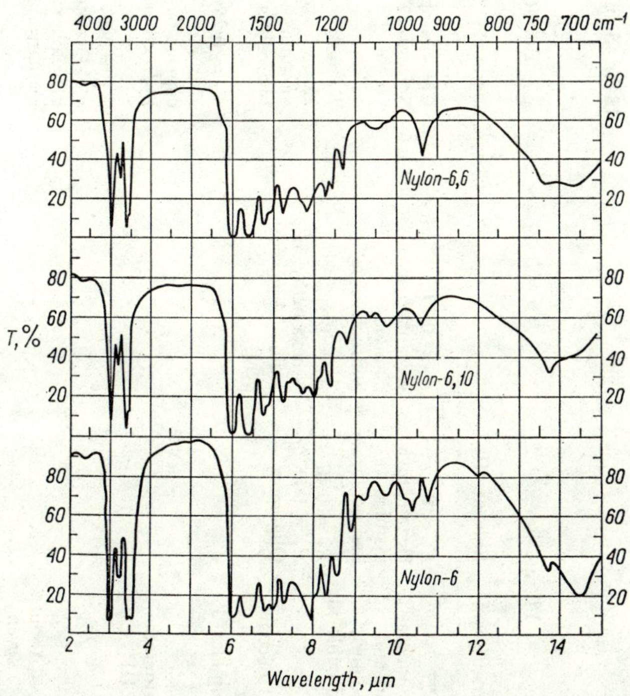

Fig. 9.1 Infrared absorption spectra of some polyamides [50] (by permission of the copyright holders, John Wiley & Sons, Inc.)

other compounds containing the —NH— group, the remaining bands exhibit characteristic differences which make possible the differentiation of individual polyamide types. Infrared spectra in the range 800–200 cm^{-1} of 18 polyamides were measured by Matsubara *et al.* [51]. The spectra were taken from films cast from formic acid solutions of the polyamide. The position of the amide group absorption was established and a number of correlations of the absorption and the polyamide structure were found.

Ultraviolet spectra were used by Turska *et al.* [52] for identification of ε-caprolactam and nylon-6.

If the sample cannot be directly analysed by infrared spectroscopy it can be thermally decomposed and the soluble decomposition products examined [53]. Rothe *et al.* [54] use infrared spectra for determination of *N*-acyl-lactam groups in nylon-6.

Table 9.2 Diagram for Identification of Polyamides (from [58], by permission of Carl Hanser Verlag, München)

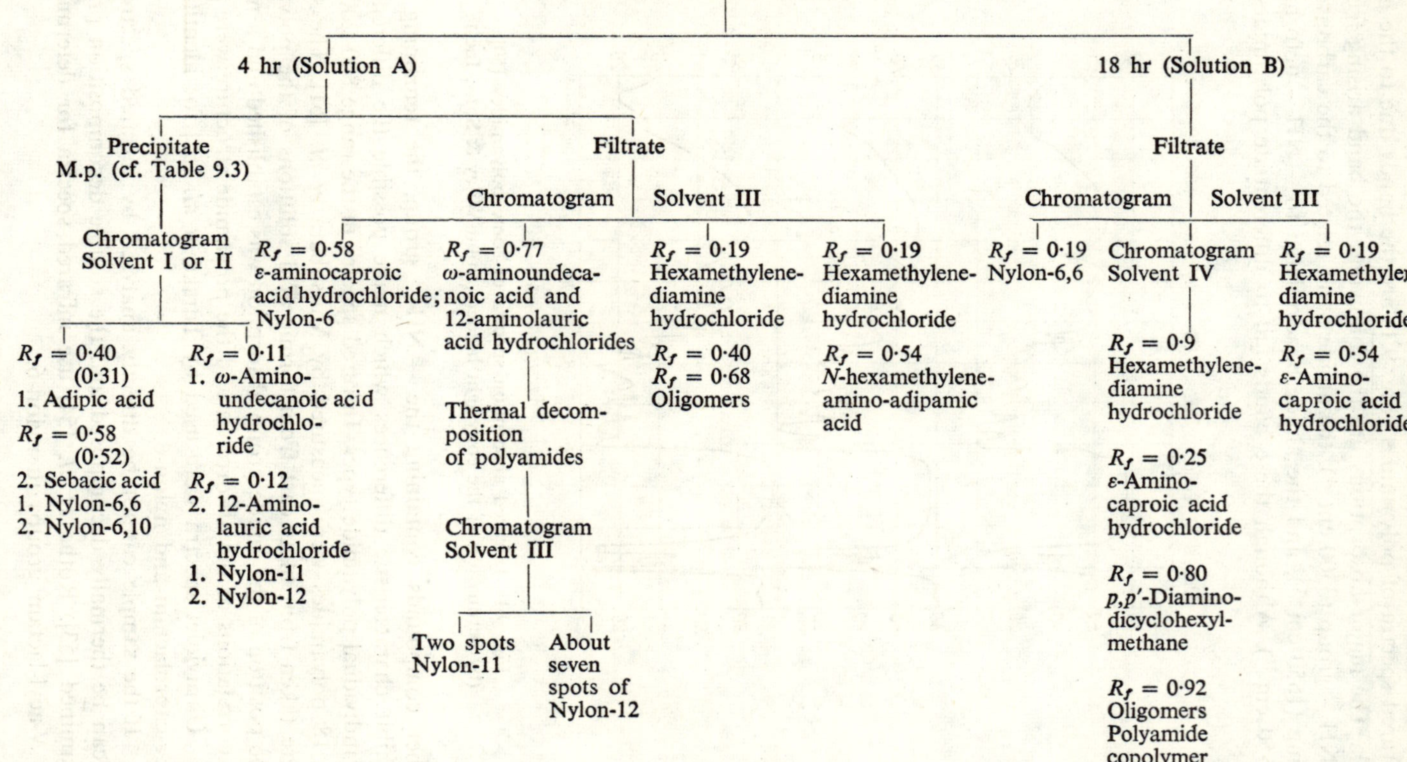

Chromatographic Methods

The following procedure for identification of polyamides by paper chromatography was reported by Ayres [55].

A 1% solution of polyamide in 90% formic acid is used; 0·25 ml of this solution is spotted on a Whatman 54 paper, and the chromatogram is developed using 88% formic acid in an atmosphere saturated with water vapour at 20±2°C. After drying the paper the chromatogram spots are detected with a 0·1% solution of Blue 2 BS in 0·1% acetic acid at 80°C, then the R_f value which is characteristic of the individual polyamide type under specified conditions, is calculated. The R_f values for nylons-6, 6,6, and 6,10 are 0·8, 0·4, and 0·3, respectively. When 90% formic acid is used as developing solvent the R_f values obtained are identical for nylons-6 and 6,6, whereas 84% formic acid fails to make nylon-6,10 move from the starting line.

Polyamides are also identified by their hydrolysis products by means of paper chromatography, as described by Zahn *et al.* [56,57].

A 50 mg sample of polyamide is hydrolysed in a sealed tube with 0·5 ml of 6*N* hydrochloric acid at 110°C for about 10 hours. Next, the hydrochloric acid is removed under vacuum at 100°C, the residue is dissolved in 2·5 ml of water and extracted with ether to isolate the acidic component of the polyamide. The aqueous layer containing amine hydrochlorides is neutralised with dilute sodium hydroxide. Ether is removed from the extract by evaporation and the acids are dissolved in a suitable volume of water to make approximately a 1% solution. Three 3 μl spots are placed on a Whatman 1 paper (20×30 cm) in a single run, and the chromatogram is developed by the ascending technique using a (60:30:10 by vol.) propanol–25% aqueous ammonia–water solution. The chromatogram is visualised with 0·04% bromothymol blue at pH 10. Under those conditions adipic acid has an R_f = 0·5 and sebacic acid, R_f = 0·7. For the separation of hexamethylenediamine and ε-aminocaproic acid a better developer is a solution of butanol–formic acid–water (75:15:10, by vol.) In both procedures the amine chromatograms are detected with ninhydrin reagent, followed by drying at 105°C for about 5 min to achieve a deep violet colour. The R_f values for ε-aminocaproic acid and hexamethylenediamine are 0·60 and 0·16 respectively.

The Ninhydrin Reagent. A 0·5% ninhydrin solution in butanol–2*N* acetic acid (95:5 by vol.).

A thin layer chromatography procedure was developed for the identification of several polyamides by Braun and Vorendohre [58]. It is illustrated by the diagram in Table 9.2, and some of the properties of the polyamide hydrolysis products are listed in Table 9.3. According to this procedure, a 2 g polyamide sample is hydrolysed by heating with 20 ml of 20% hydrochloric acid for 4 hr. Although hydrolysis is incomplete, a sample of the solution is withdrawn for identification (Solution *A*). The remainder

(Solution *B*) is further hydrolysed for another 14 hr, when nylon-6,6 and its copolymers react. Solution *A* is chilled, or even left overnight to allow the acids to crystallise before analysing the copolymers. The precipitate is filtered off, washed with ether, and the unhydrolysed polyamide is separated. The remaining clear filtrate is examined for the nylon-6 content.

Table 9.3 Properties of the Polyamide Hydrolysis Products
(from [58], by permission of Carl Hanser Verlag, München)

Polyamide	*M.p. of the crystalline hydrolysis product,* °C	*Appearance of the hydrolysate after* 4 hr *hydrolysis with conc. HCl at 120*°C
Nylon-6	no crystals	clear solution
Nylon-6,6	149–150	a few crystals of adipic acid
Nylon-6,10	132-134	many sebacic acid crystals
Nylon-11	172–178	the polyamide does not undergo hydrolysis
Nylon-12	150–155	the polyamide does not undergo hydrolysis
Polyamide copolymer 6/6,6	149–150	adipic acid crystals
Polyamide copolymer 6/6,6/DCHM	149–150	adipic acid crystals

Identification of Acids. A small quantity of the precipitate is dissolved in 1 ml of hot methanol and on cooling a 2 μl portion of the solution is applied to a silica-gel G covered chromatographic plate. Two solvent systems areused in developing the chromatogram, namely

I. (90:16:8, by vol.) benzene–methanol–acetic acid
II. (90:25:4, by vol.) benzene–dioxan–acetic acid

The coating length is 10 cm and the developing time 20 min.

To remove acetic acid the plate is heated in a drier at 120°C for 30 min. On cooling a portion of the plate is sprayed with bromocresol green reagent (see p. 284) and the other portion with a modified ninhydrin solution (see p. 284). It is dried in a hot air stream. The ninhydrin-sprayed portion of the plate exhibits pink to violet spots which appear near the starting line and indicate (the hydrochlorides of) ω–aminoundecanoic and 12-aminolauric acids with the same R_f value. Other polyamides do not produce spots in that area. If, on the bromocresol green-sprayed portion of the plate, a yellow spot appears against a blue background, with $R_f = 0{\cdot}40$ for the developer I, or with $R_f = 0{\cdot}31$ for developer II, the presence of adipic acid is indicated. Sebacic acid gives rise to spots with $R_f = 0{\cdot}58$ or 0·52 for developers I and II respectively, but these are unstable and disappear in time. The difference between the R_f values for adipic and sebacic acids is large enough for identification. For an additional check on the analysis simultaneous chromatograms of standard samples of adipic and sebacic acids may be run, or else the chromatograms obtained in the analysis may be compared with relevant literature data†. The presence of the two acids just discussed indicates nylons-6,6 and 6,10 or their copolymers with polyamides based on adipic or sebacic acids. The products of hydrolysis of polyamides-11 and 12 are found on the portion of the plate sprayed with bromocresol green reagent, wheret hese should appear as light yellow or blue spots near the starting line.

Identification of the Filtrate from Hydrolysates A and B. Solution *A* (1 ml) is diluted with 0·5 ml of methanol. A 2 μl aliquot of this solution is applied to a chromatographic plate coated with silica gel G previously dried in air. The chromatogram is developed for 70–90 min, to a length of 10 cm, with *n*-butanol, acetic acid, and water (80:20:20 by vol.)—developer III. Next, the plate is sprayed with ninhydrin solution and dried in an air stream. The spots produced correspond to ε-aminocaproic acid hydrochloride ($R_f = 0{\cdot}58$), ω-aminoundecanoic acid hydrochloride and 12-aminolauric hydrochloride ($R_f = 0{\cdot}77$). Three spots with $R_f = 0{\cdot}19$, 0·40 (weak) and 0·68 arise from nylon-6,10. The lower spot corresponds to hexamethylenediamine hydrochloride, whereas the other two are due to oligomers resulting from prolonged hydrolysis. Two spots of $R_f = 0{\cdot}19$ and 0·54 indicate nylon-6,6 or its copolymer. The lower spot is from hexamethylenediamine hydrochloride and the upper one from ε-aminocaproic acid or from *N*-hexamethyleneamino-adipamic acid, $NH_2(CH_2)_6NH—CO—(CH_2)_6—COOH$, which does not appear on prolonged hydrolysis (Solution *B*). If these two spots also appear on the chromatogram of Solution *B*, the lower spot and the upper spot correspond to hexamethylenediamine hydrochloride and ε-caproic acid, as before. This indicates the presence of 6/6,6 copolymer.

In the case of polyamides it has to be additionally ascertained whether or not *p,p'*-diaminodicyclohexylmethane as well as hexamethylenediamine is present. To do this, 1 μl of Solution *B* is applied to a silica-gel G coated plate in two places for development with two different eluting agents. A 10 cm length of the chromatogram is developed with a solvent system consisting of chloroform, methanol, 25% aqueous ammonia, and water (45:45:7:3 by vol.) (Solvent IV). The plate is dried in a warm air stream, then part of it is sprayed with ninhydrin reagent and dried until violet spots appear. The spot with $R_f = 0{\cdot}10$ is due to hexamethylenediamine hydrochloride, that with $R_f = 0{\cdot}25$ corresponds to ε-aminocaproic acid, and that with $R_f = 0{\cdot}95$ to oligomers. Subsequently the whole plate is sprayed with Blue B salt solution, dried in warm air, sprayed with 0·1*N* sodium hydroxide and dried again in warm air. The first spots to appear are those on the part of plate sprayed with ninhydrin, and are due to *p,p'*-diaminodicyclohexylmethane with $R_f = 0{\cdot}80$. This procedure has the advantage of allowing the latter diamine to be easily distinguished from possible

† Hummel, D. O., *Atlas der Kunststoff-Analyse*, Vol. 1, *Hochpolymere und Harze*, Carl Hanser Verlag, Munich, 1968, pp. 1582, 1586.

oligomers with $R_f = 0{\cdot}92$, since only this diamine develops a colour with Blue B salt reagent as intense as that of the spots on the previously unsprayed portion of the plate. Exact R_f values can be measured only after several hours, as the compound gives rise to a 2 cm tail that gradually vanishes. In this procedure ε-caprolactam/polyamide-6,6 copolymer (copolyamide 6/6,6) can be distinguished from its copolymer analogue with *p,p′*-diamino-dicyclohexylmethane (copolymer 6/6,6/DCHM).

Differentiation of Nylon-11 from Nylon-12. In acidic hydrolysates ω-aminoundecanoic and 12-aminolauric acids cannot be distinguished, and an additional test is required. This involves pyrolysis of a 70–80 mg sample of polyamide by heating it in a beaker over a small flame. Some vapours may evolve in the process. The pyrolysate is dissolved in 0·5 ml of methanol and warmed for a short period; 1·5 μl of this solution is applied to a thin layer plate coated with silica gel G and a 10 cm chromatogram is developed using Solvent III. After drying, the plate is sprayed with the ninhydrin reagent and dried in warm air. Nylon-11 gives rise to two spots, one of lower intensity than the other. Several spots appear from nylon-12.

REAGENTS FOR DETECTION OF CHROMATOGRAMS

Bromocresol Green. The indicator (40mg) is dissolved in 100 ml of ethanol and 0·1*N* sodium hydroxide solution is added until a blue colour develops.

Modified Ninhydrin Reagent. Solution I: 0·1 g of ninhydrin is dissolved in a mixture containing 50 ml of ethanol, 10 ml of glacial acetic acid and 2 ml of 2,4,6-collidine. Solution II: 0·5 g of cupric nitrate is dissolved in 50 ml of ethanol. Solutions I and II are combined prior to their use in quantities 50:3 parts by vol. respectively.

Blue B Salt. A 5% solution in water.

The technique of gas-liquid chromatography was used in the analysis of homo- and copolyamides by Tengler [59].The preparation of the material for the analysis involves hydrogen chloride hydrolysis, followed by conversion of the hydrolysate components into their trimethylsilyl derivatives.

9.3 CHEMICAL COMPOSITION OF THE POLYMER

Determination of Constituents

Quantitative analysis consists of the hydrolysis of polyamides and the subsequent determination of the polyamide components in the hydrolysate. A variety of hydrolysing agents and reaction times are employed by diverse authors for hydrolysis of a specific polymer type.

For the hydrolysis of nylons-6 and 6,6 26% hydrochloric acid and reaction times of several hours are used by Haslam [46], whereas for nylon-11 the same investigator uses 18% hydrochloric acid for 40 hr. For hydrolysis of nylons-6 and 6,6 Frey [36] uses 3*N* hydrochloric acid for ten or more hours and 6*N* hydrochloric acid for polyamide-6,10. Nylon-6,6 is hydrolysed with 20% hydrochloric acid for 22 hr by Clasper [45]. Schröder [60] recommends using conc. hydrochloric acid for 5—6 hr for the hydrolysis of nylons-6, 6,6, and 6,10, and for 12 hr at increased pressure for nylon-11. Thionyl chloride is used for the hydrolysis of nylon-6 by Schaffgen [61]. Concentrated hydrochloric acid for 5 hr, or for 2 hr in a sealed ampoule at 120°C, is recommended by Majewska [39] for the hydrolysis of nylons-6, 6,6, 6,10, and for 10 hr in a sealed ampoule at 120°C in the analysis of nylon-11.

The separation and determination of components in hydrolysate solutions may be effected in various ways. A conventional method involves isolation of dicarboxylic acids by ether extraction, removal of ether by distillation, recrystallisation of the acid from water, followed by drying and weighing of the crystals. After the ether extraction the solution which contains amine hydrochlorides is evaporated to dryness; the residue is dissolved in hot ethyl alcohol, purified by refluxing over activated carbon, filtered, and allowed to stand to obtain the amine hydrochloride crystals which, after filtering and drying, are determined gravimetrically.

The dicarboxylic acid content in the hydrolysis products may also be determined without previous extraction by direct potentiometric titration [55,62]. Hexamethylenediamine hydrochloride does not affect the determination. ε-Aminocaproic acid produced in the acid hydrolysis of nylon-6 may be determined either by potentiometric or visual titration [63].

The methods described above are used in the analysis of simple polyamides. On the other hand, in the analysis of mixed polyamides containing different dicarboxylic acid and amino acid components, chromatographic [39,60,62,64], spectrophotometric [30,44] and colorimetric [65] techniques are of value.

A simple procedure for quantitative separation and determination of the hydrolysate products, using ion-exchange chromatography [39], is given below. The method involves the separation of acidic from basic components and their subsequent determination in the eluates by titration. For the separation of the components, chromatographic columns 50 cm long and 12 mm in diameter are connected to a distilled water reservoir, and packed with the following ion-exchange resins of grain size 0·15–0·30 mm.

1. Amberlite IR 45, a weakly basic anion-exchange resin, capable of retaining strong acids, is used for the separation of free hydrochloric acid (employed in the hydrolysis reaction) and ε-aminocaproic acid in the form of the hydrochloride from free ε-aminocaproic acid.
2. Amberlite IR 120, a strongly acidic cation-exchange resin, capable of retaining amino acids and hexamethylenediamine, is used for the separation of amines and amino acids from hydrochloric acid and dicarboxylic acids (adipic, sebacic).
3. Amberlite IRA 400, a strongly alkaline anion-exchange resin, capable of retaining weak and strong acids and amino acids, is used for the separation of hydrochloric acid, dicarboxylic acid, and amino acids from hexamethylenediamine.

Preparation of Ion-Exchange Columns. Ten g of ion exchanger in a beaker is first mixed with distilled water, then transferred to the column, and washed with 400 ml of the solution used for column regeneration, that is, 10% sodium hydroxide solution in the case of anion-exchange resins (Amberlite IR 45 and IRA 400), and 10% hydrochloric acid in the case of cation-exchange resins (Amberlite IR 120). Next, the columns are washed with distilled water until neutral to methyl orange. Desorption of individual components from the columns is effected using a variety of solutions, the volumes of which, at a flow rate level of 2 ml per min, are as follows:

1. adipic acid: 200 ml distilled water,
2. sebacic acid: 200 ml distilled water–ethyl alcohol mixture (1:1),
3. hexamethylenediamine: 300 ml distilled water,
4. ε-aminocaproic and ω-aminoundecanoic acids are first eluted with 10 ml 0·1*N* sodium hydroxide, followed by 200 ml water.

The quantities specified here are valid only for an ion-exchanger of a given type, grain size, column length, and eluent flow rate, and should be corrected whenever any of these parameters is changed.

PROCEDURE. The hydrolysate of 1 g of the polyamide with 4 ml of conc. hydrochloric acid is transferred quantitatively to a 100 ml volumetric flask and the contents are diluted to the mark with distilled water; 10 ml aliquots are introduced into individual columns and eluted as described above.

Total hydrochloric and adipic acids in the eluate from the Amberlite IR 120 column are determined by titration with 0·1*N* sodium hydroxide with phenolphthalein as indicator, and the hydrochloric acid content is established by the Volhard method. The adipic acid content $X(\%)$ is found from the difference of these results, from the formula

$$X = \frac{(a-b)0{\cdot}1 \times 58{\cdot}06}{m}$$

where:

a = volume (ml) of 0·1*N* sodium hydroxide used for the titration of hydrochloric and adipic (sebacic) acids,

b = volume (ml) of 0·1*N* silver nitrate used for the titration of hydrochloric acid,

m = weight of polymer sample, g.

The eluate from the strongly alkaline anion-exchange column containing hexamethylenediamine (Amberlite IRA 400) is titrated with 0·1*N* hydrochloric acid with methyl orange as indicator. The hexamethylenediamine content $X(\%)$ is calculated from the formula

$$X = \frac{0{\cdot}1 \times 58\, a}{m}$$

where:

a = volume (ml) of 0·1*N* hydrochloric acid,

m = weight of polymer sample, g.

The eluate from the weakly basic anion-exchange column (Amberlite IR 45) containing ε-aminocaproic acid or ω-aminoundecanoic acid is neutralised with 0·1*N* hydrochloric acid with phenolphthalein as indicator, then treated with 20 ml of approx. 40% formalin which has been previously neutralised, and titrated with 0·1*N* sodium hydroxide by the Sörensen method [38]. The amino acid content *X*(%) is calculated from the formula

$$X = \frac{0{\cdot}1aM}{m}$$

where:

a = volume (ml) of 0·1*N* sodium hydroxide used for titration,
M = molecular weight of the amino acid under examination,
m = weight of polymer sample, g.

A column chromatography technique using silica gel was developed by Schröder for the quantitative determination of binary mixed polyamides such as adipate–sebacate–hexamethylenediamine, adipate–ketopimelate–hexamethylenediamine, and caprolactam–ω-aminoundecanoic acid, and terpolymers composed of these same components [66].

Techniques for identification may also be used as quantitative methods for the determination of polyamide components, namely those of thin layer and paper chromatography with photometric determination of the spots or elution of the spots followed by microtitration.

Determination of the Carboxyl End Groups

A number of techniques are reported in the literature for the determination of carboxyl groups in polyamides, specifically alkalimetric titrations [67–69], colorimetric titration using the methylene blue reaction [70], or esterification of COOH groups with diazomethane, followed by determination of methoxy groups [71]. The best results are obtained by titrimetric methods. Waltz and Taylor [69] take polyamide solutions in benzyl alcohol and titrate in a nitrogen atmosphere with potassium hydroxide solution in a benzyl alcohol–methanol solution with phenolphthalein as indicator, or conductometrically. *m*-Cresol polyamide solutions are recommended by Loepelman [67] and conductometric titration with potassium hydroxide solution in *n*-propyl alcohol. Schnell [72] reports a titrimetric analysis using α-phenylethanol with an indicator composed of phenolphthalein with thymol blue in ethanol.

Losev and Fedotova [65] treat a polyamide solution with sodium hydroxide solution and back titrate the excess with hydrochloric acid.

The labelled atom technique for determination of COOH groups was employed by Garman and Gibson [73].

Turska and Wolfram [68] modified the Taylor method by replacing benzyl alcohol with a (4:1) benzyl alcohol–*n*-propyl alcohol mixture to make titration at room temperature possible. Majewska *et al.* modified the Taylor method by titrating at a temperature > 75°C without a nitrogen atmosphere [74]. A titration technique for the analysis of COOH groups by protection of amino groups with formaldehyde is reported by Vaiman and Pokrovskii [75].

The Modified Waltz–Taylor Method. A 0·15–0·20 g sample of polyamide is weighed to the nearest 0·2 mg into a 100 ml flask. It is then treated with 120 ml of benzyl alcohol and heated on an oil bath until completely dissolved (ca. 15 min.); 3 ml of *n*-propyl alcohol is added to the hot (90°C) solution, and the latter is immediately titrated with 0·05*N* sodium hydroxide, using 10 drops of phenolphthalein indicator. The temperature of the titrated solution should not be lower than 75°C. A blank is run under analogous conditions. The COOH group content *X* is calculated from formula

$$X = \frac{(a-b)n}{c} \times 10^{-3} \text{ equiv. per g}$$

where:

a = volume (ml) of sodium hydroxide titrant used for titration of the the blank,
b = volume (ml) of sodium hydroxide used in titration of the unknown,
n = normality of the sodium hydroxide titrant,
c = weight of polyamide sample, g.

Determination of the Amine End Groups

The amine end groups may be determined by direct titration with acids with a coloured indicator [69], conductometrically [36,67], potentiometrically [69,76], or by the van Slyke method [78] (reaction with nitrous acid and subsequent colorimetric determination after formation of a coloured complex with dinitrofluorobenzene). A method of catalytic acylation of the amine group with benzoyl chloride, followed by back titration of its excess is reported by Litvinenko *et al.* [79]. The method is accurate but extremely laborious.

A simple technique of visual titration of amino groups was developed by Elias [80]. A 0·2–0·3 g sample of powdered polyamide is placed in a 100 ml conical flask, treated with 10–15 ml of *m*-cresol, and dissolved at a temperature not higher than 150°C. As soon as the sample dissolves (ca. 5 min) 10 ml of methanol is added, then 5 drops of indicator, 0·1% aqueous alcoholic thymol blue (80 parts water 20 parts alcohol by vol.). It is titrated using a microburette, with an accuracy of 0·02 ml, until a red colour persists. The amine group content *X* is found from formula

$$X = \frac{an}{c} \times 10^{-3} \text{ equiv. per g}$$

where:

a = volume (ml) of the acid titrant used,
n = normality of the titrant,
c = weight of polyamide sample, g.

In the potentiometric Waltz–Taylor titration *m*-cresol is used to dissolve the polyamide, and the titration is carried out with perchloric acid in methanol. In the case of alcohol-soluble polyamides, 72% ethanol is used as solvent, and the titration is performed using hydrochloric acid solution. A procedure using potentiometric titration for the determination of *m*-phenylenediamine in process liquors to control the polymerisation process is described by Kreshkov [81]. The end amidino groups in nylon-6 are determined by Schlach [82] using hydrazine hydrate.

Determination of other Additives and Contaminants

Caprolactam and Oligomers in Nylon-6

Nylon-6 usually contains monomer and oligomers in various amounts, which may be determined by a number of procedures. One of these involves Soxhlet extraction of a 5 g sample with 190 ml of distilled water for 16 hr, followed by drying the polyamide for 6 hr under reduced pressure (650 mm Hg). Water is then determined separately. From the weight loss of the sample during extraction and the moisture content the caprolactam content is calculated [77]. According to Hermans [83] cyclic dimers, trimers and tetramers are also present in the aqueous extract, though their solubility in water is much lower than that of caprolactam; for example for caprolactam, dimer, trimer and tetramer the solubility values, in g per 100 g of H_2O are, respectively: 320, 0·1, 0·093, and 0·04.

A sublimation method was worked out by Harford and Joyce [84] for the determination of low molecular impurities (oligomers). According to this procedure, about 2 g of the polyamide is heated for 30 min at 180°C under a pressure of 3 mm Hg in a sublimation apparatus. The sublimation products are collected in a receiver cooled with an acetone–"dry ice" mixture and subsequently transferred to a weighed bottle containing acetone, which is then removed by evaporation and the residue is dried in a desiccator and weighed.

A procedure for the simultaneous determination of monomer and water in nylon-6 is reported by Schenker *et al.* [85] who employed vacuum extraction at 200°C, followed by determination of the monomer by measuring the refractive index, and finding the water content from the difference in weight loss of the sample.

A polarographic method for the analysis of caprolactam was employed by Robinson [86]. The compound was extracted with water, then converted into the Schiff base (a product of the reaction of formaldehyde with ε-aminocaproic acid), and determined by the method of standard addition.

Water-extractable components of nylon-6 can be determined by differential refractometry, and determination of the caprolactam content in the aqueous extract, by a technique developed by Ongemach *et al.* [87]. The procedure involves extraction of the monomer with carbon tetrachloride from the aqueous extract containing cyclic oligomers and the monomer, followed by infrared determination of the monomer on the basis of its characteristic absorption at 1550 cm^{-1}. In the extracted solution cyclic oligomers are determined by the differential refractometry technique. The same author determines caprolactam in the aqueous extract directly by the gas chromatography method without additional chemical operations [88].

A rapid analysis of caprolactam by infrared spectrophotometry is carried out in nylon-6 by Mori [89]. The procedure involves dissolving a polyamide sample in *o*-chlorophenol and measuring the intensity of the 892 cm^{-1} band. The relative error in the measurement amounts to 1·4%.

A colorimetric method for the determination of caprolactam in nylon-6 in aqueous extract evolved by Mazur [90] involves treatment of the extract with hydroxylamine sulphate and ferric chloride in hydrochloric acid.

Water

A remarkable feature of polyamide resins is their substantial water absorption which in the case of nylon-6 is as high as or greater than 10%. The high water content has an adverse effect on both the fabricating properties and the quality of finished products. A number of recognised methods are in use for water determination in polyamides.

Heating a powdered polymer sample under vacuum, collecting the separated water in a receiver, and titrating with the Fischer reagent is recommended by Haslam and Clasper [91] for nylon-6. The method is of low accuracy for low water contents.

The above method was modified by Reid and Turner [92], who heated a sample in a dry nitrogen stream at 120°C. The water produced was trapped of in an absorption vessel under continuous titration with the Karl Fischer reagent.

Direct determination of water by the Karl Fischer method is given in Kline's monograph [77]. A 1·5–3·0 g sample of polyamide is dissolved in 100 ml of anhydrous *m*-cresol and titrated with the Fischer reagent either visually or potentiometrically.

The NMR technique was used by Slonim *et al.* [93] for determination of water in nylon-6 and is suitable for samples containing 3–17% water. At lower water contents the broadened signal of the water protons coalesces with the proton signals from water adsorbed on the solid. In the case of higher water contents the calibration curve deviates from linearity.

A manometric method for the determination of minute quantities of water in polyamides (> 0·01%) is reported by Illing and Hobe [94] and involves heating a sample at 150–200°C under high vacuum and measuring the pressure differences with an oil vacuum gauge calibrated for a low water content range.

Hydroxymethyl Groups in Formaldehyde-Modified Polyamides

Quantitative analysis of hydroxymethyl groups in formaldehyde-modified polyamide resins involves the decomposition of samples with phosphoric acid and removal of the formaldehyde produced, by absorption in a sodium sulphite solution. This is followed by acidimetric titration.

In the procedure a 1·5 g sample in a 250 ml flask is treated with a 1:1 phosphoric acid–water mixture (50 ml), and the contents are slowly brought to boiling. After addition of 200 ml of water the mixture is distilled. The distillation rate should be 3 ml per min. The liberated formaldehyde is absorbed in a receiver containing 50 ml of 25% sodium sulphite solution, and determined by titration with 1N hydrochloric acid, with thymolphthalein as indicator [95]. The hydroxymethyl groups content X(%) is found from the formula

$$X = \frac{3{\cdot}102\, vN}{m}$$

where:

v = volume (ml) of the HCl titrant used in the determination,
N = normality of the titrant,
m = weight of sample, g.

Determination of alcoholic (hydroxymethyl, hydroxyethyl) groups in modified polyamides can be made by Zeisel's method.

Plasticisers [96]

A 1 g sample of polyamide is placed in a 200 ml flask and dissolved in 5 ml of 90% formic acid (dissolution may be accelerated by heating on a water bath). Next, 10 ml of methanol is added, and, with continuous stirring, 70 ml of ether is carefully poured into the solution. After 30 min the solution is filtered through a sintered glass no. 3 funnel; the flask is washed with ether, and the polyamide filter cake is washed with four 20 ml portions of ether. The filtrate is transferred to a 500 ml beaker, the ether is removed by evaporation on a water bath until the formic acid odour disappears. The beaker is then chilled, 20 ml of ether is added again to dissolve the plasticiser, the solution is filtered as above and the beaker is washed with three 20 ml portions of ether. The combined filtrates are transferred to a weighed evaporating dish and the ether is evaporated on a water bath. The dish containing the plasticiser is dried to constant weight at 100°C. The plasticiser content is calculated as a percentage of the original sample.

Graphite

Certain commercial brands of polyamide contain several per cent of graphite as an additive to improve stability to light. A 2 g polyamide sample is hydrolysed with hydrochloric acid until soluble products are formed. The solution is filtered to separate graphite which is dried in air and subsequently determined gravimetrically [77].

A colorimetric method for the determination of manganese stabiliser in the presence of TiO_2 and an optical brightener is described by Majewska [97].

References

1. Fraenkel-Conrat, H., Cooper, M., Olcott, H. S., *J. Am. Chem. Soc.*, **67**, 950 (1945).
2. Korshak, V. V., Mozgova, Kh. I., Shkolina, N. A., *Vysokomol. Soedin.*, **1**, 1364 (1959).
3. Azimov, S. A., Ustanov, Kh., Kordub, N. V., Slepakova, S. I., *Vysokomol. Soedin.*, **2**, 1459 (1969).
4. Sumitimo, H., Hachikama, I., *Kogyo Kagaku Zasshi*, **62**, 132 (1959).
5. Schuttenberg, H., Schutz, R. C., *Makromol. Chem.*, **143**, 153 (1971).
6. Zakrzewski, L., Zieliński, W., *Mat. Plastiche*, **37**, 145 (1971).
7. Dorst, H. G., Fischer, K., *Brit. Plastics*, **44**, 27 (1971).
8. Zieliński, W., *Polimery*, **16**, 386 (1971).
9. Kagiya, T., Izu, M., Matsuda, T., Fukui, K., *J. Polymer Sci. A–1*, **5**, 15 (1967).
10. Terent'ev, A. P., Dunina, V. V., Rukhodze, E. G., *Vysokomol. Soedin.*, *Ser. A*, **9**, 599 (1967).
11. Vinogradova, S. V., Korshak, V. V., Vygodski, Ya. V., Zaitsev, V. I., *ibid.*, **9**, 658 (1967).
12. Magdeev, I. M., Shermergon, M. J., *ibid.*, **13**, 2380 (1971).
13. Guidotti, V., Russo M., Mortillaro, L., *Makromol. Chem.*, **147**, 111 (1971).
14. Stirnova, G. S., Andreev, D. N., *Vysokomol. Soedin.*, *Ser. B*, **13**, 534 (1971).
15. Khalifa, A. M. J., Kolesnikov, G. S., *ibid.* **13**, 262 (1971).
16. Petru, K., *Plasticke Hmoty Kaučuk*, **8**, 353 (1971).
17. Credali, L., Parrini, P., *Polymer*, **12**, 717 (1971).
18. de Sigy, P., *Japan Plastics*, **5**, 33 (1971).
19. Nishimura, I., *Japan Plastics Age*, **9**, 39 (1971).
20. Griehl, W., Rustem, D., *Kunststoff-Rundschau*, **17**, 2 (1969).
21. Adsett, I. M., *Brit. Plastics*, **44**, 47 (1971).
22. Butta, E., de Petris, S., *Europ. Polymer J.*, **7**, 387 (1971).
23. Overberger, G. G., Ohnishi, A., Gomes, A. S., *J. Polymer Sci.*, *A-1*, **9**, 1139 (1971).
24. Radivilova, L. A., Ratalovo, L. G., Volosova, K. N., Kavets, L. F., *Plast. Massy*, **1960**, [6], 14.
25. Cairns, T. L., Foster, H. D., Larchar, A. W., Schneider, A. K., Schreiber, R. S., *J. Am. Chem. Soc.*, **71**, 651 (1949).
26. Albrecht, W., Chrzczonowicz, W. S., Czternastek, W., Włodarczyk, M., Ziabicki, A., *Poliamidy* (*Polyamides*), WNT, Warsaw, 1964, p. 98.
27. Rafikow, S. R., Czełnokowa, G. N., Zurawlewa, I. W., Gribkowa, P. N., *J. Polymer Sci.*, **53**, 75 (1961).
28. Korshak, V. V., Chelnokova, G. N., Gribkova, P. N., *Vysokomol. Soedin.*, **1**, 208 (1959).
29. Bogdanov, V. V., Kudryavtsev, G. I., Manalrosova, F. N., Spirina, I. A., Ostromol'skii, P. E., *ibid.*, **3**, 324 (1961).
30. Champetier, G., Pied, I., *Makromol. Chem.*, **43**, 64 (1961).
31. Korshak, V. V., Mozgova, K. K., Lavrishchev, K. L., *Vysokomol. Soedin.*, **1**, 990, 1159, 1164 (1959).
32. Volokhina, A. V., Bogdanov, V. V., Kudryavtsev, G. I., *ibid.*, **2**, 92 (1960).
33. Evans, R. D., Mighton, H. R., Flory, P. J., *J. Am. Chem. Soc.*, **72**, 2018 (1950).
34. Korszak, W. W., *Chemia związków wielocząsteczkowych* (*Chemistry of High-Molecular Compounds*) (transl. from Russian), PWN, Warsaw, 1957, p. 158.
35. Thinius, K., *Analytische Chemie der Plaste*, Springer Verlag, Berlin, 1952, p. 322.
36. Frey, H. T., Knox, I. R., *Polyamides*, in *High Polymers*. Vol. XII, *Analytical Chemistry of Polymers, Part I*, Kline, G. M. (Ed), Interscience, New York, 1959, pp. 273, 279, 290.
37. Schwemmen, T., *Textilrundschau*, **11**, 4 (1956).
38. Czerepko, K., Jaroszewicz, L., *Chem. Anal.* (*Warsaw*), **2**, 173 (1957).
39. Majewska, F., Wajnryb, M., *Polimery*, **8**, 63 (1963).

40. Roff, I. V., *Analyst*, **79**, 306 (1954).
41. Jaroszewicz, L., Czerepko, K., *Chem. Anal. (Warsaw)*, **10**, 37 (1965).
42. Franc, I., *Coll. Czech. Chem. Comm.*, **22**, 1253 (1957).
43. Johnston, F. R., Wheedon, E., *J. Text. Inst. Trans.*, **55**, 162 (1964).
44. Hummel, D. O., *Kunststoff-, Lack- und Gummi-Analyse*, Carl Hanser Verlag, Munich, 1959, p. 247.
45. Clasper, M., Haslam, J., *Analyst*, **74**, 224 (1949).
46. Haslam, J., Swift, S. D., *ibid.*, **79**, 82 (1954).
47. Klacel, Z., *Plasticke Hmoty Kaučuk*, 1959, **10**, 379.
48. *Pensez Plastiques*, **14**, 19 (1967).
49. Sthal, S. S., Dmitreva, V. I., Bezuglyi, B. D., *Plast. Massy*, 1968, **9**, 61.
50. *See ref.* 36, pp. 275, 276.
51. Matsubara, I., Itoh, Y., Shinomiya, M., *J. Polymer Sci., Pt. B*, **4**, 47 (1966).
52. Turska, E., Kroh, I., Kalinowski, A. P., *Polimery*, **8**, 212 (1963).
53. Harms, D. H., *Anal. Chem.*, **25**, 1140 (1953).
54. Rothe, M., Mazanek, I., *Makromol. Chem.*, **145**, 197 (1971).
55. Ayres, C. W., *Analyst*, **78**, 382 (1953).
56. Zahn, H. B., Wolf, H., *Melliand Textilber.*, **32**, 317 (1951).
57. Zahn, H. B., Wolleman, D., *ibid.*, **32**, 927 (1951).
58. Braun, D., Vorendohre, G., *Kunststoffe*, **57**, 821 (1967).
59. Tengler, H., *Plastverarbeiter*, **22**, 329 (1971).
60. Schröder, E., *Plaste u. Kautschuk*, **5**, 3 (1958).
61. Schaffgen, I. R., Koontz, F. H., Fietz, R. F., *J. Polymer Sci.*, **40**, 385 (1959).
62. Ecochard, F., Duveau, N., *Makromol. Chem.*, **7**, 148 (1951).
63. Łada, Z., Młodecka, I., *Chem. Anal. (Warsaw)*, **9**, 353 (1964).
64. Haslam, J., Money, E., *Analyst*, **82**, 101 (1957).
65. Losev, I. P., Fedotova, O. Ya., *Praktikum po Khimii Vysokomolekularnych Soedinenii* (*Handbook of Chemistry of High Molecular Compounds*), Goskhimizdat, Moscow, 1962.
66. Schröder, E., *Plaste u. Kautschuk*, **8**, 121 (1961).
67. Loepelman, F., *Faserforsch. Textiltechn.*, **3**, 58 (1952).
68. Turska, E., Wolfram, L., *Zesz. Nauk. Polit. Łódź, Chemia*, **7**, 79 (1958).
69. Waltz, I. E., Taylor, G. B., *Anal. Chem.*, **19**, 448 (1947).
70. Nowakowski, A., Chojnacki, S., *Konferencja dotycząca chemii polimerów* (*Conference on the Chemistry of Polymers*) Łódź, 1954.
71. Staudinger, H., Nuss, O., *J. Prakt. Chem.*, **157**, 284 (1941).
72. Schnell, H., *Makromol. Chem.*, **2**, 172 (1948).
73. Garman, R., Gibson, M., *Anal. Chem.*, **37**, 1309 (1965).
74. Majewska, I., Miłosz A., Keystek, H., *Polimery*, **12**, 3 (1967).
75. Vaiman, E., Pokrovskii, L. I., *Zh. Prikl. Khim.*, **34**, 232 (1961).
76. Esnerova, I., *Plasticke Hmoty Kaučuk*, **4**, 42 (1967).
77. *See ref.* 36, p. 285.
78. Matthes, A., *J. Prakt. Chem.*, **162**, 245 (1945).
79. Litvinenko, Z., Titarenko, N., Zikrans, V., *Vysokomol. Soedin., Ser. A*, **2**, 214 (1964).
80. Elias, H. G., Schumacher, R., *Makromol. Chem.*, **76**, 23 (1964).
81. Kreshkov, A. P., Shvietsova, L. N., *Plast. Massy*, 1969, **12**, 44.
82. Schlach, P., Rieker, I., *Angew. Makromol. Chem.*, **15**, 203 (1971).
83. Hermans, P. M., *Nature*, **177**, 126 (1956).
84. Harford, W. E., Joyce, R. M., *Analyst*, **79**, 82 (1954).
85. Schenker, H. H., Casto, C. C., Mullen, P. W., *Anal. Chem.*, **29**, 825 (1957).
86. Robinson, T. A., *ibid.*, **39**, 7 (1967).
87. Ongemach, G. C., Dorman-Smith, V. A., Beier, W. E., *Anal. Chem.*, **38**, 1 (1966).
88. Ongemach, G. C., Moody, A. C., *Anal. Chem.*, **39**, 1005 (1967).
89. Mori, S., Okazaki, K., *J. Polymer Sci.*, *A-1*, **5**, 231 (1967).
90. Mazur, H., *Roczniki Zakł. Higieny*, **14**, 509 (1963).

91. Haslam, J., Clasper, M., *Analyst*, **77**, 413 (1952).
92. Reid, V. W., Turner, L., *Analyst*, **86**, 36 (1961).
93. Slonim, I. Ya., Urman, Ya. G., Konovalov, A. Ts., *Plast. Massy*, **1963**, [4], 58.
94. Illing, G., Hobe, D. v., *Z. Anal. Chem.*, **230**, 418 (1967).
95. Majewska, F., *Sprawozdanie, ITS, Kontrola analityczna poliamidu M* (*Report*, Institute of Plastics, Warsaw, *Analytical Control of Polyamide M*), 1964, 49/50 p. 5.
96. Haslam, J., Willis, H. A., *Identification and Analysis of Polymers*, 1st. Ed., Iliffe, London, 1965, p. 110.
97. Majewska, I., *Polimery*, **15**, 596 (1970).

Chapter 10

UNCURED EPOXY RESINS

Epoxy resins are compounds which contain in their molecule more than one 1,2-epoxy group capable of undergoing polyreactions, referred to as curing reactions. These reactions occur upon addition of curing agents such as amines, amides or carboxylic acid anhydrides. The uncured resins, which are highly viscous liquids soluble in organic solvents, become insoluble, infusible, hard materials due to their crosslinked structure, and are commonly referred to as epoxy resins [1,2,3,4].

In technologically advanced countries today over 20 varieties of epoxy resins, which may be classed in two main groups, are manufactured. To the first group belong the products of the polyaddition reaction of epichlorohydrin to compounds with active hydrogen atoms, and the polymers of epichlorohydrin itself. This group includes aliphatic resins made from polyhydric alcohols and polyethers (1,4-butanediol, glycerol, polypropylene glycol, etc.), and aromatic resins produced in a reaction of epichlorohydrin with phenols, phenolic resins and amines (bisphenol A, tetrachlorobisphenol, aniline, *p*-aminophenol, etc.).

The second group of epoxy resins comprises the products of epoxidation of unsaturated compounds including aliphatic and cycloaliphatic resins. The former are the products of addition of oxygen to the C=C double bonds of unsaturated hydrocarbons, unsaturated alcohols, or of esters of unsaturated fatty acids. The latter comprise the products of epoxidation of cycloaliphatic olefins and of their derivatives.

Of the various epoxy resins mentioned, the oldest known ones, notably the product of the reaction of bisphenol A and epichlorohydrin, are of prime significance. Depending on the quantitative ratio of the reagents and on the reaction conditions, the resins produced have molecular weight from ca. 400 to 4000, and range from viscous liquids to solids, with colour ranging from pale yellow to light brown.

Owing to their excellent working properties (the resins can be cured at room temperature or slightly higher) as well as superior mechanical and electrical characteristics, epoxy resins are extensively used as cast resins in electrical and electronic industries, and for the manufacture of glues, laminates and varnishes.

10.1 STRUCTURE

Epoxy resins from bisphenol A and epichlorohydrin represent a mixture of a variety of compounds differing considerably in their chemical structure. This is largely due to the fact that the reaction, between the phenolic group of bisphenol A and epichlorohydrin, is very complex. The reaction is known now to occur after both patterns (A) and (B), although the latter is predominant especially at temperatures above 50°C [1,4]:

$$\text{—ONa} + \text{ClCH}_2\text{CH—CH}_2\text{(—O—)} \rightarrow \text{—OCH}_2\text{CH—CH}_2\text{(—O—)} + \text{NaCl} \qquad \text{(A)}$$

$$\text{—ONa} + \text{CH}_2\text{—CHCH}_2\text{Cl (—O—)} \xrightarrow{\text{H}_2\text{O}} \text{—OCH}_2\text{CH(OH)CH}_2\text{Cl} \rightarrow \qquad \text{(B)}$$

$$\xrightarrow{\text{NaOH}} \text{—OCH}_2\text{CH—CH}_2\text{(—O—)} + \text{NaCl}$$

In reaction (B) unreacted chlorohydrin groups are likely to remain in certain amounts in the product. Also the product may contain chlorine in the CH_2Cl group that is not adjacent to the OH group, as a result of a possible addition of phenol to the other, inner C atom of the epoxy group [5]:

$$\text{—ONa} + \text{CH(CH}_2\text{Cl)—CH}_2\text{(—O—)} \xrightarrow{\text{H}_2\text{O}} \text{—O—CH(CH}_2\text{Cl)(CH}_2\text{OH)}$$

This chlorine atom is split off by sodium hydroxide with much more difficulty than the chlorine from chlorohydrin groups and, if it does split, epoxy groups are not formed.

Subsequent bisphenol molecules or free phenolic groups of the reaction products of bisphenol with epichlorohydrin, add to the epoxy groups (as to epichlorohydrin) to form hydroxyl alcoholic groups:

$$\text{—OCH}_2\text{CH—CH}_2\text{(—O—)} + \text{HO—} \rightarrow \text{—OCH}_2\text{CH(OH)CH}_2\text{O—}$$

Free phenolic groups of the polyaddition reaction products may remain unetherified, particularly when bisphenol A is used in excess in the reaction mixture.

The presence of compounds containing alcoholic groups capable of reacting with epoxy groups in the reaction mixture is another source of side reactions that may give rise to a variety of products. Thus, for instance, if epichlorohydrin adds to chlorohydrin groups, the groups formed

$$\text{—OCH}_2\text{CH(CH}_2\text{Cl)—O—CH}_2\text{CH—CH}_2\text{(—O—)}$$

contain chlorine that can be eliminated by sodium hydroxide only with difficulty as in the other groups mentioned above.

As with most other resins, it is difficult to obtain complete knowledge of the chemical composition of epoxy resins. However, much valuable information on the chemical composition of resins based on bisphenol A and epichlorohydrin may be acquired by chemical methods. The major functional groups of the resins in question have been detected and determined (with the exception of ether groups) and they are mainly epoxy groups, alcoholic and phenolic hydroxyl groups, α-glycol groups, active chlorine associated with chlorohydrin groups, inactive chlorine, and even olefinic C=C double bonds.

A considerable amount of information on the structure of epoxy resins can be obtained from their infrared spectra. This method is increasingly replacing chemical methods. Among the main reasons why chemical methods have a more limited application may be the increasing number of varieties of resins. The method of infrared spectrophotometry is especially useful in studying cured resins, and in following the curing process of epoxy resins. A particularly wide range of epoxy resins has been investigated by this method by Serboli [6]. The infrared spectra of 24 major types of commercial epoxy resins were recorded over a range of 4000–675 cm^{-1} along with a detailed interpretation of the spectra, and particular attention was given to the epoxy group bands in the range of 950–750 cm^{-1}.

Of the older studies devoted to infrared investigations on the epoxy resin structure, mention should be made of the Dannenberg paper [7]. A near infrared spectral range of 0·6–2·5 μm was used to identify functional groups both in liquid and solid uncured resins and in cured resins. The process of hardening of Epon 828 epoxy resin with ethylenediamine was studied by following the varying contents of epoxy, hydroxyl and amino groups. Likewise the studies by infrared spectrophotometry by Shimazaki and Kojima [8], Lee [9] Gurman *et al.* [10], Feazel and Verchot [11], Mak and Rogers [12] and by Dannenberg and Harp [13] concerned the epoxy resin curing process. These and the research by Slonim *et al.* [14] by NMR spectroscopy, all shed some light on the structure of both cured and uncured epoxy resins.

Recently Wiesner [15] has succeeded in identifying low molecular transitory intermediates of the reaction of epichlorohydrin with bisphenol A, using column and paper chromatography techniques. The intermediates are compounds that are terminated largely with chlorohydrin, α-glycol groupings, epoxy and phenolic hydroxyl groups. The complete composition of bisphenolic epoxy resins was identified by Ghosh *et al.* [16] using thin layer chromatography. According to Biesenberger *et al.* [17] oligomers of bisphenol epoxy resins of up to five bisphenol units were resolved by recycle gel permeation chromatography.

It should be said, however, that chromatographic techniques can furnish only approximate information on the complete composition of epoxy resins, and then only for resins of low molecular weight.

10.2 QUALITATIVE ANALYSIS

Chemical Methods

Uncured epoxy resins may be identified by characteristic tests for epoxy groups (the Feigl test for aldehyde, lepidine test, tests with dinitroarylsulphonic acids). Furthermore, bisphenol resins may be detected by means of the colour reactions of groupings characteristic of bisphenol A (see other tests).

The Feigl Aldehyde Test

For analytical procedure see p. 28.

Lepidine Test

Pyridine bases diluted with alcohols were found by Lohmann [18] to give colour reactions with alkylene oxides. Colour reactions such as those were used by Swann [19] to detect epoxy groups in epoxy resins, using lepidine as reagent. To a small sample of resin placed in a test tube are added 3 ml of 2-ethoxyethanol and 5 drops of lepidine. The contents are heated on an oil bath at 120°C, and the developing colour is observed. A blue colour indicates the presence of free epoxy groups in the resin.

m-Dinitroarenesulphonic Acids Tests

Epoxy resins undergo cleavage when treated with *m*-dinitroarenesulphonic acids, as they do with hydrogen halides [20,21]. The esters thus formed give colour reactions with tetrabutylammonium hydroxide in aqueous dioxan, or with primary and secondary amines in dimethylformamide. The reactions are highly characteristic of epoxy compounds, and are highly sensitive as they enable epoxy groups to be detected in polymers down to a concentration of 0·001%.

PROCEDURE [22]. A small quantity (ca. 5–50 mg) of the resin is dissolved in 0·5 ml dioxan, and 0·5 ml of 0·4% dioxan solution of 2,4-dinitrobenzenesulphonic acid is added. After 30 min 2 ml of 0·5% tetrabutylammonium hydroxide solution in a (9:1) dioxan-water mixture is added. If a red colour immediately appears this indicates the presence of free epoxy groups.

PROCEDURE [21]. Several mg of the resin are dissolved in 0·5 ml of dioxan and to the solution is added 0·5 ml of a 0·5% solution of 2-methyl-3,5-dinitrobenzenesulphonic acid in the same solvent. After 30 min 2 ml of 5% *n*-butylamine solution in dimethylformamide is added. Immediate appearance of a bluish-green colouration indicates the presence of free epoxy groups.

Cured epoxy resins or those modified with fatty acids, which do not therefore contain free epoxy groups, fail to give the above-mentioned colour reactions.

Foucry Test

A sample of finely divided resin (0·25 g) [23] is dissolved, with vigorous shaking, in 98% sulphuric acid at room temperature and the following tests are carried out.

a. One ml of the solution in a test tube is treated with 1 ml of 63% nitric acid; the mixture is shaken and allowed to stand for 5 min. Subsequently the contents of the test tube are poured, with continuous stirring, into 100 ml of 5% aqueous sodium hydroxide. An orange-red colour immediately appears.

b. Ten ml of 98% sulphuric acid is mixed with 50 ml of distilled water and in the still warm solution is dissolved 2·5 g of mercuric oxide; 5 ml of this reagent is poured into 1 ml of the sulphuric acid resin solution, the mixture is shaken, and allowed to stand for 30 min. When epoxy resins are present an orange-red precipitate forms.

Neither free bisphenol A nor free epoxy groups give colour reactions in the Foucry test [23]. Most likely therefore the colour reactions observed are due to the presence of ether groupings produced by a reaction of bisphenol A with epichlorohydrin. The resins derived from resorcinol fail to give these reactions.

p-Phenylenediamine

In a test tube 30 mg of *p*-phenylenediamine is dissolved in 8 ml of distilled water and 0·5–1·0 g of finely divided resin is added. A pink colour appears immediately and grows in intensity after 12 hr. Modified epoxy resins, containing ester components, are subjected to prior hydrolysis with alcoholic potassium hydroxide. The water-insoluble residue is examined. Urea-melamine, and phenol-formaldehyde resins present interfere with the determination and should be separated from the epoxy resin.

The Feigl Test for Phenol

For procedure refer to p. 29.

Swann's Filter Paper Test

Two drops of the resin solution [19] or an equivalent amount of the resin is dissolved in a quantity of sulphuric acid so as to produce the intensity of the colour of a solution of 0·1N potassium dichromate. If necessary the mixture is warmed to 40–50°C to facilitate dissolution. One or two drops of this solution are applied to a piece of filter paper previously washed vertically with a dilute hydrochloric acid solution. In the presence of epoxy resins derived from bisphenol A a light-purple colour appears within 1 min and sometimes turns blue.

The test is characteristic of epoxy resins. Bisphenol-formaldehyde resins do not give any colour in this test. Down to 1% of epoxy resin may be detected in paint finishes with the help of the filter paper test. Sensitivity, however, is lower when substances are present which give rise to darkening solutions in concentrated acids.

Chromatography

The foregoing chemical tests are inadequate for identification of a large variety of commercial cured epoxy resins, especially in the presence of additives and fillers. According to Braun and Lee [24] components of those resins can be conveniently identified by analysing higher boiling pyrolysis products by thin layer or gas-liquid chromatography.

Infrared Spectroscopy

The following characteristic bands that are due to major functional groups are observed in the infrared spectrum of uncured epoxy resins (Fig. 10.1) at 2·9 μm (OH groups), 8 μm (aromatic ether groups), 9·6 μm (aliphatic ether groups), 12·1 μm (*para*-substituted benzene ring), 10·9 μm (main

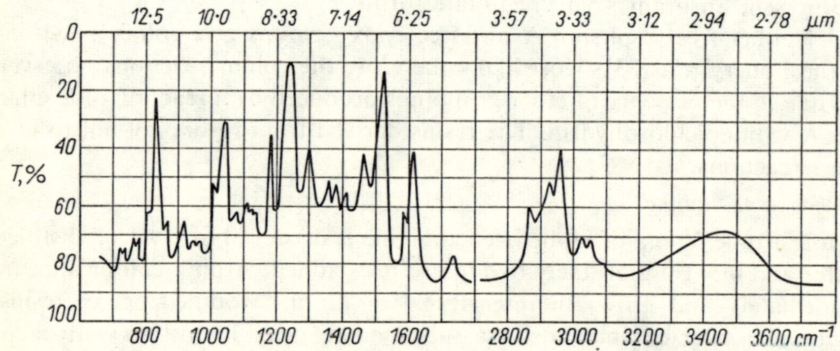

Fig. 10.1 Infrared absorption spectrum of an epoxy resin derived from bisphenol A and epichlorohydrin; capillary film on sodium chloride plate from acetone solution

band of the terminal epoxy groups) [25]. In the near infrared region of epoxy groups there are bands due to CH_2 groups of the oxirane ring at 2·205 μm (stronger) and 1·159 μm (weaker), and also bands of OH groups at 1·428 μm [7].

The spectra of epoxy resins are similar to those of phenol-formaldehyde resins. The latter, however, do not have a band at 9·6 μm, hence this band is suitable for the identification of epoxy resins in the presence of phenol-formaldehyde resins.

The spectra of cured epoxy resins derived from bisphenol A and epichlorohydrin are not substantially different from those of uncured resins. The main difference is the lack of the 10·9 μm epoxy group band and the presence of new bands from the crosslinking agents [8,25] (Fig. 10.2).

A particularly large amount of experimental data, of assistance in the identification of a number of epoxy resin varieties, is to be found in the paper by Serboli [6] (spectra of 24 varieties of epoxy resins), and in the book by Lee and Neville [2].

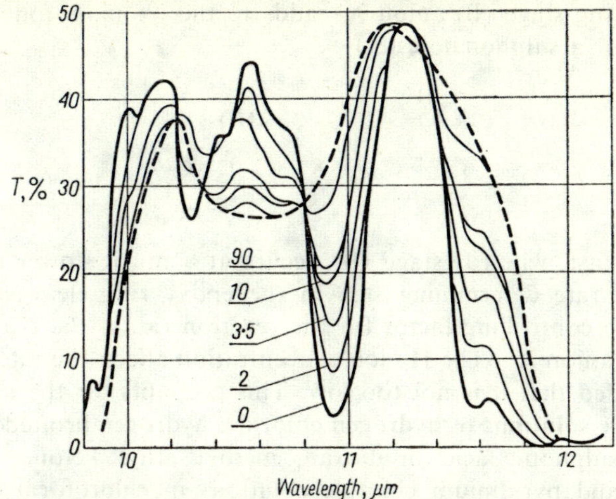

Fig. 10.2 Variation in the infrared absorption spectrum of epoxy groups in the Epon type resin in the curing process; numbers at the curves denote curing time (hr) at 65°C; — uncured resin; — partly cured resin — — — completely cured resin (reprinted with permission from *Anal. Chem.*, 1956, **28**, 86. Copyright by the American Chemical Society)

10.3 CHEMICAL COMPOSITION OF THE POLYMER

For the chemical characterisation of a resin, its elemental composition is determined, and also epoxy groups, alcoholic and phenolic hydroxyls, α-glycol groupings, and the bisphenol component content. Separate determinations are made of chlorine in chlorohydrin groupings (active chlorine), and the organically bound chlorine incorporated in bisphenol A resin groupings other than chlorohydrin (inactive chlorine).

Determination of Epoxy Groups

Chemical Methods

Of the large number of existing methods for the determination of epoxy groups in a variety of epoxy compounds [26], the commonest techniques used in polymer analysis involve the addition of hydrogen halides (almost exclusively hydrogen chloride and hydrogen bromide) to epoxy groups. For the colorimetric determination of minor quantities of epoxy groups in polymers, use is also made of 2,4-dinitrobenzenesulphonic acid [27] which reacts with the epoxy groups nearly as rapidly as do hydrogen halides [20]. The addition of the acid to an epoxy group proceeds in two stages. In the first stage the epoxy group oxygen becomes protonated to form the oxonium ion:

$$-\overset{\overset{\displaystyle O}{/\ \ \backslash}}{C\!-\!\!-\!C}- + H^{+} \rightarrow -\overset{\overset{\displaystyle \overset{H}{O^{\oplus}}}{/\ \ \backslash}}{C\!-\!\!-\!C}- \qquad \text{(A)}$$

In the following stage (B) anion X^- adds to the oxonium ion to produce halohydrin (or a sulphonate ester):

$$\underset{\diagdown\;\overset{\displaystyle\overset{H}{O^{\oplus}}}{}\;\diagup}{-C\text{——}C-} + X^- \rightarrow -\overset{\overset{\displaystyle HO}{|}}{C}-\underset{\underset{\displaystyle X}{|}}{C}- \qquad \text{(B)}$$

Stage (A) is fast, whereas stage (B) occurs at a much slower rate and is therefore the rate determining step in the epoxy ring cleavage reaction. Therefore, the controlling factor for the reaction rate is the type and concentration of anion X^-. The H^+ ion concentration affects the rate to a lesser extent, provided that it is not too low. This accounts for the use both of strongly acidic solutions of hydrogen chloride, hydrogen bromide or 2,4-dinitrobenzenesulphonic acid in dioxan, methyl ethyl ketone, or glacial acetic acid, and pyridinium chloride solutions in chloroform or even in pyridine, as reagents for the determination of epoxy groups in epoxy resins. Also for this reason epoxy groups can be directly titrated in a neutral solution of quaternary ammonium bromides or iodides with perchloric acid, the anion of which does not take part in the reaction.

Pyridinium Chloride-Pyridine Method

This is one of the oldest methods known to be used for determination of epoxy groups in epoxy resins [28]. The method involves heating a resin sample for a short time to boiling with a standard pyridine hydrochloride solution in pyridine, followed by back-titration of the unreacted hydrochloric acid with alcoholic sodium hydroxide using phenolphthalein as indicator.

Pyridinium Chloride-Chloroform Method

Replacement of the excess pyridine by chloroform in the method described above actually lowers the reactivity of the reagent, but it increases the precision and accuracy of the analysis and suppresses undesirable side reactions [26].

REAGENTS

Pyridine Hydrochloride. A 1*N* solution in chloroform: 75 g of anhydrous pyridine is weighed into a 2 l. measuring cylinder and 400 ml of anhydrous chloroform is added, after which the cylinder is weighed to the nearest gram. The cylinder is cooled with ice-water, and gaseous hydrogen chloride is introduced into the solution through a glass tube reaching to the bottom. About 35 g of hydrogen chloride is introduced during which the contents of the cylinder are allowed to warm up to room temperature, and the excess hydrogen chloride vapours are removed from the air space of the cylinder by passing a stream of dry air. A 10 ml aliquot is then withdrawn with a pipette, and after addition of 5 ml of water it is titrated with 0·5*N* alcoholic sodium hydroxide with phenolphthalein as indicator. The amount of pyridine *X* (g) needed to neutralise any free hydrogen chloride plus its 5% excess is calculated from the formula

$$X = \frac{79{\cdot}1 \times 1{\cdot}05 v_1 n v_2}{1000 v_3} - p$$

where:

v_1 = the volume (ml) of 0·5*N* sodium hydroxide used for titration of the solution aliquot examined,
n = normality of the sodium hydroxide solution,
v_2 = total volume (ml) of the prepared mixture after withdrawal of the aliquot,
v_3 = volume (ml) of the portion of the mixture withdrawn,
p = weight (g) of pyridine initially taken,
79·1 = mol. wt. of pyridine.

The amount of pyridine thus found is added to the mixture and diluted with chloroform to a volume equal to $v_1 n v_2 / v_3$.

To make sure that the solution was prepared correctly, 25 ml of the mixture is titrated with 0·5*N* alcoholic sodium hydroxide, using phenolphthalein as indicator. A parallel titration is carried out with the same volume of the mixture after cautiously refluxing it in a fume cupboard for 15 min. The results obtained from both determinations should coincide within 0·1 ml of the titrant.

0·2*N Pyridine Hydrochloride Solution in Chloroform.* A 200 ml portion of the 1*N* solution prepared as described is diluted with anhydrous chloroform to exactly 1 litre.

PROCEDURE. The 0·1*N* pyridine hydrochloride solution in chloroform (0·25 ml) is pipetted into a glass-joint 250 ml conical flask equipped with a reflux condenser, and a sample of the resin, containing 2–3 mequiv. of epoxy groups, is added. The mixture is first warmed to 40°C to dissolve the sample, then it is refluxed for 2 hr. After cooling to room temperature, 10 ml of distilled water is added followed by 0·2 ml of 1% ethanolic phenolphthalein solution, and it is titrated with 0·5*N* alcoholic sodium hydroxide until pink. Near the end point the mixture is vigorously shaken following the addition of each portion of the titrant. If the sodium hydroxide solution used is less than 15 ml, the determination is repeated with a smaller amount of the resin. A blank is run analogously, and the acid content of the resin is separately determined. A resin sample is dissolved in 25 ml of chloroform, followed by addition of 25 ml of methanol and 10 ml of distilled water and titrated with sodium hydroxide solution as described above.

CALCULATION. The epoxy group content in the resin under analysis (in equiv. per 100 g) is calculated from the formula

$$\alpha\text{-Epoxide content, equiv. per 100 g} = \frac{(v_1 - v_2)n}{10m} + K$$

where:

v_2 and v_1 = the volume (ml) of 0·5*N* sodium hydroxide solution used for the titration of the resin sample and the blank, respectively,

n = normality of the sodium hydroxide solution,

m = weight of the resin sample, (g),

K = the acid content in the resin, equiv. per 100 g.

To calculate the results as a percentage of epoxide groups, the α-epoxide content (in equiv. per 100 g) should be multiplied by the factor 43·0.

Commercial epoxy resins are commonly characterised by the value of their epoxide equivalent weight. This is defined as the amount of the resin, in g, which contains one equivalent of epoxy groups, namely

$$\text{epoxide equivalent weight, } g = M/N$$

where:

M = the molecular weight of the resin,

N = the number of epoxy groups in the molecule.

Methods Using Hydrogen Chloride in Neutral Solvents

Dioxan, which is an excellent solvent for epoxy resins, was used for the first time as a reaction medium by King [29]. It was established that epoxy groups react rapidly with a solution of hydrogen chloride in dioxan, and the solvent need not be absolutely anhydrous. The reaction is therefore carried out at room temperature. Since, however, hydrogen chloride fails to react instantly, the analysis is conducted with about a twofold excess of the acid solution and after some time the unreacted acid is back-titrated with a suitable reagent. The King method is unquestionably one of the best methods available for determination of epoxy groups in epoxy resins [28,30,31]. Methyl ethyl ketone and dimethylformamide were used as solvents by Jung and Kleeberg [32] and Tanaka and Kakiuchi [33], respectively.

As a rule, the excess of unreacted hydrogen chloride is determined by titration with alcoholic sodium hydroxide. If the resin under examination contains acidic compounds or substances that undergo hydrolysis in hydrochloric acid solution, better results are achieved by argentometric [34] or mercurimetric [35] titrations on condition, of course, that the resin does not contain chlorides in large quantities.

Alkalimetric Method [28]

REAGENTS

Indicator Solution. Cresol red (0·1 g), as its sodium salt, is dissolved in 100 ml of 50% ethanol (stock solution); 1 ml of this solution is added to 100 ml of 95% ethanol and is neutralised with 0·1*N* methanolic sodium hydroxide until the solution turns violet (the neutralised solution is prepared immediately before use).

Hydrogen Chloride Solution in Dioxan (0·2*N*). Concentrated hydrochloric acid (1·6 ml) is added to 100 ml of purified dioxan. The reagent must be carefully mixed. The required quantity is prepared immediately before the determination.

Purified Dioxan. Dioxan is purified by refluxing for 3 hr with solid potassium hydroxide (about 3% by wt. relative to dioxan) in a nitrogen atmosphere followed by distillation over potassium hydroxide under nitrogen (the initial fraction boiling up to 98°C is discarded). Purified dioxan is stored in a dark glass bottle under nitrogen.

Sodium Hydroxide. A 0·1*N* methanolic solution.

PROCEDURE. A sample of the resin, containing 2–4 mequiv. of epoxy groups, is weighed into a 250 ml conical flask containing 25 ml of purified dioxan. To facilitate solution, the contents are warmed to 40°C. When the sample is dissolved the flask is cooled to room temperature and 25 ml of the dioxan solution of hydrogen chloride is added; it is stoppered with a glass stopper, thoroughly mixed and allowed to stand for 15 min. Then 25 ml of neutralised indicator solution is added with a measuring cylinder, and the excess acid is titrated with 0·1*N* methanolic sodium hydroxide until a violet colour appears. A blank is run in an identical manner, and the content of acidic or basic compounds is determined. (A resin sample is dissolved in 25 ml of neutralised indicator solution and titrated with standard methanolic acid or alkali solution to an end point indicated by a change in colour from yellow to violet).

CALCULATION. The epoxy group content in the resin under test (in equiv. per 100 g) is evaluated from a formula analogous to the one given above.

Mercurimetric Method [35]

REAGENTS

Diphenylcarbazone. A 0·5% ethanolic solution.

Mercuric Nitrate. A 0·2*N* solution in methanol; 11 g of mercuric oxide is dissolved in 15 ml of dilute (1:1) nitric acid, and diluted with methanol to 1 litre.

PROCEDURE. A sample of the resin, containing about 1 mequiv. of epoxy groups (0·1–0·5 g) is weighed into a 100 ml conical flask and 10 ml of 0·2*N* hydrochloric acid in dioxan is added. After dissolving, the solution is allowed to stand for 15 min. Subsequently, 5 drops of the indicator solution are added and the excess hydrochloric acid is titrated with mercuric nitrate solution until a blue-violet colour appears. The epoxy group content is calculated as described above.

Direct Titration

The ease with which the epoxy groups incorporated in epoxy resins undergo cleavage in the presence of hydrogen halides, permits their determination by direct titration. The first reagent to be used was hydrogen chloride in acetic acid solution with a potentiometric titration [36].

Another method consists in visual titration using hydrogen bromide solution in acetic acid [37]. Hydrogen bromide is a much more potent epoxy-splitting agent but the most active agent is hydrogen iodide which, however, cannot be used because of its reducing properties. In the potentiometric titration using hydrogen bromide the equilibrium at the end point was found by Dijkstra and Dahmen [38] to be reached rather slowly. Hence, the reaction time must be considerably prolonged to obtain accurate results. Another disadvantage of this procedure is that because of the volatility of hydrogen bromide the titre of its solution in acetic acid is unstable, and preparation of the reagent is inconvenient. Among the advantages of the procedure are versatility, rapidity, and simplicity. Moreover, according to Budyak *et al.* [31] the procedure may give more precise and more accurate results in the presence of larger quantities of organic acids (or chlorides) than the King method. According to the analytical procedure reported in the monograph by Lee and Neville [2], a resin sample is dissolved in chlorobenzene and titrated with 0·1*N* hydrogen bromide solution in glacial acetic acid using methyl violet as indicator until a bluish-green colour appears.

In the above-mentioned paper [38] a method is described in which the epoxy compounds are directly titrated with hydrogen bromide or hydrogen iodide generated *in situ*. These hydrogen halides are generated during titration with a standard perchloric acid solution in acetic acid, when the reaction medium consists of a solution of tetraalkylammonium bromides or iodides in glacial acetic acid. As the hydrogen halide is generated it reacts instantly in quantities exactly equivalent to the quantity of perchloric acid added. The potentiometric titrations performed demonstrate that in the presence of sufficient excess of the ammonium salt the reaction is so rapid that the titration curves obtained are of the correct shape. Another advantage of the Dijkstra method is the stability and easy preparation of the reagent. However, a shortcoming is the rather high price of the ammonium salts.

A method based on the same principle for determination of epoxy compounds (including epoxy resins) was developed by Jay [39]. A detailed description of this method is given below. With the help of the above reagent aziridines can also be titrated. (Aziridines contain a nitrogen atom, in place of the epoxy oxygen, in a three-membered ring).

REAGENTS

Tetraethylammonium Bromide. A solution of 100 g of the bromide in 400 ml of glacial acetic acid.

Tetraethylammonium Iodide. A 10% solution in chloroform.

Methyl Violet. A 0·1% solution in glacial acetic acid.

Perchloric Acid. A 0·1*N* solution in glacial acetic acid.

PROCEDURE. A sample of the resin, containing 0·6–0·9 mequiv. of epoxy groups, is weighed into a 50 ml beaker, dissolved in 10 ml of chloroform*

* According to Jay [39] acetone, benzene, or chlorobenzene may well be used instead of chloroform. On the other hand acetic acid alone is recommended by Dijkstra and Dahmen, since benzene or chlorobenzene decreases the reaction rate.

and 10 ml of the ammonium salt solution is added; 2 or 3 drops of the indicator solution are added and the mixture is immediately titrated with 0·1*N* perchloric acid solution in glacial acetic acid until a light-green colour appears. A blank should be run separately.

INSTRUMENTAL METHODS

Colorimetric Determination with 2,4-Dinitrobenzenesulphonic Acid [27]

The epoxy groups contained in epoxy resins react in dioxan solution at much the same rate with 2,4-dinitrobenzenesulphonic acid as with hydrogen chloride to form hydroxy-esters. The esters react with piperazine (and also with piperidine) in dimethylformamide to produce coloured solutions that exhibit a maximum absorbance at 390 nm [20].

REAGENTS

2,4-*Dinitrobenzenesulphonic Acid.* A 1% solution in dioxan
Piperazine Hydrate. A 1% solution in dimethylformamide

PROCEDURE. A polymer sample is dissolved in dioxan and diluted so that the resulting solution contains about 0·3 mequiv. of epoxy groups in 1 ml; 5 ml of this solution is pipetted into a 10 ml volumetric flask and treated with 2 ml of 2% 2,4-dinitrobenzenesulphonic acid solution in dioxan. After 1 hr the volume is made up to the mark with dioxan; 1 ml of this solution is added to 2 ml of 1% piperazine hydrate in dimethylformamide. After 10 min its absorbance is measured at 390 nm with respect to an analogous reference containing the same quantities of all the reagents except the sample under analysis.

If the polymer analysed contains less than 15 mequiv. of epoxy groups per 1 g, the sample (not exceeding 0·3 g) is dissolved in 1 ml of dioxan and 1 ml of 0·5% 2,4-dinitrobenzenesulphonic acid solution is added. After 1 hr 4 ml of piperazine hydrate in dimethylformamide is added and the procedure described previously followed. In the latter case, and also whenever coloured materials are analysed, a blank is run which differs from the actual determination in that to both the polymer solution and to the reference, 1 ml of dioxan is added instead of 2,4-dinitrobenzenesulphonic acid solution. The value of the absorbance measured in the blank is subtracted from the value of the absorbance found in the test.

The epoxy group content in the sample under examination, expressed in mequiv., is found from a calibration curve prepared as described above, using a standard reference of the same type of epoxy compound as the compounds under analysis. 3-Phenoxy-1,2-epoxypropane is used as the standard reference in the case of epoxy resins. The method described is especially suitable for the determination of minute quantities of epoxy groups.

Infrared Spectroscopy

Epoxy groups in uncured epoxy resins may be identified on the basis of a band at 10·9 μm which is the main band of these groups terminating the polymer molecules [25,40]. However, since more precise chemical methods are available, infrared spectroscopy is used merely in examination of the degree of cure of the resins.

Epoxy groups, when present in minor quantities, may be determined by the strong band around 2·2 μm which is due to the —CH_2— groups within the three-membered epoxy ring [41,7,40].

Determination of Hydroxyl Groups

The presence of reactive epoxy groups in epoxy resins means that only a few of the many methods available to determine hydroxyl groups can be successfully used. Specifically, common acetylation methods fail, as the epoxy group reacts with acetic acid which is present in acetic anhydride or takes part in the OH acetylation reaction:

$$\underset{\diagdown\,O\,\diagup}{CH_2\text{—}CH}\text{—} + CH_3COOH \rightarrow CH_2\text{—}\overset{\displaystyle OH}{\overset{|}{CH}}\text{—}$$
$$\qquad\qquad\qquad\qquad\qquad\quad \overset{|}{OOCCH_3}$$

The hydroxyl group formed reacts with acetic anhydride to generate more acetic acid. The reaction, although slow, proceeds to complete conversion of the epoxy groups. Additionally, epoxy groups react with pyridine to form coloured salt-like compounds, which restricts the use of this base as a reaction medium.

In the determination of hydroxyl groups the best results are achieved with lithium aluminium hydride reagent, which does not react at all with epoxy groups. The method is inconvenient, however, and its use is limited to better equipped analytical laboratories. Meanwhile, in industrial practice suitably modified acetylation methods have found wide application. Of some significance also is acylation with stearyl chloride.

Gasometric Lithium Aluminium Hydride Method

The method developed by Stenmark and Weiss [42] for the determination of the total hydroxyl content in epoxy resins, which involves the reaction with lithium aluminium hydride, allows very accurate results to be achieved. Lithium aluminium hydride reacts with hydroxyl groups, as with other groups and compounds containing active hydrogen, according to the reaction

$$4ROH + LiAlH_4 \rightarrow LiAl(OR)_4 + 4H_2$$

A sample of the resin to be analysed is dissolved in tetrahydrofuran at 0°C and treated with anisole or tetrahydrofuran solution of lithium aluminium hydride after the removal of air with nitrogen. The volume of liberated hydrogen is measured gasometrically or determined by the gas-liquid chromatography technique [43]. In calculating results allowance must be made for the water content, as well as for other active hydrogen-containing compounds present in the resin under examination [28].

Acylation

Detailed comparative studies of several acetylation procedures, the common feature of which is the simultaneous determination of epoxy groups, were made by Bring and Kadleček [44]. The hydroxyl group content is found as the difference between the total epoxy and hydroxyl group contents and the epoxy group content alone which is determined independently. The best results were found by using a reagent composed of pyridine perchlorate and acetic anhydride in pyridine solution. The reaction of pyridine perchlorate with epoxy groups follows the equation:

$$\text{—}\underset{\diagdown\ O\ \diagup}{CH\text{—}CH_2} + HClO_4 \cdot C_5H_5N \rightarrow \text{—}\underset{HO}{\underset{|}{CH}}\text{—}\underset{OClO_3}{\underset{|}{CH_2}} + C_5H_5N$$

Pyridine perchlorate has the advantage of being a crystalline solid, non-hygroscopic, and stable in air, and can easily be prepared with adequate purity by recrystallisation from hot water.

REAGENTS

Pyridine Perchlorate. Concentrated perchloric acid (144 g) is added dropwise with cooling to 120 g of pyridine. The precipitated crystals are filtered off, recrystallised twice from hot water and dried in air.

Acetylating Mixture. Acetic anhydride (12 g) is dissolved in 88 g of pyridine.

PROCEDURE. A sample of the resin (2·5–3·0 g) and about 0·3 g of pyridine perchlorate per mmole of epoxy groups contained in the sample, are weighed into a 250 ml ground-glass conical flask equipped with a reflux condenser, and 25 ml of acetylating mixture is added. The contents of the flask are gently warmed under reflux on a moderately warm water bath with continuous stirring until the resin is completely dissolved, (otherwise the mixture will turn dark brown at a later stage). Then the mixture is warmed to boiling, gently refluxed for 30 min. On cooling, 2 ml of water is added to the flask and the condenser is flushed with 10–15 ml of pyridine. On cooling again to room temperature a few drops of 1% alcoholic phenolphthalein solution are added and the mixture is titrated, with vigorous stirring, with 1*N* methanolic potassium hydroxide until a permanent pink colour appears. A blank is carried out in a similar manner.

CALCULATION. The total OH group content X (in equiv. per 100 g of resin) is calculated from formula

$$X = \frac{5{\cdot}569a+(v_1-v_2)n}{10m} - 2E$$

where:

a = weight of the pyridine perchlorate, g,

v_1 = the volume (ml) of 1*N* potassium hydroxide solution used in the blank titration,

v_2 = the volume (ml) of 1*N* potassium hydroxide solution used in the titration of the sample of the resin analysed,

n = normality of the potassium hydroxide solution,
m = weight of the resin sample, g,
E = the epoxy group content, mmole per g of resin.

A method based on a similar principle was developed by Levkovich and Kuvichinskaya [45]. This involves prior destruction of epoxy groups with hydrogen chloride, removal of the excess hydrogen chloride with a nitrogen stream, and determination of the hydroxyl groups, both originally contained in the resin and those formed in the reaction with hydrochloric acid, by acetylation in pyridine. The acetylation methods, though inferior to the Stenmark and Weiss method or to the stearyl chloride procedure as regards accuracy, are most suitable for analytical process control in industry, since they are rapid and simple.

Acylation with Stearyl Chloride

Determination of hydroxyl groups by this procedure involves warming a resin sample to boiling with a solution of stearyl chloride in benzene or carbon tetrachloride [46] or in chloroform [44] and passing nitrogen through the reaction mixture. Hydrogen chloride is generated as a result of the reaction:

$$ROH + CH_3(CH_2)_{16}COCl \rightarrow CH_3(CH_2)_{16}COOR + HCl$$

The hydrogen chloride evolved is absorbed in water and determined by titration with $1N$ aqueous sodium hydroxide, with methyl orange as indicator. Of all the acylation methods for determining hydroxyl groups, this procedure ensures the most accurate results [44].

Infrared Spectroscopy

The total hydroxyl group content in epoxy resins was determined by Adams [47] on the basis of the infrared absorption spectrum of pyridine solutions. A single band at 3·08 μm occurs in the infrared spectrum, due to associated OH groups, irrespective of the kind of the resin.

Determination of Phenolic Hydroxyl Groups

Separate determination of the phenolic hydroxyl groups as well as alcoholic hydroxyls in epoxy resins is not a simple task. This is because the phenolic hydroxyls are usually present in minor quantities, and other groupings and additives contained on the resins interfere with the analysis. Essentially only three methods for the determination of these groups are available, and none provides accurate results. The oldest known is the Seidl and Makeš method [48] based on the acidic properties of phenol hydroxyls. This involves potentiometric titration of a resin solution in a mixture of dimethylformamide and isopropyl alcohol, with potassium methoxide. A method involving amperometric titration of a resin sample with a bromide-bromate solution, that is, one that is essentially similar to the method of determination of dihydroxyphenylpropane in these resins [49], was developed by

Gintsburg and Flegontova [50]. The author of this chapter has evolved a method of colorimetric determination of minor quantities of phenol hydroxyls in polymers, including epoxy resins, which is based on a colour reaction with 2,4-dinitrobenzenesulphonyl chloride and piperazine [51]. This is the only method available for the determination of very small quantities of phenol hydroxyls in epoxy resins. The titanium tetrachloride method [52], suitable for polycarbonates, fails here because epoxy resins are insoluble in a methylene chloride-acetic acid mixture.

Acidimetric Method [48]

Into a dry three-necked Quickfit 100 ml flask is weighed 1–2 g of resin (the weight depends on the phenolic hydroxyls content in the resin) and 20 ml of dimethylformamide and 10 ml of isopropyl alcohol are added. On complete dissolution of the resin an antimony electrode and salt bridge filled with a saturated solution of lithium chloride in butanol are attached to the flask. The bridge is connected to a saturated calomel electrode, and the titration is carried out under nitrogen with $0{\cdot}1N$ potassium methoxide in a methanol-benzene mixture, with continuous stirring with a magnetic bar. Portions of 0·5 ml are added, and the potentiometer readings are noted when the value of the potential difference does not vary by more than ± 2 mV during 30 sec.

Colorimetric Method

The analytical reaction [51] underlying this method consists of two stages. First, the condensation reaction is carried out, in which 2,4-dinitrobenzenesulphonyl chloride is condensed with the resin phenolic hydroxyls in the presence of triethylenediamine in an acetone-benzene solution to form the ester of 2,4-dinitrobenzenesulphonic acid. In the second stage, the ester is treated with a solution of piperazine in dimethylformamide. A yellowish-green product is formed with a broad absorbance maximum at 390 nm. A detailed analytical procedure is given in Chapter 8, p. 262.

Determination of α-Glycol Groups

1,2-Glycols are oxidised with periodic acid according to the reaction:

$$RCHOHCH_2OH + IO_4^- \rightarrow IO_3^- + RCHO + HCHO + H_2O$$

For the determination of these groups when they are present in minor quantities at the ends of the epoxy resin molecules, Stenmark [53] used as oxidant benzyltrimethylammonium periodate in a neutral solution. The choice of such conditions was justified by the presence of a large number of epoxy groups capable of reacting with periodic acid. Moreover, the insolubility of epoxy resins in water necessitates a nonaqueous reaction medium and ammonium salts with large organic substituents are readily soluble in organic solvents.

REAGENTS

Benzyltrimethylammonium Periodate. Periodic acid ($HIO_4.2H_2O$) (2·6 g) is dissolved in 475 ml of methanol. The solution is neutralised with benzyltrimethylammonium hydroxide (a 35% solution in methanol), then 15 ml of glacial acetic acid and 5 ml of water are added, and the contents are thoroughly mixed.

PROCEDURE. A sample of the resin (0·3–0·5 g containing not more than 0·3 mmole of α-glycol groups) is weighed into a 300 ml bromination flask, and 25 ml of chloroform is added. The flask is chilled with water, then 25 ml of the periodate solution is added from a pipette. The mixture is allowed to stand in an ice bath for 2·5 hr. Next, 100 ml of water at 0°C is added, the flask is stoppered, the contents are vigorously shaken for 30 sec, and 5 ml of 10% sulphuric acid is added, followed by 15 ml of 20% aqueous potassium iodide. The liberated iodine is titrated with 0·1*N* sodium thiosulphate solution, and some starch solution is added near the end point. A blank is run separately. The α-glycol group content *X* (in moles per 100 g) is calculated from the formula

$$X = \frac{(v_1 - v_2)n}{20m}$$

where:

v_1 = the volume (ml) of 0·1*N* sodium thiosulphate solution used for titration of the blank,

v_2 = the volume (ml) of 0·1*N* sodium thiosulphate solution used for titration of the resin sample,

n = normality of the sodium thiosulphate solution,

m = weight of resin sample, g.

Determination of Chlorine Bound in Chlorohydrin Groups (Active Chlorine)

Chlorine contained in chlorohydrin groups is easily split off by the action of alkalis. This chlorine is determined either by treatment of a dioxan solution of the resin with 3*N* alcoholic potassium hydroxide at room temperature [28], or by heating the resin solution to boiling for 2 hr with 0·5*N* alcoholic potassium hydroxide [54]. After acidification, the cleaved chlorine in the form of chloride ions is determined by potentiometric titration with 0.05*N* silver nitrate [54]. In calculating the results of the analysis allowance should be made for the chloride content (ionic chlorine), which is found as described below.

Total Organically Bound Chlorine

The total organically bound chlorine content in epoxy resins is usually determined by methods that involve oxidation of the organic material either in a Parr bomb [5] or by the Grote method. In the latter method the bound chlorine is oxidised to free chlorine which, in turn, is reduced to hydrogen chloride and the chloride ions are determined by argentometric titration [55]. On the other hand, according to the Czechoslovak standard (ČSN 640 333) [55] total organic chlorine is determined by hydrolytic abstraction as a result of heating for 2 hr with 0·5*N* potassium hydroxide in an ethylene glycol-dioxan mixture. The author of this chapter has worked out a procedure, based on the latter method, involving determination of cleaved chlorine by potentiometric titration with a silver nitrate solution in a non-aqueous medium containing acetic acid [56]. The procedure is given below. The method may prove particularly useful for minor quantities of organic chlorine or when the material under analysis is available in restricted amounts. In calculating the results in the determination of total organic chlorine, allowance should be made, as in the case of active chlorine, for the chloride (ionic chlorine) content.

REAGENTS AND SOLUTIONS

Silver Nitrate. A 0·003*N* solution in 85% acetic acid.

Dioxan, A.R. grade. Commercial reagent is purified as described in Vogel's text-book [57].

Potassium Hydroxide. A 0·5*N* solution in a dioxan-ethylene glycol mixture (3:2).

PROCEDURE. A sample of the resin, containing not more than 2 mg of organically bound chlorine, is weighed with an accuracy of 0·0001 g into a flat bottomed 100 ml flask with a large ground joint. The sample is dissolved in 5 ml of dioxan, and 5 ml of potassium hydroxide solution in dioxan-ethylene glycol mixture is added. The contents are heated for 2 hr on an oil bath under mild reflux. On cooling the condenser is flushed out with 10 ml dioxan, and 10 ml of glacial acetic acid is added. A silver and a glass electrode are immersed and the conventional potentiometric titration is carried out. A blank is run in analogous manner.

CALCULATION. The organically bound chlorine content *X*(%) is calculated from formula

$$X = \frac{35{\cdot}45(v_2 - v_1)\,n}{10\,m} - \%\mathrm{Cl}^-$$

where:

v_2 = the volume (ml) of 0·003*N* silver nitrate used in the titration of the sample,

v_1 = the corresponding volume (ml) for the blank,

n = normality of the silver nitrate solution,

$\%\mathrm{Cl}^-$ = ionic chlorine content, %,

m = weight of the resin sample, g.

Inactive Chlorine

Organically bound chlorine contained in groups other than chlorohydrin groups in bisphenol resins is determined as the difference between total organic chlorine content and active chlorine content.

Bisphenol Component

A colorimetric method of determining the bisphenol component based on the reaction of bisphenol A with formaldehyde was developed by Swann and Esposito [58]. The method is suitable for analysis of both unmodified epoxy resins and those modified with fatty acid resins, including mixtures of epoxy resins with silicones, alkyds, or in the presence of colophony. The method fails, however, when melamine resins are present, as the solutions obtained are turbid and interfere with the colorimetric determination.

REAGENTS

Paraformaldehyde. Reagent (0·15g) is dissolved in 9 ml of conc. sulphuric acid and 1 ml of water is added. Freshly prepared solutions should be used.

PROCEDURE. A resin sample of 0·5 g is dissolved in a 100 ml volumetric flask in methyl ethyl ketone and the volume is made up to the mark; 3 ml (which corresponds to 15 mg of the resin) is evaporated to dryness in a 25 ml conical flask at 105–110°C. On cooling 3 ml of conc. sulphuric acid is added to the flask which is then closed with a tube filled with calcium chloride, and the contents are heated at 40°C for 30 min until the resin dissolves completely; 2 ml of the paraformaldehyde solution is then added and the heating is continued for a further 30 min. The warm solution is poured into about 15 ml of water with vigorous stirring, the contents are quantitatively transferred to a 200 ml volumetric flask and the volume is made up to the mark with water. The blue colour that appears attains a maximum intensity after 2 hr. Absorbance is measured at 650 nm, and the bisphenol content is evaluated from the calibration curve made with pure bisphenol A.

10.4 DETERMINATION OF CONTAMINANTS

Commercial epoxy resins contain certain amounts of low molecular weight contaminants such as water, sodium chloride, free bisphenol A, and sometimes also organic solvents.

Determination of Water

Water in epoxy resins is determined by titration with the Karl Fischer reagent. The presence of free epoxy groups does not interfere with the determination.

Determination of Chlorides

Chloride ions contained in epoxy resins may be determined by the method developed by the author of this chapter [56] which involves potentiometric titration of a resin solution in dioxan-acetic acid mixture with 0·003*N* (or even 0·001*N*) silver nitrate in 85% acetic acid. The analytical procedure is given below. Other methods available in the literature are less convenient. For instance, a method involving titration in an acetone-water mixture [59] is unsuitable for resins of higher molecular weight, whereas the polarographic technique [60] is much more troublesome, because among other reasons the resin must be digested to convert it into inorganic ions prior to the determination.

REAGENTS AND SOLUTIONS

Silver Nitrate. A 0·003*N* (or 0·001*N*) solution in 85% acetic acid.

Dioxan, A.R. grade. The commercial reagent is purified as described in Vogel's text-book [57].

Sodium Acetate. A solution (ca. 0·3%) in acetic acid; 2 g of sodium hydroxide is dissolved in 8 ml of water and diluted to 1000 ml with glacial acetic acid.

PROCEDURE. A sample of the resin, containing not more than 1 mg of Cl^-, is weighed to the nearest 0·0001 g into a 200 ml beaker, and dissolved in 50 ml of hot dioxan. After cooling to room temperature, 25 ml of sodium acetate solution in acetic acid is added with vigorous stirring. This solution is titrated potentiometrically by conventional means using a glass and a silver electrode immersed directly in the solution under examination.

CALCULATION. The ionic chlorine content *X*(%), is evaluated from the formula

$$X = \frac{35{\cdot}45\,vn}{10\,m}$$

where:

v = the volume (ml) of 0·003*N* (or 0·001N) silver nitrate solution,
n = normality of silver nitrate solution,
m = weight of the resin sample, g.

Determination of Free Bisphenol A (Dian)

If free phenolic groups are present in epoxy resins in minor quantities, or whenever a very high accuracy is not required, free dian may be determined directly in the resin. The method developed by Ulbrich [49] involves amperometric titration with a potassium bromide-bromate solution using a platinum rotating electrode, whereas the procedure reported by Silin [61] is based on a colour reaction of bisphenol A with 4-aminoantipyrine.

A more precise and reliable method for quantitative analysis of free bisphenol A calls for prior isolation of this component from the resin either by steam distillation or by precipitation from the resin solution in a suitable solvent with water, methanol or aliphatic hydrocarbons [55]. Unconverted bisphenol A may also be isolated from epoxy resin by extraction of the bisphenol A from the resin by using a suitable solvent immiscible with water, using aqueous sodium hydroxide; a mixture of 2-ethoxyethanol with toluene (6:4) is recommended as solvent. Liberated bisphenol A is determined colorimetrically by coupling it with diazotised *p*-nitroaniline [55].

Determination of Sodium

Buscás *et al.* [62] determined sodium in certain varieties of epoxy resins by atomic absorption spectroscopy. The analysis was conducted by the internal standard technique using a 5% solution of the resin in a cyclohexanone-xylene mixture.

Determination of Volatile Matter

Heat 5 g of the resin in a stream of air at 140°C for 3 hr and determine the loss in weight (after DIN 16945) [55].

10.5 ANALYSIS OF CURED EPOXY RESINS

Finished products made of epoxy resins, such as cast resins, laminates, coats, and joints produced with the use of epoxy glues, contain cured resins that are built from enormous, crosslinked macromolecules.

In the curing process the linear build up of the polymer chain is accompanied by the coupling of linear elements to form three dimensional crosslinked structures.

The degree of cure directly affects the properties of the resins. Degree of cure can be determined from degree of conversion and degree of crosslinking. Degree of conversion is found from the quantity of reactive groups of the resins or of the curing agent which underwent reaction. Generally the epoxy groups are determined for that purpose. The degree of crosslinking is found on the basis of the physical distortion properties of cured products: heat resistance and insolubility. Heat resistance is determined by measuring the distortion of a resin under load at elevated temperatures (e.g. by the Martens method after DIN 16 945), whereas insolubility is measured as a swelling value in solvents. The degree of crosslinking of an amine-cured epoxy resin was examined also by chemical analysis by Bell [63].

DETERMINATION OF CONVERSION

Chemical Methods

The epoxy group content in cured epoxy resins is determined by the King method [29] based on the reaction with a solution of hydrogen chloride in dioxan. The analysis is essentially the same as the one described above for uncured resins (p. 304). The only differences are a longer reaction time, and fine division of the resin prior to the reaction. For instance, according to Dannenberg and Harp [13] the resin is ground in a vibrating ball mill. Finely divided material, with a particle size of the order of several μm, is subsequently treated with a solvent in which the resin swells but does not dissolve. According to Anderson [64] the unreacted excess hydrogen chloride should be determined by argentometric titration, because the amines present as curing agents in the resins interfere with the determination by the acidimetric method in the original procedure given by King.

Instrumental Methods

The process of curing of epoxy resins may be followed either by the method of infrared spectrophotometry [8–11,13,65,66] measuring absorbance of the band due to the three-membered epoxy ring vibration (10·92 μm) [13] (Fig. 10.2) or in the near infrared region [7]. In the latter case the unreacted epoxy group content is determined on the basis of the absorption bands at 2·205 and 1·159 μm, whereas the hydroxyl group content is determined on the basis of absorbance at the isosbestic point at 1·456 μm [7,13].

Chemical transformations occurring during the process of curing of epoxy resins, which results in changes in structure, were also investigated by NMR spectroscopy [14,67].

References

1. Brojer, Z., Hertz, Z., Penczek, St., *Żywice epoksydowe* (*Epoxy Resins*), WNT, Warsaw, 1972.
2. Lee, H., Neville, K., *Handbook of Epoxy Resins*, McGraw-Hill, New York, 1967.
3. Lidařik, M., Kincl, J., Roth, V., Bring, A., *Epoxydové pryskyřice*, SNTL, Prague, 1961.
4. Paquin, A. M., *Epoxydverbindungen und Epoxydharze*, Springer Verlag, Berlin-Heidelberg–Göttingen, 1958.
5. Belanger, W. J., Shulte, S. A., *Modern Plastics*, **37**, [3], 154 (1959).
6. Serboli, G., *Kunststoffe-Plastics*, **13**, 106, 150 (1966).
7. Dannenberg, H., *SPE Transactions*, **3**, 78 (1963).
8. Shimazaki, A., Kojima, M., *Kogyo Kagaku Zasshi*, **66**, 1610 (1963).
9. Lee, H., *Plastics Technol.*, **7**, [2], 47 (1961).
10. Gurman, I. M., Zalkind, G. I., Akutin, M. S., *Plast. Massy*, **1966**, [8], 69.
11. Feazel, C. E., Verchot, E. A., *J. Polymer Sci.*, **25**, 351 (1957).
12. Mak, H. D., Rogers, M. G., *Anal. Chem.*, **44**, 837 (1972).
13. Dannenberg, H., Harp, W. R., *Anal. Chem.*, **28**, 86 (1956).
14. Slonim, I. Ya., Lyubimov, A. N., Kovarskaya, B. M., *Chem. Průmysl*, **13**, 606 (1963).
15. Wiesner, I., *Coll. Czech. Chem. Comm.*, **32**, 448 (1967).
16. Ghosh, P. K., Bandyopadhyay, C., Saha, A. N., *J. Polymer Sci., A-1*, **6**, 3418 (1968).
17. Biesenberger, J. A., Tan, M., Duvdevani, I., *J. Appl. Polym. Sci.*, **15**, 1549 (1971).
18. Lohmann, H., *Angew. Chem.*, **52**, 407 (1939).
19. Swann, M. H., *Official Digest*, **30**, 1277 (1958).
20. Urbański, J., *Mikrochim. Acta*, **1965**, 60.
21. Urbański, J., *Polimery*, **15**, 241 (1970).
22. Urbański, J., *Plaste u. Kautschuk*, **15**, 260 (1968).
23. Foucry, M. J., *Peintures, Pigments, Vernis*, **30**, 925 (1954).
24. Braun, D., Lee, D. W., *Kunststoffe*, **62**, 571 (1972).
25. Haslam, J., Willis, H. A., *Identification and Analysis of Plastics*, 1st Ed., Iliffe Books, London, 1965.
26. Jungnickel, J. L., Peters, E. D., Polgar, A., Weiss, F. T., in *Organic Analysis*, *Vol. I*, Mitchell J., Jr., Kolthoff, I. M., Proskauer, E. S., Weissberger A., (Eds), Interscience, New York, 1953, p. 127.
27. Urbański, J. *Chem. Anal.* (*Warsaw*), **13**, 73 (1968).
28. Dannenberg, H., Harp, W. R., Jr., *Anal. Chem.*, **28**, 86 (1956).
29. King, G., *Nature*, **164**, 706 (1949).
30. Ulbrich, V., *Plaste u. Kautschuk*, **15**, 546 (1968).
31. Budyak, N. F., Zil'bershtein, F. S., Ogienko, R. A., *Plast. Massy*, **1969**, [5], 66.
32. Jung, G., Kleeberg, W., *Kunststoffe*, **51**, 714 (1961).
33. Tanaka, Y., Kakiuchi, H., *Kobunshi Kagaku*, **20**, 629 (1963).
34. Stenmark, G. A., *Anal. Chem.*, **29**, 1367 (1957); Gawłowski J., *Chem. Anal.* (*Warsaw*), **10**, 1141 (1965).

35. Vorobjov, V., *Chem. Průmysl*, **13**, 381 (1963).
36. Durbetaki, A. J., *J. Am. Oil Chem. Soc.*, **33**, 221 (1956).
37. Durbetaki, A. J., *Anal. Chem.*, **28**, 2000 (1956).
38. Dijkstra, R., Dahmen, E. A. M. F., *Anal. Chim. Acta*, **31**, 38 (1964).
39. Jay, R. R., *Anal. Chem.*, **36**, 667 (1964).
40. Loefgren, P. A., *Farg Lack*, **15**, 199 (1969).
41. Goddu, R. F., Delker, D. A., *Anal. Chem.*, **30**, 2013 (1958).
42. Stenmark, G. A., Weiss, F. T., *Anal. Chem.*, **28**, 1784 (1956).
43. Norton, E. J., Turner, L., Salmon, D. G., *Analyst*, **95**, 80 (1970).
44. Bring, A., Kadleček, Fr., *Plaste u. Kautschuk*, **5**, 43 (1958).
45. Levkovich, G. A., Kuvichinskaya, T. V., *Lakokrasoch. Mater. i ikh Primen.*, **1966**, [6], 54.
46. Raymond, E., Bouvetier, E., *Compt. Rend.*, **209**, 439 (1939).
47. Adams, M. R., *Anal. Chem.*, **36**, 1688 (1964).
48. Seidl, J., Makeš, J., *Syntetické Pryskyřice*, **5**, 363 (1955).
49. Ulbrich, V., *Chem. Listy*, **50**, 743 (1956).
50. Gintsburg, V. I., Flegontova, L. M., *Zavodsk. Lab.*, **27**, 392 (1961).
51. Urbański, J., *Chem. Anal. (Warsaw)*, **15**, 615 (1970).
52. Horbach, A., Veiel, U., Wunderlich, H. *Makromol. Chem.*, **88**, 215 (1965).
53. Stenmark, G. A., *Anal. Chem.*, **30**, 381 (1958).
54. Kasterina, G. N., Kalinina, L. S., *Khimicheskie metody issledovaniya sinteticheskikh smol i plasticheskikh mass* (*Chemical Methods of Analysis of Synthetic Resins and Plastics*), Goskhimizdat, Moscow, 1963.
55. Rybnikář, F., Ditrych, Z., Klácel, Z., Ordelt, O., *Analyza a zkoušeni plastickych hmot* (*Analysis and Testing of Plastic Materials*), SNTL, Prague, 1965.
56. Urbański, J., Iwańska, St., *Chem. Anal. (Warsaw)*, **9**, 11 (1964).
57. Vogel, A. I., *A Text-Book of Practical Organic Chemistry*, Longmans, London, 1961.
58. Swann, M. H., Esposito, G. G., *Anal. Chem.*, **28**, 1006 (1956).
59. Kalinina, L. S., Barulina, M. V., *Plast. Massy*, **1961**, [8], 62.
60. Seidl, J., Ditrych, Z., *Syntetické Pryskyřice*, **5**, 310 (1955).
61. Silin, Ł., *Polimery*, **6**, 284 (1961).
62. Buscás, M. M., Obiols, J., Puigjaner, J. M., *Afinidad*, **24**, 185 (1967).
63. Bell, J. P., *J. Polymer Sci.*, *A-2*, **8**, 417 (1970).
64. Anderson, H. C., *Plastics Technol.*, **5**, [7], 40, (1959).
65. Feltzin, J., Longenecker, D. M., Petker, J., *SPE Transactions*, **5**, 111 (1965).
66. Krejcar, E., Klaban, J., *Plaste u. Kautschuk*, **17**, 513 (1970).
67. Chistyakov, V. A., Khozin, V. G., Prokop'ev, V. P., Kostochko, A. V., Agishev A. Sh., *Vysokomol. Soedin.*, *Ser. B.*, **14**, 699 (1972).

Chapter **11**

POLYURETHANES

Polyurethanes are formed in polyaddition reactions between isocyanate groups (—NCO) and alcoholic hydroxyl groups. These polymers contain the characteristic urethane link, $>N-\overset{\overset{O}{\|}}{C}-$, the name of which is derived from urethans which are esters of carbamic acid [1]. Polyurethanes are obtained from aliphatic or aromatic isocyanates and diols or polyols [2–6] according to the reaction:

$$n\text{OCN—R—NCO} + n\text{HO—R}'\text{—OH} \rightarrow -\left[-\text{O—R}'\text{—O—}\overset{\overset{O}{\|}}{C}\text{—NH—R—NH—}\overset{\overset{O}{\|}}{C}-\right]_n-$$

Polyester or polyether resins containing hydroxyl groups may be used in this reaction to give polyester- or polyether-polyurethanes, respectively [7]. Polyurethanes can also be synthesised from diols and alkyl- or arylphosphoryl di-isocyanates (polyphosphorylurethanes) [8], or from bisphenols or their dichloroformates and *N,N'*-dialkyl- or diaryl-arylenediamines to yield poly(carbonate-urethanes) [9], or from chloro-substituted aromatic di-isocyanates, fluoro-substituted aliphatic di-isocyanates, or fluoro-substituted diols and bisphenols [10], nitrodiols, or *p,p'*-di-isocyanatodiphenylmethane [11]. Depending on the raw materials used, either linear or crosslinked polyurethanes are produced. Linear polyurethanes are formed from substrates containing at least four methylene groups between the functional groups (e.g. from hexamethylene di-isocyanate and 1,4-butanediol).

Polyurethane resins are soluble in common organic solvents such as dioxan, tetrahydrofuran, dimethylformamide, cyclohexanone, *m*-cresol, and also in dimethylsulphoxide, formic acid and 60% sulphuric acid. They are not, however, resistant to oils, petrol fractions, and weak acid. Mixed polyurethanes, that is, those prepared from mixed glycols, exhibit a lower melting point and better flexibility. Polyurethanes undergo hydrolysis under the action of acid and alkalis. Substrates with hydroxyl groups of increasing hydrophobic character and those with longer polyether or polyester chains have a higher resistance to hydrolysis [12]. A process of gradual hydrolysis occurs under the action of light, as found from infrared evidence (the presence of OH absorption bands and the disappearance of the —NCO band in the spectrum). The photolytic decomposition process proceeds faster in a humid atmosphere.

Polyurethanes are widely used for the manufacture of both hard and soft foamed plastics [13–16] which are in great demand in a large number of industries (refrigeration, building engineering, electronics, shipbuilding [3, 17, 18]) and for manufacturing binding agents, varnish coats, and synthetic rubbers. Linear polyurethanes are used for the manufacture of fibres and moulding mixtures [3, 19].

11.1 QUALITATIVE ANALYSIS

The following methods have been found useful in identification of polyurethanes: infrared and NMR spectroscopy, chromatographic techniques, and colour reactions.

Colour Reaction of the —NCO *Group with p-Dimethylaminobenzaldehyde* [20].

A 0·5 g sample is dissolved in 5–10 ml of glacial acetic acid at room temperature and 0·1 g of *p*-dimethylaminobenzaldehyde is added. In the presence of the —NCO group the solution turns yellow after several minutes. If the sample fails to dissolve at room temperature it should be slightly warmed or if insoluble it should be dissolved first in a suitable solvent and then acetic acid should be added. The reaction can be used for the detection of free isocyanate and polyurethane resin.

Colour Reaction of the —NCO *Group with Malachite Green* [21]

REAGENT. This contains 0·20 g of the *n*-butylamine–malachite green complex in 50 ml of benzene.

A 0·1–0·2 g polymer sample is treated with 0·5–1 ml of the reagent. In the presence of isocyanate a green colour appears after 20–30 sec.

Colour Reaction with p-Nitrobenzenediazonium Fluoroborate

This reaction is used for differentiating tolylene di-isocyanate (TDI), methylene bis(*p*-phenylene di-isocyanate) (MDI) and napthylene di-isocyanate (NDI) in polyurethane foams [22].

Filter paper is moistened with a 1% *p*-nitrobenzenediazonium fluoroborate solution. A glass rod heated to red heat is applied to the foam sample, and the filter paper is inserted in the fumes produced from decomposing polymer. A red-brown colour on the paper is indicative of the presence of TDI, whereas yellow or violet colours point to the presence of MDI or NDI respectively.

Colour Reaction to Differentiate Polyester- from Polyether-Polyurethanes [23]

REAGENTS

Sodium Hydroxide Solution. A 2*N* methanolic solution with a little added phenolphthalein.

Hydroxylamine Hydrochloride. A saturated methanolic solution.

Hydrochloric Acid. A 1*N* solution.

Ferric Chloride. A 3% solution.

A weighed sample of 50 mg of finely powdered material is treated with a few drops of the sodium hydroxide solution, then with a few drops of the hydroxylamine hydrochloride solution. The mixture must be alkaline. For highly crosslinked polymers the mixture is heated for 20–30 sec. The mixture is then acidified with the hydrochloric acid, and each drop of the acid is followed by an addition of the ferric chloride reagent until a colour develops. In the presence of a polyester-urethane the mixture turns violet. In the case of a polyester-urethane on a castor oil or dimerised fatty acid base the colour developed is brown or violet-brown. Polyether-urethanes do not give the characteristic colour reaction.

Infrared Spectroscopy

Infrared spectroscopy by the ATR technique is the simplest method for the identification of polyurethanes, but this is of limited range. The main uses are related to the following:

1. classification test for polyurethanes,
2. differentiation of polyester- from polyether-urethanes,
3. check on the presence of aromatic components on the basis of absorption at 1600 cm^{-1} (6·25 μm),
4. check on the use of water or a diamine to extend the polymer chain, on the basis of the 1640 cm^{-1} (6·1 μm) absorption, the amine itself, however, being unidentifiable. Lack of absorption at 1667–1640 cm^{-1} (6·0–6·1 μm) indicates the absence of diamine or of water.

On the other hand infrared spectroscopy is inadequate for the identification of

1. polyester components, diols and diamines used to extend the polymer chain,
2. di-isocyanates and their isomers. Aromatic di-isocyanates only can be detected.

The presence of the urethane structure is detected on the basis of the bands at 1538 cm^{-1} (6·5 μm; amide II —NH deformation), 1724 cm^{-1} (5·8 μm; amide I —C=O stretch), and 3330 cm^{-1} (3·0 μm; —NH stretch). Polyester or polyether links in polyurethanes may be detected from the difference in intensity of the absorption bands at 1250 cm^{-1} (8·0 μm; $-\overset{\overset{\large O}{\|}}{C}-O-C-$ group) in the case of polyester-urethane, and at about 1110 cm^{-1} (9·0 μm; —C—O—C— group) in the case of polyether-urethane. Polyester-polyurethanes give rise to a stronger and broader band than do their polyether counterparts as illustrated by the spectra in Figs. 11.1 and 11.2. Better spectra can be obtained by the FMIR (Frustrated Multiple Internal Reflection) technique.

Soft polyurethane elastomers were identified by Boyarchuk *et al.* [24] and Zharkov [25] by infrared spectroscopy on the basis of the characteristic hydrogen bonds.

Adipic and sebacic acids were also identified in polyester-polyurethanes, which were made to crystallise by stretching, by the infrared technique by Corish [26].

An infrared identification procedure for polyurethane elastomeric joints and coatings is reported by Leukroth [27]. A complete identification can be achieved by hydrolysis followed by separation and identification of the hydrolysis products.

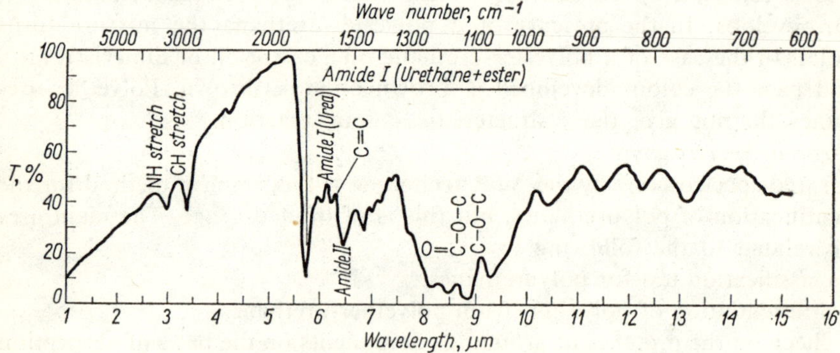

Fig. 11.1 ATR spectrum of polyester-urethane

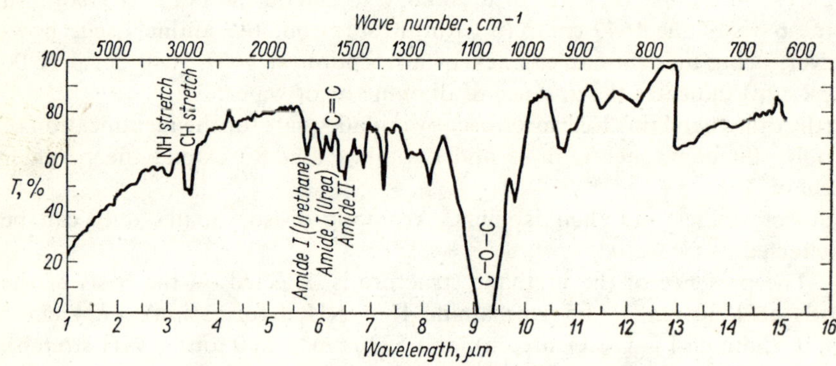

Fig. 11.2 ATR spectrum of polyether-urethane

Methods based on hydrolysis for identification of polyurethane foams and for linear polyurethanes are described by Corish [28] and Mulder [29], respectively.

The method of identification developed by Dawson *et al.* [30] to be used for a variety of polyurethanes is described below.

Hydrolysis Followed by Separation into Fractions

A 1 g sample of resin is heated in a stainless steel ampoule under pressure with 15 ml of 3*N* sodium hydroxide for 16 hr at 150°C. The mixture is allowed to cool to room temperature, and the hydrolysate is transferred to a funnel with a Teflon stopcock and extracted with four 20 ml portions of ether. In the case of polyester-polyurethanes, on evaporation of the ether, amines formed from di-isocyanates and those used to extend the polymer chain are obtained. The extracted hydrolysate is acidified with hydrochloric acid and again extracted with ether. Dibasic acids from the polyester are isolated by the removal of ether. The acidic solution is made alkaline with solid potassium hydroxide and is extracted in a liquid-extraction apparatus for at least 6 hr. The glycol components of the polyester along with other hydroxy compounds used to extend the polymer chain are isolated on evaporation of the ether.

In the case of polyether-polyurethanes, in the first ether extraction amines and polyether are obtained jointly. After evaporation of the ether hydrochloric acid is added to convert the amines into hydrochlorides, and the polyether is extracted with ether. The amines may also be separated from the polyether by evaporation of the mixture to dryness, followed by addition of potassium hydroxide to liberate amines, which are then extracted with ether.

Aliphatic amines are more readily soluble in water than aromatic amines, and thus are more difficult to extract with ether. Dilution of the hydrolysate with water should therefore be avoided, and the ether extraction should be carried out with at least six 10 ml portions of ether.

Separation of the hydrolysate into fractions follows the pattern given below (reprinted by permission of the copyright holders, John Wiley & Sons, Inc.).

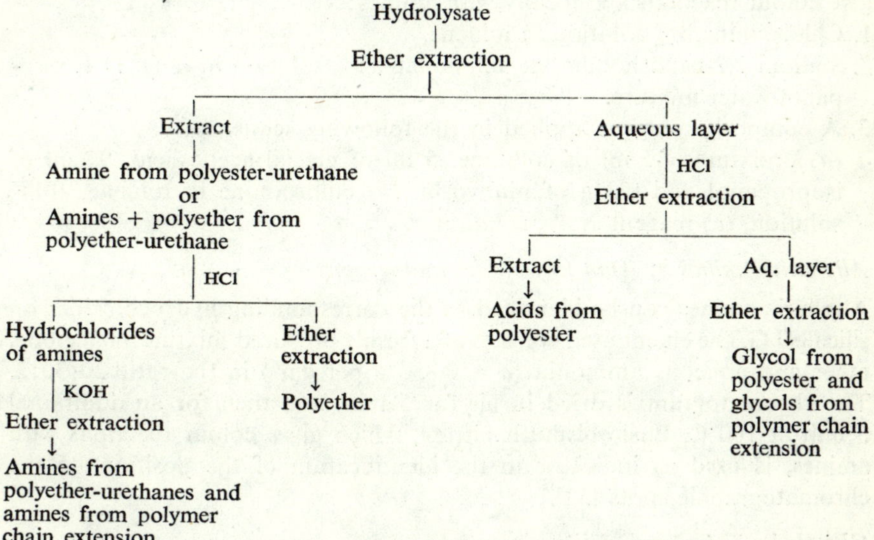

Each fraction is examined by infrared spectroscopy by which dicarboxylic acids, certain amines [31] and glycols can easily be identified. On the other hand, a mixture of glycols with amines is analysed by using NMR spectroscopy and chromatographic techniques.

Aromatic Amines by Thin Layer Chromatography

Silica gel is used as adsorbent. As eluents four solvent compositions are used, which are of varied separating power, as shown in Table 11.1.

Table 11.1

Diamine	R_f *in the solvent system*			
	A	B	C	D
2,4-Tolylenediamine	0·36	0·43	0·30	0·18
2,6-Tolylenediamine	0·43	0·63	0·47	0·24
Methylenedianiline	0·40	0·55	0·39	0.20
1,5-Naphthylenediamine	0·47	0·60	0·52	0·39
p-Phenylenediamine	0·22	0·51	0·31	0·08
m-Phenylenediamine	0·36	0·41	0·30	0·07
o-Phenylenediamine	0·40	0·44	0·35	0·19
Diethyltolylenediamine	0·53	0·54	0·40	0·30
Dichlorobenzidine	0·58	0·64	0·57	—
Methylenedi-*o*-chloroaniline	0·58	0·66	0·61	0·52

A—toluene–pyridine (40:10).
B—methylene chloride–cyclohexane–diethylamine (30:15:5).
C—cyclohexane–ethyl acetate–diethylamine (25:20:5).
D—toluene–acetone–aqueousammonia (d = 0·880 g per cm^3) (40:10:2).

The chromatograms are visualised with a number of reagents which give colour reactions specifically with amines:

1. Chloroimine, 1% solution in toluene.
2. Sodium 1,2-naphthoquinone sulphonate, 1% solution in a (1:1) isopropanol-water mixture.
3. A composite reagent applied in the following sequence: (*a*) a mixture of 5 ml of collidine, 5 ml of glacial acetic acid, 95 ml of isopropanol and 0·3 g of ninhydrin, (*b*) chloroimine in toluene, 0·1% solution, (*c*) reagent 2, 4% solution.

Aliphatic Amines by Thin Layer Chromatography

Aliphatic amines can be identified as the corresponding hydrochlorides on silica gel G.The eluent used consists of a freshly prepared mixture of acetone, ether and aqueous ammonia (d = 0·880 g per cm^3) in the ratio 20:30:2. The chromatogram is dried in air for 5 min, and then for an additional 5 min at 100°C. Basic bismuth nitrate, which gives colour reactions with amines, is used as indicator in the identification of the position of the chromatographic spots [32].

Glycols by Thin Layer Chromatography

Glycols are separated on silica gel PF or on alumina G, with a variety of eluents to develop the chromatograms and using rosaniline periodate [33] and chromium trioxide for detection of the spots. The glycols listed in Table 11.2 have been identified by this method.

Table 11.2

Glycol	R_f in the solvent system				
	A	B	C	D	E
Ethylene glycol	0·36	0·36	0·24	0·25	0·25
1,2-Propanediol	0·51	—	0·33	—	0·44
1,4-Butanediol	0·45	0·53	0·48	—	—
Diethylene glycol	0·27	0·29	0·10	—	—
Triethylene glycol	0·20	0·20	0·15	—	—
1,6-Hexanediol	0·60	0·64	—	0·30	—
1,2,6-Hexanetriol	0·27	0·23	—	—	0·15
1,1,1-Trimethylolpropane	0·48	0·56	0·15	—	—
Dipropylene glycol	0·49	0·57	—	0·20	—
2,2,4,4-Tetramethyl-1,3-cyclobutane-diol	0·89	—	—	0·71	—
Glycerol	—	—	—	—	0·07

A—silica gel PF; ethyl acetate–acetone (50:50)
B—silica gel PF; diethyl ether–methanol (90:10) [29].
C—alumina; ethyl acetate.
D—silica gel PF; ethyl acetate.
E—alumina G; chloroform–toluene–acetic acid (77:17:6).
Adsorbents were activated by heating at 110°C for 1 hr.

Other procedures for identification of glycols, using thin layer chromatography with a variety of eluents and identification reagents, are reported by Knappe [34] and Braun [35].

Glycols by NMR Spectroscopy

The method consists in dissolving the glycol fraction in pyridine with tetramethylsilane as internal reference. Glycols were directly identified in samples of polyurethanes soluble in formic acid (foams, elastomers) by Brame *et al.* [36]. As a solvent for identification of glycols by the NMR technique a mixture of dimethylacetamide and butylamine was used by Okuto [37].

Glycols by Gas Chromatography

Glycols produced during the hydrolysis of polyurethanes are dissolved in acetone and separated on a 1 m × 2 mm column packed with Porapak Q [30]. Individual glycols exhibit characteristic retention times as shown in Table 11.3. Glycols may also be identified by gas-liquid chromatography after prior esterification of hydroxyl groups with acetic anhydride [38, 39, 40].

11.2 QUANTITATIVE ANALYSIS

Quantitative analysis consists in hydrolysis of polyurethanes, followed by determination of the hydrolysate components.

Various kinds of polyurethanes undergo alkaline [41] or acid hydro-

Table 11.3 (reprinted from [30] by permission of the copyright holders, John Wiley & Sons, Inc.)

Glycol	*Retention time* min
Ethylene glycol	5·6
1,2-Propanediol	8·0
1,3-Propanediol	11·2
2,3-Butanediol	11·5
1,3-Butanediol	14·3
1,4-Butanediol	17·5
2,2-Dimethyl-1,3-propanediol	18·9
Glycerol	19·1
Diethylene glycol	19·8
2-Methyl-2,4-pentanediol	20·7
Dipropylene glycol	25·8
2,2,4,4-Tetramethyl-1,3-cyclobutanediol	29·0
1,6-Hexanediol	29·6
Triethylene glycol	33·5
1,2,6-Hexanetriol	35·9

lysis [42, 43], aminolysis with aniline [44], or hydrolysis with water under high pressure in the presence of magnesia [45].

The hydrolysate can be separated by a number of procedures. Conventional methods of separation and identification of the products of hydrolysis of polyester-urethane and of a polyurethane synthesised from 1,4-butanediol and hexamethylene di-isocyanate are used by Kasterina and Kalinina [46]. In the analysis of the polyester-urethanes alkaline hydrolysis was carried out with alcoholic potassium hydroxide solution, whereas in the other case 60% sulphuric acid was used in the hydrolysis. Separation of the components was carried out by extraction and precipitation, and the compounds were identified gravimetrically and titrimetrically.

A paper chromatography technique is employed by Schröder [47] for the separation of the hydrolysis products in the analysis of polyurethanes derived from various glycols and hexamethylene di-isocyanate. The aromatic amines produced from isocyanates are determined by gas chromatography or ion-exchange chromatography [48].

The technique of ion-exchange chromatography is utilised by Majewska [42] in the analysis of tolylene di-isocyanate polyester-urethane and of polyurethane formed from 1,4-butanediol and hexamethylene di-isocyanate. The method involves hydrolysis of a 2 g sample with conc. hydrochloric acid for 5 hr, followed by separation of the hydrolysis products on ion-exchange columns and their subsequent determination in the eluate. Two kinds of ion-exchange resins are used: an anionite—strongly basic—and a cationite—strongly acidic. To determine dicarboxylic acids a hydrolysate sample is passed through the cationite bed. Total acids are determined alkalimetrically in the eluate (dicarboxylic and hydrochloric acids), and hydrochloric acid is separately determined argentometrically. The dicarboxylic acid content is calculated from the difference between the two determinations. To determine the aliphatic amine, a hydrolysate sample is passed through the anionite, and the amine is determined in the eluate by the Sörensen method [49]. To determine the glycols, a hydrolysate sample is passed through both columns; it is immaterial which column is used first. The glycol is determined in the eluate by potassium dichromate oxidation. Glycols may also be determined by oxidation with periodic acid [50–52] or with potassium permanganate [43].

Isocyanate Groups

For the quantitative determination of isocyanate groups volumetric, gravimetric, colorimetric, and instrumental techniques can be employed. Underlying the first two methods is the reaction of the isocyanate group with amines to yield disubstituted derivatives of urea:

$$RNCO + RNH_2 \rightarrow RNH—CO—NHR$$

In the gravimetric determination aniline is used as reagent. The disubstituted urea derivatives produced are sparingly soluble in water, and can be isolated and determined gravimetrically [53].

Volumetric determination involves treatment of the sample with excess amine and back-titration of the excess with hydrochloric acid solution. The most reactive are the amines in the order: dibutylamine (5–10 min), dimethylamine (30 min) and butylamine (45 min). Titration can be carried out in a variety of inert solvents such as chlorobenzene [54], acetone [53], dioxan [55]. Ammonia reacts more rapidly than amines with the isocyanate group in acetone solution. The procedure is given below.

A 0·2–0·5 g polyurethane sample is placed in a 250 ml conical flask, mixed with 25 ml of acetone, and then with 25 ml of 0·2*N* ammonia in acetone. The mixture is allowed to stand at room temperature for 10 min, then the excess ammonia is back-titrated with 0·1*N* hydrochloric acid, using methyl red indicator, until the solution is a distinct red.

Free Di-isocyanate in Prepolymers and Elastomers

Spectrophotometric methods have proved adequate in these analyses. A procedure for the analysis of various isocyanates in foamed polyurethanes, using malachite green, is reported by Kubitz [21]. Another procedure for the determination of TDI in elastomers and polyurethane foams is given by Remington [22]. This involves dissolving the polymer sample, distillation of the solvent together with the isocyanate under reduced pressure, formation of a coloured complex with a sodium nitrite–cellosolve reagent, and subsequent determination of its absorption with a colorimeter. The method allows TDI to be determined with an error of 0·2%.

A procedure using 1-*N*-naphthylethylenediamine for analysis of TDI in foams is described by Kalman [56]. A 0·5 g sample is extracted with 25 ml of ethyl alcohol, the solution is treated with 5 ml of 0·1*N* hydrochloric acid, then cooled and diazotised with 5 ml of sodium nitrite solution prepared from 0·3 g of sodium nitrite and 0·5 g of potassium bromide in 10 ml of water. After 5 min 3 ml of 10% sulphanilic acid solution is added to fix the excess nitrite, and after another 10 min 0·5 ml of the 1-*N*-naphthylethylenediamine reagent (50 mg of the compound dissolved in 59 ml of water) is added, and the absorbance is measured colorimetrically.

Isocyanate Groups in Prepolymers [57]

About 0·5 g of the prepolymer is weighed into a conical flask, 25 ml of toluene is added, and the mixture is allowed to dissolve completely. Next, the solution is treated with 25 ml of 0·1*N* *n*-butylamine in toluene, the whole is shaken for 15 min, 100 ml of isopropyl alcohol is added, then 4–6 drops of 0·1% bromophenol blue indicator solution, and the solution is titrated with 0·1*N* hydrochloric acid. A blank is run independently.

The isocyanate group content (%) is calculated from the formula

$$\% \text{—NCO} = \frac{4{\cdot}202(v_1 - v_2)n}{m}$$

where:

v_1 and v_2 = volumes (ml) of the hydrochloric acid consumed in the titration of the blank and the sample respectively,

n = normality of the hydrochloric acid,

m = the weight of the sample, g.

n-Dibutylamine is used in the analysis of —NCO groups in prepolymer by Malec and David [58]. The sample is dissolved in trichlorobenzene at a temperature not higher than 70°C. Prior to the titration some methanol is added the solution.

Isocyanate Groups in Elastomers

The following procedure has been developed by Klacel and Vaculikova [59] for the determination of isocyanates. This involves reaction with dibutylamine, followed by potentiometric titration of the excess amine in a non-aqueous medium.

REAGENTS AND APPARATUS

Dimethylformamide. Of a moisture content less than 0·01%.

Di-n-butylamine. A solution in dimethylformamide (3 ml of the amine in 500 ml of dimethylformamide).

Perchloric acid. A 0·02*N* solution in anhydrous methanol.

Magnetic stirrer.

Potentiometric Titration Apparatus with Glass and Calomel Electrodes. Prior to use the glass electrode is immersed in dimethylformamide for 3 days.

PROCEDURE. A 2–3g sample of powdered polymer of a grain size not larger than 3 mm, is weighed into a dry conical flask; 25 ml of the dibutylamine solution is added by pipette, and the flask is tightly stoppered. The contents are agitated until dissolution of the sample is complete. Then 25 ml of dimethylformamide is added, the contents are quantitatively transferred to a potentiometric cell, and titrated with perchloric acid. The titration end point may also be determined with bromocresol green, the colour changing from blue to yellow. One ml of 0·02*N* perchloric acid corresponds to 0·84 mg of NCO groups.

A correction should be made in the calculation of the NCO group content for the possible presence of acidic compounds derived from polyesters or alkaline compounds in the sample analysed.

Acid compounds are determined by titration of the polymer solution in dimethylformamide with 0·01*N* potassium chlorate in methanol, with bromophenol blue indicator, and the alkaline compounds by titration with 0·1*N* alcoholic potassium hydroxide (thymol blue). In either case potentiometric titration can be used.

The error in the determination by this procedure amounts to 2·5% for the presence of NCO groups in about 0·4% quantity.

Instrumental Methods

Determinations of free tolylene di-isocyanate in prepolymer by gas-liquid chromatography are given by Neubaur *et al.* [60]: 4·5 g of prepolymer sample is treated with 0·5 g of trichlorobenzene, and dissolved in 10 ml of ethyl acetate; 1 μl of this solution is injected into a 100″ × 1/8″ copper column coated with a 2·5% solution of silicone rubber SE-30. The column temperature is maintained below the polymer decomposition temperature, which is ca. 160°C.

Infrared Spectroscopy [61–63]

Determination of the NCO group in polyurethanes consists in measurement of the relation between the intensity ratio of the band at 2270 cm^{-1} (characteristic of the isocyanate group) and that of the band at 2950 cm^{-1} (CH stretching vibrations, assumed as internal reference). The spectra are taken by the potassium bromide disc technique. The method is rapid and simple but a calibration curve has to be prepared for each polymer type. The mean error of the determination is not higher than ±10%, and sensitivity of the method is 0·05%.

NMR Spectroscopy

Structures of the above prepolymers were studied by establishing the urethane and diurethane linkages by Sumi *et al.* [64], and specifically in the following pairs:

a. polypropylene glycol/MDI,
b. polypropylene glycol/2,4-TDI,
c. polypropylene glycol/a mixture of 2,4- and 2,6-TDI.

A prepolymer solution in a polar solvent such as *N,N*-dimethylacetamide or dimethylsulphoxide was used in the determination. The MDI prepolymer gives rise to a single NH signal at 9·47 ppm. In the case of 2,4-TDI two signals from the NH proton of 2,4-diurethane are observed at 9·43 and 8·55 ppm, and another signal due to 4-urethane at 9·60 ppm.

A mixture of 2,4- and 2,6-TDI exhibits a spectrum with five signals, of which three are identical with the 2,4-TDI prepolymer spectrum, and the additional two signals at 8·92 and 8·72 ppm are due to the 2-urethane NH proton. The method is good for the quantitative analysis of isocyanates.

References

1. Olczyk, W., *Poliuretany* (*Polyurethanes*), WNT, Warsaw, 1966.
2. Frisch, K. C., Reegen, S. L., Floutz W., *J. Polymer Sci. A-1*, **5**, 35 (1967).
3. Grochowski, M., *Polimery*, **9**, 507 (1964).
4. Otey, F. B., Zagorew, B. L., Mehltretter, C. L., *J. Appl. Polymer Sci.*, **9**, 3337 (1965).
5. Saotome, K., Komoto, H., *J. Polymer Sci.*, **5**, 119 (1967).
6. Ryan, P. W., *Brit. Polymer J.*, **3**, 145 (1971).
7. Buist, I. M., Lowe, A., *ibid.*, **3**, 104 (1971).
8. Hashimoto, S., Furukawa, I., *Kobunshi Kagaku*, **25**, 11 (1968)
9. Metzner, M., Marclay, R., Merrlam, C. N., *J. Appl. Polymer Sci.*, **9**, 3337 (1965).
10. Hollander, L., Trischler, F. D., Gosnell, R. B., *J. Polymer Sci.*, *A-1*, **5**, 2757 (1967).
11. Kozlov, L. M., Solodova, N. L., and Akhmetova, Z. U., *Vysokomol. Soedin.*, *Ser. B.*, **13**, 370 (1971).
12. Müller, E., *Angew. Makromol. Chem.*, **14**, 75 (1970).
13. Bulatov, G. A., *Plast. Massy*, **1967**, [5], 72.
14. Goujon, I., *Pensez Plast.*, **14**, 25 (1967).
15. Kuryla, W. C., *J. Appl. Polymer Sci.*, **9**, 1019 (1965).
16. Bedoit, W. C., *J. Cell. Plastics*, **7**, 110 (1971).
17. Allan, G. W., *Appl. Plastics*, **8**, 23 (1965).
18. Frentsch, G., *Plaste u. Kautschuk*, **18**, 801 (1971).
19. Mitchell, D., *Brit. Plastics*, **40**, 105 (1967).
20. Swann, M. H., Esposito, G. G., *Anal. Chem.*, **30**, 107 (1958).
21. Kubitz, K. A., *Anal. Chem.*, **29**, 884 (1957).
22. David, D. J., Staley, H. B., *High Polymers*, Vol. XVI, *Analytical Chemistry of the Polyurethanes*, Part III, Wiley-Interscience, New York, 1969, pp. 327, 333, 360.
23. Baumann, G. F., Steingiser, S., *J. Appl. Polymer Sci.*, **1**, 251 (1959).
24. Boyarchuk, Yu. M., Rappoport, L. Ya., Nikitin, V. N., Apucimina, N. P., *Vysokomol. Soedin.*, **7**, 778 (1965).
25. Zharkov, V. V., Rudneskii, N. K., *ibid.*, **10**, 29 (1968).
26. Corish, P. J., *Anal. Chem.*, **31**, 1298 (1959).
27. Leukroth, G., *Kunststoffe*, **23**, 1118 (1970).
28. Corish, P. J., Davison, W. H. T., *J. Chem. Soc.*, **1955**, 2431.
29. Mulder, I. L., *Anal. Chim. Acta*, **38**, 563 (1967).
30. Dawson, B., Hopkins, S., Sewell, P. R., *J. Appl. Polymer Sci.*, **14**, 35 (1970).
31. Brügel, W., *Introduction to Infrared Spectroscopy*, Methuen, London, 1962.
32. Kirchner, I. I., *Thin Layer Chromatography* in *Techniques of Organic Chemistry*, Vol. XII, Interscience, New York, 1967, p. 160.
33. Conacher, H. B. S., Rees, D. I., *Analyst*, **91**, 55 (1966).
34. Knappe de Peteri, E., *Z. Anal. Chem.*, **199**, 270 (1964).
35. Braun, D., Mai, E., *Kunststoffe*, **58**, 637 (1968).
36. Brame, E. G., Ferguson, R. C., Thomas, G. J., *Anal. Chem.*, **39**, 517 (1967).
37. Okuto, H., *Makromol. Chem.*, **98**, 148 (1966).
38. Esposito, G. G., Swann, M. H., *Anal. Chem.*, **33**, 1854 (1961).
39. Percival, D. F., *ibid.*, **35**, 236 (1963).

40. Wittendorfer, R. E., *Anal. Chem.*, **36**, 930 (1964).
41. Winterscheid, H., *Seifen, Öle, Fette, Wachse*, **80**, 404 (1954).
42. Majewska, F., *Sprawozdanie I.T.S., Analiza tworzyw poliuretanowych metodą chromatografii jonitowej* (*I.T.S. Report, Analysis of Polyurethanes by Ionite Chromatography*), 1962.
43. Schröder, E., *Plaste u. Kautschuk*, **9**, 186 (1962).
44. Fanica, L., *Chim. Anal.* (*Paris*), **38**, 360 (1956).
45. Rinkie, H., *Ger. Pat.* 981268 12, 101 (1951).
46. Kasterina, T. N., Kalinina, L. S., *Khimicheskie metody issledovaniya sinteticheskikh smol i plasticheckikh mass* (*Chemical Methods of Analysis of Synthetic Resins and Plastics*), Gos. Nauch.-Tekhn. Izd. Khim. Lit., Moscow, 1963.
47. Schröder, E., *Plaste u. Kautschuk*, **8**, 121 (1961).
48. Schröder, E., *ibid.*, **10**, 25 (1963).
49. Schröder, E., *ibid.*, **5**, 3 (1958).
50. Desmelle, P., Nandet, M., *Bull. Soc. Chem. France*, **5**, 112 (1945).
51. Johanson, M. I., *Ind. Eng. Chem., Anal. Ed.*, **16**, 626 (1944).
52. Warhowsky, B., Enrug., P. I., *ibid.*, **18**, 263 (1946).
53. Stagg, H. E., *Analyst*, **71**, 557 (1946).
54. Spitelberger, G. I., *Liebigs Ann.*, **562**, 99 (1949).
55. Siggia, S., Gardon, A., *Anal. Chem.*, **20**, 1084 (1948).
56. Kalman, M., *ibid.*, **29**, 552 (1957).
57. *Pensez Plast.*, **12**, 7 (1965).
58. David, D. J., Staley, H. B., *High Polymers*, Vol. XVI, *Analytical Chemistry of the Polyurethanes*, Part III, Wiley-Interscience, New York, 1969, p. 357.
59. Klacel, Z. and Vaculikova, I., *Plasticke Hmoty Kaučuk*, **7**, 334 (1970).
60. Neubaur, N. R., Skreckoski, G. R., White, R. G., Kane, A. I., *Anal. Chem.*, **35**, 1647 (1963).
61. Murphy, E. B., O'Neil, W. A., *SPE Journal*, **18**, 191 (1962).
62. Greth, G. G., Smith, R. G., Rudkin, G. O., *J. Cell. Plastics*, **1**, 159 (1965).
63. Zhokhova, F., Zharkov, V. V., *Plast. Massy*, **1967**, [8], 63.
64. Sumi, M., Hokki, Y., Nakai, Y., Nakabayashi, M., Kauzawa, T., *Makromol. Chem.*, **78**, 146 (1964).

Chapter **12**

POLYFORMALDEHYDE

High molecular weight formaldehyde polymers [1,2] are made by polymerisation of gaseous formaldehyde in organic solvents in the presence of ionic catalysts. The crude polymerisation product from this process has, like paraformaldehyde itself, a linear structure with hemiacetal hydroxyl end groups. Replacement of the hydroxyl by some other groups improves the heat-resistance of crude polyformaldehyde. For this purpose acetylation with acetic anhydride is commonly used.

Thermostable polyformaldehyde, manufactured by Du Pont and known as Delrin, is a solid of specific gravity 1·425 g per cm^3, melting at ca. 180°C, insoluble in water and most organic solvents except dimethylsulphoxide, and affected only by strong acids and alkalis. Delrin products exhibit excellent mechanical properties, and therefore are used in the manufacture of parts (such as gears and valves) for pumps, engine parts and the like.

A high polymer of formaldehyde with properties similar to those of Delrin can also be obtained by polymerisation or copolymerisation of trioxan. Such commercial products (Celcon, Hostaform) contain in addition to $—CH_2—O—$, oxyethylene $—CH_2CH_2O—$ repeating units, and are thus ethylene oxide- or dioxolan-trioxan copolymers.

12.1 STRUCTURE

Polymerisation of formaldehyde may lead to the formation of products of varying structure. Polymerisation in the presence of alkalis yields carbohydrate type polymers $(—CHOH—)_n$. Another structure occurs in paraformaldehyde and α-polyoxymethylene, as demonstrated by detailed investigations by Staudinger [3], namely the acetal or polyoxymethylene unit $(—CH_2O—)_n$. Investigations made by Koch and Lindvig [4] have shown that polymeric polyformaldehyde has the same structure. The evidence is based on the fact that on heating to > 200°C, $99{\cdot}6 \pm 0{\cdot}3\%$ formaldehyde is produced and that the product is completely hydrolysed to formaldehyde with strong acids and alkalis. The absence of C—C bonds, in conjunction with a high degree of crystallinity, suggests a linear structure. The unmodified polymer molecules are terminated with OH groups. This may be inferred from the participation in the acetylation reaction, by the formation of water on depolymerisation (to be determined by the Fischer method), and by the presence of some methoxyl groups. The acetylated products contain largely acetyl groups at the ends of the molecules.

Recently NMR spectroscopy was used in studies of the structure of polyformaldehyde [5,6] and trioxan-dioxolan copolymers [7,8]. Also pyrolysis gas chromatography was used in similar structural studies on formaldehyde-dioxolan copolymers [9].

12.2 QUALITATIVE ANALYSIS

Flame Test

Polyformaldehyde melts with decomposition and burns with a bluish flame. The evolving fumes are neutral and smell of formaldehyde.

Chromotropic Acid Test

A small polymer sample is heated at 60–70°C for 10 min with ca. 2 ml conc. sulphuric acid and a few crystals of chromotropic acid. An intense violet colour indicates the presence of formaldehyde.

Infrared Spectroscopy

The infrared spectrum of non-stabilised polyformaldehyde (Fig. 12.1) contains a number of characteristic bands of high intensity in the 8–11 μm (900–1250 cm^{-1}) range, of which the strongest are due to CH_2 groups and to the C—O bond [4,10,11]. Additional bands occur in the spectrum of Delrin from the carbonyl group at 1755 and 1740 cm^{-1} [4] and from the amino groups of the stabiliser added to this plastic, at 1547, 1642, and 3320 cm^{-1} [12].

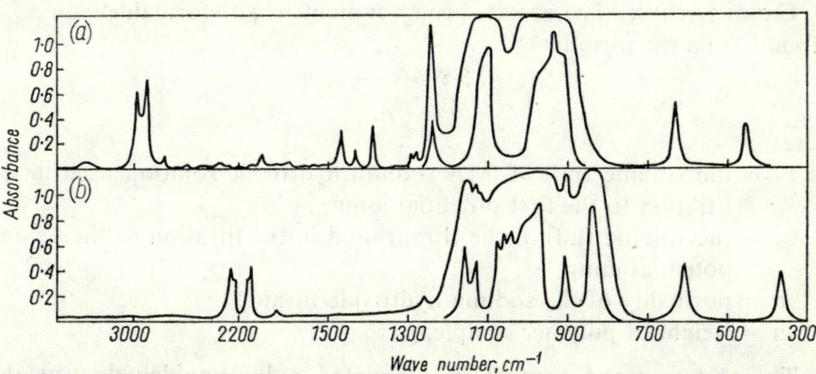

Fig. 12.1 Infrared spectra of polyformaldehyde [10]: *a*—suspended in Nujol and in hexachlorobutadiene, *b*—by the KBr disc technique (by permission of the authors and the Chemical Society)

12.3 CHEMICAL COMPOSITION OF THE POLYMER

Acetate Groups

The method for the determination of acetate groups in polyformaldehyde has been worked out by Koch and Lindvig [4]. The method consists in gradual decomposition of the polymer with methanolic hydrochloric acid, removal of the methyl acetate by distillation, and hydrolysis of the ester with standard sodium hydroxide. The effect of the medium on the quantitative methyl acetate hydrolysis was investigated by Feigin and Kosinska [13] who also established exact conditions for the determination of acetate groups by the Koch and Lindvig procedure.

REAGENTS

Hydrochloric Acid Solution. In methanol, 2*N*.

Sodium Hydroxide. Aqueous 1*N* and 0·1*N* solutions.

Hydrochloric Acid. A 1*N* solution.

PROCEDURE. A 20 g polymer sample is weighed into a round-bottom flask equipped with a dropping funnel and a reflux condenser, and 30 ml of 2*N* hydrochloric acid in methanol is added. The flask is heated on a water bath and methanol is distilled off along with the methyl acetate formed. The receiver is chilled to −10°C. Three 5 ml portions of methanol are then dropped into the flask from the funnel, and the methanol is distilled off. The fraction boiling below 70°C is collected. The distillate is neutralised, and 2 ml of 1*N* sodium hydroxide is added, and the whole is heated in a 100 ml round-bottom flask under reflux on a water bath for 1 hr. The condenser is cooled to −10°C. The mixture containing methyl acetate hydrolysis products is cooled to −5°C, neutralised with 1*N* hydrochloric acid and 2 ml excess of this acid is added. The solution is titrated potentiometrically with 0·1*N* sodium hydroxide solution, using a glass-calomel electrode system.

CALCULATION. The acetate group content in polyformaldehyde $X(\%)$ is found from the formula

$$X = \frac{5{\cdot}904(v_1 - v_2)n}{m}$$

where:

v_1 = the volume (ml) of 0·1*N* sodium hydroxide solution used in the titration to the first potential jump,

v_2 = the volume (ml) of the titrant used in the titration to the second potential jump,

n = normality of the sodium hydroxide titrant,

m = weight of polymer sample, g.

The acetate group content in acetylated polyformaldehyde was determined by Doerffel *et al.* [14] by infrared spectrophotometry, and on this basis they calculated the number-average molecular weight of the polymer. The analysis was made by the potassium bromide disc technique, by measuring the absorbance at 1760 cm^{-1}, which is due to carbonyl in the acetate end groups.

Methoxyl Groups

Methoxyl groups in the thermostable high polymer polyformaldehyde are best determined by the Zeisel method. It must be said, however, that the conventional method is not feasible for the analysis of these polymers, as they do not dissolve in conc. hydriodic acid. This problem was overcome by Koch and Lindvig [4] who used phenol as a solvent for polyformaldehyde. Lebedeva and Pisarenko [15] similarly analysed methoxyl groups in a variety of polymers.

From detailed studies by Matlina and Vishnyak [16] the Koch and Lindvig method was found to yield low results on account of the considerable heat from the exothermic reaction of phenol with formaldehyde in the initial stage of the determination, and because of the formation of a phenolic resin which hinders decomposition of polyformaldehyde. These authors succeeded in obtaining accurate results by replacing hydriodic acid with a potassium iodide–phosphoric acid mixture.

REAGENTS

Absorption Solution. Sodium acetate (hydrated) (100 g) is dissolved in glacial acetic acid and the volume is made up to 1 litre; 1 ml of bromine is added to 200 ml of this solution.

Sodium Thiosulphate. Aqueous solution, 0·01*N*.

PROCEDURE [16]. A polymer sample is weighed into a 50 ml flask (Fig. 12.2) in such a quantity as to keep the thiosulphate titrant consumption within 10 ml (which corresponds to 0·050 g of paraformaldehyde, 0·070 g of α-polyoxymethylene, or 1 g of Delrin). Phenol is added (3 g for powdered samples, and a 10-fold excess relative to the sample weight for a granulated

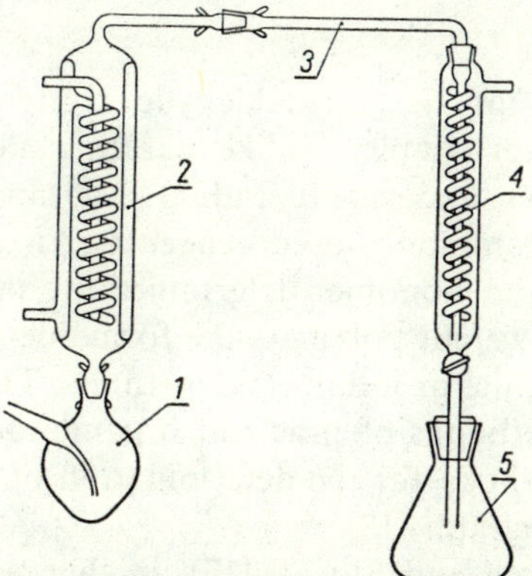

Fig. 12.2 Apparatus for determination of methoxyl groups in formaldehyde polymers: *1*—reaction flask, *2*—reflux condenser, *3*—Ascarite-filled tube, *4*—absorber, *5*—titration flask

product, but not less than 5 g). A mixture of 2·4 g of potassium iodide and 4·0 ml of phosphoric acid (d = 1·70 g per cm^3) is then added. The flask is connected to a reflux condenser. The absorber is filled with 8 ml of sodium acetate-bromine solution in acetic acid. The flask is heated on an air bath at 140–160°C for an hour under a nitrogen stream passed at a rate of 14–15 ml per min. The absorber contents are transferred to a titration flask (previously filled with 10 ml of 20% sodium acetate solution) and the absorber and tube connecting the absorber with the condenser (see Fig. 12.2), are rinsed with 10 ml of water. To remove excess bromine 1·8 ml of 2% formic acid is introduced. The flask is tightly stoppered and allowed to stand for 5 min, 5 ml of 10% sulphuric acid and 0·1 g potassium iodide are added, the flask is stoppered again and allowed to stand in the dark for 10 min. The liberated iodine is titrated with 0·01*N* thiosulphate solution with 0·5% starch solution (2 ml) added as indicator towards the end of titration. A blank is run separately. The volume of the reagent used in the blank should not exceed 0·06 ml. Whether or not the determination is successful depends on how complete is the dissolution of the polymer in the reaction medium [16].

Methoxyl groups in paraformaldehyde were determined by Přibyl and Slovák [17] by the method of Fischer and Schmidt [18]. This involves conversion of methyl alcohol (formed here from hydrolytic splitting of the polymer) into methyl nitrite. Iodometric determination of methyl nitrite (originally used in the Fischer and Schmidt method) was replaced by ultraviolet spectrophotometry. The ester isolated from the reaction mixture was first hydrolysed to methyl alcohol, which was then isolated and oxidised with potassium permanganate to formaldehyde. This was converted into a colour complex with chromotropic acid. Absorbance was measured at 570 nm.

Formaldehyde

Precise determination of the formaldehyde content of high polymeric polyformaldehyde is not simple. Unlike paraformaldehyde or α-polyoxymethylene this polymer undergoes gradual decomposition with difficulty under the action of commonly used reagents. An additional problem is that the quantities of the monomer determined are in the vicinity of 100%.

In low molecular weight polymers the formaldehyde content is usually determined by the sulphite or iodometric methods. The methods are treated at greater length in textbooks of practical organic analysis. A comprehensive review of the methods for the determination of formaldehyde is covered in Walker's monograph [19].

According to Přibyl and Slovák [17], in the case of higher polymers (but of the paraformaldehyde type) the sulphite method provides too low results on account of incomplete dissolution of the polymer. These authors recommend instead a method that involves the reaction with hydroxylamine hydrochloride in 1*N* sodium hydroxide solution. In high polymers formaldehyde was determined by Koch and Lindvig [4], who examined the pyrolysis products of the sample under study.

Dioxolan Content

1,3-Dioxolan units incorporated into the polyoxymethylene structure were determined by the NMR technique by Pfab and Bernhart [8]. Pyrolysis gas chromatography was employed by Minin *et al.* [9].

Water

The water contained in formaldehyde polymers can occur both in a chemically combined form as hemiacetal end groups

$$HO \; CH_2O \cdot CH_2O \ldots CH_2O \; H$$

and as free water referred to as moisture. In low polymers of formaldehyde (paraformaldehyde, α-polyoxymethylene) determination of both combined and free water offers no particular problems. The combined water can be abstracted from the polymer with strong alkalis in the process of complete degradation of the polymer, and it can be determined jointly with free water by the Karl Fischer reagent titration. According to Lyubomilov and Usevich [20] methanolic sodium hydroxide may be used as an alkaline reagent. As a hydrolytic and solvolytic agent sodium methoxide in methanol was used in the analysis of this polymer by Přibyl and Slovák [17]. In this procedure the result from the blank determination was lowered, and the reaction time was substantially shortened so that the side reactions which might lead to the formation of water could be avoided. Water was determined by the same method in the pyrolysate of the polymer in 1,3-propanediol by Muroi and Ogawa [21]. The same authors [22] also followed the Reid and Turner procedure (see p. 58). Polyformaldehyde was decomposed by heating to 200–250°C and the volatile products were absorbed in a methanol-ethylene glycol mixture at 10°C.

Determination of combined water in thermostable polymer by the techniques involving the Karl Fischer reagent titration in solution is not feasible because of the insolubility of the polymer. The Muroi and Ogawa procedure [21] can be used here by using a suitably higher decomposition temperature. Thus total water content in the polymer is established.

Free water can be determined in thermostable polymer as in paraformaldehyde by titration of the powdered polymer suspended in methanol with the Karl Fischer reagent. The chemically combined water can be found from the difference between the results from both determinations.

Antioxidants

Schröder *et al.* [23] determined antioxidants and thermal stabilisers in polyformaldehyde by successive extraction of the polymer with chloroform, heptane and methanol. In the individual fractions obtained these authors identified individual compounds by chemical methods, by thin layer chromatography, or by infrared spectrophotometry.

References

1. Enikolopyan, N. S., Vol'fson, S. A., *Khimiya i tekhnologiya poliformal'degida* (*Chemistry and Technology of Polyformaldehyde*), Khimiya, Moscow, 1968.
2. Schweitzer, E., MacDonald, R. N., Punderson, J. O., *J. Appl. Polym. Sci.*, **1**, 158 (1959).
3. Staudinger, H., *Die hochmolekularen organischen Verbindungen*, Springer-Verlag, Berlin, 1960.
4. Koch, T. A., Lindvig, P. E., *J. Appl. Polymer Sci.*, **1**, 164 (1959).
5. Sobottka, J., Keller, F., *Plaste u. Kautschuk*, **16**, 92 (1969).
6. Sobottka, J., Uffrecht, H. H., *ibid.*, **18**, 737 (1971).
7. Fleischer, D., Schulz, R. C., *Makromol. Chem.*, **152**, 311 (1972).
8. Pfab, W., Bernhart, K. H., *Z. Anal. Chem.*, **251**, 25 (1970).
9. Minin, V. A., Berlin, A. A., Varshavskaya, A. I., Kovtun, T. S., Karmilova, L. V., Enikolopyan, N. S., *Vysokomol. Soedin.*, *Ser. A.*, **14**, 9 (1972).
10. Novak, A., Whalley, E., *Trans. Faraday Soc.*, **55**, 1484 (1959).
11. Hummel, D. O., Scholl, F., *Atlas der Kunststoff-Analyse*, Carl Hanser Verlag, Munich; Verlag Chemie, Weinheim, 1968.
12. Stejný, J., Majer, J., *Chem. Průmysl*, **12**, 53 (1962).
13. Feigin, E., Kosinska, V., *Plast. Massy*, **1961**, [6], 8.
14. Doerffel K., Friedrich, H., Grohn, H., Wimmers, D., *Plaste u. Kautschuk*, **12**, 524 (1965).
15. Lebedeva, A. I., Pisarenko, E. S., *Zh. Analit. Khim.*, **17**, 636 (1962).
16. Matlina, E. S., Vishnyak, Yu. I., *Plast. Massy*, **1965**, [2], 58.
17. Přibyl, M., Slovák, Z., *Z. Anal. Chem.*, **202**, 23, 112 (1964).
18. Fischer, W. M., Schmidt, A., *Ber.*, **59B**, 679 (1926).
19. Walker, J. F., *Formaldehyde*, Reinhold, New York, 1964.
20. Lyubomilov, V. I., Usevich, T. D., *Plast. Massy*, **1961**, [2], 67.
21. Muroi, K., Ogawa, K., *Bull. Chem. Soc. Japan*, **36**, 965 (1963).
22. Muroi, K., Ogawa, K., *ibid.*, **36**, 1278 (1963).
23. Schröder, E., Hagen, E., Helmstedt, M., *Plaste u. Kautschuk*, **14**, 560 (1967).

Chapter 13

POLYOLEFINS

Polyolefins represent the products of polymerisation of unsaturated aliphatic hydrocarbons. Of the highest technical and commercial interest are polyethylene, polypropylene, ethylene-propylene copolymers, and also polybutenes. The following copolymers are also known: ethylene-vinyl acetate, ethylene-acrylates or methacrylates, or ethylene with acrylic, methacrylic, itaconic, maleic and fumaric acids [1-3].

Polyethylene is manufactured commercially by high pressure and low pressure polymerisation. In the high pressure process, which occurs by a free-radical mechanism, the catalysts used are minor quantities of oxygen or peroxides. This process is used only for the manufacture of polyethylene since polypropylene made by this process is of a very low molecular weight. The low pressure process is now more widely used and operates by an ionic mechanism in the pressure range 0–10 atm. The catalysts used here are referred to as Ziegler catalysts and constitute mixtures of organometallic compounds (such as trialkylaluminium or haloalkylaluminium) and salts of variable-valence heavy metals (e.g. $TiCl_4$).

Also of considerable commercial interest is the Phillips process based on the polymerisation of olefins at medium range pressures (up to 100 atm) with the use of supported oxide catalysts (chromium oxides).

The products made by individual processes from the same feedstock differ substantially in their properties. High pressure polyethylene has a molecular weight of the order of 20,000–30,000, density of about 0·92 g per cm^3, and a softening point above 100°C. This is a soft and transluscent material, and thus finds application, among others, in the manufacture of films. Low pressure polyethylene usually has a higher molecular weight of the order of 100,000, density of 0·93–0·95 g per cm^3, and a softening point of about 120°C. This product is much tougher and less translucent, and for these reasons it is fabricated by extrusion and moulding.

Currently, polyethylene grades are distinguished according to their density and not by the manufacturing process by which they are made. Thus three grades of polyethylene are commonly recognised: polyethylene of low density (0·910–0·925 g per cm^3), of medium density (0·926–0·940 g per cm^3), and of high density (0·941–0·965 g per cm^3).

Polyethylene at room temperature is resistant to the majority of chemical agents. In benzene or carbon tetrachloride at room temperature this polymer swells only slightly.

Polypropylene is made exclusively by the ionic polymerisation process. It is of a still lower density than polyethylene (0·91 g per cm³) but its softening point is much higher (140–155°C). A disadvantage of polypropylene compared with polyethylene is its higher susceptibility to oxidants. Polypropylene is fabricated into fibres, films, tubes and pipes and a variety of technical finished products.

Ethylene-propylene copolymers are manufactured by ionic polymerisation with certain complex catalysts. Copolymers containing ethylene in minor quantities are aimed at improving the flexibility of polypropylene. Copolymers having about equal proportions of ethylene and propylene are used as elastomers. These copolymers usually contain certain amounts of a third monomer which is generally a diene. The double bond present in the diene plays an important role in the vulcanisation process.

Polyisobutylene is made by ionic polymerisation with boron trifluoride as calatyst. This polymer is resilient and in many respects not inferior to natural rubber. With regard to chemical resistance and electrical properties polyisobutylene is superior to the latter. Its applications range from lining for chemical plant and pipelines to insulating electric cables. An isobutylene copolymer with a minor quantity of isoprene or butadiene is known under the name of butyl rubber.

13.1 STRUCTURE

Polyethylene

Ethylene polymers have the very simple general formula $(CH_2)_n$. No functional groups enter their structure with the exception of minor quantities of oxygen-containing groups which result from side reactions accompanying the polymerisation process. These are mainly carboxylic, carbonyl (ketonic) and preoxide groups. Various polyethylene grades also contain minor quantities of olefinic bonds. These polymers may, however, incorporate considerable branching, which varies in degree according to the polymerisation process used.

Chemical methods of analysis are only suitable for determining the quantity of carboxylic groups, peroxide groups and olefinic bonds. The most valuable information about the structure of polyethylenes can be obtained almost exclusively by instrumental methods. It is seen from the infrared spectra of polyethylenes that the oxygen contained in these materials may be present in a wide variety of forms such as hydroperoxide groups C—O—O—H (band at 2·81 μm), anhydride groups, and ester, aldehydic, ketonic, and carboxylic groups (bands due to the C=O group at about 5·8 μm) [4–6]. In addition, as demonstrated by the investigations of Luongo [7], peracid groups occur in polyethylene heated for a prolonged period in air.

The infrared spectrum of polyethylene exhibits bands due to unsaturated groups, specifically at about 6·1 μm (all types of olefinic bonds), 10·07 and 11·00 μm (—CH=CH_2 vinyl groups), 10·35 μm (olefinic —CH=CH— groups within the polymer chain), and at 11·25 μm ($>$C=CH_2 vinylidene groups) [4,6,8].

In the spectra of polyethylene numerous bands due to the methylene and methyl groups also appear. From the quantitative measurements of absorbance in the bands due to CH_3 groups (in the first place at 7·25 μm) conclusions can be drawn as to the presence of short-chain branches in polyethylene macromolecules, as the side chains are largely composed of alkyl groups. Short branched chains in polyethylene may also be revealed by examination of the absorption bands due to —CH_2— rocking modes in the 13 μm region.

Infrared spectrophotometry was used in studies on branching in polyethylene by a number of authors [9–13] (see also [14] and the references therein).

On the other hand, long-chain branching in polyethylene can be determined by light-scattering [15], measurements of intrinsic viscosity [15–19], and by gel-permeation chromatography [17,18].

Poor solubility of polyethylenes restricts the use of NMR spectroscopy to the study of more soluble products, primarily those of low density polyethylene. On the basis of the NMR spectrum of low density polyethylene examined at an elevated temperature (35–130°C), Nishioka and Kato [20] established that the methyl group content coincides with the data from infrared spectrophotometry. In the NMR spectra CH_3 triplets are observed, substantiating the evidence from infrared spectroscopy for the presence of ethyl and butyl branches. Chain branching of polyethylenes was also studied by Lissac and Berticat by the NMR technique [21]. These results concerning chain branching in polyethylenes have been further proved by radiation methods [15,22,23].

Trestianu and Sandulescu [24] examined the structure of polyethylene (and polypropylene) by a pyrolytic gas chromatography method.

On the basis of the above data, polyethylene molecules are assumed to consist of long chains with minimum branching, the branches being mostly short chain alkyl groups of 2 and 4 carbon atoms. The ratio of ethyl to butyl groups in these molecules is about 2:1. Low density polyethylene has one such short branch for about every 50 carbon atoms and the branches are randomly distributed along the main chain. Furthermore this variety of polyethylene contains a certain quantity of long chain branches which are several tens of carbon atoms in length. High density polyethylene has few branches (one per 300–400 carbon atoms), distinguishing it from low density polyethylene. In addition to alkyl groups in the polyethylene molecules minor quantities of oxygen-containing groups are also present. These are mostly ketonic groups which may be distributed at various sites in the molecules. At the end of certain polyethylene molecules there may be unsaturated groups; in low density polymer these are mainly vinylidene groups at a concentration of 0·3 olefinic bonds per macromolecule, whereas in high density grades these are mainly vinyl groups at a concentration of about one olefinic bond per macromolecule.

Polymers of Ethylene Homologues

From the previous section it is seen that the two principal types of polyethylenes are distinguished by their branching frequency. The high density polymers contain few branch points, while the low density polymers have a relatively high branching frequency, the branches being randomly distributed along the main chains.

The structure of polymers of ethylene homologues is far more complex since their molecules, irrespective of the method of polymerisation, invariably incorporate branches in a quantity at least equal to the degree of polymerisation. These branches are, of course, alkyl substituents of the ethylene homologues. If the orientation of these alkyl groups is random, the polymer is atactic. It is now common practice, however, to perform the polymerisation process in the presence of stereospecific catalysts, resulting in polymers with a regular disposition of alkyl groups and without additional branches, that is, a stereoregular structure. The way these alkyl groups are regularly arranged may be diverse, and hence we have a variety of stereoregular homopolymers. These varieties differ considerably in physical and chemical properties.

In the case of copolymers the structure of the macromolecules is still more complex, as the individual monomer molecules may combine in a random or regular way, and the sequence of such linking may be varied.

The determination of stereoregularity in polyolefins is a problem of considerable theoretical and practical interest. The problem has been dealt with in recent years by numerous investigators who employed a range of instrumental methods.

The structure of polypropylene polymers was studied by infrared spectroscopy [25–30], NMR spectroscopy [31–37], and X-ray spectroscopy [38].

The structure of ethylene-propylene copolymers was studied by infrared spectroscopy [39–51], by NMR spectroscopy [36,37,51–53], X-ray spectroscopy [50], and by pyrolysis mass spectrometry [54], while the structure of copolymers of ethylene with ester type monomers was investigated by NMR spectroscopy [55] and pyrolysis gas chromatography [56].

In addition to the branching alkyl groups, molecules of the ethylene homologue polymers contain quantities of the same functional groups that are present in polyethylene (see above). These polymers usually have more oxygen-containing groups because of their higher susceptibility to oxidation. The presence of these groups may be detected and determined by infrared spectroscopy [26,57,58] as well as by chemical methods.

13.2 QUALITATIVE TESTS

Chemical Methods

Flame Test

When placed in a burner flame polyolefins melt, catch fire easily and burn with a somewhat sooty flame. The burning of polyolefins is accompanied by the odour of paraffin wax. The fumes exhibit a neutral reaction. When the flame is extinguished the odour of higher fatty acids appears.

For identification of polyethylenes use may also be made of their chemical stability and their behaviour towards organic solvents. Polyethylenes do not dissolve in the common solvents. They do dissolve, however, in boiling decalin and tetralin and in higher aromatic hydrocarbons at temperatures above 100°C.

Mercury Salt Test

A polymer sample [59] is heated in a small test tube of heat resistant glass. The mouth of the tube is covered with a filter paper moistened with a mercury oxide solution in sulphuric acid (5 g of yellow mercury oxide, 15 ml of conc. sulphuric acid, and 80 ml of water). A yellow spot suggests the presence of polypropylene and polyisobutylene or butyl rubber. Polyethylene gives a negative reaction. In the presence of natural rubber, nitrile rubber, or polybutadiene a dark-brown colour develops.

To differentiate polypropylene and polyisobutylene [60] the pyrolysis products are introduced into a 5% methanolic mercuric acetate solution. The resulting solution is evaporated to dryness. The residue is extracted with petroleum ether and the extract is filtered free of solids. From the concentrated filtrate methoxyisobutylene mercury acetate crystallises as long needles of m.p. 55°C. In the case of polypropylene the crystals do not form.

Wijs Reagent Test

A polymer sample is dissolved in *p*-dichlorobenzene and the Wijs reagent is added dropwise. Decolourisation is indicative of double bonds present in large amounts. Polydienes—polybutadiene and polyisoprene—give a positive test.

Instrumental Methods

Infrared Spectroscopy

All polyolefins [14,61,62] exhibit strong absorption bands at about 3·4 and 6·8 μm due to the $—CH_2—$ methylene groups. On the basis of these bands polyolefins can be distinguished from other polymers. Polyolefins can be differentiated among themselves on the basis of the following characteristic bands: polyethylene (strong doublet at 13·7 and 13·9 μm), polypropylene (sharp bands of medium intensity at 8·6, 10·0, 10·3, and 11·9 μm), polyisobutylene or butyl rubber (a strong band at 8·1 μm), and polybutene-1 (a strong band at 13·1 μm).

Polyethylene grades of different density and ethylene-propylene copolymers may be distinguished on the basis of the intensity of the absorption bands in the range 850–1050 cm^{-1}. The bands are due to diverse unsaturated groups such as $RR'C{=}CH_2$ (888 cm^{-1}), $RCH{=}CH_2$ (909 cm^{-1}), and $RCH{=}CHR'$ (965 cm^{-1}), which are present in variable quantities and are characteristic of the individual polymer types mentioned earlier.

The absorbance for these polymers at sample thickness of 0·55 mm are listed in Table 13.1. Typical spectra are given in Figs. 13.1–13.4.

Table 13.1 Absorbance (*A*) of Polyolefins at 888, 909 and 965 cm^{-1}

Polymer Type	A_{888}	A_{909}	A_{965}
High density polyethylene: from medium pressure			
Phillips process	0·09	1·55	0·08
from low pressure Ziegler process	0·14	0·35	0·09
Low density polyethylene:	0·15	0·10	0·07
Ethylene-propylene copolymer:			
from medium pressure process	0·25	0·90	0·22
from low pressure process	0·19	0·19	0·16

Individual polyethylene grades may be identified by measurement of the absorption intensity at 7·23 μm (methyl groups). The methyl group content is a measure of the degree of branching. Low density polyethylene contains 20–30 methyl groups per 1000 carbon atoms, whereas high density

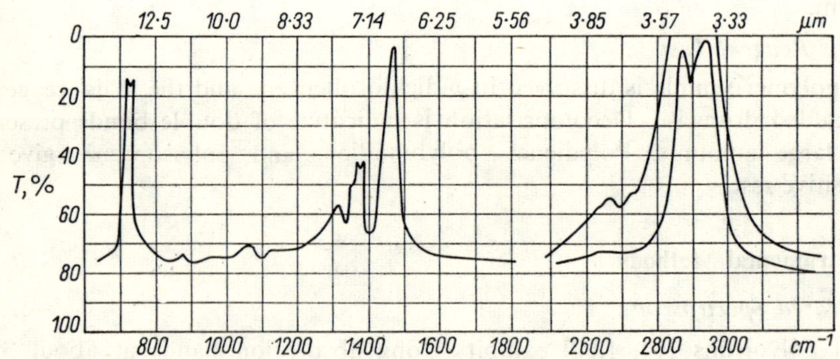

Fig. 13.1 Infrared absorption spectrum of a low density polyethylene; hot-formed film

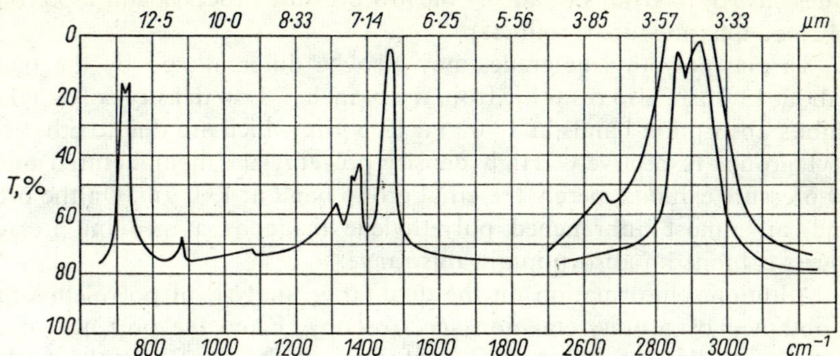

Fig. 13.2 Infrared absorption spectrum of a low pressure and low density polyethylene; pressed film

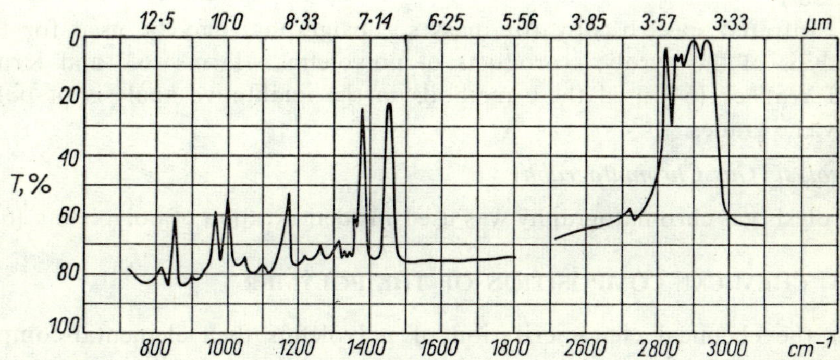

Fig. 13.3 Infrared absorption spectrum of isotactic polypropylene; pressed film

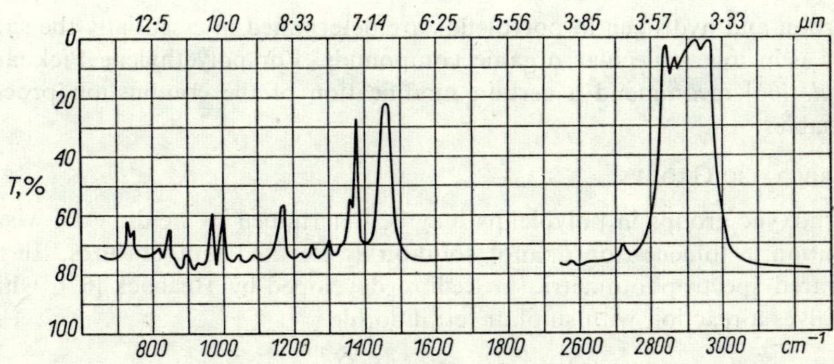

Fig. 13.4 Infrared absorption spectrum of ethylene-propylene copolymer (20:80); pressed film

grades have 5–10 groups (made by the low pressure process), and 1–3 groups (the medium pressure products).

Various polyethylene grades may also be differentiated by the bands at about 13·0 μm, also resulting from the branches. Low density polyethylene exhibits absorption bands at 13·0 and 13·5 μm, which are due to ethyl and butyl groups respectively. High density polyethylene manufactured under low pressure exhibits merely the ethyl group band at 13·0 μm. On the other hand, an almost unbranched polyethylene made by a medium pressure process exhibits no absorption in this range.

Additional information on the qualitative analysis of polyolefins may be obtained by atomic emission spectroscopy. From the presence of the characteristic lines of the residual catalyst traces (Ti, Al, Cr, B) the method of preparation may be inferred. Titanium and aluminium are present in low pressure polyolefins, and chromium in polyolefins made under medium pressure. The lack of characteristic lines of these elements indicates that the polymer was manufactured by the free radical process, that is, under high pressure.

Infrared spectroscopy (or mass spectrography) may be used for the analysis of the pyrolysis products of polyolefins. Harms [63] and Kruse and Wallace [64] used these methods in the qualitative analysis of polyethylene grades.

Pyrolysis Gas Chromatography

Pyrolysis gas chromatography was used in identification of polyolefins [65].

13.3 CHEMICAL COMPOSITION OF THE POLYMER

For the chemical characterisation of polyolefins their elemental composition, content of constituent monomers (in copolymers), alkyl groups, and functional groups such as carboxyls, olefinic bonds, carbonyl groups, as well as active oxygen and proportions of stereoisomers are determined.

Elemental Analysis

Carbon and hydrogen in polyolefins are determined in essentially the same way as in low molecular organic compounds. For polyethylene Pickhardt *et al.* [66] recommend a certain modification of the combustion process variables.

Carboxylic Groups

Carboxylic groups in polyolefins may be determined by acidimetric visual titration in toluene-isopropanol solution at elevated temperatures. In the infrared spectrophotometric procedure developed by Heacock [67], which involves a reaction with sulphur tetrafluoride

$$R{-}C\begin{matrix}\nearrow O\\ \searrow OH\end{matrix} + SF_4 \rightarrow R{-}C\begin{matrix}\nearrow O\\ \searrow F\end{matrix} + HF + SOF_2$$

the absorbance which results from the acyl fluoride produced in the reaction is measured at 5·45 μm (C—F bonds) and 5·8 μm (C=O).

Ester Structures

In oxidised polyethylene, ester structures are determined by infrared spectroscopy on the basis of the C—O absorption band at 1175 cm^{-1} [57].

Olefinic Double Bonds

The double bonds in polyolefinic plastics were determined by the method based on bromine addition by Szewczyk *et al.* [68]. The addition products were analysed by the Schöniger combustion procedure, and the bromide ions were determined by mercuric nitrate titration. Tunnicliffe [69] developed a method for determination of double bonds in polyolefins, based on bromination with pyridinium bromide perbromide.

The minor unsaturation in polyisobutene may be determined by a radiochemical method based on the addition of labelled chlorine [70].

Haslam and Willis [61] describe a method for determination of unsaturated groups in polyethylene, involving infrared spectroscopic measurements, prior to and after bromination, at 10·30, 11·00, 11·27 μm, and at 11·27 μm respectively. This procedure is aimed at eliminating the interfering effect of the alkyl group band at 11·18 μm. Likewise Szewczyk *et al.* [71] determined the double bond content by infrared spectrophotometry before and after the bromination of polymer samples.

The question of the determination of a wide range of olefinic groupings by infrared spectrophotometry has already been treated along with the subject of identification of individual varieties of polyethylene (see p. 344).

Carbonyl Groups

Carbonyl groups in polyethylenes are usually determined by infrared spectroscopy on the basis of the band at 5·8 μm [5,7,61].

Nitzl [72] determined these groups in polyethylene films by a radiochemical method or titrimetrically after treatment with ^{14}C-labelled 2,4-dinitrophenylhydrazine.

Active Oxygen

Peroxide groups in polyethylenes may be determined as in low molecular weight compounds, by iodimetric titration.

In polypropylene the active oxygen was determined by Giuffre and Cassani [73], Jadrniček [74] and Boček [75] by colorimetric methods based on oxidation of leuco forms of certain dyes; Boček [76] also made use of the colour reaction with Fe^{2+} and phenanthroline.

Pokorný [77] determined peroxide groups in polypropylene, using a titrimetric procedure based on oxidation of Fe^{2+} to Fe^{3+} followed by potentiometric titration with EDTA solution.

Hydroperoxide groups in polyolefins may be detected by infrared spectrophotometry on the basis of the absorption band at 2·8 μm.

Lisovskii *et al.* [78] analysed oxygen in polyethylene by fast-neutron activation and measurement of the quantity of ^{16}N produced.

Estimation of Tacticity of Homopolymers

Homopolymers of different tacticity differ considerably in physical properties, including solubility in organic solvents. This fact has been exploited for analytical purposes.

Russell [79] developed a method for the determination of tacticity of polypropylene by extraction of the polymer under nitrogen, first with ethyl ether then with *n*-heptane.

Examination of the spectra of the extracts demonstrated that the ethereal extract is atactic, whereas the *n*-heptane extract contains 20–40% of the isotactic form.

Czechoslovak investigators [80,81] developed a method for determination of the polypropylene structures of different tacticity by extraction with other solvents, namely naphthalene, ethyl ether or chloroform. Since results for an extraction procedure are affected by the size of the macromolecules, in this method the polymer is first dissolved in molten naphthalene, and the resulting melt is cooled. During the ether extraction of the cool mixture the atactic form passes into the ether layer in the largest proportion; when ether is replaced by chloroform, the atactic and stereo-block forms both pass to the chloroform layer. The residue from the naphthalene-chloroform extraction is mainly the isotactic form, but this may also contain some less regular structures.

Ungureanu and Firoiu [82] determined the quantity of isotactic polypropylene by extraction of the atactic polymer with boiling *n*-heptane in a reaction vessel of special design.

The proportion of the structures of different tacticity in polyolefins can also be evaluated on the basis of their infrared spectra [25,26,27,30,83], but these methods are of lesser practical significance compared to the ones discussed above.

Tacticity of polypropylene was estimated by X-ray spectroscopy by Natta *et al.* [38], and by Tincher [33] by NMR spectroscopy on the basis of the methylene proton signals. The determination of the structure and composition of polybutenes [84] was also attempted on the basis of their NMR spectra and with the help of pyrolysis gas chromatography.

Methyl Groups

Haslam and Willis [61] give a detailed account of the determination of CH_3 groups in polyethylene by infrared spectrophotometry. One methyl group per 1000 carbon atoms gives an absorbance of 0·85 for measurement at 7·26 μm with a 1 cm path-length.

PROPORTIONS OF MONOMERS IN COPOLYMERS

The proportion of both monomers in ethylene-propylene copolymers has been estimated by infrared spectrophotometry. Wei [45] developed a method of quantitative analysis based on the measurement of the absorbance ratio, measured at 8·7 and 13·9 μm, the logarithm of which is a linear function of the quantity of the propylene component. Also, according to Drushel and Iddings [41] the ratio of the absorbance measured at 968 cm^{-1} to the absorbance value at 1150 cm^{-1} (both bands are due to CH_3 groups) is a function of the content of the propylene components, whereas the ratio of the absorbance measured at 745 cm^{-1} to that found at 720 cm^{-1}, corrected for the amount of the stereoblock sequences, is a measure of the relative quantity of isolated ethylene components. The quantity of ethylene component in the same copolymers was determined on the basis of near infrared bands at 2·275 and 2·312 μm by Bly *et al.* [39], while the propylene component was estimated on the basis of the absorbance ratio measured at 2·466 and 2·35 μm by Takeuchi *et al.* [85]. Lomonte and Tirpak [43] quantitatively determined the ethylene homopolymer, in block ethylene-propylene copolymers, by measuring the absorbance at 720 cm^{-1} at 20°C and 180°C. These polymers were also analysed by infrared spectroscopy by Key *et al.* [86], Natta *et al.* [87], Majer [88], and Brame *et al.* [89].

Brown *et al.* [40] and Takeuchi *et el.* [90] analysed the propylene components in these copolymers by examining the infrared spectra of the copolymer pyrolysis products. The method is based on the observation that the ratio of vinyl to vinylidene groups in the pyrolysate is a function of the proportion of propylene component present in the polymer.

Blockiness of these copolymers was examined by pyrolysis mass spectrometry by Nencini *et al.* [54].

Ethylene-propylene copolymers were identified by pyrolysis gas chromatography by Androsova *et al.* [91]. The same method was used by Barton *et al.* [92], who analysed mixtures of the above copolymers with polypropylene and polybutene.

Ethylene-methyl acrylate copolymers were examined by Bombaugh *et al.* [56] who used pyrolysis gas chromatography and differential thermal analysis.

The composition of ethylene-propylene copolymers was determined by high-resolution NMR by Kimmer *et al.* [93]. The same technique was employed by Porter *et al.* [55] for analysis of copolymers containing ester groups and for ethylene homopolymers. This method was also found to be useful for analysis of certain homopolymers in which branching and molecular weight can be determined by measuring the area under the CH_3 group signals. In ethylene-vinyl acetate copolymers and in ethylene-ethyl acrylate copolymers the two monomer constituents may be determined qualitatively and quantitatively even at a ratio in excess of 20:1.

The monomers in ethylene terpolymers were determined by NMR spectroscopy by Altenau *et al.* [94].

13.4 DETERMINATION OF CONTAMINANTS AND ADDITIVES

Water and Carbon Black

Water in polyolefins is determined by the Reid and Turner method described on p. 58.

Carbon black is determined by a procedure involving pyrolysis of the polymer in a nitrogen stream at 500°C (Fig. 13.5) [61]. The results are too low by about 5%. If the polymer contains no carbon black, the porcelain boat has no residue after the pyrolysis. Under the conditions of the analysis the polymer undergoes complete decomposition.

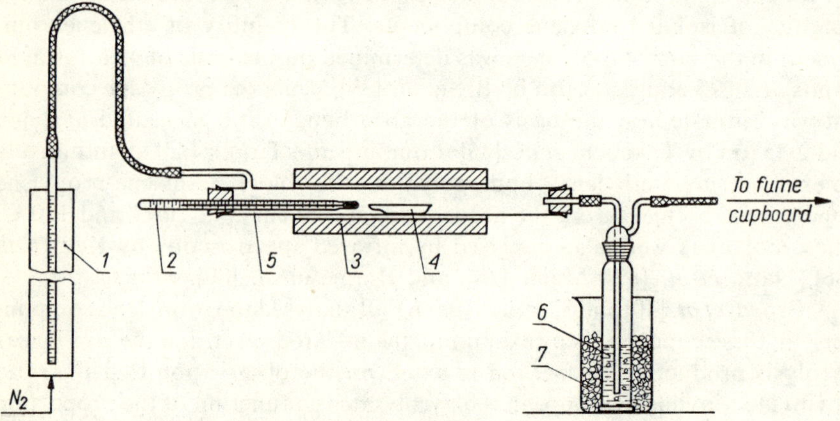

Fig. 13.5 Apparatus for determination of carbon black in polyethylene; *1*—rotameter, *2*—thermometer, *3*—electric furnace, *4*—porcelain boat, *5*—Pyrex tube, *6*—trichloroethylene, *7*—dry ice (from [61] by permission of the copyright holders).

PROCEDURE. The apparatus is set up as shown in Fig. 13.5 and 1 ± 0·1 g of polymer is weighed into a porcelain boat previously ignited in the burner flame. The nitrogen stream flow rate is controlled with a rotameter at a level of 1·7 ± 0·3 l. per min. The sample in the boat is placed in the middle of a Pyrex glass tube, which is then closed with a stopper equipped with a thermometer passing through it along with a tube for the nitrogen. The tube is put in a furnace preheated to 300–350°C for about 10 min. From that moment further heating is controlled so that a temperature of 480°C is achieved after 10 min, and 500°C after another 10 min. This temperature is maintained for 10 min. Next, the tube is removed from the furnace and allowed to cool in a nitrogen stream for 5 min. Then the boat is taken out of the tube and cooled in a desiccator for 30 min and weighed.

Ash

Polymer (50 g) [14] is weighed in a weighed platinum evaporating dish previously ignited at 800°C. Initially 15–25 g of the sample is put in the dish, which is then placed on a hot plate. The heating is interrupted when the sample is completely molten. The vapours are burnt in the burner flame. The material should be incinerated gently. As soon as the first portion is burnt another 15–25 g portion of the sample is placed in the dish, and the procedure is followed as with the first portion. When the entire sample is incinerated the residue is ignited at 800±50°C for 15–20 min. The dish is allowed to cool and is weighed again.

Metals and Other Elements

Trace impurities, which are largely residual catalysts from the polymerisation process, are determined as follows.

After Ashing of the Polymer. Ashing of the polymer may be effected in a variety of ways. Bolleter [95] heated the polymer with potassium sulphate and nitrate. Anduze [96] burned a polymer sample in a high-pressure oxygen bomb. Takeuchi and Suzuki [97] incinerated a polymer sample by heating it in an open crucible with 10% (relative to the sample weight) of sodium carbonate to avoid loss of ash. Smith [98] burnt a sample in the usual manner in an open platinum crucible. Novák and Mika [99], Mische *et al.* [100] and also Bolleter [95] wet-ashed the polymer sample in conc. sulphuric acid in a Kjeldahl flask with addition of nitric acid [95] or hydrogen peroxide [99,100]. Slovák and Přibyl [101] incinerated a polymer sample by burning and fusion with sodium hydrogen sulphate in an open platinum crucible, while Fratkina [102] heated it with cupric oxide and carbon black.

The elements contained in the solution after wet or dry ashing may be determined titrimetrically (chlorine and sulphur), by polarography or fluorimetry [100], or colorimetry, by use of colour reactions of iron with 1,10-phenanthroline or α,α'-bipyridyl, of titanium with hydrogen peroxide and sulphuric acid or with chromotropic acid [101], of aluminium with 8-hydroxyquinoline, of phosphorus with methylene blue, and of vanadium with 3,3′-diaminobenzidine [98]. In the solution from ashing a large number of metals were determined by emission spectrography [102].

Trace amounts of volatile sulphur compounds and chlorine compounds in polyethylene were determined by combustion of very large samples (up to 25 g) in a quartz tube by Negina *et al.* [103].

After Leaching the Impurities from Polymer Samples with Solvents. Baram and Grigor'eva [104] determined the residual complex catalysts in polyolefins by a method based on the analysis of the filtrate from washing of the polymer sample. Total titanium and aluminium was determined, using the reaction with EDTA. Titanium only was determined colorimetrically with hydrogen peroxide, and chlorine by potentiometric titration.

Directly in a Polymer Sample. Aluminium along with titanium (Glużińska [105]) and titanium alone (Slovák and Fischer [106]) were determined directly in a polymer sample by emission spectroscopy. The method of X-ray fluorescence was used in the analysis of such elements as Al, Ti, Cr, Fe, Cu, and Cl in polyethylene by Hayashi *et al.* [107,108]. X-ray emission spectroscopy was first used for this purpose by Smith and Maute [109] who determined Al, Ti, Cl, and Fe. Barium and calcium present as stearates were analysed by flame photometry by Schröder *et al.* [110]. Trace amounts of carbon in polypropylene were determined by atomic absorption spectrophotometry by Yanagisawa *et al.* [111]. Trace impurities were estimated by the neutron activation technique by Sorantin and Patek [112].

Antioxidants

Antioxidants are low molecular weight compounds and, unlike polyolefins, are readily soluble in common organic solvents. Therefore they are usually determined after separation from polymers by extraction with suitable organic solvents and occasionally with water [113]. In the extract antioxidants can be determined by a wide variety of methods.

Phenol-type antioxidants were estimated by alkalimetric titration with sodium methoxide [114] or with alcoholic potassium hydroxide, or else with tetraethylammonium hydroxide solution in a benzene-methanol mixture [115]. Organic compounds of bivalent sulphur and of phosphorus, being reducing agents, were determined by oxidation with an aromatic peracid. The excess acid was titrated iodimetrically [116].

Numerous antioxidants are polarographically active and can be determined in the extract by polarographic methods.

Colorimetric methods are widely used in the analysis of antioxidants. For example amines (derivatives of naphthylamines and of *p*-phenylenediamine) in polyethylene [117] were determined in this way. Ultraviolet spectroscopy is very often used in the analysis of antioxidants. This method was employed in the analysis of 2,6-di-*tert*-butyl-*p*-cresol (Ionol) and 4,4′-thio-bis(6-*tert*-butyl-*m*-cresol) (Santonox) after extraction with carbon tetrachloride (Ionol at 248 nm and Santonox at 277 nm) [118].

Antioxidants present in polyethylene were determined by extraction with chloroform and chromatographic separation on an alumina-packed column [119].

Ionol in polyolefins in the presence of Santonox was also estimated by the above method. The antioxidants were first extracted with cyclohexane, then Ionol was oxidised. Absorbance was measured at 365 nm [120].

Phenolic antioxidants in polypropylene were determined by the same method after extraction with heptane [121].

Antioxidant compositions in polyolefins are conveniently analysed by thin layer chromatography. This technique was used for their analysis in polyethylene after alkane extraction [122–125] or after dissolving the polymer in toluene, followed by precipitation of the polymer with ethyl alcohol [126].

Volatile and thermally stable antioxidants can be isolated from the polymer by heating [127] and subsequently determined by one of the methods mentioned above, for instance ultraviolet spectroscopy [127], or in the same operation by gas-liquid chromatography [128–130].

Antioxidants in polyethylene were also determined by gas-liquid chromatography after chloroform extraction [122].

Volatile antioxidants and stabilisers in polyolefins have also been determined by mass spectrometry [131].

Finally, these additives, having substantially different molecular or electronic spectra from polyolefins, may be determined directly in solid polymer samples either by ultraviolet [122] or infrared spectroscopy [132].

References

1. Hagen, H., Domininghaus, H., *Polyäthylen und andere Polyolefine*, Brunke Garrels, Hamburg, 1961.
2. Bojarski, J., Lindeman, J., *Polietylen* (*Polyethylene*), WNT, Warsaw, 1963.
3. Ritchie, P. D., (Ed), *Vinyl and Allied Polymers, Vol. I, Aliphatic Polyolefins and Polydienes: Fluoro-olefin Polymers*, Iliffe Books, London, 1968.
4. Aggarwal, S. L., Sweeting, O. J., *Chem. Revs.*, **57**, 665 (1957).
5. Rugg, F. M., Smith, J. J., Bacon, H. C., *J. Polymer Sci.*, **13**, 535 (1954).
6. Smith, D. C., *Ind. Eng. Chem.*, **48**, 1161 (1956).
7. Luongo, J. P., *J. Polymer Sci.*, **42**, 139 (1960).
8. Popov, V. P., Voropaeva, R. M., *Plast. Massy*, **1969**, [5], 68.
9. Janicka, K., Wiecheć, L., *Polimery*, **11**, 153 (1966).
10. Tirpak, G. A., *J. Polymer Sci., Pt. B*, **4**, 111 (1966).
11. Willbourn, A. H., *J. Polymer Sci.*, **34**, 569 (1959).
12. Bădilescu, S., Toader, M., Oprea, H., Badilescu, I. I., *Materiale Plastice*, **7**, 468 (1970).
13. Baltá Calleja, F. J., Hidalgo, A., *Kolloid-Z. Z. Polym.*, **229**, 21 (1969).
14. Voter, R. G., in *High Polymers, Vol. XII, Analytical Chemistry of Polymers*, Kline, G. M., (Ed), *Part I, Analysis of Monomers and Polymeric Materials*, Interscience, New York, 1959.
15. Billmeyer, F. W., Jr., *J. Am. Chem. Soc.*, **75**, 6118 (1953).
16. Völker, H., Luig, F. J., *Angew. Makromol. Chem.*, **12**, 43 (1970).
17. Cote, J. A., Shida, M., *J. Polymer Sci., Pt. A-2*, **9**, 421 (1971).
18. Drott, E. E., Mendelson, R. A., *J. Polymer Sci., Pt. A-2*, **8**, 1373 (1970).
19. Tung, L. H., *J. Polymer Sci., Pt. A-2*, **9**, 759 (1971).
20. Nishioka, A., Kato, Y., *Kogyo Kagaku Zasshi*, **68**, 1452 (1965).
21. Lissac, P., Berticat, P., *J. Macromol. Sci.*, **5**, 901 (1971).
22. Harlen, F., Simpson, W., Waddington, F. B., Waldron, J. D., Baskett, A. C., *J. Polymer Sci.*, **18**, 589 (1955).
23. Lawton, E. J., Zemany, P. D., Balwit, J. S., *J. Am. Chem. Soc.*, **76**, 3437 (1954).
24. Trestianu, S., Sandulescu, D., *Rev. Chim.* (*Bucharest*), **18**, 419 (1967).
25. Grant, I. J., Ward, I. M., *Polymer*, **6**, 223 (1965).
26. Luongo, J. P., *J. Appl. Polymer Sci.*, **3**, 302 (1960).
27. Miyazawa, T., Fukushima, K., Ideguchi, Y., *J. Polymer Sci., Pt. B*, **1**, 385 (1963).
28. Okamoto, M., Miyamichi, K., Ishizuka, O., *Kobunshi Kagaku*, **21**, 218 (1964).
29. Peraldo, M., Natta, G., Zambelli, A., *Makromol. Chem.*, **89**, 273 (1965).
30. Quynn, R. G., Riley, J. L., Young, D. A., Noether, H. D., *J. Appl. Polymer Sci.*, **2**, 166 (1959).
31. Chierico, A., Del Nero, G., Lanzi, G., Mognaschi, E. R., *Eur. Polymer J.*, **2**, 339 (1966).
32. Ohnishi, S., Nukada, K., *J. Polymer Sci., Pt. B*, **3**, 179, 1001 (1965).
33. Tincher, W. C., *Makromol. Chem.*, **85**, 34 (1965).
34. Woodbrey, J. C., *J. Polymer Sci., Pt. B*, **2**, 315 (1964).
35. Ando, I., Nishioka, A., *Polymer J.* (*Japan*), **2**, 161 (1971).
36. Ferguson, R. C., *Macromolecules*, **4**, 324 (1971).

37. Crain, W. O., Jr., Zambelli, A., Roberts, J. D., *Macromolecules*, **4**, 330 (1971).
38. Natta, G., Mazzanti, G., Crespi, G., Moraglio, G., *Chim. Ind.* (*Milan*), **39**, 275 (1957).
39. Bly, R. M., Kiener, P. E., Fries, B. A., *Anal. Chem.*, **38**, 217 (1966).
40. Brown, J. E., Tryon, M., Mandel, J., *Anal. Chem.*, **35**, 2172 (1963).
41. Drushel, H. V., Iddings, F. A., *Anal. Chem.*, **35**, 28 (1963).
42. Gössl, T., *Makromol. Chem.*, **42**, 1 (1960).
43. Lomonte, J. N., Tirpak, G. A., *J. Polymer Sci.*, *Pt. A*, **2**, 705 (1964).
44. Smith, W. E., Stoffer, R. L., Hannan, R. B., *J. Polymer Sci.*, **61**, 39 (1962).
45. Wei, P. E., *Anal. Chem.*, **33**, 215 (1961).
46. Ciampelli, F., Valvassori, A., *J. Polymer Sci.*, *Pt. C*, **16**, 377 (1967).
47. Kimmer, W., Schmolke, R., *Plaste u. Kautschuk*, **15**, 807 (1968).
48. Ciampelli, F., Tosi, C., *Spectrochim. Acta*, **24A**, 2157 (1968).
49. Tosi, C., Valvassori, A., Ciampelli, F., *Eur. Polymer J.*, **5**, 575 (1969).
50. Tosi, C., Catinella, G., *Makromol. Chem.*, **137**, 211 (1970).
51. Miyamoto, T., Inagaki, H., *J. Polymer Sci.*, *Pt. A-2*, **7**, 963 (1969).
52. Chierico, A., Del Nero, G., Lanzi, G., Mognaschi, E. R., *Eur. Polymer J.*, **5**, 115 (1969).
53. Takegami, Y., Suzuki, T., *J. Polymer Sci.*, *Pt. B*, **9**, 109 (1971).
54. Nencini, G., Giuliani, G., Salvatori, T., *J. Polymer Sci.*, *Pt. B*, **3**, 483 (1965).
55. Porter, R. S., Nicksic, S. W., Johnson, J. F., *Anal. Chem.*, **35**, 1948 (1963).
56. Bombaugh, K. J., Cook, C. E., Clampitt, B. H., *Anal. Chem.*, **35**, 1834 (1963).
57. Lomonte, J. N., *Anal. Chem.*, **36**, 192 (1964).
58. Adams, J. H., Goodrich, J. E., *J. Polymer Sci.*, *Pt. A–1*, **8**, 1269 (1970); Adams, J. H., *J. Polymer Sci.*, *Pt. A-1*, **8**, 1279 (1970).
59. Burchfield, H. P., *Anal. Chem.*, **17**, 806 (1945).
60. Krause, A., Lange A., *Kunststoff-Bestimmungsmöglichkeiten*, Carl Hanser Verlag, Munich, 1965.
61. Haslam, J., Willis, H. A., *Identification and Analysis of Plastics*, 1st Ed., Iliffe Books, London, 1965.
62. Hummel, D. O., *Kunststoff-, Lack- und Gummi-Analyse*, Carl Hanser Verlag, Munich, 1958; Hummel, D. O., Scholl, F., *Atlas der Kunststoff-Analyse*, Carl Hanser Verlag, Munich, Verlag Chemie, Weinheim, 1968.
63. Harms, D. L., *Anal. Chem.*, **25**, 1140 (1953).
64. Kruse, P. F., Wallace, W. B., Jr., *Anal. Chem.*, **25**, 1156 (1953).
65. Cobler, J. G., Samsel, E. P., *SPE Transactions*, **2**, 145 (1962).
66. Pickhardt, W. P., Safranski, L. W., Mitchell, J., Jr., *Anal. Chem.*, **30**, 1298 (1958).
67. Heacock, J. F., *J. Appl. Polymer Sci.*, **7**, 2319 (1963).
68. Szewczyk, H., Jaworski, M., Zielasko, A., Jerzykiewicz, B., Gąsior, K., *Chem. Anal.* (*Warsaw*), **12**, 967 (1967).
69. Tunnicliffe, M. E., *J. Appl. Polymer Sci.*, **12**, 827 (1970).
70. McGuchan, R., McNeill, I. C., *J. Polymer Sci.*, *Pt. A-1*, **5**, 1425 (1967).
71. Szewczyk, H., Jaworski, M., Zielasko, A., Richter, S., *Polimery*, **14**, 16 (1969).
72. Nitzl, K., *Papier*, **21**, 393 (1967).
73. Giuffre, L., Cassani, F., *Chim. Ind.* (*Milan*), **46**, 1183 (1964).
74. Jadrniček, B., *Chem. Průmysl*, **15**, 681 (1965).
75. Boček, P., *Chem. Průmysl*, **17**, 216 (1967).
76. Boček, P., *ibid.*, **17**, 439 (1967).
77. Pokorný, B., *Chem. Průmysl*, **18**, 315 (1968).
78. Lisovskii, I. P., Smakhtin, L. A., *Zh. Analit. Khim.*, **24**, 749 (1969).
79. Russell, C. A., *J. Appl. Polymer Sci.*, **4**, 219 (1960).
80. Menčik, Z., *Chem. Průmysl*, **12**, 518 (1962).
81. Rybnikář, F., Ditrych, Z., Klácel, Z., Ordelt, O., *Analýza a zkoušeni plastických hmot* (*Analysis and Testing of Plastic Materials*), SNTL, Prague, 1965.
82. Ungureanu, N., Firoiu, E., *Mat. Plastice*, **1**, 148 (1964).

83. Hughes, R. H., *J. Appl. Polymer Sci.*, **13**, 417 (1969).
84. Barrall, E. M., Porter, R. S., Johnson, J. F., *J. Chromatogr.*, **11**, 177 (1963).
85. Takeuchi, T., Tsuge, S., Sugimura, Y., *Anal. Chem.*, **41**, 184 (1969).
86. Key, L. C., Trent, F. M., Lewis, M. E., *Appl. Spectroscopy*, **20**, 330 (1966).
87. Natta, G., Mazzanti, G., Valvassori, A., Pajaro, G., *Chim. Ind.* (*Milan*), **39**, 733 (1957).
88. Majer, J., *Chem. Průmysl*, **21**, 237 (1971).
89. Brame, E. G., Jr., Barry, J. E., Toy, F. J., Jr., *Anal. Chem.*, **44**, 2022 (1972).
90. Takeuchi, T., Tsuge, S., Ito, H., *Japan Analyst*, **18**, 383 (1969).
91. Androsova, V. M., Seidov, N. M., Shabaeva, T. M., *Zavodsk. Lab.*, **34**, 668 (1968).
92. Bartoň, J., Ďuďrovič, V., Stehliková, E., *Plaste u. Kautschuk*, **14**, 316 (1967).
93. Kimmer, W., Kuzay, P., Schmolke, R., *Plaste u. Kautschuk*, **17**, 261 (1970).
94. Altenau, A. G., Headley, L. M., Jones, C. O., Ransaw, H. C., *Anal. Chem.*, **42**, 1280 (1970).
95. Bolleter, W. T., *Anal. Chem.*, **31**, 201 (1959).
96. Anduze, R. A., *Anal. Chem.*, **29**, 90 (1957).
97. Takeuchi, T., Suzuki, M., *Kogyo Kagaku Zasshi*, **66**, 1541 (1963).
98. Smith, A. J., *Anal. Chem.*, **36**, 944 (1964).
99. Novák, K., Mika, V., *Chem. Průmysl*, **13**, 360 (1963).
100. Mische, W., Schröder, E., Thinius, K., *Plaste u. Kautschuk*, **16**, 23 (1969).
101. Slovák, Z., Přibyl, M., *Chem. Průmysl*, **15**, 610 (1965).
102. Fratkina, G. P., *Zavodsk. Lab.*, **24**, 1373 (1958).
103. Negina, V. R., Krashennikova, E. P., Mikhailova, A. T., Ratnikova, K. S., Dokuchaeva, M. T., *Zh. Analit. Khim.*, **22**, 1552 (1967).
104. Baram, A. A., Grigor'eva, N. F., *Plast. Massy*, **1965**, [2], 66.
105. Gluzińska, M., *Chem. Anal.* (*Warsaw*), **7**, 451 (1962).
106. Slovák, Z., Fischer, J., *Chem. Průmysl*, **15**, 725 (1965).
107. Hayashi, S., Sugahara, K., Teranishi, K., *Japan Analyst*, **16**, 1328 (1967); **17**, 1502 (1968).
108. Hayashi, S., Sugahara, K., Shimojo, M., *Japan Analyst*, **18**, 862 (1969).
109. Smith, G. D., Maute, R. L., *Anal. Chem.*, **34**, 1733 (1962).
110. Schröder, E., Hagen, E., Zysik, M., *Plaste u. Kautschuk*, **13**, 720, (1966).
111. Yanagisawa, M., Suzuki, M., Takeuchi, T., *Talanta*, **14**, 933 (1967).
112. Sorantin, H., Patek, P., *Z. Anal. Chem.*, **211**, 99 (1965).
113. Yushkevichyute, S. S., Shlyapnikov, Yu. A., *Plast. Massy*, **1967** , [1], 54.
114. Kapišinska, V., Šmid, J., *Plast. Hmoty Kauč.*, **7**, 144 (1970).
115. Gurvich, Ya. A., Gal'pern, G. M., Il'ina, V. A., *Plast. Massy*, **1967**, [12], 65.
116. Kellum, G. E., *Anal. Chem.*, **43**, 1843 (1971).
117. Paskvitovskaya, S. E., Medvedeva, M. D., Gusev, V. I., *Plast. Massy*, **1967**, [7], 55.
118. Maltese, P., Clementini, L., Frittelli, G., *Mat. Plastiche*, **31**, 226 (1965).
119. Campbell, R. H., Wise, R. W., *J. Chromatogr.*, **12**, 178 (1963).
120. Stafford, C., *Anal. Chem.*, **34**, 794 (1962).
121. Souček, J., Vašátkova, J., Čadersky, I., *Chem. Průmysl*, **16**, 348 (1966).
122. Schröder, E., Rudolph, G., *Plaste u. Kautschuk*, **10**, 22 (1963).
123. Slonaker, D. F., Sievers, D. C., *Anal. Chem.*, **36**, 1130 (1964).
124. Dobies, R. S., *J. Chromatogr.*, **40**, 110 (1969).
125. Sokołowska, R., *Roczn. Państw. Zakł. Hig.*, **20**, 661 (1969).
126. Woggon, H., Korn, O., Jehle, D., *Nahrung*, **9**, 495 (1965).
127. Yushkevichyute, S. S., Shlyapnikov, Yu. A., *Plast. Massy*, **1966**, [12], 62.
128. Long, R. E., Guvernator, III G. C., *Anal. Chem.*, **39**, 1493 (1967).
129. Long, R. E., Christian, G., *Anal. Chem.*, **39**, 1493 (1967).
130. Lappin, G. R., Zannucci, J. S., *Anal. Chem.*, **41**, 2076 (1969).
131. Yoshikawa, T., Ushimi, K., Nimura, K., Tamura, M., *J. Appl. Polymer Sci.*, **15**, 2065 (1971).
132. Majer, J., Kocmanová, V., *Chem. Průmysl*, **17**, 372 (1967).

Chapter **14**

POLYMERS OF CHLORINE–CONTAINING OLEFINS

The major representative of this polymer group is poly(vinyl chloride). Other chlorine-containing homopolymers of interest are chlorinated poly(vinyl chloride), neoprene and poly(vinylidene chloride). Among the copolymers of commercial value are vinyl chloride-vinyl acetate and vinyl chloride-vinylidene chloride copolymers.

14.1 POLY(VINYL CHLORIDE)

Poly(vinyl chloride) (PVC) [1,2] is commercially manufactured largely by polymerisation of vinyl chloride in the presence of peroxides, that is, in a free-radical polymerisation process. The grades of highest purity are made by bulk polymerisation. However, the most widespread processes in the commercial manufacture of PVC are emulsion and suspension polymerisation. Emulsion polymerisation is carried out usually with ammonium or alkali persulphates as catalysts. The product is isolated from the reaction mixture by driving off water which is the reaction medium, and therefore the polymer contains the salt present in the solution, emulsifiers, and other contaminants. Nevertheless, emulsion PVC has a purity that is adequate for the majority of end uses. In suspension polymerisation the most commonly used catalysts are organic peroxides and hydroperoxides. The precipitated polymer is isolated by filtration or centrifuging, followed by washing. The product is of high purity and finds numerous applications including electrical insulation.

Poly(vinyl chloride) has a rather low solubility in common organic solvents. Cyclohexanone, tetrahydrofuran, dimethylformamide and nitrobenzene are among the few available solvents. This polymer has no definite melting or softening points, and above 100° it decomposes, the rate of decomposition increasing with temperature. To prevent degradation of PVC during its fabrication and use stabilisers are added. These usually include cadmium, barium, lead, and calcium salts of higher fatty acids, epoxy compounds, and organotin compounds which are potent antidarkening agents.

In practice both plasticised and unplasticised PVC is used. Unplasticised or lightly plasticised PVC finds application in the manufacture of tubes, pipes, flooring, as well as lining and parts of chemical plants exposed to corrosive agents. From plasticised PVC (the amount of plasticisers added reaches 50%) are made tubing, hoses, stoppers, footwear, raincoats, packaging and the like. The usual plasticisers include cresyl phosphate, and esters of phthalic acid and of other dicarboxylic acids, mainly adipates and sebacates.

Structure

Poly(vinyl chloride) molecules are composed of recurring links of —CH_2—CHCl— arranged in a head-to-tail sequence [3–5]. From infrared spectroscopic studies [6,7–13] and from an analysis of molecular weight distribution curves [14] the presence of short and long side chains was established. According to Lyngaae–Jorgensen [12] PVC molecules contain 4–5 branches in total including 0·6 long chain branches per 1000 structural units.

The studies of Caraculacu *et al.* [13] performed by selective phenolysis have demonstrated, contrary to previous results, that no chlorine atoms are attached to the branching points in PVC molecules.

β-Chlorovinyl and vinyl groups were also detected by infrared spectroscopy [15] in this polymer.

By the same method the tacticity of PVC was studied [16–22] and also the effects of thermal and mechanical factors on the stereoregularity of this polymer [17]. Tacticity in PVC was also investigated by high-resolution NMR [16,22–28] and laser Raman spectroscopy [29].

Qualitative Analysis

Chlorine

Chlorine present in PVC is detected by the Beilstein test [30] (see p. 26) or by chromate oxidation [31]. In the latter procedure a polymer sample is subjected to pyrolysis. The evolving fumes are passed over a filter paper strip previously moistened with a solution of a sexivalent chromium salt. Bleaching of the paper is indicative of the presence of chlorine. However, the test is not sensitive enough nor characteristic.

Colour Reactions

Colour Reactions after Cold Pyridine Treatment [32]. First the plasticiser is removed from the polymer by ether extraction. The ether-insoluble residue is dissolved in tetrahydrofuran, the solid is removed by filtration or centrifugation, and the polymer is precipitated with methanol from the resulting solution. The precipitate is filtered off and dried at 80°C. The plasticiser-free polymer is treated in the cold with 1 ml of pyridine, allowed to stand for several minutes, and 2 or 3 drops of 5% methanolic sodium hydroxide are added. The developing colour is observed after 5 min and after about an hour the chlorine-containing polymer is identified by using the data in Table 14.1.

Colour Reaction after Boiling Pyridine Treatment [32]. The polymer freed from plasticiser is heated to boiling with 1 ml of pyridine for 1 min.

Table 14.1 Colour Reactions of Chlorine-Containing Polymers with Pyridine at Room Temperature
(from [32], by permission of Kunststoff Verlag)

Polymer	*Colour*		
	Immediately	*After* 5 min	*After* 1 hr
Natural rubber hydrochloride	nil	nil	nil
Chlorinated natural rubber	olive-green to olive-brown	brown	brown
Vinyl chloride-vinyl acetate copolymers	colourless to pale yellow	pale yellow to yellowish-green	yellow to brown
Poly(vinylidene chloride)	dark brown	dark brown	black
Chlorinated PVC	blood red	blood red	blood red to brown
Powdered PVC	nil or yellow	light yellow or brown	light brown to dark red
PVC finished products	nil	clear solution, surface of undissolved material yellow	

The solution is divided into two portions (see Table 14.2). One portion is heated again to boiling and carefully treated with 2 drops of 5% methanolic sodium hydroxide. The colour is observed immediately and after 5 min. The other portion is allowed to cool and is treated in the same way with 2 drops of 5% methanolic sodium hydroxide. The colour is observed immediately and 5 min later.

The chlorine-containing polymers may also be detected by colour reactions with mono- and dichloroacetic acid [33] (see p. 389).

Table 14.2 Colour Reaction of Chlorine-Containing Polymers with Boiling Pyridine
(from [32], by permission of Kunststoff Verlag)

Polymer	*Colour*			
	in boiling solution		*in cooled solution*	
	immediately	5 min *later*	*immediately*	5 min *later*
Chlorinated natural rubber	brown	brown	olive-green	olive-brown
Natural rubber hydrochloride	no apparent colour			
Vinyl chloride-vinyl acetate copolymers	yellowish-brown	brown-red	nil	light brown
Poly(vinylidene chloride)	dark-brown precipitate			
PVC	olive-green	brown	no colour or yellowish	olive-green
Chlorinated PVC	blood red or brown		brown	brown

Instrumental Methods

The infrared spectrum of PVC is very characteristic (Fig. 14.1) and can be used for differentiating between PVC and certain other polymers of this group (such as chlorinated rubber). However, for interpretation of the spectrum, results from elemental, qualitative, and occasionally also quantitative analyses are usually required [34].

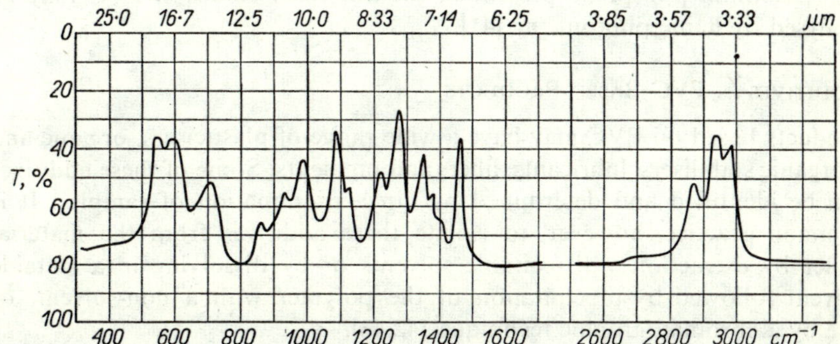

Fig. 14.1 Infrared absorption spectrum of poly(vinyl chloride); film from tetrahydrofuran solution

For identification of chlorine-containing polymers (including PVC) Tsuge *et al.* [35] used pyrolysis gas chromatography.

Quantitative Analysis

Determination of Structural Elements of the Polymer

For chemical characterisation of poly(vinyl chloride) total chlorine is determined first. Quantitative determination of chlorine in allyl groups and tertiary chlorine is made by infrared spectroscopy [36], while the double bonds in PVC are estimated by bromination [37,38]. The same method is used for determination of the —C=C— double bonds in vinyl chloride-vinyl acetate copolymers [39]. Peroxide groups in irradiated polymer may be estimated by the ferrous thiocyanate method [40].

Water

Moisture content in emulsion PVC is determined by drying in a drier or with an infrared lamp. Total volatile matter in suspension PVC is determined by this method.

Ash

Ash in PVC is determined (as sulphates) by the following procedure.

In a porcelain crucible previously ignited to a constant weight is placed a 2 g PVC sample weighed to the nearest 0·0002 g. The sample is moistened with sulphuric acid and carefully heated over a burner flame or on a hot plate. As soon as most of the vapours cease to evolve, the crucible is allowed to cool, another few drops of sulphuric acid are added, and the heating is resumed. After incineration of the sample the residue is ignited in a muffle at 800°C. The alternate adding of sulphuric acid and igniting is repeated until a light coloured ash results. Next, a little ammonium carbonate is added, and the whole is heated first on a hot plate then ignited in a muffle to constant weight.

Disodium phosphate present in the ash from emulsion PVC may be analysed by flame photometry [41].

Additives in PVC-Based Products

Products based on PVC may have a wide range of plasticisers, organic and inorganic stabilisers, lubricants, fillers and pigments. Some of these additives can be identified and determined by direct examination of samples. It is common practice, however, to isolate these additives from the material either by extraction with suitable solvents or by dissolving in a suitable solvent followed by precipitation of the polymer with a non-solvent, or else by a chromatographic technique [42,43].

Fillers

Carbon black in PVC was determined by extraction in a Soxhlet apparatus by Erdos [44], who used extraction thimbles of special design.

Stabilisers

Determination of stabilisers present in PVC is not a simple task, as these additives belong to a wide variety of classes of both organic and inorganic compounds. An additional problem is the fact that mixtures are often used.

Some of the stabilisers may be determined in PVC without resorting to ashing of the material under analysis. Organotin stabilisers were determined specifically by a method involving titration with sodium methoxide in a tetrahydrofuran-benzene-dimethylformamide solution [45]. Stabilisers of the same kind were also determined by another titrimetric procedure in which 0·001*M* EDTA titrant was used in combination with catechol violet as indicator [46].

Urea and thiourea derivatives and related compounds were identified specifically by thin layer chromatography [47] and by paper chromatography [48,49]. The latter was used in identification of a composite stabiliser of higher fatty acids (stearic, lauric, palmitic, and myristic acids) [50]. Likewise dibutyltin in the presence of dioctyltin was detected by this method [51].

Araki *et al.* [52] employed a gas chromatography technique for identification of stabilisers (and plasticisers) in PVC.

Phenolic stabilisers (2,6-di-*tert*-butyl-*p*-cresol and bisphenol A) were analysed polarographically [53]. Dialkyltin was determined colorimetrically using the colour reaction with 4-(2-pyridylazo)resorcinol and EDTA [54]. Diphenylthiourea was determined from ultraviolet spectra at 300 nm [55]. Organotin stabilisers were analysed by X-ray fluorescence [56]. Sulphur-containing dialkyltin stabilisers were determined by infrared spectroscopy directly in the sample under analysis, as well as after isolation by extraction. The stabilisers thus isolated were also identified by NMR spectroscopy and by gas chromatography [57]. The metals contained in stabilisers such as zinc, lead, and calcium were determined by atomic absorption spectrophotometry after dissolution of the unknown in dimethylacetamide [58].

Stabilisers in PVC may also be detected by infrared [59] or ultraviolet spectra [60].

Finally, a wide variety of stabilisers, plasticisers, and lubricants were determined after chromatographic separation (chloroform-methanol extract) by infrared and X-ray spectroscopy [61].

The mixture of metals used in stabilisers, however, is usually determined, after ashing of the sample, by such methods as polarography [50,62,63], oscillopolarography [48], flame photometry [62,64], or colorimetry [65–67].

Plasticisers

The plasticisers present in PVC are identified and quantitatively determined either by ethereal extraction, or by precipitation from cyclohexanone or tetrahydrofuran solutions of PVC products, using a non-solvent for the polymer (ethanol). The latter procedure can be used for simultaneous determination of inorganic fillers which remain as a residue when dissolving PVC products.

The plasticisers thus isolated are dried to a constant weight at 105°C after removal of the solvent from the extract, weighed, and identified or quantitatively analysed by the methods described in various monographs concerned with the analysis of polymers and plasticisers [34,68,69]. Chromatographic techniques are popular, with quite a profusion of pertinent literature available to date. Thus Gude [70] determined a wide range of commercial plasticisers used in the PVC industry, using paper chromatography with a 4:1 acetone-water mixture as the mobile phase.

Thin layer chromatography was used for identification of plasticisers by Campbell *et al.* [71], after dissolving the material (a PVC composition) in tetrahydrofuran followed by precipitation of the polymer with methanol. By the same procedure Steuerle and Pfab [72] identified plasticisers in PVC tubes on extraction with dimethoxymethane, while Sokołowska [73] used this procedure for tritolyl phosphate in miscellaneous PVC products. Dotreppe-Grisard and Dumoulin [74] used a procedure in which the plasticisers were isolated from PVC by dissolving it in tetrahydrofuran, precipitating the polymer with methanol, and extracting the filtrate with ether, and finally separating by thin layer chromatography. The plasticisers thus separated were identified by a number of methods such as determination of R_f value, ultraviolet or infrared spectrophotometry, and colorimetry.

A review of the available results of experimental techniques of thin layer and gas-liquid chromatography, as well as of infrared spectrophotometry as used in the analysis of the PVC plasticisers has been made by Fischer and Leukroth [75]. According to these authors the best results are achieved when several techniques are used for the same analytical problem. For the separation of plasticiser compositions the most suitable is thin layer chromatography, whereas for their identification infrared spectrophotometry is best.

The phthalate plasticisers were analysed quantitatively by Fischer and Jaehn [76] by pyrolysis gas chromatography. A polarographic method was found satisfactory for determination of 2-ethylhexylphthalate in PVC by Woggon and Köhler [77].

Instrumental analysis occasionally allows the determination of plasticisers to be made without having to resort to the laborious isolation procedures described above. Svoboda [78] isolated plasticisers from PVC products by simply heating them in a furnace with a temperature gradient, followed by infrared spectroscopic identification. Mansfield [79] determined total plasticisers directly in a sample by broad line NMR spectroscopy.

Recently Bareich [80] carried out interesting studies on the effect of components of PVC composition on physico-chemical and rheological properties. The results of these studies, when used to work out a computer routine, enabled an automatic analysis of these compositions to be performed.

14.2 CHLORINATED POLY(VINYL CHLORIDE)

Chlorinated PVC is made by treatment of poly(vinyl chloride) with chlorine. As a result, further substitution of chlorine for hydrogen atoms occurs to yield a product of higher chlorine content than PVC (specifically 64–68% instead of 55–56%).

Chlorinated PVC is more soluble in organic solvents than PVC. Because of this and its excellent adhesion to metals, this polymer is used primarily in corrosion-resistant paints.

Qualitative Analysis

Chlorinated PVC gives a colour reaction both with pyridine (Table 14.1) and morpholine [see poly(vinylidene chloride)]. Other properties of chlorinated PVC of use in identification purposes are its good solubility in ethyl acetate and its high chlorine content. Compared with the infrared spectrum of PVC, that of chlorinated PVC exhibits bands at 930 cm^{-1}, considerable absorption at 1282 cm^{-1}, and a stronger band at 1204 cm^{-1} [34,81].

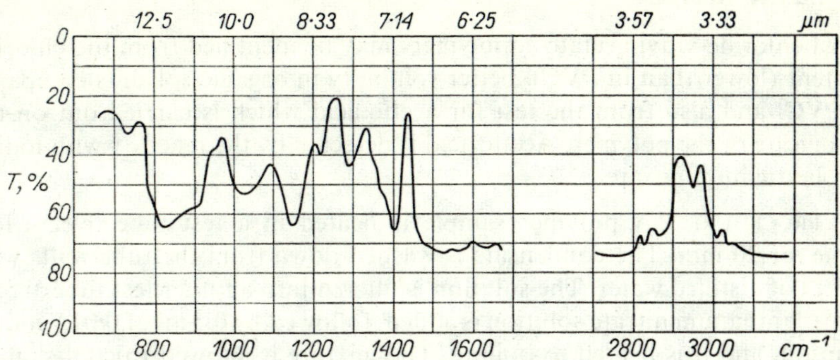

Fig. 14.2 Infrared absorption spectrum of chlorinated poly(vinyl chloride) (66% Cl); film from CH_2Cl_2 solution

14.3 POLY(VINYLIDENE CHLORIDE)

Poly(vinylidene chloride) is manufactured by polymerisation of vinylidene chloride in the presence of peroxide catalysts. The homopolymer is of rather little interest, but its copolymers, especially its copolymer with vinyl chloride, is of more use commercially.

Qualitative Analysis

As distinct from other chlorine-containing polymers, poly(vinylidene chloride) gives a characteristic colour reaction with morpholine [30]. A small polymer sample is immersed in 1 ml morpholine. Poly(vinylidene chloride) gradually darkens, and several hours later the solution becomes turbid and almost black (chlorinated PVC dissolves in morpholine to yield a solution at most red-brown, whereas PVC gives no colour reaction with this reagent). This polymer also gives a colour reaction with pyridine (Table 14.1). An additional aid in the identification of poly(vinylidene chloride) is chlorine analysis, since this polymer contains an exceptionally high amount of chlorine (theor. 73·15%). However, most information can be derived from its infrared spectrum [82].

14.4 VINYL CHLORIDE—VINYL ACETATE COPOLYMERS

Some of the properties of PVC, such as its low flow properties, poor solubility and inadequate adhesion, may be considerably improved by copolymerisation of vinyl chloride with certain other vinyl derivatives such as vinyl acetate. Of commercial significance are vinyl acetate copolymers which contain 65–95% vinyl chloride. These copolymers find a wide range of application in the manufacture of synthetic fibres, films, linings, adhesives, varnishes, as well as in miscellaneous injection fabricated products. The tacticity of these copolymers was examined by high-resolution NMR by Takeuchi *et al.* [83].

Qualitative Analysis

Vinyl chloride-vinyl acetate copolymers may be identified from the chlorine content (lower than in PVC), better solubility in organic solvents compared to PVC, and also from the test for acetic acid which is carried out on the pyrolysate of the polymer. Acetic acid is detected by the reaction with iodine and lanthanum nitrate.

PROCEDURE. A polymer sample is heated in a test tube over a low flame for 20 min. The condensate is washed down from the tube walls with 1–2 ml of distilled water. The solution is filtered into another test tube, 0·5 ml of 5% lanthanum nitrate solution is added, followed by 0·5 ml of 0·01*N* iodine solution, and it is heated to boiling. The mixture is allowed to cool slightly and 1*N* ammonia solution is added carefully with a pipette so as to make a well-defined separate layer. In the presence of acetic acid a blue ring forms.

These copolymers may be identified on the basis of their infrared spectra, specifically by the vinyl acetate bands at 1026, 1212, 1370, and 1739 cm^{-1} [34,81,84,85] (Fig. 14.3).

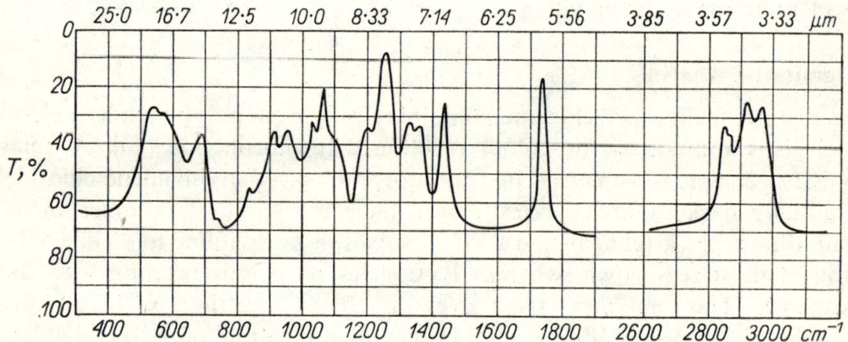

Fig. 14.3 Infrared absorption spectrum of vinyl chloride–vinyl acetate copolymer (90:10); film from ethylene dichloride solution

Quantitative Analysis

Chemical Methods

The vinyl chloride content is calculated from the chlorine analysis.

Gurvich, Balandina and Kosmakova [86] discussed the earlier methods of determination of vinyl acetate content in its copolymers with vinyl chloride and concluded that neither the procedure of Lardera *et al.* [87] nor that reported by Thinius *et al.* [88], in which the acetic acid produced from the hydrolysis of copolymers is distilled off, provides accurate results. These authors recommend a rapid analytical procedure, in which an acetone solution of the copolymer is hydrolysed with 0·1*N* ethanolic potassium hydroxide. After addition of 15–20 ml of 0·1*N* hydrochloric acid, the liberated acetic acid is determined by conductimetric titration with 0·1*N* alkali solution. Alternatively alkaline hydrolysis and potentiometric titration may be used [89,90].

However, vinyl acetate in the copolymers considered can best be determined by the procedure described below [66].

Table 14.3 Hydrolysis Conditions for Vinyl Chloride-Vinyl Acetate Copolymers
(by permission, from *Kunststoff-Bestimmungsmöglichkeiten*, Carl Hanser Verlag, München, 1970)

Percentage of vinyl acetate	*Sample wt.* g	*Concentration*		*Hydrolysis time*, hr		
		of ethanolic KOH	*of* H_2SO_4 *and* $AgNO_3$ *solutions*	*thermostat* 30°C	*room temperature* 20–25°C	*room temperature* 25–30°C
0–5	0·4–0·5	0·2*N*	0·05*N*	2	3·5	2·5
5–10	0·18–0·2	0·2*N*	0·05*N*	2	3·5	2·5
10–30	0·18–0·2	0·5*N*	0·1*N*	1·5	2·5	2
30–60	0·18–0·2	0·5*N*	0·1*N*	2	3·5	2·5
>60	0·13–0·15	0·5*N*	0·1*N*	3	6	4

A polymer sample (according to the data of Table 14.3) is weighed into a Quickfit 100 ml conical flask and 20 ml of tetrahydrofuran is added. The sample is dissolved by vigorous shaking. The flask is then placed in a thermostat at 30°C for 10 min, then 5·00 ml of alcoholic potassium hydroxide is added (the concentration of potassium hydroxide solution is shown in Table 14.3) to prevent precipitation of the polymer. Hydrolysis is carried out at 20–25°C or 25–30°C (reaction time is given in Table 14.3). Finally, with adequate stirring, 30 ml of a 1 : 1 ethanol-water mixture is slowly added to precipitate the polymer in the form of fine powder. The excess hydroxide is titrated with sulphuric acid (of concentration as in Table 14.3) using 1 ml of 0·1% alcoholic thymol blue as indicator. After the titration 1 ml of 1*N* sulphuric acid is added and, with stirring, the chloride ions produced from partial hydrolysis of vinyl chloride are titrated potentiometrically with silver nitrate solution (concentration as in Table 14.3). Other esters, if present, interfere with the determination. The vinyl acetate content *X*(%) is found from the formula

$$X = k\,\frac{v_1 - v_2 - v_3}{m}$$

where:

v_2 and v_1 = the volumes (ml) of the sulphuric acid used in the titration in the unknown and in the blank respectively,

v_3 = the volume (ml) of the silver nitrate titrant used in titration of the unknown,

m = weight of copolymer sample, g,
k = 0·8609 for 0·1N titrants, 0·4304 for 0·05N titrants.

INSTRUMENTAL METHODS

The composition of the copolymers may be rapidly and accurately determined by methods based on infrared spectroscopy [85,91–93]. The methods are most frequently based on the measurement of absorbance at ca. 1740 cm^{-1}. According to Haslam and Willis [34] absorbance should be measured at 2·15 μm. Free vinyl acetate present in these polymers is determined by measuring absorbance due to the $—C{=}CH_2$ vinyl group at 1·63 μm.

The composition of vinyl acetate-vinyl chloride copolymers may also be analysed by NMR spectroscopy by measuring the ratio of the area under the signals resulting from tertiary protons of both kinds of monomer [94]. Two other NMR spectroscopy procedures were shown to be suitable for determination of the composition of these copolymers. By these procedures homopolymer compositions can be differentiated from the corresponding copolymers [95].

The composition of these copolymers may also be determined by pyrolysis gas chromatography [96].

14.5 VINYL CHLORIDE—VINYLIDENE CHLORIDE COPOLYMERS

The disadvantages of poly(vinylidene chloride)—its tendency to crystallisation and a high softening point—can be eliminated by copolymerisation of vinylidene chloride with up to 20% vinyl chloride. These copolymers find a range of applications in the chemical industry in view of their nonflammability and chemical resistance, properties which are characteristic of PVC. The structure of these copolymers was investigated by such methods as NMR [97,98], infrared spectroscopy [99] and pyrolysis gas chromatography [100]. From the evidence of the NMR spectra the way the structural units are linked depends on the quantitative ratio of the monomers used [97].

Vinyl chloride-vinylidene chloride copolymers may be identified from their high chlorine content (57–72%), their insolubility in cyclohexanone and tetrahydrofuran, and their infrared spectra [81,85].

References

1. Krekeler, K., Wick, G., *Kunststoff-Handbuch, Band II, Polyvinylchlorid*, Carl Hanser Verlag, Munich, 1963.
2. Matthews, G., *Vinyl and Allied Polymers, Vol. II, Vinyl Chloride and Vinyl Acetate Polymers*, Iliffe Books, London, 1972.
3. Fuller, C. S., *Chem. Revs.*, **26**, 162 (1940).
4. Marvel, C. S., *An Introduction to the Organic Chemistry of High Polymers*, Wiley, New York, 1959.
5. Horner, R. C. A., *Chem. Ind.*, (*London*), **1960**, 308.
6. Boccato, G., Rigo, A., Talamini, G., Zilio-Grandi, F., *Makromol. Chem.*, **108**, 218 (1967).
7. Grisenthwaite, R. J., Hunter, R. F., *Chem. Ind.*, (*London*), **1958**, 719.

8. George, M. H., Grisenthwaite, R. J., *Chem. Ind.* (*London*), **1958**, 1114.
9. Freeman, M., Manning, P. P., *J. Polymer Sci., Pt. A*, **2**, 2017 (1964).
10. Krasovec, F., *Rep. Josef Stefan Inst.*, **3**, 203 (1959).
11. Cotman, J. D., *Ann. N. Y. Acad. Sci.*, **57**, 417 (1953).
12. Lyngaae-Jorgensen, J., *J. Chromatogr. Sci.*, **9**, 331 (1971).
13. Caraculacu, A. A., Bezdadea, E. C., Istrate, G., *J. Polymer Sci., Pt. A-1*, **8**, 1239 (1970).
14. Talamini, G., Vidotto, G., *Chim. Ind.* (*Milan*), **46**, 16 (1964).
15. Baum, B., Wartman, L. H., *J. Polymer Sci.*, **28**, 537 (1958).
16. Enomoto, S., Asahina, M., Satoh, S., *J. Polymer Sci., Pt. A-1*, **4**, 1373 (1966).
17. Krimm, S., Enomoto, S., *J. Polymer Sci., Pt. A*, **2**, 669 (1964).
18. Krimm, S., Folt, V. L., Shipman, J. J., Berens, A. R., *J. Polymer Sci., Pt. B*, **2**, 1009 (1964).
19. Štokr, J., Schneider, B., Kolinský, M., Ryska, M., Lím, D., *J. Polymer Sci., Pt. A-1*, **5**, 2013 (1967).
20. Millan, J., Smets, G., *Makromol. Chem.*, **121**, 275 (1969).
21. Pohl, H. U., Hummel, D. O., *Makromol. Chem.*, **113**, 203 (1968).
22. Glazkovskii, Yu. V., Kol'tsov, A. I., Pyrkov, L. M., *Vysokomol. Soedin., Ser. B*, **12**, 715 (1970).
23. Hoffmann, W., Kimmer, W., *Plaste u. Kautschuk*, **13**, 519 (1966).
24. Satoh, S., *J. Polymer Sci., Pt. A*, **2**, 5221 (1964).
25. Tincher, W. C., *Makromol. Chem.*, **85**, 20 (1965).
26. Wilkes, C. E., *Macromolecules*, **4**, 443 (1971).
27. Pham-Quang Tho, *J. Polymer Sci., Pt. B*, **7**, 103 (1969).
28. Carman, C. J., Tarpley, A. R., Jr., Goldstein, J. H., *Macromolecules*, **4**, 445 (1971).
29. Liebman, S. A., Foltz, C. R., Reuwer, J. F., Obremski, R. J., *Macromolecules*, **4**, 134 (1971).
30. Brauer, G. M., Newman, S. B., in *High Polymers, Vol. XII, Analytical Chemistry of Polymers*, Kline, G. M., (Ed), *Part III, Identification Procedures and Chemical Analysis*, Interscience, New York, 1962.
31. Roberts, D., *Rubber Age*, **26**, 22 (1945).
32. Hummel, D. O., *Kunststoff-Rundschau*, **5**, 85 (1958).
33. Winterscheidt, H., *Seifen, Öle, Fette, Wachse*, **80**, 239 (1954).
34. Haslam, J., Willis, H. A., *Identification and Analysis of Plastics*, 1st Ed., Iliffe Books, London, 1965.
35. Tsuge, S., Okumoto, T., Takeuchi, T., *Bull. Chem. Soc. Japan*, **42**, 2870 (1969).
36. Bengough, W. I., Onozuka, M., *Polymer*, **6**, 625 (1965).
37. Thinius, K., Hagen, E., *Plast. Inst. Trans.*, **34**, 11 (1966).
38. Zil'berman, E. N., Perepletchikova, E. M., Getmanenko, E. N., *Plast. Massy*, **1972**, [4], 67.
39. Hagen, E., *Plaste u. Kautschuk*, **14**, 565 (1967).
40. Zeppenfeld, G., *Makromol. Chem.*, **90**, 169 (1966).
41. Schröder, E., Hagen, E., Zysik, M., *Plaste u. Kautschuk*, **14**, 318 (1967).
42. Veres, L., *Kunststoffe*, **59**, 241 (1969).
43. Fassy, H., Lalet, P., *Plast. Mod. Elast.*, **21**, [2], 131 (1969).
44. Erdos, G., *Magy. Kem. Lap.*, **24**, 513 (1969).
45. Kapišinská, V., *Plasticke Hmoty Kaučuk*, **7**, 236 (1970).
46. Bergner, K. G., Rudt, U., Mack, D., *Dt. Lebensmitt. Rdsch.*, **63**, 180 (1967).
47. Korn, O., Woggon, H., *Nahrung*, **8**, 351 (1964).
48. Korn, O., Woggon, H., *Ernährungsforschung*, **10**, 57 (1965).
49. Kula, H., *Roczn. Państw. Zakł. Hig.*, **20**, 307 (1969).
50. Klácel, Z., Bednaříková, B., *Plasticke Hmoty Kaučuk*, **2**, 323 (1965).
51. Visintin, B., Pepe, A., Giuseppe, S. A., *Ann. Inst. Super. Sanita*, **1**, 767 (1965).
52. Araki, S., Suzuki, S., Kitano, M., *Nippon Kagaku Kaishi*, **1**, 590 (1972).
53. Vasil'eva, A. A., Vodzinskii, Yu. V., Korshunov, I. A., *Zavodsk. Lab.*, **34**, 1304 (1968).

54. Sawyer, R., *Analyst*, **92**, 569 (1967).
55. Hagen, E., *Plaste u. Kautschuk*, **14**, 158 (1967).
56. Havranek, E., Bumbalova, A., Kapišinská, V., *Chem. Průmysl*, **20**, 536 (1970).
57. Udris, J., *Analyst*, **96**, 130 (1971).
58. Musha, S., Munemori, M., Nakanishi, Y., *Japan Analyst*, **13**, 330 (1964).
59. Chevalier, P., *Ind. Plast. Mod.*, **16**, [3], 70 (1964).
60. Uhde, W. J., Zydek, G., *Z. Anal. Chem.*, **239**, 25 (1968).
61. Crowo, J. A., *Brit. Plastics*, **40**, 84 (1967).
62. Zilio-Grandi, F., Libralesso, G., Perazzolo, A., Sassu, G., Svegliado, G., *Mat. Plastiche*, **30**, 643 (1964).
63. Fassy, H., Lalet, P., *Chim. Anal.* (*Paris*), **52**, 1281 (1970).
64. Schröder, E., Hagen, E., Zysik, M., *Plaste u. Kautschuk*, **13**, 720 (1966).
65. Gude, A., *Kunststoffe*, **52**, 472 (1962).
66. Krause, A., Lange, H., *Kunststoff-Bestimmungsmöglichkeiten*, Carl Hanser Verlag, Munich, 1965.
67. Ochsenbein, P., *Kunststoffe*, **58**, 366 (1968).
68. Kasterina, G. N., Kalinina, L. S., *Khimicheskie metody issledovaniya sinteticheskikh smol i plasticheskikh mass* (*Chemical Analysis of Synthetic Resins and Plastics*), Goskhimizdat, Moscow, 1963.
69. Wandel, M., Tengler, H., Ostromow, H., *Die Analyse von Weichmachern*, Springer-Verlag, Berlin, 1967.
70. Gude, A., *Kunststoffe*, **52**, 679 (1962).
71. Campbell, G. B., Foxton, A. A., Worsdall, R. L., *Lab. Pract.*, **19**, 369 (1970).
72. Steuerle, H., Pfab, W., *Dt. Lebensmitt. Rdsch.*, **65**, 113 (1969).
73. Sokołowska,R., *Roczn. Państw. Zakł. Hig.*, **20**, 281 (1969).
74. Dotreppe-Grisard, N., Dumoulin, J., *Rev. Belge Mat. Plast.*, **9**, 529 (1968).
75. Fischer, W., Leukroth, G., *Plastverarbeiter*, **20**, 107 (1969).
76. Fischer, W., Jaehn, L., *Plastverarbeiter*, **17**, 117 (1966).
77. Woggon, H., Köhler, U., *Kunststoffe*, **57**, 583 (1967).
78. Svoboda, P., *Plaste u. Kautschuk*, **17**, 560 (1970).
79. Mansfield, P. B., *Chem. Ind.* (*London*), **1971**, 792.
80. Bareich, G., *SPE-Journal*, **26**, [4], 71 (1970).
81. Hummel, D., *Kunststoff-, Lack- und Gummi-Analyse*, Carl Hanser Verlag, Munich, 1958; Hummel, D. O., Scholl, F., *Atlas der Kunststoff-Analyse*, Carl Hanser Verlag, Munich, Verlag Chemie, Weinheim, 1968.
82. Krimm, S., *Fortschr. Hochpolymer-Forsch.—Adv. Polymer Sci.*, **2**, 51 (1960).
83. Takeuchi, T., Yamazaki, M., Mori, S., *J. Polymer Sci.*, *Pt. B*, **4**, 695 (1966).
84. Caroti, G., *Poliplasti e Plast. Rinf.*, **8**, [37], 4 (1960).
85. Grisenthwaite, R. J., *Plastics*, **27**, [291], 117 (1962).
86. Gurvich, A. B., Balandina, V. A., Kosmakova, R. V., *Plast. Massy*, **1961**, [12], 51.
87. Lardera, M. R., Cernia, E., Mori, A., *Ann. Chim.*, (*Rome*), **46**, 194 (1956).
88. Thinius, K., Schröder, E., Waurick, U., *Plaste u. Kautschuk*, **4**, 409 (1957).
89. Kvasha, N. M., Sviridova, L. N., Strukova, M. P., *Zavodsk. Lab.*, **34**, 676 (1965).
90. Getmanenko, E. N., Perepletchikova, E. M., *Zh. Analit. Khim.*, **25**, 1832 (1970).
91. Takeuchi, T., Mori, S., *Kogyo Kagaku Zasshi*, **68**, 643 (1965).
92. Wiberley, S. E., Sprague, J. W., Campbell, J. E., *Anal. Chem.*, **29**, 210 (1957).
93. Zakrzewski, L., *Polimery*, **6**, 169 (1961).
94. Grassie, N., McNeill, I. C., McLaren, I. F., *J. Polymer Sci.*, *Pt. B*, **3**, 897 (1965).
95. Takeuchi, T., Yamazaki, M., *Kogyo Kagaku Zasshi*, **68**, 931 (1965).
96. Okumoto, T., Takeuchi, T., Tsuge, S., *Bull. Chem. Soc. Japan*, **43**, 2080 (1970).
97. Chûjô, R., Satoh, S., Nagai, E., *J. Polymer. Sci.*, *Pt. A*, **2**, 895 (1964).
98. McClanahan, J. L., Previtera, S. A., *J. Polymer Sci.*, *Pt. A*, **3**, 3919 (1965).
99. Germar, H., *Makromol. Chem.*, **84**, 36 (1965).
100. Tsuge, S., Okumoto, T., Takeuchi, T., *Makromol. Chem.*, **123**, 123 (1969).

Chapter 15

FLUOROETHYLENE POLYMERS

15.1 POLYTETRAFLUOROETHYLENE

Polytetrafluoroethylene, like poly(vinyl chloride), is made by a free radical polymerisation process [1,2]. This polymer is remarkable for its outstanding resistance towards the most aggressive chemical agents, including concentrated acids and alkalis, and organic solvents even at their boiling points. The material is also resistant to both low and high temperatures. It is extremely valuable in chemical and electrical applications for manufacturing gaskets, seals, parts of equipment, and high voltage insulation to name but a few.

The structure of polytetrafluoroethylene was studied by Bunn and Howells [3] by X-ray techniques, and by Smith [4] by NMR spectroscopy.

Qualitative Analysis

Polytetrafluoroethylene does not melt. In a flame this polymer decomposes with depolymerisation to yield products of pungent odour. The pyrolysis products give no precipitate with silver nitrate. For identification of polytetrafluoroethylene use is made of the presence of fluorine, and of its resistance to concentrated acids and alkalis.

No characteristic chemical reactions are available for this polymer. However its infrared spectrum is characteristic and simple, and contains

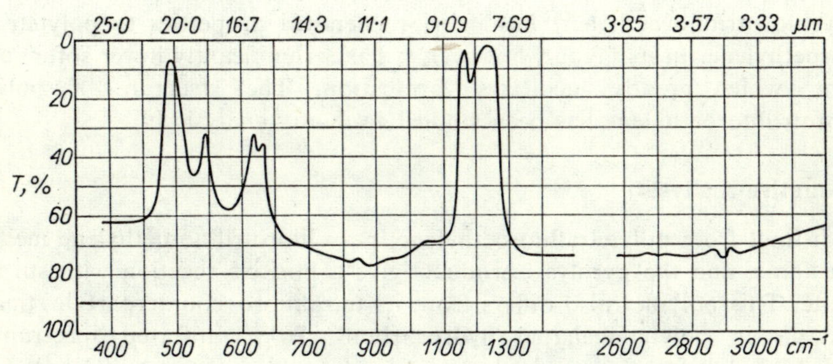

Fig. 15.1 Infrared absorption spectrum of polytetrafluoroethylene; KBr disc

bands due to the CF_2 group at 4·3 μm [10], in the 8–9 μm range (very strong bands, Fig. 15.1), and at 15·7, 16·0, 18·1, and 19·4 μm [5,6]. The intensity of the bands in the range 12–15 μm increases with the quantity of amorphous phase in the polymer [7]. The mass spectrum of the pyrolysis products is also useful for analysis of this plastic material [6].

Chemical Composition of the Polymer

For characterisation of polytetrafluoroethylene the fluorine content is usually determined. The methods for determination of this element have been treated on p. 42–45.

Additives and Contaminants

Water

Moisture content in polytetrafluoroethylene is determined either by titration with the Karl Fischer reagent or by drying to a constant weight at 150°C under reduced pressure.

Inorganic Fillers

Inorganic fillers in this polymer are determined by the following procedure. A 2–3 g sample of powdered polymer is weighed to the nearest 0·0002 g into a nickel crucible. The crucible is placed in a muffle furnace heated to 600°C for 4 hr, then allowed to cool in a desiccator and weighed.

Metals

Minor quantities of metals present in polytetrafluoroethylene have been determined by emission spectroscopy [8]. The analysis was performed in an electric arc since on incineration possible formation of volatile iron fluoride must be taken into consideration.

15.2 POLYCHLOROTRIFLUOROETHYLENE

Polychlorotrifluoroethylene has inferior chemical properties to polytetrafluoroethylene. In its favour, however, it has a significantly lower softening point, which greatly facilitates fabrication. The structure of polychlorotrifluoroethylene has been studied by X-ray analysis [9].

Qualitative Analysis

As distinct from polytetrafluoroethylene, polychlorotrifluoroethylene melts in a flame, and the pyrolysis products give a positive reaction with silver nitrate. This polymer also differs from its tetrafluoro counterpart in that it swells in polyhalogenated hydrocarbons. In its infrared spectrum, (Fig. 15.2) besides the bands due to CF_2 (notably very strong bands in the range of 8–9 μm), a very strong band appears at 10·3 μm from the

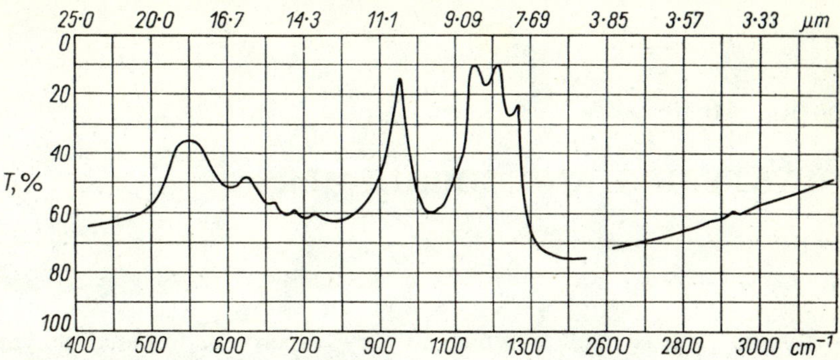

Fig. 15.2 Infrared absorption spectrum of polychlorotrifluoroethylene; KBr disc

C—F bond [9,10]. Like polytetrafluoroethylene, this polymer may also be identified by the mass spectrum of its pyrolysis products [11,12].

Chemical Composition of the Polymer

Fluorine and chlorine in polychlorotrifluoroethylene are determined, as in other polymers or low molecular weight compounds, usually by combustion in a Parr micro bomb.

Metals

Minor quantities of metals present in this polymer are determined as in polytetrafluoroethylene by emission spectroscopy [8].

References

1. Lazar, M., Rado, R., Kliman, N., *Ftoroplasty* (*Fluorinated Polymers*), Energiya, Moscow, 1965.
2. Sittig, M., *Fluorinated Hydrocarbons and Polymers*, Noyes Dev. Corp., New Jersey, 1966.
3. Bunn, C. W., Howells, E. R., *Nature*, **174**, 549 (1954).
4. Smith, J. A. S., *Discussions Faraday Soc.*, **19**, 207 (1955).
5. Hummel, D. O., *Kunststoff-, Lack- und Gummi-Analyse*, Carl Hanser Verlag, Munich, 1958.
6. Krimm, S., *Fortschr. Hochpolymer-Forsch.–Adv. Polymer Sci.*, **2**, 51 (1960).
7. Voter, R. C., in *High Polymers, Vol. XII, Analytical Chemistry of Polymers*, Kline, G. M., (Ed), *Part I, Analysis of Monomers and Polymeric Materials*, Interscience, New York, 1959.
8. Waring, C. L., Annell., C. S., *Anal. Chem.*, **25**, 1174 (1953).
9. Liang, C. Y., Krimm, S., *J. Chem. Phys.*, **25**, 563 (1956).
10. Haslam, J., Willis, H. A., *Identification and Analysis of Plastics*, 1st Ed., Iliffe Books, London, 1965.
11. Harms, D. L., *Anal. Chem.*, **25**, 1140 (1953).
12. Kruse, P. F., Jr., Wallace, W. B., *Anal. Chem.*, **25**, 1156 (1953).

Chapter **16**

POLYSTYRENE AND COPOLYMERS

16.1 POLYSTYRENE

Polystyrene [1,2] is made commercially by a peroxide-initiated polymerisation process or simply by a heat-initiated process. Hydrogen peroxide or benzoyl peroxide is used as initiator, and the raw material for any of the polymerisation processes is synthetic styrene. Polymerisation can be effected in bulk, in solution, in emulsion or in suspension. Of major practical interest are the methods of continuous bulk polymerisation of pure monomer, and the suspension method.

The molecular weight of commercial grades ranges from 50,000 to 150,000 or more, and the specific gravity from 1·040 to 1·065 g per cm^3.

Polystyrene is a clear, hard, rigid thermoplastic, soluble in aromatic hydrocarbons, halogenated hydrocarbons, esters, and ketones. It is insoluble in petroleum and in such solvents as alcohols and organic acids.

Polystyrene is fabricated mainly by injection, extrusion and blow moulding. It is extensively used for a variety of articles in daily use and for electrical and electronic equipment. Polystyrene has found wide application in its expanded form, sometimes known as "styrofoam".

Another polystyrene grade—high-impact polystyrene—comprises polystyrene modified with butadiene-styrene rubber or with *cis*-1,4-polybutadiene rubber. Other varieties of high-impact polystyrene include styrene-acrylonitrile copolymers (SAN) and acrylonitrile-butadiene-styrene (ABS) terpolymers, the latter enjoying ever increasing popularity.

Structure

The structure of polystyrene was first investigated by Staudinger and Steinhofer [3], who subjected the polymer to thermal decomposition and identified low molecular weight degradation products. They concluded that polystyrene molecules are built on the head-to-tail pattern. However, irregular structures present in minor quantities in the polymer could not be ascertained by this procedure. Krimm [4] studied the structure of polystyrene by X-ray techniques. Crystalline polystyrene isolated by selective extraction was found by Natta *et al.* [5] to be isotactic. Recently the tacticity of polystyrene was examined by infrared spectroscopy by Pokrovskii and Fedorova [6], who made use of bands at 557 and 540 cm^{-1}. Tacticity in poly (α-methylstyrene) was investigated by Braun *et al.* [7].

Based on the results of investigations to date the polystyrene molecules are assumed to be essentially linear, as shown by the formula:

$$\sim\sim CH_2-\underset{|}{CH}-\left[-CH_2-\underset{|}{CH}-\right]_n-\sim\sim$$

Polystyrene macromolecules also contain some branches and some of the molecules may be substituted at their ends by the initiator fragments.

Qualitative Analysis

Flame Test

Polystyrene burns with a very sooty flame. After removal from the flame it continues to burn with a characteristic smell of styrene.

Indophenol Test after Feigl [8]

The sample is evaporated to dryness with four drops of fuming nitric acid (d = 1·5 g per cm^3). The residue is heated in a small heat-resistant test tube over a low flame. The tube is covered with filter paper previously soaked in an ethereal solution of 2,6-dibromoquinono-4-chloroimide then dried in air. The pyrolysis is carried on for not more than one minute. After removal from the test tube the paper is placed in ammonia vapour or is treated with 1–2 drops of dilute ammonia solution. A blue colour develops in the case of polystyrene and its copolymers (for instance with butadiene or with acrylonitrile and butadiene). The test is characteristic of the polymers mentioned.

Analysis of the Pyrolysis Products

A small polymer sample is heated in a test tube fitted with a glass wool plug. The volatile products should condense in the plug. After cooling, the plug is extracted with ether and styrene is detected as follows.

a. The ethereal extract is treated with excess bromine (until just yellow). The ether is allowed to evaporate on a watch glass and the dibromostyrene crystals produced are recrystallised from petroleum, m.p. 74°C.

b. Treatment with nitric acid yields $C_6H_5CHNO.CH_2NO_2$ (styrene pseudonitrosite) of m.p. 93–94°C. This compound can be identified polarographically by the value of its half-wave potential, $E_{1/2}$ = 0·27 V.

When more unknown polymer is available the pyrolysate may be condensed in a receiver and styrene identified from its b.p. (146°C), index of refraction (1·5470), or by infrared spectroscopy.

Polystyrene is also easily identified by its very characteristic infrared spectrum (Fig. 16.1) in a direct analysis [9,10].

Both polystyrene alone and its copolymers can be identified by pyrolysis gas chromatography [11–15] coupled with mass spectrometry [15] or by thin layer chromatographic analysis [16] of the pyrolysate.

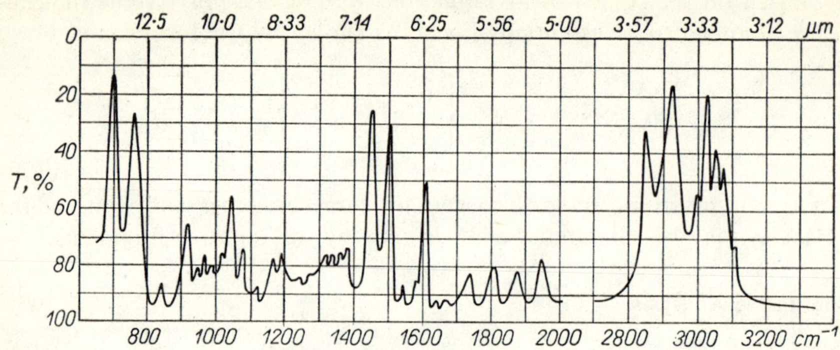

Fig. 16.1 Infrared absorption spectrum of polystyrene; film from ethylene dichloride solution

High impact polystyrene can be identified by its absorbance in carbon tetrachloride solution measured at 320 nm [17].

Quantitative Analysis

High molecular weight peroxides present in polystyrene were determined polarographically in a supporting electrolyte of 0·2*M* lithium chloride in a (2:1) benzene-methanol mixture by Novák and Malinsky [18].

Minor quantities of olefinic double bonds in polystyrene may be evaluated by a radiochemical method developed by McNeill and Haider [19] based on the addition of labelled chlorine.

Free Styrene

Free styrene monomer can be determined by both chemical and instrumental methods, including polarography, coulometric and potentiometric titrations, gas chromatography and ultraviolet spectroscopy.

Chemical Methods

Styrene is usually determined by a method based on the reaction with Wijs reagent. Iodine monochloride adds to the styrene double bond, and the unreacted excess reagent is back-titrated iodimetrically (see p. 51).

Instead of the Wijs reagent the Hanus reagent (iodine bromide) or the Kaufmann reagent (a potassium bromide-saturated methanolic bromine solution) may be used. These reagents all react in a similar manner with styrene.

Instrumental Methods

Coulometric Titration. The method for determining the content of monomers at various stages in the manufacture of polymers, and also in finished products involves a reaction with bromine produced electrolytically [20].

PROCEDURE. A 0·15–0·3 g sample of polymer or copolymer is weighed to $\pm 0{\cdot}0002$ g into a Quickfit flat-bottomed flask, and dissolved in 5 ml of benzene, using a magnetic stirrer. After dissolution 20 ml of methanol and 40 ml of electrolyte (0·1*N* potassium bromide solution in 0·3*N* hydrochloric acid) are added in small portions with vigorous stirring. The contents of the flask are transferred quantitatively (flushing the flask with a total of 30 ml of electrolyte) to the anodic section of the coulometric cell, and the styrene is titrated with the bromine produced electrolytically with a constant current intensity of 2·5–5 mA.

Potentiometric Titration. The method [21] is based on potentiometric titration of the ethanol-soluble compounds with bromine in glacial acetic acid after precipitation of the polymer with alcohol.

Another method developed for the determination of styrene involves the reaction with mercuric acetate in methanol solution, followed by titration of the acetic acid liberated in equivalent quantity to the styrene content [22].

Polarography. Styrene in polystyrene can be determined polarographically either directly, or indirectly when styrene derivatives obtained by its chemical transformation are examined. Among the direct methods [23–25] of interest is the procedure developed by Paściak [25], reported below, in which the polystyrene solution in benzene-dimethylformamide mixture is examined polarographically.

REAGENTS

Tetrabutylammonium Iodide. Commercial reagent is purified by recrystallisation from anhydrous ethyl acetate and dried at 60°C under reduced pressure. M.p. of the pure compound is 141·0°C.

PROCEDURE. In a polarographic cell are placed 2 ml of 0·1*N* tetrabutylammonium iodide in dimethylformamide solution and 0·50 ml of a previously prepared 2·5 or 5·0% benzene solution of the polystyrene sample under analysis. After removal of oxygen from the solution the polarogram curve is recorded from $-1{\cdot}6$ V with respect to the inner electrode (mercury pool electrode). The $E_{1/2}$ value relative to this electrode is $-2{\cdot}46$ V. The concentration of styrene is estimated from a calibration curve (for a large series of samples) or by other methods (comparison with a standard, standard addition) with the use of a previously prepared reference solution of a similar concentration of styrene and polystyrene.

The direct methods are, however, inconvenient, as the ammonium salts used must be very thoroughly purified. The use of these salts is indispensable in view of the strongly negative half-wave potential of styrene. Thus considerable attention has been paid to indirect methods, especially to the procedure in which styrene is converted into the pseudonitrosite by treatment with nitrous acid:

$$C_6H_5.CH{=}CH_2 \xrightarrow{HNO_3} C_6H_5.\underset{\displaystyle NO}{\underset{|}{CH}}{-}\underset{\displaystyle NO_2}{\underset{|}{CH_2}}$$

The value of the half-wave potential of styrene pseudonitrosite is −0·27 V. Styrene and its homologues were determined in the respective homopolymers by this procedure by Šedivec and Flek [26], Klácel and Štrauf [27], and Alekseeva *et al.* [28,29].

Styrene and α-methylstyrene in the presence of each other were determined [21,30] by both polarography and oscillopolarography after isolation of the monomers from the copolymers and conversion into mercuri-acetate derivatives. The method of oscillopolarography was found to be of higher precison.

Gas Chromatography. Styrene and alkyl benzenes in styrene polymers were determined by gas chromatography after isolation of the low molecular compounds by either of the following two procedures [31]:

a. by distillation of a polymer solution in *o*-dichlorobenzene; styrene and other volatile hydrocarbons collect in the first fraction.

b. by precipitation of the polymer with methanol from a dichloromethane solution.

Both procedures yield results in accord with the determination by chemical methods (using Wijs reagent) and by polarography and ultraviolet spectroscopy.

The above method was used for the determination of styrene in emulsion polystyrene preceded by ether extraction (Pozdeeva *et al.* [32]), or in polymer dispersions (Schmötzer [33]) after isolation of the monomer from the dispersion by steam distillation.

Gas chromatography techniques were used for the analysis of styrene and volatile matter in styrene polymers by Ragelis and Gajan [34], Crompton *et al.* [35], Rose [36], Bright *et al.* [37] Groebel [38], and Kleshcheva *et al.* [39,40].

Ultraviolet Spectroscopy. Styrene monomer in polystyrene was determined by ultraviolet spectroscopy by McGovern *et al.* [41], Newell [42] and Bright *et al.* [37] Absorbance is measured at 282 and 291 nm [37,41] or at 251 nm [42]. Klácel and Štrauf [27] followed the same method for the analysis of free styrene in polystyrene modified with styrene-butadiene rubber, after prior solubilisation of the polymer in toluene and subsequent precipitation of the polymer with methanol. Absorption of the supernatant solution was measured at 291 nm.

Ash and Inorganic Components

To determine the ash content, about 5 g of divided polystyrene is weighed into a crucible to the nearest 0·002 g. The crucible is placed in a muffle furnace heated to 600–800°C and is ignited to constant weight.

Minor quantities of chlorides in polystyrene were determined by Carroll [43], who used coulometric titration in alcoholic solution after removing the polymer by dissolving the sample in 2-butanone then precipitating with 95% ethanol. Trace amounts of iron, zirconium and magnesium in expanded polystyrene were determined by Toma and Crisan [44] by atomic absorption spectrometry on a wet-ashed polymer sample. Finally Crompton [45] determined trace quantities of a range of elements in polystyrene (and also in polyolefins) either directly in the polymers by such methods as emission spectroscopy, X-ray fluorescence and neutron activation analysis, or after their isolation from polymers by treatment with suitable solvents using a variety of instrumental methods.

Volatile Matter

Luce [46] reports two procedures for determination of volatile matter.

1. A 0·5–0·7 g sample of divided polymer is dissolved in a covered Petri dish in 10 ml of methyl ethyl ketone. The lid is removed and the solvent carefully evaporated to obtain a thick but still not too viscous liquid. The dish is placed in a vacuum drier and dried at 138–140°C for 2 hr. After cooling in a desiccator the residue is weighed.

2. Divided polymer is heated in a vacuum drier for 2 hr at 140°C, keeping the pressure at a level below 25 mm Hg. After cooling in a desiccator the residue is weighed.

Volatile matter in expanded polystyrene can be determined by one of the following procedures.

1. Removal of the isopentane hydrocarbons (residual pneumatogen) from foamed polystyrene by heating to 140°C then burning them in an oxygen stream. The carbon dioxide produced from the combustion is absorbed in an Ascarite-filled vessel which is then weighed.

2. Determination of total volatile matter, including pneumatogen, unreacted monomer, and water, by heating with an infrared lamp.

Individual volatile compounds are determined as follows:

a. unreacted styrene—by potentiometric titration with bromine solution of the filtrate after precipitation of the polymer with a non-solvent,
b. water—by the Karl Fischer reagent titration of the polymer in methanol-methylene chloride solution,
c. blowing agent—as the difference between total volatile matter and total styrene and water.

n-Pentane or the C_4—C_6 hydrocarbons present as blowing agents in expanded polystyrene were determined by gas chromatography [47]. This method was also used for the analysis of other volatile compounds in polystyrene [48].

Initiators

Bis-α,α'-dimethylbenzyl peroxide present in styrene polymers was determined by Brammer *et al.* [49] after acetone extraction, by thin layer chromatography.

Stabilisers

Thin layer chromatography was used by Uhde and Zydek [50] for the determination of 2-hydroxybenzophenone derivatives added to polystyrene as ultraviolet radiation absorbers, while Simpson and Currell [51] employed this technique in the analysis of certain antioxidants, ultraviolet absorbers and stabilisers, after their isolation from polymers by treatment with suitable solvents. Polygard stabiliser and dioctyl phthalate plasticiser in a high-impact polystyrene were analysed by ultraviolet spectrophotometry [52].

METHANOL-SOLUBLE MATTER

The procedures in use for the determination of methanol-soluble matter involves solubilisation of the polymer in methyl ethyl ketone, dioxan or some other suitable solvent, followed by precipitation of the polystyrene with methanol [46].

Monomer, oligomers, plasticisers, and other methanol-soluble compounds remain in solution. The precipitated polymer is washed with methanol and dried at 65–70°C to constant weight.

Composition of High-Impact Polystyrene

The content of uncombined polybutadiene rubber in high-impact polystyrene was determined turbidimetrically by Sul'zhenko [53], using benzene as solvent and an acetone-methanol mixture as the non-solvent, whereas Armitage [54] determined uncombined polybutadiene by pyrolysis gas chromatography. The content of synthetic rubber in these plastics was also determined by infrared spectroscopy [55] on the basis of absorbance in carbon tetrachloride solution at 967 cm^{-1}. Balandina *et al.* [56] determined uncombined polybutadiene rubber in these plastics by titration with the Hanus reagent and uncombined polystyrene was evaluated gravimetrically by precipitation with suitable non-solvents.

16.2 STYRENE-BUTADIENE COPOLYMERS

Styrene-butadiene copolymers are manufactured by the free-radical polymerisation process with peroxide catalysts such as potassium persulphate or benzoyl peroxide, and certain polymerisation regulators which are usually sulphur compounds. The process is run as an emulsion polymerisation.

Of greatest commercial interest are copolymers containing about 30% styrene. These materials are used for the manufacture of automobile tyres. These copolymers swell or dissolve in aromatic hydrocarbons and in many halogenated hydrocarbons. Their molecular weights range from 25,000 to 500,000. Finished products made from this rubber contain up to 40% carbon black, zinc oxide, antioxidants (such as phenyl-β-naphthylamine) and 1–2% sulphur as a vulcanising agent.

Copolymers containing 50–85% styrene find many applications including the manufacture of strong rubber grades for shoe heels and soles, flooring and electrical insulation. These polymers are more soluble in organic solvents than those poorer in styrene. The high-styrene materials are fabricated by methods common to thermoplasts.

Structure

It follows from the physical properties of low-styrene styrene-butadiene copolymers that their macromolecules are not regular in structure. Based on studies by both chemical and instrumental techniques, the following structural features are present.

1. Some of the styrene units are arranged in blocks [57], since in the pyrolysis products a certain amount of free styrene can be detected.
2. Butadiene adds both at 1,4 and 1,2 positions to yield vinyl side groups. This was established by the reaction of double bonds with perbenzoic acid, and from infrared spectroscopic studies, and NMR [58].
3. Branches and crosslinks are present in the copolymer. Materials made at higher temperatures are more highly branched and crosslinked [59].

Styrene-butadiene copolymers may therefore be represented by the following structure:

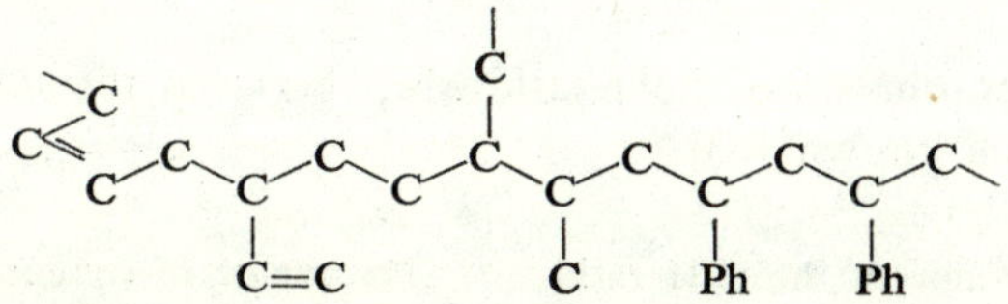

```
                                  |
                                  C
     ╲C                           |
C⩽        C       C               C       C        C       C
    C╱  ╲  C ╱  ╲  C ╱      ╲ C ╱  ╲ C ╱  ╲  C ╱  ╲
           |                        |        |        |
          C=C                       C        Ph       Ph
```

The structure of the block styrene-butadiene copolymers was recently studied by methods such as the spin-probe technique [60], NMR [61], and electron microscopy [62]. The styrene blocks in such polymers were found to be spherical and to incorporate up to 400 styrene units.

Qualitative Analysis

The styrene component of styrene-butadiene copolymers may be detected by the indophenol test after Feigl or as free styrene produced from thermal decomposition of the copolymer. Styrene is identified as described previously. The butadiene component may be identified by the reaction with the Wijs reagent. The reagent becomes decolorised because of iodine chloride addition to the C=C double bonds of the butadiene monomers.

Characteristic of these copolymers is a test in which the material is subjected to nitration with simultaneous oxidation, followed by reduction of the nitro derivative formed, and coupling with β-naphthol. A red dye is produced in this reaction. With polystyrene yellow solutions result.

Styrene-butadiene copolymers may also be identified by gas chromatography [13], or infrared spectrophotometry (from a characteristic butadiene band at 972 cm^{-1}) [10] (Fig. 16.2).

Structural Elements of the Polymer

DETERMINATION OF STYRENE AND BUTADIENE BY ELEMENTAL ANALYSIS

In the case of pure styrene-butadiene copolymers the contents of the styrene and butadiene components may be found by elemental analyses using

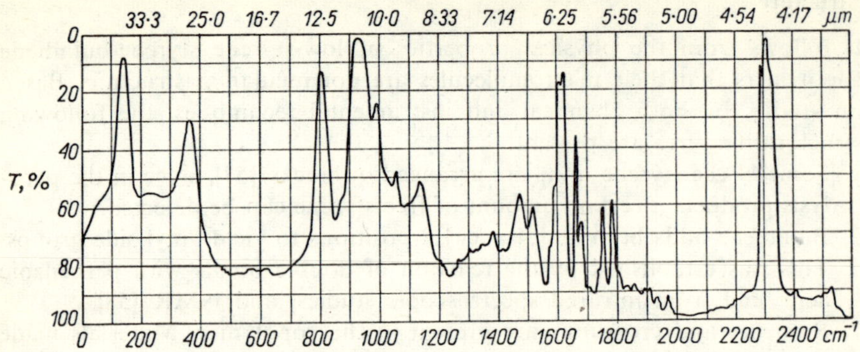

Fig. 16.2 Infrared absorption spectrum of styrene-butadiene (24:76) copolymer [10]; film from benzene solution

a calibration graph, since the relationship between the composition and the C/H ratio is known [63].

Value of the C/H ratio	Percentage of styrene
11·915	100
11·38	90
10·88	80
10·42	70
10·00	60
9·58	50
9·20	40
8·84	30
8·50	20
8·20	10
7·94	0

Butadiene

The butadiene content in styrene-butadiene copolymers may be determined from the iodine number of a *p*-dichlorobenzene solution of the polymer [63].

PROCEDURE. A 0·10 g copolymer sample is heated with 50 g of *p*-dichlorobenzene at 175–185°C until dissolution is complete (20–180 min). The flask contents are cooled to room temperature, 50 ml of chloroform is added, followed by 25 ml of Wijs reagent in carbon tetrachloride. The flask is tightly closed with a ground-glass stopper and allowed to stand in the dark for 1 hr. Next, 25 ml of 15% aqueous potassium iodide and 50 ml of water are added, and the liberated iodine is titrated with 0·1*N* sodium thiosulphate. At the end of the titration 25 ml of ethanol is added to prevent the formation of an emulsion. A blank is run independently. The butadiene content X(%) is found from the formula

$$X = 2{\cdot}705\,\frac{(v_1 - v_2)n}{m}$$

where:

v_1 and v_2 = the volumes (ml) of 0·1N thiosulphate solution used in titrations of the blank and the unknown respectively,

n = normality of the titrant,

m = weight of sample, g.

The butadiene content in these copolymers may also be determined by infrared spectrophotometry [9] or by thermogravimetric analysis [64].

STYRENE

The method for the determination of combined styrene in its copolymers with butadiene worked out by Hilton [65] consists in oxidation and nitration to *p*-nitrobenzoic acid, which is determined in an alkaline solution by ultraviolet spectroscopy. Butadiene present does not interfere with the determination. Combined styrene in the copolymers may also be estimated by ultraviolet spectroscopy [66]. The method is convenient, but satisfactory results are achieved only in the absence of stabilisers, which cannot always be removed easily.

On the other hand, good results are obtained from the method based on the measurement of the refractive index with the Abbé refractometer to determine combined styrene in the polymers [67].

Combined styrene was determined by infrared spectrophotometry from the absorbance bands at ca. 2·2 and 2·4 μm, which are due to the aromatic and aliphatic C—H bonds [68], or from the ratio of absorbance measured at 10·3 μm to that measured at 13·25 μm [69,70] or from the ratio of the absorbance at 10·3 μm to that at 6·3 μm [9]. Structural elements of styrene-butadiene copolymers were studied by NMR spectroscopy by measuring the intensity of the signals due to the aromatic and olefinic protons [58]. The structural elements can also be determined in these polymers by pyrolysis gas chromatography [71–74].

FREE POLYSTYRENE

Free styrene homopolymer may be determined by a procedure worked out by Kolthoff *et al.* [75], in which the polymer is subjected to degradation by oxidation with *tert*-butyl hydroperoxide in *p*-dichlorobenzene solution in the presence of osmium tetroxide. The copolymer is split at the double bonds into ethanol-soluble products. Polystyrene molecules do not undergo cleavage, and since polystyrene is insoluble in ethanol, it can be removed from the mixture by filtration and determined gravimetrically.

Additives and Contaminants

Water, ash, stabilisers, fillers, trace metals and free styrene monomer are also determined in styrene-butadiene copolymers.

WATER

Water in styrene-butadiene rubber may be determined by the Dean–Stark method [46,76]. A sample of rubber (ca. 100 g) cut into pieces of thickness not greater than 13 mm is treated with 500 ml of anhydrous toluene, and the water present in the polymer comes off as the azeotrope during a 75 min distillation.

The pertinent US Standard [77] provides a procedure in which a sample not smaller that 450 g is rolled in a hot mill at 93–104°C to constant weight.

Ash

A 2–5 g sample of the copolymer, after hot milling as described above, is wrapped in filter paper and burnt. It is then ignited in a porcelain crucible placed in a muffle furnace heated to 550±25°C [46].

Carbon Black

A 0·3–0·5 g sample of the polymer rolled into a sheet and dried in a drier is placed in a porcelain boat inside a medium-size crucible and inserted into a combustion tube. The tube is thoroughly flushed with oxygen-free carbon dioxide and then heated to 550°C in a carbon dioxide stream. The hydrocarbon portion of the rubber is thus volatilised. After this operation the crucible is allowed to cool and is weighed. The residue of carbon black in the crucible is subsequently burnt in the muffle furnace heated to 550°C in an air or oxygen atmosphere [46].

Stabilisers

Stabilisers of the aromatic amine type, such as *N*-phenyl-*β*-naphthylamine, are determined, after dissolving the polymer in a suitable solvent, by ultraviolet spectrophotometry [46]. Alkyl-phenol type stabilisers are analysed in the same manner. In this instance however, copolymer solutions are not used directly for spectrophotometry; instead the copolymer is extracted with ethanol-toluene azeotrope and the extract, after dilution with a suitable solvent, is studied [46].

Additives used as chain stoppers in styrene-butadiene latexes, notably 4,4′-dihydroxydiphenyl sulphide [78] and sodium dimethyldithiocarbamate [79], were determined polarographically by Paściak.

Phosphorus-containing stabilisers, such as tris(nonyl phenyl)phosphite, are analysed by a procedure in which the copolymer under test is first heated with zinc oxide at 500°C, then dissolved in sulphuric acid. On addition of ammonium vanadate and ammonium molybdate the solution becomes coloured. Its absorbance at 345 nm is measured [46].

Trace Metals

Trace metals are determined in the ash, using colorimetric techniques, after incineration of the sample in a muffle furnace at 550°C. Copper is specifically determined on the basis of the colour reaction with dithiocarbamate. Manganese is evaluated most frequently by periodate oxidation to permanganate, and iron is best determined by the reaction of ferrous ions with *o*-phenanthroline [46].

FREE STYRENE

Free styrene in styrene-butadiene copolymers, as in polystyrene, may be determined by electroanalytical methods or by gas chromatography. Free styrene and α-methylstyrene in styrene- or α-methylstyrene-butadiene latexes can be determined by ultraviolet spectroscopy after extraction with alkanes or methanol [80,81].

INITIATORS

The polymerisation initiator α,α-dimethylbenzyl hydroperoxide (cumene hydroperoxide) is determined in styrene-butadiene latexes by polarography [82].

16.3 ACRYLONITRILE-BUTADIENE-STYRENE (ABS) TERPOLYMERS

Qualitative Analysis

Styrene combined in ABS polymers may be detected by Feigl's indophenol test, and also by successive nitration, reduction and coupling with β-naphthol. It can also be identified as liberated styrene on pyrolysis by the methods covered below. These styrene terpolymers may be analysed by gas chromatography [13]. Infrared spectroscopy based on the bands characteristic of polystyrene at 700 and 752 cm^{-1}, butadiene at 970 cm^{-1} and polyacrylonitrile at 2273 cm^{-1} [9,10] is also useful. The method is, however, unsuitable for differentiation of the terpolymer from a mixture of the homopolymers. ABS-type terpolymers may also be identified by pyrolysis gas chromatography [83].

Structural Elements of the Polymer

ELEMENTAL ANALYSIS [84]

a. Acrylonitrile content $X(\%)$ is calculated from the nitrogen analysis

$$X = 3{\cdot}787\ N$$

where N is the percentage of nitrogen found.

b. The proportions of carbon and hydrogen (%) of the acrylonitrile constituent are calculated as

$$C = 2{\cdot}572\ N \qquad \text{and} \qquad H = 0{\cdot}216\ N$$

where N is again the percentage of nitrogen.

The values thus found are subtracted from the results of elemental combustion analysis for carbon and hydrogen for the terpolymer under test, and the C/H ratio of the styrene-butadiene part of the terpolymer is thus calculated. Using this value the amounts of styrene and butadiene are found in the same manner as for styrene-butadiene copolymers.

SPLITTING WITH OXIDATIVE AGENTS [84]

A quantity (at most 0·5 g) of ground polymer is heated to the boil with 20–30 ml of methyl ethyl ketone in a round-bottomed flask (non-crosslinked polymers dissolve, while crosslinked samples swell). Now at a temperature of 60°C, 5 ml of *tert*-butyl hydroperoxide is added followed by 1 ml of osmium tetroxide solution (80 mg in 100 ml of benzene; if a black precipitate is present this reagent is unsuitable for use) and refluxed for 2 hr. If this time is not sufficient to solubilise the polymer, a further 5 ml of the hydroperoxide and 1 ml of the osmium tetroxide solution are added and the refluxing is continued for a further 2 hr.

The solution is diluted with 20 ml of acetone and the fillers are removed by filtration with a porosity 2 fritted glass funnel. The precipitate is washed with acetone, dried and weighed.

The filtrate is added dropwise into a 5–10-fold volume of methanol, and the styrene-acrylonitrile part of the terpolymer is precipitated by heating, and cooling, or by addition of a few drops of ethanolic potassium hydroxide. The precipitate is filtered off on a porosity 2 fritted glass funnel, dried in a vacuum drier at 70°C, and weighed.

In the starting material, as well as in the styrene-acrylonitrile part precipitated, the nitrogen content is determined.

Based on the results obtained, the contents of individual components are estimated in the following manner.

a. Acrylonitrile X_a(%) is

$$X_a = 3{\cdot}787\ N_{ABS}$$

where N_{ABS} = the percentage of nitrogen in the sample.

b. Styrene X_s(%) is

$$X_s = (100a_{sa}/m) - 3{\cdot}787\ N_{sa}$$

where:

a_{sa} = the weight of the styrene-acrylonitrile part precipitated, g,
N_{sa} = the percentage of nitrogen in the styrene-acrylonitrile part,
m = weight of polymer sample, g.

c. Butadiene X_b(%) is

$$X_b = 100 - 100(a_{sa} + a_n)/m - 3{\cdot}787(N_{ABS} - N_{sa})$$

where a_n = the weight of filler precipitate, g.

Butadiene by Wijs Method

Combined butadiene in ABS terpolymers can also be quantitatively analysed by a procedure based on the reaction with Wijs reagent (see p. 51).

Structural Elements by Infrared Spectroscopy

All structural elements of ABS terpolymers (including styrene-acrylonitrile copolymers) may be determined by infrared spectrophotometry [9,85–88]. According to Haslam and Willis [9] the ABS terpolymer components may be quantitatively analysed by measuring the absorbance of 0·002 in. thick films at 4·4 μm (acrylonitrile), 6·25 μm (styrene), and 10·3 μm (butadiene). The content of individual components is evaluated from the formulae:

$$A_{4.4\mu m}/A_{6.25\mu m} = K' \text{ wt. \% acrylonitrile/wt. \% styrene}$$

$$A_{10.3\mu m}/A_{6.25\mu m} = K'' \text{ wt. \% butadiene/wt. \% styrene}$$

The values of constants K' and K'' are found by measuring the absorbances of copolymers of known composition.

On the other hand Weir *et al.* [87] evaluated the composition of ABS terpolymers by measurement of the absorbance of thin films at 2220 cm^{-1} (acrylonitrile), 1645 cm^{-1} (butadiene), and 755 cm^{-1} (styrene). The content of ABS terpolymers in compositions with PVC was also estimated by these investigators from the absorbance measured at 2220 cm^{-1}.

The composition of ABS terpolymers, as well as that of styrene-acrylonitrile copolymers, may be quantitatively analysed on the basis of the infrared spectra of their pyrolysis products [89].

Monomers

Free styrene or acrylonitrile in styrene-acrylonitrile copolymers can be determined by polarographic procedures developed by Claver and Murphy [90], and Crompton and Buckley [91]. The same procedures are suitable for the analysis of these monomers in ABS terpolymers.

References

1. Hertz, Z., Krajewski, B., Penczek, I., Płochocki, A., Wiecheć, T., *Polistyren (Polystyrene)*, WNT, Warsaw, 1962.
2. Teach, W. C., Kiessling, G. C., *Polystyrene*, Reinhold, New York, 1960.
3. Staudinger, H., Steinhofer, A., *Ann.*, **517**, 35 (1935).
4. Krimm, S., *J. Phys. Chem.*, **57**, 22 (1953).
5. Natta, G., Corradini, P., *Makromol. Chem.*, **16**, 77 (1955); Natta, G., Danusso, F., Moraglio, G., *Makromol. Chem.*, **20**, 37 (1956).
6. Pokrovskii, E. I., Fedorova, E. F., *Vysokomol. Soedin.*, **6**, 647 (1964).
7. Braun, D., Heufer, G., Johnsen, U., Kolbe, K., *Ber. Bunsenges. Physik. Chem.*, **68**, 959 (1964).
8. Feigl, F., Anger, V., *Modern Plastics*, **37**, [9], 151 (1960).
9. Haslam, J., Willis, H. A., *Identification and Analysis of Plastics*, 1st Ed., Iliffe Books, London, 1965.
10. Hummel, D., *Kunststoff-, Lack- und Gummi-Analyse*, Carl Hanser Verlag, Munich, 1958.
11. Jones, C. E. R., Moyles, A. F., *Nature*, **189**, 222 (1961).
12. Lehmann, F. A., Brauer, G. M., *Anal. Chem.*, **33**, 673, (1961).
13. Voigt, J., *Kunststoffe*, **51**, 18, 314 (1961).
14. Fuchs, P., Szepesy, C., *Chromatographia*, **1**, 310 (1968).
15. Feuerberg, *Glas-Instrum.-Techn. Fachzeit. Lab.*, **13**, 1185, 1191 (1969).
16. Braun, D., Nixdorf, G., *Gummi Asbest Kunststoffe*, **22**, 183 (1969).
17. Rao, K. V. C., *Angew. Makromol. Chem.*, **19**, 113 (1971).
18. Novák, V., Malinsky, J., *Plaste u. Kautschuk*, **15**, 722 (1968).
19. McNeill, I. C., Haider, S. I., *Eur. Polymer J.*, **3**, 551 (1967).
20. Gurvich, D. B., Balandina, V. A., Paikina, L. M., *Zavodsk. Lab.*, **30**, 278 (1964).
21. Balandina, V. A., Gurvich, D. B., Kleshcheva, M. S., Nikitina, V. A., Nikolaeva, A. P., Novikova, E. M., *Analiz polimerizatsionnykh plastmass*, (*Analysis of Polymerisation Plastics*), Khimiya, Leningrad, 1967.
22. Kreshkov, A. P., Balyatinskaya, L. N., Tur'yan, Ya. I., *Plast. Massy*, **1965**, [2], 52; Kreshkov, A. P., Balyatinskaya, L. N., *Zavodsk. Lab.*, **32**, 141 (1966).
23. Bezuglyi, V. D., Dmitrieva, V. N., *Zavodsk. Lab.*, **24**, 941 (1958).
24. Gintsberg, E. G., Igonin, L. A., *Khim. Prom.*, **1954**, 355.

25. Paściak, J., *Chem. Anal.* (*Warsaw*), **5**, 477 (1960).
26. Šedivec, V., Flek, J., *Coll. Czech. Chem. Comm.*, **25**, 1293 (1960).
27. Klácel, Z., Štrauf, M., *Kauč. Plast. Hmoty*, **1963**, [1], 19; Rybnikář, F., Ditrych, Z., Klácel, Z., Ordelt, O., *Analýza a zkoušeni plastických hmot* (*Analysis and Testing of Plastic Materials*), SNTL, Prague, 1965.
28. Alekseeva, T. A., Kruglyak, L. P., Bezuglyi, V. D., *Zavodsk. Lab.*, **29**, 657 (1963).
29. Alekseeva, T. A., Usikova, L. G., Bezuglyi, V. D., *Zh. Analit. Khim.*, **18**, 520 (1963).
30. Bezuglyi, V. D., Ponomarev, Yu. P., Dmitrieva, V. N., *Zh. Analit. Khim.*, **19**, 881 (1964).
31. Pfab, W., Noffz, D., *Z. Anal. Chem.*, **195**, 37 (1963).
32. Pozdeeva, R. M., Lukhovitskii, V. I., Karpov, V. I., *Zavodsk. Lab.*, **37**, 160 (1971).
33. Schmötzer, G., *Z. Anal. Chem.*, **260**, 10 (1972).
34. Ragelis, E. P., Gajan, R. J., *J. Assoc. Offic. Agr. Chemists*, **45**, 918 (1962).
35. Crompton, T. R., Myers, L. W., Blair, D., *Brit. Plastics*, **38**, 740 (1965).
36. Rose, R. L., *Plastics*, **30**, [338], 69 (1965).
37. Bright, K., Farmer, B. J., Malpass B. W., Snell, P., *Chem. Ind.* (*London*), **1965**, 610.
38. Groebel, W., Z., *Lebensm.-Untersuch. u. -Forsch.*, **130**, 180 (1966).
39. Kleshcheva, M. S., Balandina, V. A., Usacheva, V. T., Koraleva, L. B., *Vysokomol. Soedin.*, *Ser. A*, **11**, 2595 (1969).
40. Kleshcheva, M. S., Usacheva, V. T., Pozharova, V. N., *Plast. Massy*, **1971**, [7], 57.
41. McGovern, J. J., Grim, J. M., Teach, W. C., *Anal. Chem.*, **20**, 312 (1948).
42. Newell, J. E., *Anal. Chem.*, **23**, 445 (1951).
43. Carroll, W. R., *Anal. Chem.*, **42**, 144 (1970).
44. Toma, O., Crisan, T., *Materiale Plastice*, **7**, 83 (1970).
45. Crompton, T. R., *Eur. Polym. J.*, **4**, 473 (1968).
46. Luce, E. N., in *High Polymers*, *Vol. XII*, *Analytical Chemistry of Polymers*, Kline G. M. (Ed), *Part I*, *Analysis of Monomers and Polymeric Materials*, Interscience New York, 1959.
47. Moslé, H. G., Wolf, W., Bode, W., *Kunststoffe*, **56**, 760 (1966); Kleshcheva, M. S., Balandina, V. A., Smirnova, T. V., *Plast. Massy*, **1972**, [8], 68.
48. Rohrschneider, L., *Z. Anal. Chem.*, **255**, 345 (1971).
49. Brammer, J. A., Frost, S., Reid, V. W., *Analyst*, **92**, 91 (1967).
50. Uhde, W. J., Zydek, G., *Z. Anal. Chem.*, **239**, 25 (1968).
51. Simpson, D., Currell, B. R., *Analyst*, **96**, 515 (1971).
52. Balandina, V. A., Korsakov, V. G., Malakina, N. I., Davydova, Z. F., Noskova, M. P., Shevardina, Z. I., *Plast. Massy*, **1967**, [10], 53.
53. Sul'zhenko, L. L., *Plast. Massy*, **1969**, [9], 63.
54. Armitage, F., *J. Chromatogr. Sci.*, **9**, 245 (1971).
55. Gendel'man, L. S., Vnukov, A. I., *Zavodsk. Lab.*, **34**, 1081 (1968).
56. Balandina, V. A., Malkina, N. I., Zinchenko, V. A., Gol'dshtein, I. L., Bulatova, V. M., *Plast. Massy*, **1969**, [3], 67.
57. Brüssau, R. J., Stein, D. J., *Angew. Makromol. Chem.*, **12**, 59 (1970).
58. Senn, W. L., Jr., *Anal. Chim, Acta*, **29**, 505 (1963).
59. Morton, M., Salatiello, P. P., *J. Polymer Sci.*, **6**, 225 (1951).
60. Kovarskii, A. G., Burkova, S. G., Vasserman, A. M., Morozov, Yu. L., *Dokl. Akad. Nauk SSSR*, **196**, 383 (1971).
61. Anderson, J. E., Kang-Jen Liu, *Macromolecules*, **4**, 260 (1971).
62. Fischer, E., J. *Macromol. Sci.*, *Pt. A*, **2**, 1285 (1968).
63. Kemp, A. R., Peters, H., *Ind. Eng. Chem.*, *Anal. Ed.*, **15**, 453 (1943).
64. Sircar, A. K., Voet, A., *Rubb. Chem. Technol.*, **43**, 1327 (1970).
65. Hilton, C. L., *Rubber Age*, **85**, 783 (1959).
66. Meehan, E. J., *J. Polymer Sci.*, **1**, 175 (1946).
67. Arnold, A., Madorsky, I., Wood, L. A., *Anal. Chem.*, **23**, 1656 (1951).
68. Miller, R. G. J., Willis, H. A., *J. Appl. Chem.*, **6**, 385 (1956).
69. Post, M. A., *J. Paint Technol.*, **38**, 336 (1966).

70. Post, M. A., *J. Appl. Chem.*, **17**, 203 (1967).
71. Alishoev, V. R., Berezkin, V. G., Markovich, Z. P., Talalaev, E. I., Sitnikov, L. V., Malyshev, A. I., *Zavodsk. Lab.*, **34**, 1188 (1968).
72. Alekseeva, K. V., Khramova, L. P., Strel'nikova, I. A., *Zavodsk. Lab.*, **36**, 1304 (1070).
73. Zinin, V. G., Berdina, L. Kh., Avdeeva, M. P., *Zavodsk. Lab.*, **36**, 1307 (1970).
74. Cianetti, E., Pecci, G., *Industria Gomma*, **15**, 39, 42 (1971).
75. Kolthoff, I. M., Lee, T. S., Carr, C. W., *J. Polymer Sci.*, **1**, 429 (1946).
76. Tryon, M., *J. Research Natl. Bur. Standards*, **45**, 362 (1950).
77. ASTM, D, 1416—56 T.
78. Paściak, J., *Chem. Anal. (Warsaw)*, **12**, 787 (1967).
79. Paściak, J., *ibid.*, **13**, 263 (1968).
80. Drugov, Yu. S., *Kauch. Rezina*, **1964**, [1], 51.
81. Maciejowski, F., Ružička, B., *Chem. Anal. (Warsaw)*, **11**, 45 (1966).
82. Paściak, J., Wójcik, Z., *Chem. Anal. (Warsaw)*, **10**, 355 (1965).
83. Takeuchi, T., Murase, K., *Kogyo Kagaku Zasshi*, **70**, 454 (1967).
84. Krause, A., Lange, A., *Kunststoff-Bestimmungsmöglichkeiten*, Carl Hanser Verlag, Munich, 1970.
85. Janik, A., *Chem. Anal.*, *(Warsaw)*, **12**, 87 (1967).
86. Takeuchi, T., Tsuge, S., Sugimura, Y., *J. Polymer Sci., Pt. A-1*, **6**, 3415 (1968).
87. Weir, A. P., Williams, D. A., Woodcock, J. D., *Chem. Ind. (London)*, **1971**, 990.
88. Kimmer, W., Schmolke, R., *Plaste u. Kautschuk*, **19**, 260 (1972).
89. Gross, D., *Z. Anal. Chem.*, **248**, 40 (1969).
90. Claver, G. C., Murphy, M. E., *Anal. Chem.*, **31**, 1682 (1959).
91. Crompton, T. R., Buckley, D., *Analyst*, **90**, 76 (1965).

Chapter 17

POLY(VINYL ALCOHOL) AND ITS DERIVATIVES

17.1 POLY(VINYL ACETATE)

Like most of the vinyl polymers poly(vinyl acetate) [1,2] is made commercially by polymerisation of monomeric vinyl acetate in the presence of peroxide catalysts (hydrogen peroxide, organic peroxides) or azo-bis-isobutyronitrile in a bulk or emulsion polymerisation.

Molecular weights of commercial grades range from 3500 to 50,000. Poly(vinyl acetate) is a colourless transparent material. It dissolves readily in a variety of organic solvents but not in petroleum. The solubility of this polymer depends on its molecular weight.

Poly(vinyl acetate) finds numerous applications in the manufacture of emulsion paints, varnishes and adhesives. About half of the polymer produced is used as stock for the manufacture of poly(vinyl alcohol).

Structure

As a polymerisation product of vinyl acetate, poly(vinyl acetate) can be assigned the following formula:

$$\left(\begin{array}{l} -CH_2-CH- \\ \qquad\quad | \\ \qquad\quad OCOCH_3 \end{array} \right)_n$$

In view of the studies on poly(vinyl acetate) and poly(vinyl alcohol) performed by Staudinger *et al.* [3], and by Marvel and Denoon [4], it can be assumed that the repeating units in poly(vinyl acetate) are linked "head-to-tail" after the pattern:

$$\begin{array}{l} \sim\sim -CH_2-CH-CH_2-CH-CH_2-CH-\sim\sim \\ \qquad\qquad\quad | \qquad\qquad\quad | \qquad\qquad\quad | \\ \qquad\quad OCOCH_3 \quad OCOCH_3 \quad OCOCH_3 \end{array}$$

According to McLaren and Davis [5], and Flory and Leutner [6] poly(vinyl alcohol) prepared by hydrolysis of poly(vinyl acetate) may contain up to 2% of 1,2-glycol groupings. This was found by periodic acid oxidation. The same proportion of abnormal units is therefore present in poly(vinyl acetate).

Stereoisomerism of poly(vinyl acetate) was investigated by X-ray analysis [7], NMR spectroscopy [8,9,10,11] and by infrared spectrophotometry [12]. Branching in this polymer has been studied by the dye partition technique [13].

Qualitative Analysis

Flame Test

Poly(vinyl acetate) melts and burns with a luminous sooty flame. The fumes smell of acetic acid.

Pyrolysis

On heating, poly(vinyl acetate) melts and darkens. The evolving volatile products exhibit an acid reaction.

Analysis of the Hydrolysis Products

1. A 100 mg sample of the polymer is dissolved in 20 ml of methanol, an equal volume of 1*N* methanolic sodium hydroxide is added and the solution is refluxed for 30 min. The poly(vinyl alcohol) precipitate is filtered off and identified by one of the tests given below.

The filtrate is evaporated to dryness. The residue is dissolved in a little water, twice its volume of 25% sulphuric acid is added, and half of the resulting solution is distilled into a receiver. The distillate is neutralised with 2*N* sodium hydroxide (using multi-range indicator paper) and concentrated to a volume of several ml. To this solution is added 1 or 2 drops of 5% lanthanum nitrate followed by 1 or 2 drops of 0·01*N* iodine solution in potassium iodide, and a drop of dilute aqueous ammonia. A dark blue colour indicates the presence of acetic acid. A dark blue precipitate forms if too great an excess of ammonia is used. Occasionally the colour reaction fails.

2. Detection of acetates and propionates by the Feigl method [14]. The polymer is heated for about 10 min with a few ml of dilute hydrochloric acid, then neutralised with ammonia until slightly alkaline; 1 ml of this solution is treated with 1 or 2 drops of 5% lanthanum nitrate solution and a drop of 0·1*N* iodine solution in 0·1*M* aqueous potassium iodide. Acetates give a dark blue colour whereas propionates give a brown colour.

The Liebermann–Storch–Morawski Test (see p. 28).

In this test poly(vinyl acetate) yields no colour reaction in the cold. After heating a blue-green colour develops which subsequently turns dark brown. However, the test is not specific and often gives irreproducible results.

Mono- and Dichloroacetic Acid Test

About 5 ml of mono- or dichloroacetic acid is fused [15,16], several mg of finely ground polymer is added, and the whole is heated to boiling for 1–2 min. The colours developing in this test for diverse polyvinyl polymers are listed in Table 17.1. If no colour appears after 2 min the test should be regarded as negative.

Copolymers give the colour only if the vinyl constituent comes up to two thirds of the total contents. Proteins, poly(vinyl alcohol) and salts

Table 17.1 Colour Reactions of Vinyl Polymers with Mono- and Dichloroacetic Acid

(from *Seifen, Öle, Fette, Wachse*, **80**, 239 (1954), by permission of Verlag für Chemische Industrie)

Polymer	*Colour with*	
	monochloroacetic acid	*dichloroacetic acid*
Copolymers		
Vinyl chloride/vinyl acetate, 6:4	wine red to purple	blue to purple
Vinyl chloride/vinyl acetate/vinyl alcohol, 90:4:6	wine red to purple	blue to purple
Vinyl chloride/vinyl isopropyl ether, 8:2	wine red to purple	blue to purple
Poly(vinyl methyl ether)	green	blue purple
Poly(vinyl ethyl ether)	turquoise	greenish-blue
Poly(vinyl isopropyl ether)	turquoise	greenish-blue
Poly(vinyl dodecyl ether)	green	greenish-blue
Poly(vinyl cetyl ether)	green	greenish-blue
Poly(vinyl octadecyl ether)	green	greenish-blue
PVC	blue	red-purple
Chlorinated PVC	none	none
Poly(vinyl chloroacetate)	blue-purple	blue-purple
Poly(vinyl acetate)	red-purple	blue-purple

of polyacrylic acid sometimes interfere. Other polymers do not react in this test.

Infrared Spectroscopy

Poly(vinyl acetate) can be identified by infrared spectrophotometry by the bands occurring at 714, 943, 1020, 1235, 1370, and 1739 cm^{-1} [14,17]. This polymer can be differentiated from other esters by the bands at 714 and 943 cm^{-1} [14], and it can be distinguished from cellulose acetate by the band at 794 cm^{-1} [18] (Fig. 17.1).

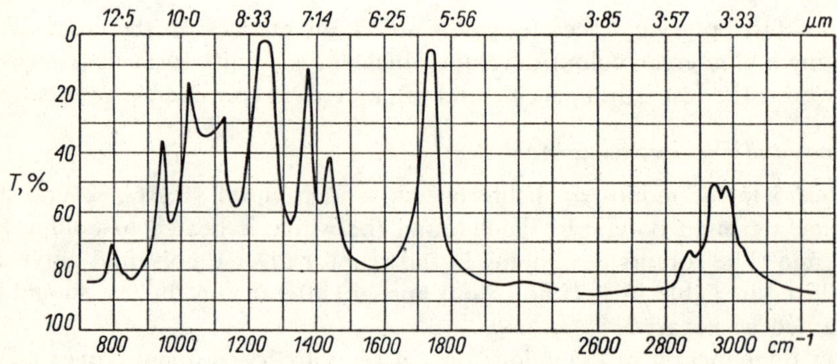

Fig. 17.1 Infrared spectrum of poly(vinyl acetate); film from acetone solution

Acetate Groups

The acetate group content in poly(vinyl acetate) is determined in the usual manner (see determination of saponification number, Section 1.4) by hydrolysing 0·1–0·2 g samples of the polymer with 0·5*N* alcoholic potassium hydroxide. The saponification value of pure poly(vinyl acetate) is 650.

Jakubiec and Truschke [19] determined the poly(vinyl acetate) content of a polyester substrate by the ATR technique by the absorbance ratio measured at 1375 and 720 cm^{-1}.

Hayashi *et al.* [20,21] examined the suitability of the colour reaction of poly(vinyl acetate) with iodine for the determination of this polymer in mixtures.

The dry weight content of poly(vinyl acetate) emulsion is determined by drying to a constant weight 3–5 g of emulsion for 15–20 hr at 40–45°C.

Additives and Contaminants

Moisture Content

The moisture content in poly(vinyl acetate) is determined by titration of a methanolic solution of the polymer with the Karl Fischer reagent.

Residual Monomer

Free vinyl acetate in poly(vinyl acetate) may be determined iodometrically by bromination either with the Kaufmann reagent [22] or directly in acetic acid solution [23], by coulometric titration [24], by polarography [25,26] and by gas-liquid chromatography [27].

Polarography. The preliminary operation in the polarographic determination [26] of free vinyl acetate is hydrolysis with lithium hydroxide in aqueous ethanol. The solution serves as the supporting electrolyte at the same time. The resulting acetaldehyde is reduced on the dropping mercury electrode. The polymer, which forms an insoluble precipitate, does not interfere with the determination, but this is not the case with other aldehydes, alkali metal and alkaline earth ions.

PROCEDURE. Into a 100 ml volumetric flask is weighed 0·5 g of poly(vinyl acetate) (varnish or emulsion); 10 ml of methanol is added and the contents are stirred with a magnetic stirrer. When polymer dissolution is complete, 5 ml of 1*N* lithium hydroxide and 5 ml of 0·5% gelatin solution are added dropwise with stirring and the volume is made up to the mark with distilled water. The contents of the flask are thoroughly mixed and allowed to stand for 5 min to allow hydrolysis of the vinyl acetate. Next, the solution is poured into a polarographic cell and a polarogram is recorded in the range from −1·5 to −2·2 V. The quantity of vinyl acetate contained in the sample is found from a calibration curve made under analogous conditions.

17.2 POLY(VINYL ALCOHOL)

Poly(vinyl alcohol) is one of the few polymers made not by polymerisation but as a result of chemical conversion of another polymer. Commercial grades are manufactured by hydrolysis of poly(vinyl acetate). Generally, acidic hydrolysis is employed, most commonly with sulphuric acid. Catalytic alkaline alcoholysis is also used. The hydrolysis process slightly affects the degree of polymerisation. Depending on the degree of polymerisation the hydrolysis products exhibit a variable solubility in water.

Poly(vinyl alcohol) is mainly used for the manufacture of poly(vinyl-acetals). In the chemically unchanged form it is a valuable material for commercial goods resistant to hydrocarbons (gasoline, mineral oils). After suitable modification certain poly(vinyl alcohol) grades are used for synthetic fibres. Poly(vinyl alcohol) is fabricated in plasticised form, using water, phosphoric acid and polyols as plasticisers.

Structure

In addition to hydroxyl groups poly(vinyl alcohol) may also contain variable quantities of acetate groups originating from vinyl acetate units which did not undergo hydrolysis. This polymer also contains minor quantities of carbonyl groups, which have been detected by chemical methods [28,29] and by ultraviolet spectrophotometry [29]. A few 1,2-glycol groups, which can be detected and determined quantitatively by periodate oxidation [30], and some α-keto-olefinic groupings [31] are also present. The arrangement of the acetyl groups remaining in poly(vinyl alcohol) may be studied by the colour reaction with iodine in the presence of boric acid [32].

Stereoisomerism of poly(vinyl alcohol) was investigated by NMR spectroscopy [9,33].

Qualitative Analysis

A notable feature of poly(vinyl alcohol) that differentiates it from most other synthetic polymers is its ready solubility in water. Poly(vinyl alcohol) is insoluble in organic solvents (with the exception of formamide).

Flame Test

When set on fire poly(vinyl alcohol) continues to burn with a luminous flame. The evolving fumes have a pungent odour but exhibit a neutral reaction.

Iodine and Borax Test

A neutral aqueous solution (5 ml) [16] is treated with 2 drops of $0{\cdot}1N$ iodine solution in potassium iodide and is diluted with water until the developing colour (blue, green or yellowish-green) can still be observed; 5 ml of this solution is treated with about 10 mg of borax and, after mixing, 5 drops. of conc. hydrochloric acid are added. An intensely green colour (especially visible on the still undissolved borax crystals) indicates the presence of poly(vinyl alcohol).

A positive reaction also occurs with poly(vinyl acetals), the colour being dependent on the degree of substitution of the polymer with acetal groups. Vinyl acetate-vinyl chloride copolymer yields a reddish complex. A variety of other compounds such as starch, pyrone and flavone derivatives, and cholic acid give a blue colour with iodine.

To differentiate poly(vinyl alcohol) from starch and to eliminate the interfering effect of the latter in spite of high dilution, the test is carried out as follows.

A few drops of 0·1*N* solution of iodine in potassium iodide solution are added to an aqueous solution of poly(vinyl alcohol). Starch gives a blue colour, whereas poly(vinyl alcohol) does not react with iodine in such a dilute solution. Starch is subsequently hydrolysed by heating with hydrochloric acid to boiling. The excess acid is neutralised with sodium bicarbonate, then one drop of 0·1*N* iodine solution in potassium iodide is added. If a blue colour now develops it is due to poly(vinyl alcohol) present in the original polymer sample.

The iodine test was also used successfully for the detection of poly(vinyl alcohol) in fabrics [34]. A sample is treated with 2 drops of aqueous boric acid and 2 drops of iodine in potassium iodide solution. A blue colour suggests the presence of poly(vinyl alcohol). Starch gives rise to a purple colour. Recently Kikukawa *et al.* [35] examined the effect of tacticity of poly(vinyl alcohol) and of the 1,2-glycol group content on the sensitivity of this colour reaction. Hayashi *et al.* [36] studied the colour reaction of partly acetalised poly(vinyl alcohol) with iodine.

Borax Test

A fairly concentrated solution of poly(vinyl alcohol) is prepared. One drop is transferred on to a spot plate and one drop of saturated borax solution is added. Poly(vinyl alcohol) forms as a gluey gel.

Infrared Spectroscopy

Polyvinyl alcohol may be identified spectrophotometrically by its infrared bands at about 3·0 μm (3340 cm^{-1}) (OH group) and 9·1 μm (1100 cm^{-1}) (secondary alcoholic C—O bond) [17]. The residual acetate groups in the

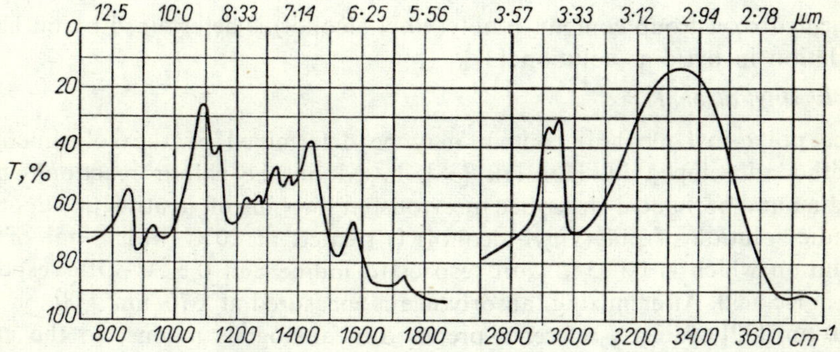

Fig. 17.2 Infrared spectrum of poly(vinyl alcohol); film on AgCl plate from aqueous solution

polymer exhibit bands at 1739 cm^{-1} (C═O group), 1370 cm^{-1} (CH_3 group) and 1250 cm^{-1} (C—O bond) [14] (Fig. 17.2).

Chemical Composition of the Polymer

Acetate Groups

The acetate group content is determined by hydrolysis with potassium hydroxide in aqueous ethanol solution [22] or with aqueous sodium hydroxide [24] after determination of the acid number.

REAGENTS

Sodium Hydroxide. Aqueous solutions, 0·1*N* and 0·5*N*.

Hydrochloric Acid. A 0·5*N* solution.

Phenolphthalein. A 1% ethanolic solution.

PROCEDURE. About 1·5 g of polymer dried to a constant weight is weighed to the nearest 0·0002 g into a 250 ml glass-joint conical flask. The sample is dissolved in 70–80 ml of water by refluxing. The solution is neutralised with 0·1*N* sodium hydroxide, using phenolphthalein. Then 20 ml of 0·5*N* sodium hydroxide is added and the solution is refluxed for 30 min. After cooling, the solution is titrated potentiometrically or visually with 0·5*N* hydrochloric acid in presence of phenolphthalein. A blank is run separately using 20 ml of 0·5*N* sodium hydroxide.

CALCULATION. The acetate group content $X(\%)$ is found from the formula

$$X = \frac{(v_1 - v_2)5{\cdot}904\, n}{m}$$

where:

v_2 and v_1 = the volumes (ml) of 0·5*N* hydrochloric acid solution used for titration of the unknown and of the blank, respectively,

n = normality of the hydrochloric acid solution,

m = weight of polymer sample, g.

Hydroxyl Groups

The hydroxyl group content in poly(vinyl alcohol) is determined by phthaloylation in pyridine solution [37].

Poly(vinyl alcohol)

The poly(vinyl alcohol) content may be determined by the colorimetric method developed by Horáček [38], based on the colour reaction with iodine and boric acid described previously. Two ml of neutral or slightly acidic solution of poly(vinyl alcohol) is treated at 20°C with 8 ml of a solution which is 0·003*M* with respect to iodine and 0·32*M* with respect to boric acid. After mixing, absorbance is measured at 670 nm [38] or at 690 nm [39] against a reference prepared in analogous manner to the unknown. Using this method Brown *et al.* [40] estimated the amount of poly(vinyl alcohol) in fabrics. Acetic acid, tartaric acid, chlorides and ethanol do not interfere in the determination, whereas peroxides, acetone, ascorbic acid, ferric chloride and certain other compounds do.

ADDITIVES AND CONTAMINANTS

Ash

About 2 g of polymer is carefully incinerated in a platinum crucible until organic matter is completely oxidised. On cooling, the residue is treated with a few ml of dilute (1:1) sulphuric acid, and heated until acid fumes stop evolving. The residue is ignited at 800°C in a muffle furnace for 30 min.

pH Determination

The acidity of 4% aqueous poly(vinyl alcohol) is measured with a glass electrode.

Sodium Acetate

Sodium acetate in poly(vinyl alcohol) was determined by conductometric titration using hydrochloric acid by Gurvich *et al.* [41]. A polymer sample (4–5 g) is weighed into a 250 ml conical flask, 150 ml of water is added, and the mixture refluxed until dissolution is complete. The cooled solution is transferred quantitatively (using 120 ml of water) to a beaker, a few ml of 0·1*N* sodium chloride is added, and after insertion of platinum electrodes, the solution is titrated conductometrically with 0·1*N* hydrochloric acid. The consumption of the hydrochloric acid in the titration of the unknown is found from the curve as the distance on the abscissa between the second and the third inflection points.

17.3 POLY(VINYL ETHERS)

Poly(vinyl ethers) are made by cationic polymerisation of vinyl ethers $CH_2{=}CH{-}OR$, where R is an alkyl or aryl group. The most common polymerisation catalyst is boron trifluoride. The process is conducted either in bulk or in dioxan solution.

Poly(vinyl methyl ether) is water-soluble but also dissolves in most organic solvents. The polymer precipitates from aqueous solution on heating. The material is used as a dressing in the textile industry and, as an emulsifier. Poly(vinyl ethyl ether) is water insoluble, but it dissolves in all organic solvents. This polymer is employed in the manufacture of varnishes and adhesives. Higher alkyl ethers have the properties of elastomers.

Qualitative Analysis

Phenyl-lithium Test

Poly(vinyl ethers) react with phenyl-lithium in tetrahydrofuran solution to yield a red colour initially which subsequently turns blue, then purple, and finally, after some time, a black gel precipitates [42].

Ethers of poly(vinyl alcohol) are not hydrolysed by alkalis.

The Liebermann–Storch–Morawski Test

In this test (see p. 28) poly(vinyl ethers) give rise to the following colours: methyl ether—green to blue, low-molecular ethyl ether—bluish-green, high-molecular ethyl ether—blue, isobutyl ether—pale green.

Poly(vinyl ethers) react with mono- and dichloroacetic acid giving rise to different colours (Table 17.1).

Quantitative Analysis of Vinyl Ethyl Ether Monomer [23]

Into a 500 ml flask is poured 50 ml of methanol and a polymer sample of 25±0·1 g is added. Next, 100 ml of 0·1*N* iodine in potassium iodide solution is added with a pipette and 1±0·1 g of sodium oxalate. The contents of the flask are shaken vigorously for 2 min, then titrated with 0·1*N* sodium thiosulphate until the iodine colour disappears. A blank is run separately using the same quantities of the reagents. The vinyl ethyl ether monomer content X (%) is found from the formula

$$X = \frac{(v_1 - v_2)3{\cdot}605n}{m}$$

where:

v_2 and v_1 = the volumes (ml) of the thiosulphate solution used in titration of the unknown and of the blank, respectively,

n = normality of the thiosulphate titrant,

m = weight of polymer sample, g.

17.4 POLY(VINYL ACETALS)

Poly(vinyl acetals), like poly(vinyl alcohol), are made not by direct polymerisation of the monomer but by chemical modification of another polymer. The substrate in this case is poly(vinyl alcohol) and the modifying compounds are aldehydes. The acetalisation reaction is conducted in the presence of acid catalysts—sulphuric, hydrochloric and other acids.

Only a few poly(vinyl acetals) are of commercial interest, namely poly(vinyl formal), poly(vinyl acetal), poly(vinyl butyral) and certain mixed acetals.

Properties of poly(vinyl acetals) depend on the degree of acetalisation, the chemical structure of the aldehyde used and the degree of polymerisation. With increasing degree of acetalisation, the polymers exhibit more hydrophobic properties, are more soluble in weakly polar solvents and have improved electrical insulating properties. The higher the degree of acetalisation the lower the softening point. Aldehydes with hydrocarbon residues enhance hydrophobic properties, whereas hydroxy-, alkoxy- and amino-aldehydes reduce these properties. The molecular weight of a poly(vinyl acetal) differs little from that of the parent poly(vinyl alcohol).

Poly(vinyl acetals) find miscellaneous applications in the manufacture of varnishes, safety glass interlayers, adhesives, films, and in the fabrication of a miscellany of goods by extrusion and moulding techniques.

Structure

According to Flory, the maximum degree of substitution of hydroxyl groups in poly(vinyl alcohol) in converting into poly(vinyl acetals) amounts to 86·47 mole per cent. The polymers should therefore invariably contain not less than 13·53 mole per cent of OH groups. In practice a higher degree of substitution is achieved but 100% acetalisation can never be attained. Thus, poly(vinyl acetals) always contain significant amounts of hydroxyl groups, and in addition quantities of acetyl groups originating from vinyl acetate units which did not undergo hydrolysis in the process of manufacturing poly(vinyl alcohol).

Poly(vinyl acetals) may be assigned the following formula:

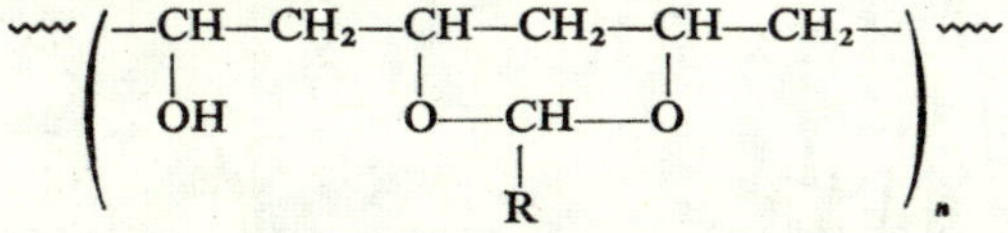

$$\sim\left(\begin{array}{c}-\underset{\displaystyle OH}{\underset{|}{CH}}-CH_2-\underset{\displaystyle O}{\underset{|}{CH}}-CH_2-\underset{\displaystyle O}{\underset{|}{CH}}-CH_2-\\ \qquad\qquad\quad \text{O—}\underset{\displaystyle R}{\underset{|}{CH}}\text{—O}\end{array}\right)_n\sim$$

The tacticity of poly(vinyl formal) was investigated by NMR spectroscopy [43,44], infrared spectroscopy [44], and electron microscopy [45].

Qualitative Analysis

Flame Test

When set on fire poly(vinyl acetals) keep on burning. Poly(vinyl butyral) burns with melting with a bluish flame having yellow edges.

Pyrolysis

Poly(vinyl acetals) melt and darken on decomposition. The evolving fumes are acidic. During pyrolysis the constituent aldehydes are given off. Aldehydes may be identified in the pyrolysate either by their odour or by characteristic reactions, including colour tests [15].

Analysis of the Hydrolysis Products

On heating to boiling with 25–50% sulphuric acid, constituent aldehydes or ketones (from ketals) are formed. The aldehyde vapours are introduced into a 2,4-dinitrophenylhydrazine solution in dilute hydrochloric acid. The 2,4-dinitrophenylhydrazones precipitated as insoluble crystals are filtered off, dried, recrystallised from alcohol, and identified from their m.p. and nitrogen content.

Iodine Test

REAGENT. Ten ml of 50% acetic acid is combined with 7 ml of iodine in potassium iodide solution (1 g of potassium iodide plus 0·9 g of iodine plus 40 ml of water plus 2 ml of glycerine).

The polymer free of plasticiser [46] is treated with 1 or 2 drops of the reagent. After 1 min the reagent is washed off and the colour is observed. Poly(vinyl acetals) yield the following colours in this test:

poly(vinyl formal)—blue to black-purple,
poly(vinyl acetal)—green,
poly(vinyl butyral)—green,
[poly(vinyl acetate) gives rise to a red colour in this test].

Poly(vinyl acetals) may be identified by their infrared spectra [14,17,18] which show bands in the range 945–1250 cm^{-1} and at 2774 cm^{-1} [poly(vinyl formal)], 1340 cm^{-1} [poly(vinyl acetal)], and 1385 cm^{-1} [poly(vinyl butyral)]. A typical spectrum is shown in Fig. 17.3.

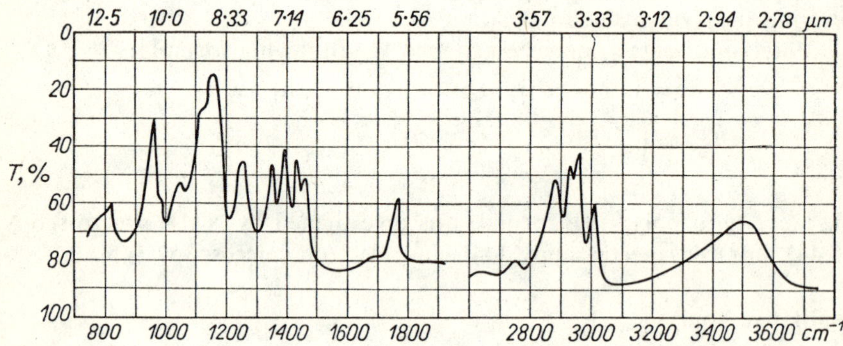

Fig. 17.3 Infrared spectrum of poly(vinyl acetal); film from ethylene dichloride solution

Chemical Composition of the Polymer

Hydroxyl Groups

Hydroxyl groups in poly(vinyl acetals) are commonly determined by acetylation with acetic anhydride in pyridine. According to the procedure reported by Kennett [23] the analysis is carried out as follows.

A 1·5 g resin sample is weighed to the nearest mg into a pressure bottle; 20 ml of a mixture of 880 ml of freshly distilled pyridine and 120 ml of acetic anhydride is pipetted into the bottle, which is then stoppered and the contents shaken to effect solution. The bottle is tied in a cloth bag and placed in a water bath at 98±2°C for 4 hr. The bag is removed and the bottle is allowed to cool in air but not in cold water, or a cold bath, or on a cold metallic plate. Then it is chilled for a short time in a cold bath, 25 ml of methyl isobutyl ketone is added, the contents are swirled to effect solution, and 5 ml of water is added. The solution is placed in a hot water bath (98±2°C) for 30 min, after which it is removed, cooled and chilled. On addition of 35 ml of chilled methyl isobutyl ketone the solution is titrated immediately with freshly standardised 0·5*N* alcoholic potassium hydroxide to a faint pink end point with phenolphthalein as indicator. A blank is run following the same procedure with the omission of the sample.

CALCULATION. The hydroxyl group content X, as wt. % of poly(vinyl alcohol), is found from the formula

$$X = 2{\cdot}933(v_1 - v_2)N$$

where:

v_2 and v_1 = the volumes (ml) of 0·5N potassium hydroxide used in the titration of the unknown and of the blank, respectively,

N = normality of the potassium hydroxide titrant.

According to another procedure [47] acetylation is run with a mixture richer in acetic anhydride (1 vol. of acetic anhydride, 3 vol. of pyridine) for ten hours or so at 55–60°C, whereupon the unreacted acetic anhydride is hydrolysed and titrated with 1N standard sodium hydroxide after the solid polymer particles have been solubilised by addition of a solvent (ethylene dichloride).

Hydroxyl groups in poly(vinyl acetals) can also be determined by phthaloylation in a manner analogous to that for poly(vinyl alcohol) (see p. 394).

Acetate Groups in Poly(vinyl formal) [23]

Into a 500 ml conical flask is weighed 2·150 g of the unknown polymer; 30 ml of pyridine and 20 ml ot methanol are added to solubilise the sample and the solution is titrated with 0·05N potassium hydroxide, using phenolphthalein as indicator, until a faint pink colour which does not disappear within 1 min is observed. A blank is run in a similar manner.

To hydrolyse the acetyl groups in the polymer 25 ml of 0·5N potassium hydroxide is added to the neutral solution and the whole is heated on a water bath under reflux for 1 hr. Next the condenser is flushed with a little water and the solution is titrated with 0·5N hydrochloric acid, using phenolphthalein as indicator. A blank is performed in analogous manner.

The acetyl group content X, as wt. % of poly(vinyl acetate), is found from the formula

$$X = 2(v_1 - v_2)$$

where v_2 and v_1 = the volumes (ml) of the 0·5N hydrochloric acid titrant used in titration of the unknown and the blank, respectively.

Acetate Groups in Poly(vinyl butyral)

A 5·0 g polymer sample is weighed into a 500 ml flask and 100 ml of methanol is added. When the sample is dissolved the solution is neutralised with 0·05N potassium hydroxide, using phenolphthalein as indicator, until a faint pink colour persists for one min.

To hydrolyse the acetyl groups introduced into the polymer, 50 ml of 0·5N ethanolic potassium hydroxide is added to the neutralised solution, and the whole is heated in a pressure bottle for 4 hr on a water bath at 98±2°C. On cooling the solution is titrated with vigorous stirring with

$0{\cdot}5N$ sulphuric acid, using phenolphthalein as indicator. A blank is run following the above procedure but omitting the unknown.

The acetate group content X, as wt. % of poly(vinyl acetate), is found from the formula

$$X = 0{\cdot}86(v_1 - v_2)$$

where v_2 and v_1 = the volumes (ml) of the $0{\cdot}5N$ sulphuric acid titrant consumed in titration of the unknown and of the blank respectively.

Combined Butyraldehyde in Poly(vinyl butyral) [23]

About 2 g of the polymer is weighed into a 300 ml conical flask, 25 ml of butanol is added followed by 50 ml of ca. $0{\cdot}5N$ hydroxylamine hydrochloride solution, and the whole is refluxed for 2 hr. On cooling, the condenser is flushed down with 50 ml of a water-methanol (1:1) mixture, a few drops of 0·04% methanolic bromothymol blue indicator are added, and the solution is titrated with $0{\cdot}5N$ sodium hydroxide. A blank is run using the same procedure but with the omission of the sample. The butyraldehyde content in poly(vinyl butyral) $X(\%)$ is found from the formula

$$X = [(v_1 - v_2)n\, 7{\cdot}21]/m$$

where:

v_2 and v_1 = the volumes (ml) of the $0{\cdot}5N$ sodium hydroxide consumed by the polymer sample and the blank respectively,

n = normality of the sodium hydroxide titrant,

m = weight of polymer sample, g.

For poly(vinyl acetals) that are insoluble in aqueous alcoholic systems, for instance poly(vinyl formals), the acetal group content is determined by procedures based on previous acidic hydrolysis of the polymer. The liberated aldehyde (formaldehyde) is driven off with steam and determined by either hydroxylamine hydrochloride (by the above method) or the Nessler reagent, or else by other available methods (for instance by polarography or colorimetrically with chromotropic acid [48]).

For polymers with a lower content of acetal groups, 20% sulphuric acid is used as a hydrolysing medium, whereas for polymers containing more acetal groups a (1:2) mixture of conc. formic acid with 20% sulphuric acid is used [24].

Acetal groups in poly(vinyl formal) were quantitatively analysed by pyrolysis gas chromatography by Takahashi *et al.* [49].

Water

Water in poly(vinyl acetals) can be determined by titration with the Karl Fischer reagent. Results in good agreement with the titration method were achieved in the determination of water by gas chromatography by Shiryaev [50].

References

1. Ushakov, S. N., *Polivinilovyi spirt i ego proizvodnye* (*Poly(vinyl alcohol) and its Derivatives*), Akad. Nauk SSSR, Moscow—Leningrad, 1960.

2. Ritchie, P. D., (Ed), *Vinyl and Allied Polymers*, Iliffe Books, London, 1968.
3. Staudinger, H., Frey, K., Starck, W., *Ber.*, **60**, 1782 (1927).
4. Marvel, C. S., Denoon, C. E., Jr., *J. Am. Chem. Soc.*, **60**, 1045 (1938).
5. McLaren, A. D., Davis, R. J., *J. Am. Chem. Soc.*, **68**, 1134 (1946).
6. Flory, P. J., Leutner, F. S., *J. Polymer Sci.*, **3**, 880 (1948); *ibid.*, **5**, 267 (1950).
7. Staudinger, H., Schwalbach, A., *Ann.*, **488**, 8 (1931).
8. Fujii, K., Fujiwara, Y., Fujiwara, S., *Makromol. Chem.*, **89**, 278 (1965).
9. Fujiwara, Y., Fujiwara, S., Fujii, K., *J. Polymer Sci., Pt. A-1*, **4**, 257 (1966).
10. Murahashi, S., Nozakura, S., Sumi, M., Yuki, H., Hatada, K., *J. Polymer Sci., Pt. B*, **4**, 65 (1966).
11. Ramey, K. C., Lini, D. C., *J. Polymer Sci., Pt. B*, **5**, 39 (1967).
12. Fujii, K., *J. Polymer Sci., Pt. B*, **5**, 551 (1967).
13. Vidyarthi, D. K., Mukhopadhyay, S., Palit, S. R., *Makromol. Chem.*, **148**, 1 (1971).
14. Hummel, D., *Kunststoff-, Lack- und Gummi-Analyse*, Carl Hanser Verlag, Munich, 1958.
15. Brauer, G. M., Newman, S. B., in *High Polymers, Vol. XII, Analytical Chemistry of Polymers*, Kline, G. M., (Ed), *Part III, Identification Procedures and Chemical Analysis*, Interscience, New York, 1962.
16. Winterscheidt, H., *Seifen, Öle, Fette, Wachse*, **80**, 239 (1954).
17. Haslam, J., Willis, H. A., *Identification and Analysis of Plastics*, 1st Ed., Iliffe Books, London, 1965.
18. Hummel, D. O., Scholl, F., *Atlas der Kunststoff-Analyse*, Carl Hanser Verlag, Munich, Verlag Chemie, Weinheim, 1968.
19. Jakubiec, R. J., Truschke, E. J., *Anal. Letters*, **2**, 25 (1969).
20. Hayashi, S., Kawamura, C., *Kogyo Kagaku Zasshi*, **72**, 2491 (1969).
21. Hayashi, S., Kawamura, C., Takayama, M., *Bull. Chem. Soc. Japan*, **43**, 537 (1970).
22. Rybnikář, F., Ditrych, Z., Klácel, Z., Ordelt O., *Analýza a zkoušeni plastických hmot* (*Analysis and Testing of Plastic Materials*), SNTL, Prague, 1965.
23. Kennett, C. J., in *High Polymers, Vol. XII, Analytical Chemistry of Polymers*, Kline, G. M., (Ed), *Part I, Analysis of Monomers and Polymeric Materials*, Interscience, New York, 1959.
24. Balandina, V. A., Gurvich, D. B., Kleshcheva, M. S., Nikitina, V. A., Nikolaeva, A. P., Novikova, E. M., *Analiz polimerizatsionnykh plastmass* (*Analysis of Polymerisation Plastics*), Khimiya, Leningrad, 1967.
25. Heitz, J., *Rev. Chim.* (*Bucharest*), **14**, 173 (1963).
26. Horáček, J., *Plaste u. Kautschuk*, **9**, 116 (1962).
27. Fossick, G. N., Tompsett, A. J., *J. Oil Colour Chem. Ass.*, **49**, 477 (1966).
28. Marvel, C. S., Inskeep, G. E., *J. Am. Chem. Soc.*, **65**, 1710 (1943).
29. Clarke, J. T., Blout, E. R., *J. Polymer Sci.*, **1**, 419 (1946).
30. Harris, H. E., Pritchard, J. G., *J. Polymer Sci., Pt. A*, **2**, 3673 (1964).
31. Sopiela, W., Majewska, J., Miłosz, A., *Polimery*, **9**, 379 (1964).
32. Hayashi, S., Takizawa, K., *Kogyo Kagaku Zasshi*, **71**, 2051 (1968).
33. Tincher, W. C., *Makromol. Chem.*, **85**, 46 (1965).
34. E. I. du Pont de Nemours, *Text. Inds.*, **134**, 140 (1970).
35. Kikukawa, K., Nozakura, S., Murahashi, S., *Polymer J.* (*Japan*), **2**, 212 (1971).
36. Hayashi, S., Takayama, M., Kawamura, C., *Kogyo Kagaku Zasshi*, **73**, 412 (1970).
37. Kasterina, G. N., Kalinina, L. S., *Khimicheskie metody issledovaniya sinteticheskikh smol i plasticheskikh mass* (*Chemical Methods of Analysis of Synthetic Resins and Plastics*), Goskhimizdat, Moscow, 1963.
38. Horáček, J., *Chem. Průmysl*, **12**, 385 (1962).
39. Chene, M., Martin-Borret, O., Clery, M., *Papeterie*, **88**, 273 (1966).
40. Brown, W. T., Olson, E. S., Keegan, H. J., *Am. Dyest. Rep.*, **56**, 36 (1967).
41. Gurvich, D. B., Balandina, V. A., Kosmakova, R. V., *Plast. Massy*, **1964**, [2], 69.
42. Fanica, L., *Chim. Anal.* (*Paris*), **38**, 353 (1956).

43. Fujii, K., Shibatani, K., Fujiwara, Y., Ohyanagi, Y., Ukida, J., Matsumoto, M., *J. Polymer Sci., Pt. B*, **4**, 787 (1966).
44. Shibatani, K., Fujiwara, Y., Fujii, K., *J. Polymer Sci., Pt. A-1*, **8**, 1693 (1970).
45. Sinitsyna, G. M., Samarina, A. V., Tarakanova, E. E., Tarakanov, O. G., Vlodavets, I. N., *Kolloid. Zh.*, **32**, 903 (1970).
46. Broockmann, K., Müller, G., *Farbe u. Lack*, **61**, 217 (1955) .
47. Krause, A., Lange, A., *Kunststoff-Bestimmungsmöglichkeiten*, Carl Hanser Verlag, Munich, 1970.
48. Ogasawara, K., Nakamura, N., Matuzawa, S., *Makromol. Chem.*, **149**, 291 (1971).
49. Takahashi, I., Yamane, T., Suzuki, T., Mukoyama, T., *Kogyo Kagaku Zasshi*, **73**, 303 (1970).
50. Shiryaev, B. V., *Plast. Massy*, **1972**, [2], 69.

Chapter **18**

ACRYLIC POLYMERS

This group of high polymers includes polymers of acrylic and methacrylic acids and their salts, esters, amides and nitriles [1–3]. Among these polymers polymethacrylates, polyacrylonitrile, and the acrylonitrile copolymers are of major commercial significance. Polymers of the acids themselves, including their alkali metal salts, have not found application as typical plastic materials, as they easily swell and soften in water. Water-soluble sodium and ammonium salts are used as emulsion and suspension thickening agents for commercial compositions and may be applied as impregnating formulations for nylon and other fabrics. The ethylene copolymers containing minor additions of these salts are known under the name of "ionomers".

Acrylic polymers of all types are manufactured by free radical polymerisation initiated by peroxide catalysts. The polymerisation process is usually effected in bulk or in suspension.

18.1 POLYMETHACRYLATES AND POLYACRYLATES

The physical properties such as solubility, density, and softening point of high polymeric acrylates and methacrylates depend to a large extent on the length and branching of the alcoholic residues. The polymeric esters in which these residues are sizeable are soluble in such solvents as aromatic hydrocarbons, esters, ketones, and halohydrocarbons. The polymers with larger alcohol residues are also soluble in alkanes. Polyacrylates exhibit lower softening points than the polymethacrylates. Among the valuable properties of polyacrylates and polymethacrylates are their clarity, transparency and considerable stability to light.

Polymethacrylates are chemically more stable and more resistant to hydrolytic agents than are polyacrylates. Poly(methyl methacrylate) is remarkably resistant to alkaline hydrolysis. When heated above the decomposition temperature polymethacrylates depolymerise almost quantitatively to monomer, whereas polyacrylates are degraded with the formation of little monomer.

Polyacrylates, in particular poly(methyl acrylate), are used in the manufacture of varnishes, where they do not require the addition of plasticisers. On the other hand polymethacrylates, and especially poly(methyl methacrylate), have found a wide range of applications as a material to replace glass. The excellent transparency of poly(methyl methacrylate) is unparalleled by other materials of organic origin.

Structure

The ester groups originating from the monomers can be identified and determined in polyacrylates and polymethacrylates by chemical methods. The alcoholic components may be determined after splitting with hydriodic acid or with alkaline hydrolytic agents. When heated, poly(methyl methacrylate) undergoes depolymerisation to produce the monomer almost quantitatively, which is evidence of the regular and linear structure of molecules of this polymer.

Commercial polymethacrylates were demonstrated by Schreyer and Szigeti [4] to be 80–85% syndiotactic.

The structure of polyacrylates and polymethacrylates has been studied by a number of methods such as dielectric measurements [5], ultraviolet [6] and infrared spectroscopy [7,8] as well as by NMR spectroscopy [9–14]. The structure of end groups was studied by the dye technique [15].

Qualitative Analysis

Chemical Methods

Analysis of the Pyrolysis Products

One g of powdered polymer is heated in a heat-resistant glass test tube with a threefold quantity of ignited sand, and the distillate is collected in a receiver chilled in ice. Alternatively a glass-wool plug may be inserted in the mouth of the test tube; the liquid pyrolysis products will condense in the fibre glass. For further examination either a portion of the distillate or part of the plug may be taken.

Determination of Physical Constants. Direct determination of physical constants of the pyrolysis product of polymethacrylates allows the polymer type to be identified.

Monomer	b.p., °C	n_D^{20}
Methyl methacrylate	100·3	1·414
Ethyl methacrylate	117	1·413
n-Propyl methacrylate	141	1·418
n-Butyl methacrylate	163	1·424
Isobutyl methacrylate	155	1·420

In the analysis of other acrylic polymers including the copolymers more pyrolysate is required (from about 100 g of the polymer). The pyrolysate is separated by fractional distillation. The isolated compounds are subsequently identified by boiling point, density, index of refraction, and saponification value (for polyacrylates and polymethacrylates the saponification values are > 200 and < 20 respectively).

Phenylhydrazine Test [16]. The crude pyrolysate is dried with a little calcium chloride and redistilled. A few drops of anhydrous distillate are refluxed for 30 min with a little freshly distilled phenylhydrazine and 5 ml of toluene. A fivefold volume of 85% formic acid is then added, followed by one drop of perhydrol, and the contents are agitated for several minutes. A dark green colour developing is indicative of polyacrylates. The test can be used for the detection of polyacrylates in the presence of polymethacrylates, as the latter fail to give a colour.

The Mano Colour Reaction [17,18]. Part of the glass wool containing the pyrolysis products is carefully heated in a test tube with 1 ml of conc. nitric acid ($d = 1{\cdot}40$ g per cm^3). The resulting yellow solution is cooled and 0·5 ml of distilled water with a few drops of 5–10% sodium nitrite solution are added. A turquoise colour, which in the chloroform extraction passes into the organic layer, indicates the presence of methyl methacrylate. By this reaction polymethacrylates can be differentiated from polyacrylates. These polymers may also be differentiated from each other by the hydroxylamine test [16].

Acid Hydrolysis

One g of the polymer is mixed with 20–30 ml of 50% sulphuric acid at 60–70°C until completely dissolved. The solution is poured into 100 ml of cold water, and poly(acrylic acid) separates from the solution as a sticky material.

Alkaline Hydrolysis in Alcoholic Potassium Hydroxide

A polymer sample is heated for 1 hr with 0·5*N* ethanolic potassium hydroxide at 60°C. Under these conditions only alkyds, fats, and poly(vinyl esters), if present, undergo hydrolysis. Polyacrylates undergo hydrolytic splitting slowly and only on boiling. Under these conditions polymethacrylates are practically unaffected.

Alcoholic Constituents

A polymer sample is hydrolysed in 0·5*N* potassium hydroxide in ethylene glycol or in ethanolamine (see p. 50). The alcohols produced are isolated by distillation and identified by determination of their physical constants, colour reactions, by their infrared spectra, by paper or thin layer chromatography (e.g. as 3,5-dinitrobenzoates), or using gas-liquid chromatography.

Instrumental Methods

Infrared Spectroscopy

Polyacrylates can be identified by the bands at 7·9 and 8·55 μm. Poly(methyl acrylate) exhibits additional bands at 8·35 and 12·0 μm. Characteristic of polymethacrylates is a series of infrared bands at 7·8, 8·1, 8·55, and 8·7 μm, and a band at 13·35 μm. The alcoholic constituents of these polymers may be differentiated by their characteristic bands [19,20,21]. Acrylic and methacrylic acids were detected in styrene-containing acrylic polymers by the absorptions at 1707 and 1700 cm^{-1}, respectively [22]. Typical spectra are shown in Figs. 18.1 and 18.2.

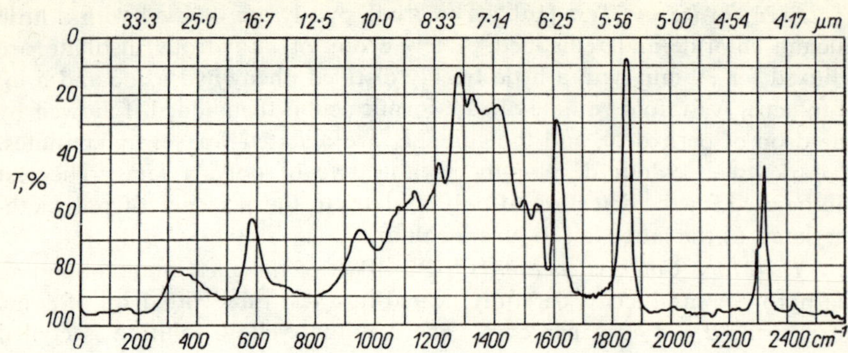

Fig. 18.1 Infrared absorption spectrum of poly(methyl acrylate); film from ethyl acetate solution

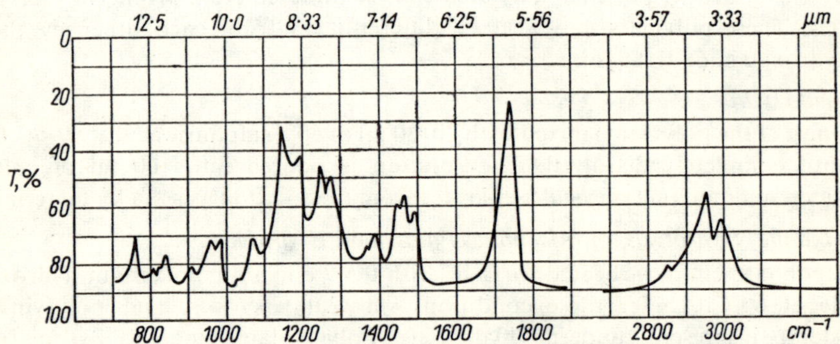

Fig. 18.2 Infrared absorption spectrum of poly(methyl methacrylate); film from acetone solution

Infrared spectroscopy was also used in the analysis of acrylic polymers in the identification of their pyrolysis products [23,24].

Gas Chromatography

Analysis of the Pyrolysis Products. As mentioned previously, polyacrylates differ in their behaviour from polymethacrylates on heating to higher temperatures. Under suitably matched conditions polymethacrylates decompose to the inherent monomers, whereas polyacrylates also form other decomposition products in addition to the monomer. Of these the alcohols derived from the alcoholic constituents of the polymer, represent a valuable analytical feature.

Both homopolymers and a wide range of copolymers can be identified by gas chromatography [25–33].

Analysis of the Solvolysis Products. The ester groups of polyacrylates and polymethacrylates are split with hydriodic acid in much the same way as are alkoxyl groups to yield alkyl iodides from corresponding alcoholic constituents. By the classical procedure the total alcoholic group content can be determined, but the groups cannot be distinguished from one another or determined individually. Haslam *et al.* [34] were the first to succeed in detecting and individually determining minor quantities of poly(ethyl acrylate) in poly(methyl methacrylate), using the Zeisel method coupled with gas-liquid chromatography. In this procedure a 20 mg polymer sample was heated for 30 min with 2·5 g phenol and 6 ml hydriodic acid. The alkyl iodides produced were absorbed in *n*-heptane chilled to very low temperatures with a methanol-dry ice mixture instead of conventional absorption of the iodides in a bromine solution. The *n*-heptane solution was then introduced into a chromatographic column. By a slightly modified procedure higher alkoxyl groups (propoxyl, butoxyl) were also detected and determined in polyacrylates and polymethacrylates [22,26,35].

Higher alkyl esters (2-ethylhexyl, dodecyl, lauryl) can also be analysed qualitatively and quantitatively after alkaline hydrolysis of polyacrylates and polymethacrylates [26]. According to this procedure 400–500 mg polymer samples were dissolved in 13 ml of tetrahydrofuran and hydrolysed in a Parr bomb with 3 ml of ca. 5*N* alcoholic potassium hydroxide by heating the bomb contents to 160°C for 4 hr. The contents were then transferred to a 25 ml volumetric flask and the volume made up to the mark with ethanol. A 10 ml aliquot of this solution was acidified with 0·5 ml of conc. hydrochloric acid. The potassium chloride precipitate settled to the bottom of the flask, and the supernatant liquid was analysed chromatographically.

Mass Spectrometry

Acrylic polymers may be identified by examination of the mass spectra of their pyrolysates [36].

Quantitative Analysis

Alcoholic Constituents

The total alcoholic residues in polyacrylates and polymethacrylates may be determined by the Zeisel method [28]. The content of individual alcoholic groups can be evaluated from gas chromatograms of the products of hydrogen iodide splitting or of alcoholic potassium hydroxide hydrolysis following the procedures described above [22,26,35].

Acidic Constituents

Total combined carboxylic groups in polyacrylates may be found from the determination of saponification values in the usual manner. Polymethacrylates are not split under these conditions. Thus in order to determine their saponification values one has to resort to potassium hydroxide hydrolysis in ethylene glycol or to aminolysis with ethanolamine [28,37].

A method for the determination of the ester groups was developed by Drinberg *et al.* [38], in which alkaline hydrolysis is effected with 2*N* potassium hydroxide in an ethanol-benzene mixture under pressure (at 120–180°C).

DETERMINATION OF ADDITIVES AND CONTAMINANTS

In the acrylic polymers moisture content, free monomer, residual catalyst, and plasticisers are usually determined.

Moisture Content

Moisture in polyacrylates and polymethacrylates is determined by titration with the Karl Fischer reagent. When dried at an elevated temperature the monomer evaporates along with water.

Free Monomers

Free monomers in polyacrylates and polymethacrylates may be determined by chemical methods based on the addition of iodine, or bromine [19,28] or mercaptans [39]. Described below is the mercaptan procedure. When the unknown contains a number of acrylic monomers (acrylates and methacrylates, acrylonitrile, and acrylamide) the method merely provides total free monomers:

$$CH_2{=}CHCOOR + C_{12}H_{25}SH \rightarrow H_{25}C_{12}S(CH_2)_2COOR$$
$$CH_2{=}CHCN + C_{12}H_{25}SH \rightarrow H_{25}C_{12}S(CH_2)_2CN$$

The excess mercaptan is determined by potentiometric titration with a silver nitrate solution in an acidic medium:

$$H_{25}C_{12}SH + AgNO_3 \rightarrow H_{25}C_{12}SAg + HNO_3$$

REAGENTS

Dodecyl Mercaptan. A 0·05*N* solution in isopropanol.

Silver Nitrate. Aqueous 0·05*N* solution.

PROCEDURE. Into a 100 ml Quickfit conical flask containing 30 ml of ethanol, 10–20 ml of the polymer solution (1–2 g of the polymer dissolved in 100 ml of dimethylformamide in a 100 ml volumetric flask) is introduced with a pipette, and from a burette are added 20 ml of 0·05*N* dodecyl mercaptan and 2 ml of 5% potassium hydroxide solution. The flask is stoppered, the contents are thoroughly mixed and allowed to stand for 1 hr when methacrylates are analysed or 10 min when other acrylic monomers are analysed. Then 2 ml of glacial acetic acid is added and the solution is titrated with 0·05*N* silver nitrate solution, using a silver electrode connected through a salt bridge to a calomel electrode immersed in a saturated potassium nitrate solution.

The monomer content X (%) is found from the formula

$$X = 10^4(v_1 - v_2)Kn/am$$

where:

v_2 and v_1 = the volume (ml) of the silver nitrate titrant used in titration of the unknown and the blank, respectively,

n = normality of the silver nitrate solution,

K = weight (g) of the monomer equivalent to 1 ml of exactly

1N silver nitrate solution (for methyl methacrylate—0·1001, methyl acrylate—0·0861 butyl methacrylate—0·1422, acrylonitrile—0·0530),

a = the volume of the sample withdrawn with the pipette,

m = weight of sample, g.

Instrumental Methods. Polarography is extensively used for the determination of free monomers in acrylic polymers. The analysis of free methyl methacrylate in its polymers may be made in such media as benzene, ethanol, or water, with the use of tetrabutylammonium iodide [40] or methyl-tri-*n*-butylammonium chloride [41,42] as the supporting electrolyte. Other reaction media include an acetone-dimethylformamide mixture with tetrabutylammonium iodide as the supporting electrolyte [43].

A colorimetric procedure for the determination of free monomer in poly(methyl methacrylate) was developed by Takayama [44]. The procedure is based on the reaction with hydroxylamine hydrochloride in alkaline solution. The hydroxamic acid formed in the reaction is subsequently converted into a coloured iron salt by adding a solution of iron in nitric acid-perchloric acid mixture in ethanol. Absorbance of the solution is measured at 532 nm. Benzoyl peroxide present in a quantity less than 0·1% does not interfere with the determination. If this compound is present in larger amounts the calibration curve is made with some addition of benzoyl peroxide to the solutions.

According to the same author [45] free monomer in poly(methyl methacrylate) may also be determined by ultraviolet spectroscopy subsequent to its isolation from the polymer by distillation with an acetic acid-water mixture. Absorbance is measured at 250 and 260 nm. Account should be taken of the acetic acid present in the distillate by titrating it with 0·5N sodium hydroxide solution. Benzoyl peroxide does not interfere with the analysis.

Free methyl methacrylate can also be analysed by infrared spectrophotometry by measuring absorbance of a polymer solution in carbon tetrachloride at 1·63 μm [46]. Up to 0·1% of the monomer can be determined by this method.

Free monomers—esters of acrylic and methacrylic acids—may be determined by gas-liquid chromatography using polymer solutions in methylene chloride [26].

Benzoyl Peroxide

Benzoyl peroxide in an acidic medium liberates iodine from iodides. The liberated iodine can be determined either by titration with sodium thiosulphate or by ultraviolet spectrophotometry [47] by measuring absorbance at 370 nm.

Benzoyl peroxide may also be analysed directly by measuring absorbance at 276 or 270 nm. The measurement is made on 1% chloroform solutions of the polymers [48].

This compound can also be determined polarographically in poly(methyl methacrylate) [49].

Sulphur

A method for the quantitative analysis of minor quantities of sulphur (down to a hundredth of 1%) in poly(methyl methacrylate) was evolved by Haslam and Squirrell [50]. The polymer was burnt in oxygen, and the gaseous sulphur combustion products were absorbed in an alkaline hydrogen peroxide solution. The sulphate ions in this solution were determined turbidimetrically, using barium chloride and measuring the absorbance at 700 nm.

Minor quantities of sulphur as SO_4^{2-} ions (from sulphuric acid) were determined colorimetrically [51] after aqueous extraction of the polymer solution in ethyl acetate.

Plasticisers

According to Haslam and Soppet [52] the plasticisers are isolated from the polymer by dissolving a sample in acetone, followed by precipitation of the polymer with petroleum ether and gravimetric determination of the plasticiser.

The plasticisers present in poly(methyl methacrylate) may be determined by ultraviolet spectroscopy [53]. Dibutyl phthalate present in this polymer is determined polarographically [49].

Antioxidants

Ionol and quinol, used as antioxidants in poly(acrylic esters), were determined polarographically by Budylina *et al.* [54].

18.2 POLYACRYLONITRILE

Polyacrylonitrile is manufactured by free radical polymerisation, as are most of the other acrylic polymers. The process is usually conducted in water in the presence of persulphate or some redox system as catalyst.

This material finds application in the manufacture of fibres, as it has no value as a thermoplast because it decomposes below its softening point. Polyacrylonitrile is soluble in few solvents, dimethylformamide being one of them.

Of the acrylonitrile copolymers, apart from the ABS terpolymers discussed elsewhere, the butadiene copolymers which incorporate 15–40% of acrylonitrile (depending on the grade), are of most interest. These materials are elastomers and find uses as petroleum- and mineral-oil-resistant rubbers.

Structure

Polyacrylonitrile contains nitrile groups. Their content may be determined indirectly by chemical methods covered in the subsequent sections of this chapter.

The structure of this polymer was studied by NMR spectroscopy [10,55–57]. The terminal —SO_3H groups in the polyacrylonitrile polymers made in a redox system (with a sulphite) may be detected by infrared spectroscopy [58].

Qualitative Analysis

Hydrolysis Products

1. When heated with concentrated aqueous solutions of alkalis, ammonia evolves almost quantitatively as a result of hydrolysis. The test is not, however, specific for polyacrylonitrile.
2. The polymer is dissolved in dimethylformamide, made strongly alkaline with conc. sodium hydroxide solution, and heated. In the presence of polyacrylonitrile a light orange-red colour develops. By this test polyacrylonitrile can be differentiated from polyamides and polyurethanes, since the latter give no colour reaction [59].
3. When heated with conc. aqueous solutions of mineral acids (such as 20% hydrochloric acid) insoluble poly(acrylic acid) precipitates [20].

Pyrolysis Products

1. A small quantity of the polymer is mixed with ca. 50 mg of a (1:1) calcium oxide-calcium carbonate mixture and heated to a temperature of 250°C in a test tube covered with filter paper soaked in a copper acetate-benzidine acetate mixture. The hydrogen cyanide evolving as a result of pyrolysis of polyacrylonitrile turns the paper blue [60].
2. In the polyacrylonitrile pyrolysate hydrogen cyanide, ammonia, nitriles and amines can be identified by infrared spectroscopy.

Nitrile Groups

A little polymer is heated in a test tube with 10 mg of sulphur. The mouth of the tube is covered with filter paper moistened with acidified 1% ferric nitrate solution. The ferric thiocyanate produced on the paper turns it red [60].

Polyacrylonitrile may be identified from the infrared spectrum of the polymer itself (bands at 1070, 1250, 1443, and 2252 cm^{-1}) (Fig. 18.3), or of its pyrolysis products [61].

Chemical Composition of the Polymer

Nitrogen in polyacrylonitrile and its copolymers can be determined by the Kjeldahl [62] or Dumas methods. Minor quantities of nitrogen may be determined by the Kjeldahl method in a procedure developed by Majewska [63].

Nitrile groups in polyacrylonitrile may be determined in the acid hydrolysate. A polymer sample is refluxed with a conc. solution of a mineral acid (for instance 20% hydrochloric acid). The poly(acrylic acid) precipi-

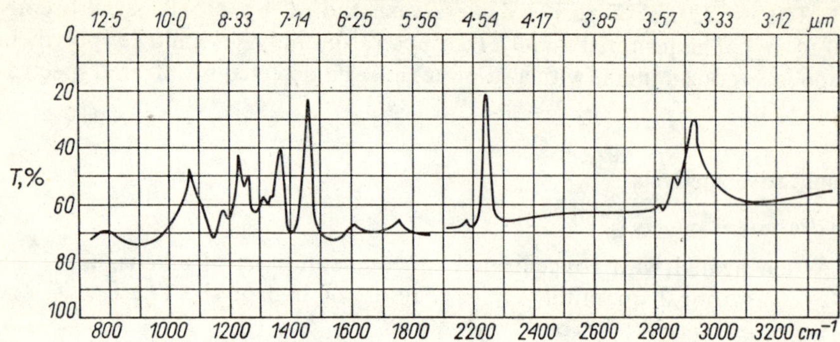

Fig. 18.3 Infrared absorption spectrum of polyacrylonitrile; KBr disc

tated as a result of hydrolysis is filtered off and washed with water. The content of carboxylic groups in the precipitate may be determined alkalimetrically. The ammonium salt present in the filtrate is decomposed by making it strongly alkaline with sodium hydroxide, and the ammonia is determined after steam distillation as in the Kjeldahl method [64].

Acrylonitrile was quantitatively determined in the acrylonitrile-butadiene copolymers by pyrolysis gas chromatography by Okumoto *et al.* [65].

Determination of Contaminants and Additives

Residual Acrylonitrile

Unreacted acrylonitrile in the acrylonitrile copolymers may be determined polarographically (see analysis of the ABS terpolymers) or by gas-liquid chromatography [66]. Free acrylonitrile in aqueous latexes was determined by a method based on the reaction with sodium sulphite [67].

Trace Sulphur

Sulphur in quantities of 0·01–0·2% can be determined in polyacrylonitrile by a method developed by Philipp and Dautzenberg [68]. The material is fused with potassium hydroxide with addition of hydrogen peroxide and subsequently reduced with a mixture of glacial acetic acid, hydriodic acid, and sodium hypophosphite (NaH_2PO_2).

The hydrogen sulphide produced is isolated from the reaction mixture, by heating to boiling in a nitrogen stream, and absorbed in a sodium hydroxide solution. Then the sulphide ions are determined in this solution by potentiometric titration with 0·001*N* mercuric chloride, or polarographically.

Dimethylformamide

Minor quantities of dimethylformamide in polyacrylonitrile fibre were determined by gas chromatography [69] following either aqueous extraction of the material or heating in a nitrogen stream and absorption in water.

References

1. Schildknecht, C. E., *Vinyl and Related Polymers*, Wiley, New York, 1952.
2. Dębski, W., *Polimetakrylan metylu* [*Poly*(*methyl methacrylate*)], WNT, Warsaw, 1969.
3. Rauch-Puntigam, H., Völker, T., *Acryl- und Methacrylverbindungen*, Springer Verlag, Berlin, 1967.
4. Schreyer, G., Szigeti, P. R., *Kunststoffe*, **53**, 703 (1963).
5. Steck, N. S., *SPE Transactions*, **4**, 34 (1964).
6. Ohara, K., *Makromol. Chem.*, **142**, 69 (1971).
7. Nagai, H., *J. Appl. Polymer Sci.*, **7**, 1697 (1963).
8. Pokrovskii, E. I., *Vysokomol. Soedin.*, **6**, 642 (1964).
9. Grassie, N., Torrance, B. D. J., Fortune, J. D., Gemmell, J. D., *Polymer*, **6**, 653 (1965).
10. Kato, Y., Nishioka, A., *Bull. Chem. Soc. Japan*, **37**, 1614 (1964).
11. Matsuzaki, K., Uryu, T., Ishida, A., Ohki, T., Takeuchi, M., *J. Polymer Sci., Pt. A-1*, **5**, 2167 (1967).
12. Ando, I., Nishioka, A., *Kogyo Kagaku Zasshi*, **73**, 375 (1970).
13. Niezette, J., Desreux, V., *Makromol. Chem.*, **149**, 177 (1971).
14. Hatada, K., Terawaki, Y., Okuda, H., Niinomi, T., Yuki, H., *Kobunshi Kagaku*, **28**, 293 (1971).
15. Palit, S. R., Mandal, B. M., *J. Macromol. Sci., Pt. C*, **2**, 225 (1968).
16. Rath, H., Nawrath, G., Schönpflug, E., *Melliand Textilber.*, **33**, 636 (1952).
17. Mano, E. B., *Anal. Chem.*, **32**, 291 (1960).
18. Dawson, T. S., Haken, J. K., *J. Chromatogr.*, **45**, 346 (1969).
19. Haslam, J., Willis, H. A., *Identification and Analysis of Plastics*, Iliffe Books, London, 1965.
20. Hummel, D., *Kunststoff-, Lack- und Gummi-Analyse*, Carl Hanser Verlag, Munich, 1958; Hummel, D. O., Scholl, F., *Atlas der Kunststoff-Analyse*, Carl Hanser Verlag, Munich; Verlag Chemie, Weinheim, 1968.
21. Katritzky, A. R., Lagowski, J. M., Beard, J. A., *Spectrochim. Acta*, **16**, 964 (1960).
22. Anderson, D. G., Isakson, K. E., Snow, D. L., Tessari, D. J., Vandeberg, J. T., *Anal. Chem.*, **43**, 894 (1971).
23. Barnes, R. B., Gore, R. C., Stafford, R. W., Williams, V. Z., *Anal. Chem.*, **20**, 402 (1948).
24. Harms, D. L., *Anal. Chem.*, **25**, 1140 (1953).
25. Davison, W. H. T., Slaney, S., Wragg, A. L., *Chem. Ind.* (*London*), **1954**, 1356.
26. Cobler, J. G., Samsel, E. P., *SPE Transactions*, **2**, 145 (1962).
27. Guillet, J. E., Wooten, W. C., Combs, R. L., *J. Appl. Polymer Sci.*, **3**, 61 (1960).
28. Kasterina, G. N., Kalinina, L. S., *Khimicheskie metody issledovaniya sinteticheskikh smol i plasticheskikh mass* (*Chemical Methods of Analysis of Synthetic Resins and Plastics*), Goskhimizdat, Moscow, 1963.
29. Lehmann, F. A., Brauer, G. M., *Anal. Chem.*, **33**, 673 (1961).
30. Strassburger, J., Brauer, G. M., Tryon, M., Forziati, A. F., *Anal. Chem.*, **32**, 454 (1960).
31. Gatrell, R. L., Mao, T. J., *Anal. Chem.*, **37**, 1294 (1965).
32. Dandoy, J., Deneubourg, C., *J. Chromatogr.*, **56**, 126 (1971).
33. Duebler, K. H., Hagen, E., *Plaste u. Kautschuk*, **16**, 169 (1969).
34. Haslam, J., Hamilton, J. B., Jeffs, A. R., *Analyst*, **83**, 66 (1958).
35. Miller, D. L., Samsel, E. P., Cobler, J. G., *Anal. Chem.*, **33**, 677 (1961).
36. Zemany, P. D., *Anal. Chem.*, **24**, 1709 (1952).
37. Brauer, G. M., Newman, S. B., in *High Polymers, Vol. XII, Analytical Chemistry of Polymers*, Kline, G. M. (Ed), *Part III, Identification Procedures and Chemical Analysis*, Interscience, New York, 1962.
38. Drinberg, A. Ya., Yakovlev, A. D., Sokolova, Z. S., *Zavodsk. Lab.*, **23**, 26 (1957).

39. Balandina, V. A., Gurvich, D. B., Kleshcheva, M. S., Nikitina, V. A., Nikolaeva, A. P., Novikova, E. M., *Analiz polimerizatsionnykh plastmass* (*Analysis of Polymerisation Plastics*), Khimiya, Leningrad, 1967.
40. Bobrova, M. I., Matveeva, A. N., *Zh. Obshch. Khim.*, **26**, 1857 (1956).
41. Stanley, E. L., in *High Polymers, Vol. XII, Analytical Chemistry of Polymers*, Kline, G. M., (Ed), *Part I, Analysis of Monomers and Polymeric Materials*, Interscience, New York, 1959.
42. Lacoste, R. J., Rosenthal, I., Schmittinger, C. H., *Anal. Chem.*, **28**, 983 (1956).
43. Paściak, J., *Chem. Anal.* (*Warsaw*), **6**, 411 (1961).
44. Takayama, Y., *Japan Analyst*, **4**, 367 (1955).
45. Takayama, Y., *Japan Analyst*, **4**, 634 (1955).
46. Miller, R. G. J., Willis, H. A., *J. Appl. Chem.*, **6**, 385 (1956).
47. Takayama, Y., Doi, K., *Japan Analyst*, **4**, 349 (1955).
48. Takayama, Y., *Japan Analyst*, **4**, 234 (1955).
49. Bezuglyi, V. D., Dmitrieva, V. N., *Zavodsk. Lab.*, **24**, 941 (1958).
50. Haslam, J., Squirrell, D. C. M., *J. Appl. Chem.*, **11**, 244 (1961).
51. Getmanenko, E. N., Perepletchikova, E. M., *Zavodsk. Lab.*, **37**, 540 (1971).
52. Haslam, J., Soppet, W., *Analyst*, **75**, 63 (1950).
53. Haslam, J., Grossman, S., *Analyst*, **79**, 238 (1954).
54. Budylina, V. V., Marinin, V. G., Vodzinskii, Yu, V., Kalinin, A. I., Korshunov, I. A., *Zavodsk. Lab.*, **36**, 1051 (1970).
55. Murano, M., Yamadera, R., *Kobunshi Kagaku*, **23**, 497 (1966); Yamadera, R., Murano, M., *J. Polymer Sci., Pt. A-1*, **5**, 1059 (1967).
56. Matsuzuki, K., Okada, M., Uryu, T., *J. Polymer Sci., Pt. A-1*, **9**, 1701 (1971).
57. Schaefer, J., *Macromolecules*, **4**, 105 (1971).
58. Yamadera, R., *J. Polymer Sci.*, **50**, Suppl. 4 (1961).
59. Ford, J. E., Roff, W. J., *J. Textile Inst.*, **45**, 580 (1954).
60. Feigl, F., Gentil, V., Jungreis, E., *Mikrochim. Acta*, **1959**, 47.
61. Burlant, W. J., Parsons J. L., *J. Polymer Sci.*, **22**, 249 (1956).
62. Majewska, J., Jordan, M., *Chem. Anal.* (*Warsaw*), **5**, 1039 (1960).
63. Majewska, J., *Chem. Anal.* (*Warsaw*), **10**, 85 (1963).
64. Winterscheidt, H., *Seifen, Öle, Fette, Wachse*, **80**, 310, 382, 404 (1954).
65. Okumoto, T., Takeuchi, T., Tsuge, S., *Japan Analyst*, **18**, 1483 (1969).
66. Reichle, A., Tengler, H., *Plastverarbeiter*, **19**, 921 (1968).
67. Taubinger, R. P., *Analyst*, **94**, 628 (1969).
68. Philipp, B., Dautzenberg, H., *Anal. Chim. Acta*, **30**, 384 (1964).
69. Nestler, H., Mai, J., *Faserforsch. Textiltechn.*, **21**, 307 (1970).

Chapter 19

CELLULOSE DERIVATIVES

Plastic materials from cellulose belong to the class of chemically modified natural polymers. We distinguish here cellulose esters and ethers, as well as hydro- and oxycelluloses. However, in the present chapter we shall be dealing only with the esters and ethers in view of their much greater commercial importance.

The most important cellulose esters include soluble cellulose nitrate, cellulose acetate, triacetate and tripropionate and cellulose acetate-butyrate. The only ether of any importance as a plastic is ethylcellulose, but other ethers are known, for example methylcellulose, ethylcellulose, benzylcellulose and sodium carboxymethylcellulose.

Cellulose derivatives find extensive application in both unplasticised and plasticised forms. In the analysis of plasticised fabricated products the determination of plasticiser content and specific gravity are of prime importance. Analysis of technical unplasticised products covers the determination of functional groups, impurities, and some physical properties. In the case of plastics containing plasticisers, the separated free cellulose derivative may be tested by the same methods as those used for technical raw material.

19.1 SEPARATION AND DETERMINATION OF PLASTICISER

Two methods are chiefly used for the separation of plasticiser from polymeric cellulose derivatives: the extraction method, and the precipitation method based on precipitation of the cellulose derivative from the solution of the plastic by means of a suitable non-solvent [1].

The precipitation method is more accurate than the extraction method in determining the polymer content. The precipitation method yields a polymer of higher purity, but it can result in low values of plasticiser content due to losses when recovering the plasticiser from the large volumes of solution.

Extraction Method

In this method, which is particularly applicable to films, a weighed quantity of the material is extracted with a liquid which acts as solvent for the plasticiser, but not for the cellulose derivative [2]. The size of the sample amounts to 3–20 g. The conventional Soxhlet apparatus is commonly used. The polymer together with the pigment, if any, remains in the extraction thimble, and the plasticiser may be recovered from the extract after evaporation of the solvent. The pigment content can be determined by ashing a portion of the residue in the thimble. *n*-Hexane is the most suitable solvent, but benzene is sometimes used. Extraction with *n*-hexane for 72 hr has been found to give a reasonably good separation of the plasticiser from any of the four cellulose derivatives: cellulose nitrate, acetate, butyrate and ethylcellulose. The error in the polymer determination does not exceed ±0·5%. Although part of the plasticiser is left unextracted in the polymer, the method, because of its simplicity and good reproducibility, is very useful for control purposes.

While working with hexane special precautions should be taken. Extraction should be conducted in a well-ventilated room and a steam bath should be used.

In the case of the plasticiser determination in cellulose nitrate, latches should be removed from the oven used for drying both the original sample and the recovered cellulose nitrate. The drying time in an oven should not exceed the time specified in the procedure. The recovered cellulose nitrate must not be stored dry. These precautions are necessary because of the risk of explosions or fires with these compounds.

The sample for extraction is prepared by finely grinding the plastic. This may be conveniently accomplished by filing a piece of the plastic with a rotary file mounted on an electric drill. If the sample is unsuitable for filing, as would be a thin film, it should be compressed into a plaque to facilitate the filing operation. The required amount of filings is dried for half an hour at 100°C in an oven and cooled in a desiccator. The thimble used for extraction should also be dried at 100°C but for at least 6 hr.

The sample weighed out in the thimble is placed in the extraction apparatus and approximately 200 ml of *n*-hexane is added to a tared round-bottomed flask. During extraction the flask is placed in a steam-heated water bath. Extraction is continued for 72 hr for samples up to 10 g, and for at least 96 hr for larger samples. A satisfactory reflux rate is 2–3 drops per sec. After extraction the thimble with its contents is removed and dried at 60–70°C in a steam oven until free of hexane odour and then for 2 hr at 100°C, after which it is weighed. Drying is then continued for 1 hr to constant weight. However, in the case of cellulose nitrate drying at 100°C should not last more than 3 hr.

For determining the plasticiser the flask is placed on a water bath and the solvent evaporated to a small volume (about 5 ml); a gentle stream of dry air is passed to facilitate evaporation. The flask is then placed in a vacuum desiccator and the rest of the solvent is removed by evacuation for 2 hr at a pressure of 1–2 mm Hg. Drying under these conditions is repeated for 1 hr periods until the difference between successive weighings is not greater than 0·002 g.

Precipitation Method

The determination is based on complete dissolution of the plastic material and precipitation of the cellulose derivative with a non-solvent. Acetone is often used as a solvent for the plastic, and hexane or isopropyl ether as non-solvent (precipitant). For some sparingly soluble ethylcellulose plastics ethyl acetate is a better solvent. Its presence, however, hinders the quantitative precipitation of ethylcellulose with hexane, and better results are obtained with heptane. The error in the precipitation method is within about ±0·5%.

Before dissolution the sample is ground or cut up and dried for 1 hr at 100°C. A known amount of the plastic is dissolved in a sufficient quantity of acetone to obtain a ca. 10% solution. For this purpose a suitable container is used, such as 2·5 l. bottle with a rubber stopper previously extracted with acetone. Dissolution is facilitated by shaking. For some samples heating to about 60°C may be necessary, but the bottle, of course, must then be opened. The size of the sample is normally about 100 g, except with cellulose nitrate when the sample should not weigh more than 10 g because of the ignition hazard during drying. After dissolution is complete, the solution is quantitatively transferred to several beakers, the walls of the bottle being washed with additional portions of acetone. The solution is diluted with hexane so that the concentration of the solution in hexane is about 5%, but without causing the polymer to precipitate. Hexane should be added slowly with constant stirring. If any precipitate appears it should be redissolved by adding a small portion of acetone. The solution should contain as much hexane as possible, as this favours in the next step the formation of a fine fibrous precipitate that is easy to filter.

Precipitation is accomplished by slowly pouring the diluted solution into a 2 l. beaker containing hot hexane. The portion of the solution added to hexane should not be too big otherwise a jelly-like precipitate is formed. If this happens, the gelatinous mass must be redissolved and reprecipitated. As a rule, the addition of 1 part of the approx. 5% solution to 4 parts of hexane yields a precipitate which separates easily from the liquid. The conditions of precipitation are not exacting and they may be suitably modified depending on the type of polymer.

The precipitated polymer is filtered off on a dried and tared filter paper placed in a large funnel. The beaker is rinsed with hot hexane, which is subsequently used for washing the precipitate. The filtrate is collected in 2 l. beakers. The precipitation and filtration steps are repeated with all the hexane-diluted acetone solutions prepared as above. The polymer remaining on the walls of the original 2 l. beaker is dissolved in acetone and precipitated separately in an appropriate excess of fresh hexane. The precipitates collected on the filter paper are washed with several portions of fresh hot hexane and the washings are combined with the filtrate.

In order to recover the plasticiser the solvent is evaporated on a steam bath. Near the end of the evaporation all the solutions are combined in one beaker. When the total volume is about 500 ml, it should be inspected for additional polymer fines that may have precipitated on evaporation of the acetone. The final evaporation requires gentler conditions and constant attention to avoid any loss of the plasticiser.

To recover the polymer the filter papers with their contents are dried at 60–70°C. To facilitate drying, the folded filter papers may be opened up and left overnight. Drying to constant weight is carried out at 100°C. The sum of the percentages of polymer, plasticiser and pigment, if any, should be within $100 \pm 0{\cdot}5\%$.

The procedure outlined above can be modified by using a smaller sample size (e.g. as small as 0·5 g) when only a determination of polymer content is required. Then a sintered glass crucible is used for filtration and vacuum drying to obtain a constant weight of the recovered polymer.

19.2 CELLULOSE ESTERS (EXCEPT CELLULOSE NITRATE)

Cellulose esters are produced on an industrial scale by esterification of cellulose, in the form of cotton linters or wood pulp, with acetic or other suitable anhydrides in the presence of catalysts such as sulphuric acid or perchloric acid.

The fully esterified products, such as cellulose triacetate, are in many cases subjected to partial hydrolysis by addition of a limited amount of water to obtain a required content of acidic groups (secondary acetates).

Cellulose acetates such as triacetate and 2–2·5 acetate are mainly utilised for making a great variety of injection-moulded, extruded, and sheet-fabricated articles. Other esters are used chiefly as moulding powders if a material tougher and more water-resistant than cellulose acetate is required.

The acetates also find application as sheet plastics, as components of lacquers and adhesives and in production of artificial fibres. They are available on the market in the form of granular material of varying plasticiser content, usually 20–30%.

Cellulose propionate is available as a clear or coloured sheeting suitable for general fabrication. The acetate-butyrate is also used for lacquers, hot-melt applications and dip-coating powders.

The solubility of cellulose esters depends on the amount of acidic groups [3]. Generally the solubility in less polar solvents such as chloroform or ethyl acetate increases with the number of these groups. There can be a solubility distinction between secondary and triacetates. The secondary acetates (yielding 51–57% of acetic acid) are soluble in acetone, in acetone–water mixture (80:20) or in conc. hydrochloric acid. Triacetates (yielding 57–62% of acetic acid) are soluble in chloroform or methylene chloride especially in the presence of a little alcohol; they swell in acetone.

Cellulose acetate butyrate is soluble in acetone-benzene mixtures and methylene chloride; cellulose propionate is soluble in benzene, methylene chloride, chloroform, esters and ketones.

Commercially available cellulose acetate contains small amounts of water—ca. 1%.

Identification

The cellulose esters are thermoplastic materials. Preliminary identification can be based on the appropriate tests given in the Section on qualitative analysis of polymers. More accurate investigation consists in saponification of esters by heating with alcoholic potassium hydroxide solution. The saponification number of the esters is above 200. The hydrolysis can also be accomplished when a sample is left to stand in a 1:1 sulphuric acid-water mixture for several hours. The distillate from a diluted solution of the hydrolysis product (after acidification in the case of alkaline hydrolysis) can be used for the detection of the acidic groups. By testing the odour of the distillate butyric acid may be easily distinguished from acetic or propionic acid. The latter acids can be detected by the reaction with lanthanum nitrate [4]. One drop of the distillate is mixed with one drop of 5% aqueous lanthanum nitrate. Then one drop of 0·1% alcoholic iodine is added, followed by several drops of 1*N* ammonia solution. In the presence of acetic or propionic acid a blue-brown colour is produced (the colour may only appear after some time).

The precipitate of hydrocellulose formed by hydrolysis is insoluble in water and organic solvents, but is soluble in Schweitzer's reagent (see p. 25). Cellulose can be detected by the Molisch test or by the reaction with aniline acetate (see p. 32–33). For this purpose the anthrone test can also be used.

To a 1 mg sample suspended in 0·5 ml of water, about 2 ml of a 0·2% solution of anthrone (9,10-dihydro-9-ketoanthracene) in conc. sulphuric acid is added until the precipitate first formed redissolves. In the presence of carbohydrate material a green colour appears and increases in intensity until a dark blue-green solution results.

The organic acids obtained by hydrolysis may be analysed using gas chromatography [5] or paper chromatography. For this purpose a sample of the distillate may be taken. In the case of alkaline hydrolysis the solution of free acids can be obtained by passing the salts through a cation exchange resin in hydrogen form [6], such as Dowex 50 (H^+).

For the paper chromatography identification [7] the solution of free monocarboxylic acids is neutralised with ammonia and concentrated to a small volume. The ammonium salts are separated by paper chromatography, using the ascending technique on Whatman No 1 filter paper, with ethyl alcohol–aqueous ammonia as a solvent. The dried chromatogram is detected by spraying with a 0·04% ethanolic bromocresol green solution. The acids are identified by comparison of R_f values of the blue spots from the test sample and from a standard solution of the pure acids.

Different cellulose esters and ethers can be identified by means of infrared spectroscopy [8,9] with the possibility of distinguishing cellulose triacetate from diacetate, the latter having a strong absorption band at 3448 cm^{-1} due to the presence of hydroxyl groups [10].

Quantitative Analysis

Cellulose esters are characterised principally by analyses of their acidic (acyl) group contents, and by determination of their viscosities. Certain other determinations such as free acid, moisture and ash contents may also be of interest for process control.

In general the methods of analysis are similar for different types of esters.

Determination of Free Acids [11]

A 5 g sample of dried cellulose acetate is placed in a 250 ml conical flask and 150 ml of cold freshly boiled distilled water is added. The contents of the flask are shaken, stoppered, and allowed to stand for 3 hr at room temperature. The cellulose acetate is filtered off and washed with distilled water. The filtrate is combined with the washings and titrated with 0·1*N* sodium hydroxide solution, using phenolphthalein as indicator. A blank determination is carried out at the same time and the free acid content calculated as acetic acid.

Determination of Combined Acetic Acid in Cellulose Acetate [11]

The determination of acetyl content in commercial cellulose acetate is based on the measurement of sodium hydroxide consumed for heterogeneous saponification of the sample. The method is also applicable to certain mixed esters of cellulose, including acetate-propionates and acetate-butyrates containing up to 40% of the propionate or butyrate groups.

About 1 g of a dried and milled sample of cellulose acetate (particles not greater than 0·5 mm) is placed in a conical flask and 40 ml of 75% ethanol is added. The contents are warmed to 50–60°C for 30 min 40 ml of 0·5*N* sodium hydroxide is added, the mixture is warmed slightly at 50°C for 15 min and then allowed to stand for 48 hr at room temperature after which the excess sodium hydroxide is back-titrated with 0·5*N* hydrochloric acid, using phenolphthalein as indicator. A slight excess of hydrochloric acid is added and several hours are allowed for the remaining traces of alkali to diffuse from the cellulose into the solution. The titration is completed with 0·5*N* sodium hydroxide solution to the appearance of a pink colour remaining after vigorous shaking. A blank test should be carried out at the same time. The method is applicable to samples containing up to 43% acetyl content. For a higher acetyl content the sample should first be agitated in the above ethanol-sodium hydroxide solution for 24 hours at 50°C.

The results are expressed as % of combined acetic acid or as acetyl groups, making a correction for the free acid content.

$$\% \text{ Acetyl} = \% \text{ combined acetic acid} \times 0{\cdot}7167$$

DETERMINATION OF COMBINED ACETIC AND BUTYRIC ACIDS IN CELLULOSE ACETATE-BUTYRATE [12]

To determine the total acyl group content of less than about 35% butyrate content, the same procedure is used as for acetyl content in cellulose acetate.

If the butyrate content is between 35 and 45%, the sample (about 0·5 g accurately weighed) is dissolved, before addition of 0·5*N* sodium hydroxide, in 100 ml of acetone-methanol mixture (1:1) by vol. If the butyrate content is above 45% a 1:1 mixture of pyridine and methanol is used as solvent. After saponification at room temperature (the samples are allowed to stand overnight) the excess sodium hydroxide is back-titrated, using phenolphthalein as indicator. The results are calculated in terms of total acyl content ($C\%$ expressed as acetic acid) by the formula

$$C = 6(nv - n_1 v_1)/W$$

where:

v = volume (ml) of sodium hydroxide added,
v_1 = volume (ml) of hydrochloric acid added,
n = normality of sodium hydroxide solution,
n_1 = normality of hydrochloric acid solution,
W = weight of sample, g.

Acetic and butyric acids are determined in the polymer on the basis of the analyses for total acetyl content (by the preceding method) and for the molar ratio at which these acids are combined in the polymer. The molar ratio is determined by saponification of the sample, acidification to liberate the acids, vacuum distillation of the acids, and measurement of their distribution ratio between *n*-butyl acetate and water.

In the case of material with a content of less than 35% butyrate groups, a sample of ca. 3 g is heated with 100 ml of 0·5*N* sodium hydroxide solution at 40°C for 48–72 hr. Then 50 ml of 1*M* phosphoric acid is added and the liberated acids are distilled off under reduced pressure. In the case of more than 35% butyrate content, a ca. 3 g sample is mixed with 100 ml of ethanol and 100 ml of 0·5*N* sodium hydroxide solution, and allowed to stand for 48–72 hours at room temperature. The regenerated cellulose is filtered off and the filtrate evaporated to dryness under vacuum. The residue is treated with 50 ml of 1*M* phosphoric acid and the released acids are vacuum-distilled.

Further procedure is the same independent of the butyrate content of the material. First, a 25 ml portion of the distillate is titrated with 0·1*N* sodium hydroxide, using phenolphthalein as indicator. (Volume of alkali required is designated *M*.) Subsequently 30 ml of the distillate is shaken thoroughly with 15 ml of acid- and water-free *n*-butyl acetate for 1 min in a small separating funnel. The volumes should be measured accurately. The mixture is allowed to stand, then the lower aqueous layer is drawn off and 25 ml of this layer is titrated with 0·1*N* sodium hydroxide. (Volume of alkali required is designated M_1.) The distribution ratio K of the mixed acids between water and *n*-butyl acetate is given by:

$$K = 100(M_1/M)$$

Distribution ratios for pure acetic and butyric acids are individually determined in the same manner, using for this purpose approximately decinormal solutions. The ratios obtained are designated K_a and K_b for acetic and butyric acids respectively.

The molar percentage B of butyric acid is given by

$$B = \frac{K - 100.K_a}{K_b - K_a}$$

and the molar percentage A of acetic acid by:

$$A = 100 - B$$

The percentage of combined acetic acid (by weight) is equal to $AC/100$, while the percentage of combined butyric acid is given by $88BC/6000$.

Besides the above conventional method [12] gas chromatography [5] may be used for the determination of acetic and butyric acids in the distillate. Another chromatographic method [13] can be applied without the vacuum distillation of free acids. After saponification of the sample and separation of the precipitated cellulose, alcohol and other solvents are removed by evaporation. The residue is dissolved in water and the solution of salts is passed through a column with a cation exchange resin in hydrogen form (Wofatite KPS 200). The effluent containing free acids is collected in a volumetric flask. By titrating a portion of the effluent the saponification number may be determined.

Separation of butyric and acetic acids is carried out by partition chromatography using a silica gel column and a mixture of chloroform and butanol for elution. The acid contents in the fractions from the column are determined by titration with 0·02*N* sodium hydroxide, using phenol red as indicator.

It is also possible to determine acetic and *n*-butyric acids by using column partition chromatography on ion exchange resins. The separation is based on the difference in sorption of the acids on a strongly acidic cation exchanger in the hydrogen form [14]. A separation technique using a cation exchanger column combined with automatic titration of eluted acids [15] seems to be suitable for analytical control work.

Another possible method for the quantitative analysis of cellulose esters involves the application of pyrolysis gas chromatography [16]. The determinations are based on the measurement of peak heights of the free acids produced by pyrolysis of the sample at 700°C.

Organic cellulose esters can be determined by gas chromatography after quantitative methanolysis to the corresponding methyl esters [17]. The methanolysis is accomplished by treatment of the sample in absolute methanol with boron trifluoride.

Determination of combined butyric and acetic acids in cellulose acetate-butyrate is described [18]. The sample is saponified and the total acidity determined by titration. Butyric acid is determined colorimetrically as cuprous butyrate (which, when taken up with chloroform, develops a blue colour) and the acetic by the difference from total acidity.

Conditions of procedures are reported [19] for the infrared spectrophotometric determination of acetate content of secondary cellulose acetate and acetate and butyrate content of cellulose acetate-butyrate esters.

Determination of Ash

A sample of dried material or of known moisture content, weighing 2–5 g, is ashed in a weighed crucible in a flame followed by ignition to constant weight in a muffle furnace at 900°C. After cooling, the crucible is reweighed and the ash content calculated.

Determination of Moisture

Cellulose acetates are hygroscopic materials and, when left exposed to the atmosphere, absorb moisture. According to a standard method [11] a sample of about 2–5 g in weight is heated for 2 hr in an oven at 100–105°C and the loss in weight is determined. A relatively fast determination of water in cellulose derivatives employs gas chromatography [20].

Determination of Viscosity [11, 21]

The viscosity of cellulose esters is a measure of their degree of polymerisation, or chain length, and is determined in solutions in suitable solvents, generally at high concentrations (12–25%). The viscosity is measured at 25°C under carefully controlled conditions, using the falling-ball method. The solution is placed in a tube of closely specified dimensions. Results are expressed in terms of the time in seconds which is required for a steel ball to fall through a 2·00 in. high column of solution indicated by two marks on the tube wall.

The composition of standard solutions for measuring the viscosity of cellulose acetates depends on the content of combined acetic acid in the tested materials. In the case of cellulose acetate-butyrate the viscosity is determined on a 20% solution of a dried sample in acetone of specified purity.

19.3 CELLULOSE NITRATE

Cellulose nitrate is prepared from dried cotton linters or paper pulp, steeped in a mixture of conc. nitric acid with either sulphuric acid or a mixture of acetic acid and acetic anhydride at 20–40°C. The properties of cellulose nitrate depend on the degree of substitution by nitro groups, their position in the macromolecule and the degree of polymerisation.

Products with 10·5–11·5% of nitrogen and with 11·5–12% of nitrogen are used for plastics (celluloid), fibres, lacquers and films. They are soluble in alcohol-ether mixtures, ketones, esters and mixtures thereof, e.g. 1:1 acetone-amyl acetate, as well as in ethylene glycol monoethyl ether (Cellosolve), diethylene glycol or glacial acetic acid. Products with more than 12% of nitrogen are insoluble in alcohol-ether mixtures and are mainly utilised for explosives. For making celluloid a product of 10·6–11·2% nitrogen is used with camphor as plasticiser.

The commercial products are usually stored in a wet state with 30–35% of a lower alcohol such as ethanol or butanol as wetting agent, which should be uniformly distributed in the whole bulk of material.

Cellulose nitrates are relatively unaffected by cold diluted acids and alkalis, but are decomposed by concentrated acids such as sulphuric acid, and concentrated alkalis.

Identification

Cellulose nitrate is thermoplastic. It is readily flammable with an exceptionally high burning rate, little smoke, and little residue. Pyrolysis gives camphor, if present, which can also be smelt on warming or on scraping. On further heating the material chars, evolves strongly acid yellow fumes of nitrogen oxides, and may ignite or explode.

Cellulose nitrate can be detected by colour reactions, such as the Molisch test (see p. 33). The sample is dissolved in acetone and 2 or 3 drops of ca. 2% ethanolic solution of α-naphthol are added. A little conc. sulphuric acid is then added slowly so as to form a bottom layer. A green ring at the interface indicates the presence of cellulose nitrate.

Cellulose nitrate produces an intense blue colour in the reaction with diphenylamine (see p. 32) and also responds positively to the anthrone test (see p. 419).

Quantitative Analysis

Among the analytical methods for soluble cellulose nitrate the determination of nitrogen content is most commonly used. The control tests also include determination of acidity, ash, viscosity and wetting agent. For special purposes the determination of stability and solubility are also carried out. The viscosity of a standard solution of specified composition is determined in the same manner as in the case of cellulose acetate.

Samples for the determination of nitrogen, ash and stability are prepared by spreading a thin layer of the material and drying, first at room temperature for 12–16 hr, and then in an oven at 100–105°C for 1 hr. This method of drying enables one to determine the content of the wetting agent at the same time.

To avoid an explosion hazard, the drying ovens should have door latches removed, and a protective face mask should be worn by the laboratory technician when samples are withdrawn.

DETERMINATION OF NITROGEN

The most frequently used is the gas-volumetric method which is accepted in many countries as a standard method [22]. The method is based on the following reaction and subsequent volumetric measurement of liberated nitric oxide:

$$2RNO_3 + 4H_2SO_4 + 3Hg \rightarrow R_2SO_4 + 3HgSO_4 + 4H_2O + 2NO$$

The determination is carried out in the Lunge nitrometer [23] or the Du Pont nitrometer.

The sample of cellulose nitrate is first dried at 90–100°C to constant weight and then 1·0–1·05 g is weighed out on an analytical balance. The sample is dissolved and transferred quantitatively into the reaction bulb of the nitrometer. When the decomposition reaction is completed the volume of the liberated nitric oxide is measured.

The procedure requires first a standardisation of the measuring part of the nitrometer which is done using a known amount of potassium nitrate to evolve a known volume of nitrogen oxide.

Besides the gasometric method a gravimetric one based on the precipitation of the insoluble nitron nitrate is recommended [24]. About 0·2 g of cellulose nitrate is weighed into a 200 ml conical flask and 5 ml of 30% sodium hydroxide and 10 ml of 3% hydrogen peroxide are added. The mixture is heated under reflux, first gently and then to boiling until a transparent liquid is obtained. Then 40 ml of water and 10 ml of 3% hydrogen peroxide are added through the condenser, and the solution is heated on a water bath at 50°C. At this temperature 40 ml of 5% sulphuric acid is slowly introduced. The mixture is heated to 80°C and 12 ml of nitron acetate solution is added. The mixture is kept at room temperature for 30–45 min then cooled in a refrigerator at a little above 0°C for at least 2 hr. The sintered glass filter and the suction flask are also cooled. The precipitate is transferred to the filter, washed with the filtrate and with several portions of ice-water (total volume 10–12 ml). The precipitate is dried at 110°C for 45 min and after cooling is weighed as $C_{20}H_{16}N_4{\cdot}HNO_3$.

$$\text{Nitrogen content, } \% = 3{\cdot}732 \frac{\text{weight of precipitate}}{\text{weight of sample}}$$

The nitron acetate solution is prepared by dissolving 10 g of nitron in 5 ml of glacial acetic acid and making up with water to 100 ml in a volumetric flask.

Another suitable method for the determination of nitrogen is a volumetric procedure based on the reduction of the nitro group to the primary amino group with excess of a ferrous salt. The ferric ions formed are titrated with titanous chloride in the presence of ammonium thiocyanate as indicator [25,26]. A polymer sample (0·8–1·0 g) is dissolved in 25 ml of a 1:1 mixture of glacial acetic acid and butyl acetate (1:1). To 20 ml of this solution are added 25 ml of 0·7*N* ferrous chloride and 25 ml of dilute (1:1) hydrochloric acid. The mixture is heated under reflux in a stream of nitrogen for 3–4 hr. After cooling, the ferric ions are titrated with titanous chloride to the disappearance of the red colour of the tervalent iron thiocyanate complex. The method is rapid and can be applied in the case of a cellulose nitrate type insoluble in sulphuric acid.

A number of other methods are available for determination of nitrogen in cellulose nitrate. These aim at rapidity, simplicity and greater accuracy. Titration methods with ferrous nitrate as a titrant are described, using both visual [27] and potentiometric [28] techniques. Colorimetric determinations are described based on the colour reaction of the nitro group with potassium hydroxide in acetone solution [29], treatment with ferrous sulphate and sulphuric acid to give a brown colour [30] and the yellow colour produced with phenoldisulphonic acid [31].

Cellulose nitrate may be determined in the presence of other cellulose derivatives by a characteristic infrared absorption band at 12·92 μm [32]. A refined calculation has been proposed [33] to achieve greater accuracy in the determination of nitrogen in cellulose nitrate by infrared spectrophotometry.

For the determination of solvents such as ether, ethanol and water in cellulose nitrate a gas chromatographic method is proposed [33].

Determination of Acidity

For the determination of acidity from inorganic acids, a sample of wet product is taken in an amount corresponding to 1 g of the dry material. The sample is mixed with 10 ml of distilled water in a conical flask, 100 ml of acetone is added and the contents of the flask are mixed until the cellulose nitrate is dissolved. After addition of a few drops of methyl red indicator and 25 ml of water the contents are titrated with 0·01*N* sodium hydroxide solution until neutral. Acidity is calculated in terms of % sulphuric acid and the correction for the blank determination is taken into account.

Total acidity, both organic and inorganic, may be determined similarly, using phenolphthalein as indicator.

Determination of Ash

The percentage of ash is determined by decomposing a dried sample with nitric acid over a water bath and then igniting at red heat to constant weight.

Heat Stability Test

The test is a measure of the resistance of cellulose nitrate to chemical degradation at elevated temperatures. A satisfactory level of stability is of particular importance with regard to safety on storage and retention of physical properties of the material.

The sample is placed in the bottom of a test tube and a strip of methyl red test paper is positioned above the sample. The tube is stoppered and heated at a temperature of 134·5±0·5°C. The colour of the test paper is noted at 5 min intervals until it changes to yellow-pink. Generally, a 25 min period is the required minimum level of stability.

19.4 CELLULOSE ETHERS

Two types of cellulose ethers are distinguished depending on their solubility in alkali and water. Methylcellulose and carboxymethylcellulose as a sodium salt belong to the alkali- and water-soluble type, whereas ethylcellulose and benzylcellulose remain insoluble.

The cellulose ethers are commonly manufactured by treatment of alkali-swollen cellulose (cotton linters or wood pulp) with alkyl chlorides or sulphates or related substances in the presence of alkali which catalyses the reaction and removes the acidic product, e.g. hydrogen chloride. Cellulose and sodium hydroxide autoclaved at 40–80°C with methyl chloride or dimethyl sulphate gives methylcellulose. Similarly ethylcellulose is prepared from ethyl chloride or diethyl sulphate. Etherification of alkali-swollen cellulose with benzyl chloride yields benzylcellulose. Carboxymethylcellulose is prepared by treatment of alkali-cellulose with monochloroacetic acid.

The solubility of methylcellulose depends on the degree of substitution: lightly etherified products are alkali-soluble, intermediate ethers are water-soluble and those with a high degree of substitution (commercially unimportant) dissolve only in organic solvents. Carboxymethylcellulose, like methylcellulose, is soluble in water, and also in ethylene glycol and ethylene chlorohydrin.

Commercial ethylcellulose containing approximately 47–48% of ethoxyl groups dissolves in aromatic and chlorinated hydrocarbons, alcohol, acetone, ether, certain esters, dioxan and pyridine.

Benzylcellulose dissolves in benzene, methylene chloride, ketones and esters and is highly resistant to water.

Methylcellulose is utilised as a permanent textile finish, size, and starch substitute, thickening and emulsifying agent, printing ink component and adhesive.

Carboxymethylcellulose (sodium salt) is also a thickening and emulsifying agent and adhesive. It has found many applications in for example, the textile and paper industry as size, in laundries as starch substitute, in pharmaceutical, and foodstuff industries, etc.

Ethylcelluloses are utilised as raw materials for lacquers and adhesives. The most important are the plastic type ethers, employed similarly to cellulose acetate but where toughness and flexibility are required. Benzylcellulose has similar, but more limited, applications.

Identification

Cellulose ethers are thermoplastics, especially at high degrees of substitution. They melt on heating, to char at higher temperatures; they burn with a yellow luminous flame. On pyrolysis methyl- and ethylcellulose evolve a smell of burning paper, and benzylcellulose a smell of bitter almonds from benzaldehyde.

Cellulose ethers give the above identification tests such as the Molisch test (in the case of ethylcellulose the colour may appear only on warming), anthrone test, and the test with aniline acetate.

A number of reactions [35] enable a distinction to be made between methylcellulose and carboxymethylcellulose. Methylcellulose is remarkable in being soluble in cold water but insoluble in hot water; its solutions form reversible gels on warming.

Carboxymethylcellulose (always marketed as the sodium salt) is ionic, the other ethers are non-ionic. Unlike methylcellulose it is soluble in both cold and hot water.

Methylcellulose gives a flocculent precipitate with 10% tannin solution. The reaction is very sensitive and even 0·1% methylcellulose in solution can be detected. Carboxymethylcellulose responds negatively but itself forms an insoluble precipitate with Zephiran (dimethylalkylbenzylammonium chloride).

Iodine-potassium iodide solution added to a solution of methylcellulose produces a violet-brown to brown colour, which disappears on addition of sodium hydroxide solution. An excess of strong acid added to a solution of carboxymethylcellulose sodium salt precipitates the free acid. Polyvalent metal salts, such as alum and copper sulphate or lead acetate, cause precipitation of metal salts of carboxymethylcellulose and can be used for gravimetric determination.

Detection of ethylcellulose is based on a specific colour reaction for ethoxyl groups [36] which are oxidised to acetaldehyde.

The sample is treated in a micro test tube with 1 drop of dichromate solution (1 g of $K_2Cr_2O_7$, 60 ml of water, 7·5 ml of conc. sulphuric acid) and then heated to 100°C.

The mouth of the test tube is covered with a filter paper disc moistened with a mixture of equal volumes of 20% aqueous morpholine solution and 5% aqueous sodium nitroprusside. A blue colour indicates the presence of ethylcellulose.

Identification of benzylcellulose is based on the detection of the benzyl groups [37].

About 2 g of a cellulose derivative is heated under reflux with 10 ml of acetic anhydride and two drops of conc. sulphuric acid for 30 min. Benzyl acetate is formed. The ester is separated by steam distillation and extraction with ether. After evaporating the ether the ester is hydrolysed with potassium hydroxide solution. Benzyl alcohol formed is then oxidised with potassium permanganate to benzoic acid, which is identified by melting point after recrystallisation.

Benzylcellulose is characterised by a specific infrared absorption band [38] and also by its ultraviolet spectrum, which can be used for both quantitative analysis and for determining the degree of substitution with benzyl groups [39].

Quantitative Analysis

Determination of Alkoxyl Content in Methyl- and Ethylcellulose

For the determination of methoxyl or ethoxyl group content in cellulose ethers a standard method [40] is used. It is based on a modified volumetric Zeisel method [41]. The determination of alkoxyl groups in cellulose ethers may also be carried out by gas chromatography [32]. The sample is decomposed with hydrogen iodide, using a modified Zeisel method, then the alkyl iodides are absorbed in 2,2,5-trimethylhexane at 80°C and finally this solution is chromatographed.

Standard methods of testing the purity of methyl- and ethylcellulose [40] also include procedures for the determination of moisture, ash, chloride, alkalinity, iron, heavy metals, viscosity, pH, solids and density.

The standard methods of testing sodium carboxymethylcellulose [43] include the determination of moisture, degree of etherification, viscosity, purity, sodium glycollate and sodium chloride.

For the determination of sodium glycollate content a colorimetric method is used based on a colour developed with dihydroxynaphthalene after the bulk of the sodium carboxymethylcellulose has been removed by precipitation of its copper salt.

The sample is dissolved in water, acidified, and dilute copper sulphate solution is added. The solution is neutralised with sodium hydroxide solution to cause the excess of copper and the copper salt of carboxymethylcellulose to precipitate. The precipitate is filtered off and the glycollate determined colorimetrically in the filtrate with 2,3-dihydroxynaphthalene. The small amount of sodium carboxymethylcellulose which remains in the filtrate will interfere. It can be determined colorimetrically with anthrone [29] and a correction made.

The carboxymethylcellulose content and the degree of substitution with carboxyl groups can be accurately determined by precipitation of its insoluble copper salt [44,45]. In the copper salt, unlike the salts of other metals e.g. aluminium, silver, lead, mercury and zirconium, the metal content is stoichiometric. The copper salt is quantitatively precipitated and weighed. Iodometric determination of copper in the precipitate can be used for estimation of the degree of substitution with carboxyl groups. On the other hand, from the weight of the precipitate and the amount of the carboxyl groups the content of carboxymethylcellulose in the tested sample can be determined.

References

1. Wandel M., Tengler, H., Ostromow, H., *Die Analyse von Weichmachern*, Springer–Verlag, Heidelberg, 1967, p. 5–11.

2. Schröder, E., *Plaste u. Kautschuk*, **7**, 167 (1960).
3. Ott, E., Spurlin, H. M., Grafflin, M. W., (Eds) *Cellulose and Cellulose Derivatives, Part III*, Interscience, New York, 1955, p. 1454.
4. Feigl, F., Anger, V., *Spot Tests in Organic Analysis*, 7th Ed., Elsevier, London, 1966, p. 454.
5. Bayars, B., Jordan, G., *J. Gas. Chromatogr.*, **2**, 304 (1964).
6. Inczédy, J., *Analytische Anwendungen von Ionenaustauschern*, Akadémiai Kiadó, Budapest, 1964, p. 235.
7. Mano, E. B., Cunha Lima, L. C. O., *Anal. Chem.*, **32**, 1772 (1960).
8. Bikales, N. M., Segal, L. (Eds), *Cellulose and Cellulose Derivatives, Part IV*, in *High Polymers, Vol. V*, 2nd Ed., Wiley-Interscience, New York, 1971, p. 60.
9. Dechant, J., *Ultrarotspektroskopische Untersuchungen an Polymeren*, Akademie-Verlag, Berlin, 1972, p. 456.
10. Hummel, D., *Kunststoff-, Lack- und Gummi-Analyse*, Carl Hanser Verlag, Munich, 1958, p. 20–23.
11. ASTM Standard D 871–54 T.
12. ASTM Standard D 817–65 T.
13. Schröder, E., Franz, J., Thinius, K., *Plaste u. Kautschuk*, **11**, 411 (1958).
14. Reichenberg, D., *Chem. Ind.* (*London*), **1956**, 956.
15. Czerwiński, W., *Chem. Anal.* (*Warsaw*), **12**, 597 (1967).
16. Groten, B., *Anal. Chem.*, **36**, 1206 (1964).
17. Emelin, E. A., Smyslova, N. F., *Zavodsk. Lab.*, **32**, 28 (1966).
18. Wandel, M., Tengler, H., *Gummi, Asbest, Kunststoffe*, **19**, 141 (1966).
19. Bockman, C. D. Jr., *Appl. Spectroscopy*, **15**, 84 (1961).
20. Keister, D. C., Harrington, R. C., *Tappi*, **50**, 81A (1967).
21. ASTM Standard D 1343–54 T.
22. ASTM Standard D 301–56.
23. Berl-Lunge, *Chemisch-technische Untersuchungsmethoden, Vol. I.*, Springer Verlag, Berlin, 1931, p. 609.
24. Kast, H., Metz, L., *Chemische Untersuchung der Spreng- u. Zündstoffe*, 2nd Ed., Vieweg, Brunswick, 1944, p. 85, 221.
25. Pierson, R. H., Julian, E. C., *Anal. Chem.*, **31**, 589 (1959).
26. Marvillet, L., Franchant, J., *Mem. Poudres*, **42**, 271 (1960).
27. Murkami, T., *Japan Analyst*, **9**, 100 (1960).
28. Saudi, S., Flaunguart, G., *Chim. Anal.* (*Paris*), **39**, 20 (1957).
29. Swann, M. H., *Anal. Chem.*, **29**, 1505, (1957).
30. Murakami, T., Ishii, E., *Japan Analyst*, **8**, 257 (1959).
31. Gardon, J. L., Leopold, B., *Anal. Chem.*, **30**, 2057 (1958).
32. Resenberger, H. M., Shoemaker, C. J., *Anal. Chem.*, **31**, 1315, (1959).
33. Clarkson, A., Robertson, C. M., *Anal. Chem.*, **38**, 522 (1966).
34. Rice, D. D., Trovell, J. M., *Anal. Chem.*, **39**, 157 (1967).
35. Neu, R., *Seifen, Öle, Fette, Wachse*, **76**, 445 (1950); *Fette u. Seifen*, **52**, 23 (1950).
36. Feigel, F., *Spot Tests in Organic Analysis*, 7th Ed., Elsevier, London, 1966, p. 692, 438.
37. Hummel, D., *Kunststoff-, Lack- und Gummi-Analyse. Textband*, Carl Hanser Verlag, Munich, 1958, p. 131.
38. Ibid, *Tafelband*, p. 20–23.
39. Houvink, R., Staverman, A. J., *Chemie und Technologie der Kunststoffe, III. Typisierung und Prüfung der Kunststoffe*, Akademische Verlagsgesellschaft, Leipzig, 1963, p. 227.
40. ASTM Standards D 1347–64 and D 914–50.
41. Samsel, E. D., Mchard, J. A., *Ind. Eng. Chem, Anal. Ed.*, **14**, 750 (1942).
42. Colber, J. G., Samsel, E. P., *Talanta*, **9**, 473 (1962).
43. ASTM Standard D 1439–64 T.
44. Corner, A. Z., Eyler, R. W., *Anal. Chem.*, **22**, 1129 (1950).
45. Timokhin, I. M., Finkelshtein, M. Z., *Zh. Prikl. Khim.*, **36**, 415 (1963).

Chapter **20**

SILICONES

Commercial polysiloxanes, commonly referred to as silicones, [1–3] are the products of hydrolytic polycondensation of monomers—low molecular weight organosilicon compounds

$$R_{4-n}SiX_n$$

in which the silicon atom is linked to two or three groups of atoms X capable of hydrolysis.

Of highest commercial interest as intermediates for the manufacture of silicones are silanes in which X may be halogens, alkoxyl or aryloxyl groups, and organic substituents R are methyl or phenyl groups. Hydrolysis of these silanes produces silanols, which subsequently react with one another to give polymers—polyorganosiloxanes—eliminating one water molecule per two reacting silane molecules. For bifunctional monomers containing two X groups the reaction proceeds according to the following equation:

$$\underset{\text{silanes}}{R_1R_2SiX_2} + 2H_2O \rightarrow \underset{\text{silanols}}{R_1R_2Si(OH)_2} + 2HX$$

$$nR_1R_2Si(OH)_2 \rightarrow \text{polyorganosiloxanes} + (n-1)H_2O$$

The structure and properties of silicones depend first on the functionality of the parent monomers, that is on the number of groups of atoms capable of hydrolysis, and secondly on the kind of organic substituents in the monomer molecule. The conditions of the polycondensation process also affect the structure and properties of the final polymer.

Bifunctional monomers give polymers which, depending on their molecular weight, range from liquids (mol. wt. ca. 1000) to solid, resilient, rubber-like materials (mol. wt. ca. 100,000). These polymers, when heated to a sufficiently high temperature, behave as thermoplasts, as they do not harden. A property peculiar to liquid silicones is their constant viscosity, almost unaffected by considerable temperature variations. These polymers find a diversity of applications as lubricants, greases and rubbers, whenever properties such as considerable heat resistance, chemical inertness, water repellence or good electrical insulation are desired.

Mixtures of tri- and bifunctional monomers are used in the manufacture of silicone resins—polymers with properties similar to those of resols or amino resins. Silicone resins are used for varnishes, silicone moulding powders, and laminates. When heated the resins become insoluble and infusible. Properties of thermosetting silicone resins depend to a considerable extent on both the R/Si ratio and the structure of substituents.

20.1 STRUCTURE

Silicones, when treated with hydrogen halides undergo splitting to yield halosilanes [4,5]. These reactions, combined with the presence of a very strong band at 9–10 μm [6,7] in the absorption spectrum of the polymers are, among others, indicative of siloxane bondings in the molecules:

$$-\overset{|}{\underset{|}{Si}}-O-\overset{|}{\underset{|}{Si}}-$$

Based on this evidence the backbone of the silicone polymers produced by hydrolytic condensation is believed to consist of recurring siloxane links. These polymers have therefore a polysiloxane structure and comprise mixtures of polyorganosiloxanes of variable degree of condensation and of diverse structure (crosslinked, linear and cylic). Under certain conditions of hydrolytic condensation the main reaction products of dialkyldichlorosilanes are low molecular cyclic polyorganosiloxanes. The structure of cyclic polyorganosiloxanes has been established by several of methods including both conventional techniques (elemental analysis and molecular weight determination) and X-ray analysis and electron diffraction [8]. In the infrared absorption spectrum of cyclic molecules the bands characteristic of the end R_3Si-groups are not observed [9,10]. Cyclic molecules are undesirable components of siloxane polycondensates.

Low molecular weight components of silicone resins can be isolated and identified by conventional methods [1] and by gas chromatography [11]. As for high polymer silicones, only the elemental constitution and the major structural elements can be identified and quantitatively determined. Among the determinable substituents are hydroxyl, alkoxyl and aryloxyl groups, halogens (both directly linked to silicon and incorporated into organic substituents), acetoxyl groups, Si—H and Si—Si groups, and the average value of the R/Si ratio.

Chlorine linked to silicon can be detected only in the products obtained from halosilanes, and even there it is present in minute quantities on account of its high reactivity with water. On the other hand, in the products from fluorine-containing monomers fluorine is present in larger quantities. This is also true of ether groups in silicone resins made from alkoxy and aryloxysilanes. Si—H groups are present in larger quantities only in resins to be applied at low curing temperatures.

For a better characterisation of the structure of polyorganosiloxanes the polymer functionality is occasionally determined. This property is related to the content of mono-, di-, and trifunctional elements. The resin is treated with anhydrous hydrogen fluoride in conc. sulphuric acid. Under these conditions the siloxane bonds are broken. The method fails, however, for aryl-substituted polymers, since aromatic substituents are abstracted to form the Si—F bond and the parent arene in the reaction with hydrogen fluoride. The Si—H bond reacts similarly.

More detailed information on the structure of silicone polymers can be derived from the infrared absorption spectrum in the range 9–10 μm [6,7]. When ring-substituted phenyl groups are present (for instance with a methyl group), the position of the substituent can be established from the infrared spectrum of the polymer [12]. However, the infrared method is unsuitable for the analysis of the composition of low molecular polyorganosiloxane mixtures, as the spectra of polysiloxane homologues do not differ sufficiently [9,10]. Instead gas chromatography is often used [13,14].

Valuable information on the chemical structure of silicone polymers can be obtained by such methods as gas chromatography after splitting with boron trifluoride [15], mass spectrometry (after pyrolysis) [16], or gas chromatography and thermal analysis [17].

20.2 QUALITATIVE ANALYSIS

GENERAL

The presence of silicon in the molecule of an organic compound imparts peculiar properties to that molecule. Thus analytical methods [18] for silicones differ substantially from analytical methods used for other organic compounds. This refers to both low molecular compounds and silicone polymers.

The thermal stability of Si—O bonds is remarkable. Polyorganosiloxanes usually undergo decomposition at considerably higher temperatures than those required for pyrolysis of conventional organic polymers.

Unlike common organic compounds, many organosilicon compounds burn with difficulty and undergo complete decomposition only at high temperatures. Thus, for combustion of such compounds for analytical purposes special conditions must be used for their pyrolysis and oxidation. Because of the considerable stability of organosilicon compounds towards chemical agents, and in view of their hydrophobic properties, special conditions are also required for other methods of chemical analysis.

SILICON

Fusion Methods

A small amount of the material to be analysed is mixed in a platinum or nickel crucible with a 5–6-fold excess of a sodium carbonate-sodium peroxide mixture, and heated, initially over a low flame then with a strong flame to effect a homogeneous melt. On cooling the melt is dissolved in dilute (1:1) hydrochloric acid, the resulting solution is transferred to a porcelain dish and evaporated to dryness. The solid residue is treated with few ml of dilute hydrochloric acid. The undissolved silica precipitate is filtered off, washed once or twice with distilled water, moistened on the filter paper with 0·1% methylene blue in 10% acetic acid, then washed again once or twice with water. The residue on the filter paper turns intense blue due to adsorption of the dye by the silica. Other dyes such as malachite green, rhodamine and safranine, may also be used [18].

For the detection of organosilicon compounds in commercial materials Kreshkov and Bork [19] fused the material with sodium carbonate and sodium peroxide, using a tiny platinum-wire loop. The melt was subsequently dissolved in a few drops of water and treated with ammonium molybdate and benzidine. A blue colour develops in the presence of organosilicon compounds.

After incineration of the sample silicon may also be detected in a test involving the formation of gaseous silicon fluoride on heating the ash with calcium fluoride and sulphuric acid. The evolving silicon fluoride makes a drop of water turn turbid due to the formation of silica.

When devising tests for silicon it should be borne in mind that in plastics containing fillers and pigments this element may be present in the inorganic form as silicates or silica. In such cases it is necessary to isolate polysiloxanes from the material by extraction, for instance with an acetone-toluene mixture, prior to the analysis for silicon.

Pyrolytic Method

As a result of thermal decomposition of polyorganosiloxanes in an inert atmosphere (such as a nitrogen stream or under reduced pressure) cyclic oligomers are formed in which silicon can be detected by combustion.

HALOGEN BONDED TO SILICON

Halogen bonded to silicon can easily be detected as hydrogen halide in the aqueous layer after hydrolysis of the sample. However, this reaction is not unambiguous, as hydrogen halide may also be formed from the hydrolysis of certain derivatives containing halogen in the organic part.

The reaction with silver nitrate is a very sensitive and characteristic test for halogens (except fluorine) directly linked to silicon [18]. In this test a 0·05–0·1 g polymer sample in a test tube is dissolved in a few ml of methanol, ethanol, or acetone, and shaken with 1–2 ml of 0·05*N* aqueous silver nitrate. Silver halide immediately precipitates in the presence of a halogen linked to silicon.

Chlorine directly bound to silicon may be detected by the method described below [18].

Powdered lead oxide (50–100 mg) alone or in a mixture with copper oxide is placed in a micro test tube, keeping the test-tube neck clean. Then a few drops of the material under test or its benzene solution are added. The test tube is covered with ashless filter paper moistened with *o*-toluidine hydrochloride solution (1 g of *o*-toluidine in 1 l. of 10% hydrochloric acid). In the presence of chlorine directly bound to silicon the paper turns tan.

The minimum detectable quantity of chlorine by this method is 2–0·13 mg.

Instrumental Methods

Polyorganosiloxanes in a mixture with other organic compounds can easily be identified by infrared spectrophotometry, because polysiloxanes, unlike other organic compounds, absorb weakly in the range 2–7 μm, but strongly in the 7–15 μm range. Particularly characteristic is a very strong broad band in the range 9–10 μm which is due to the Si—O bond. Methyl groups directly linked to silicon exhibit absorption at 7·8–8·0 μm, whereas phenyl groups can be identified by a strong absorption maximum at 7·0 μm (Typical spectra are shown in Figs. 20.1 and 20.2).

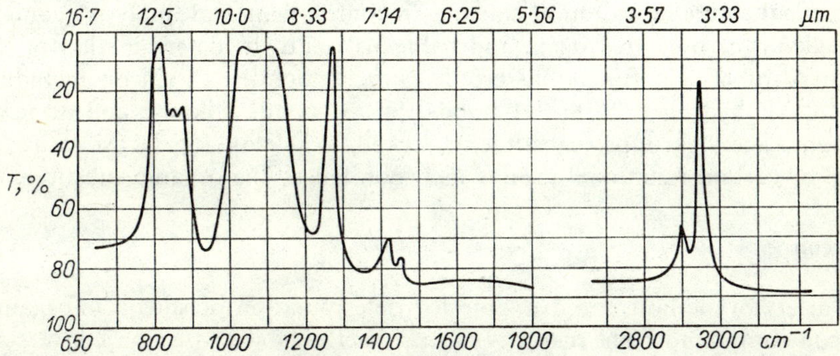

Fig. 20.1 Infrared absorption spectrum of methylsilicone resin; capillary film

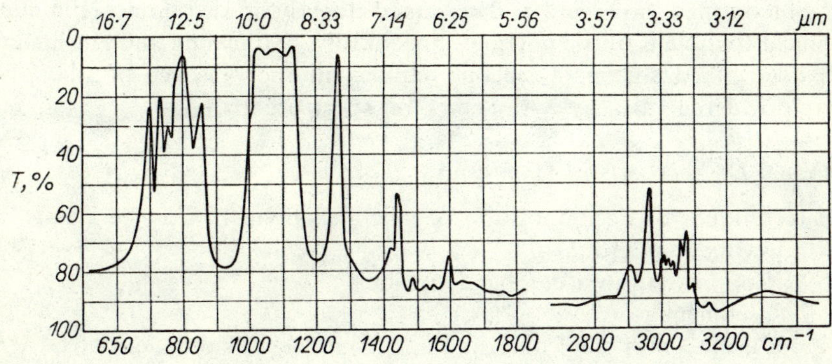

Fig. 20.2 Infrared absorption spectrum of a methylphenyl silicone resin; capillary film

Si—OH groups are detected from the band at 2·78 μm, Si—F from the band at 8·25 μm, Si—H groups from the fairly strong bands in the range 4·5–4·8 μm, Si—O—CH_3 groups at 3·5 and 8·4 μm, Si—O—Et at 8·5 μm, and Si—O—Ph from bands in the range 10·3–10·9 μm.

Much data on the characteristic absorption bands of a variety of substituents present in polyorganosiloxanes are recorded in the Hummel monograph [20] and in papers by Kriegsmann [21,22].

The oligomers formed from pyrolysis of polyorganosiloxanes in an oxygen-free atmosphere can be identified by mass spectrometry [16] which also provides some indication as to the presence of various functional groups and other substituents in polyorganosiloxanes. Similar information can also be obtained by gas chromatography [17] as well as by pyrolysis gas chromatography or gas chromatography following boron trifluoride cleavage [15].

20.3 CHEMICAL COMPOSITION OF THE POLYMER

Chemical characterisation of silicones includes elemental analysis (content of silicon, carbon, hydrogen, and halogens), and the determination of the content of certain functional groups directly bonded to silicon including hydroxyl, hydrogen of Si—H bonds, Si—Si bonds, alkoxyl and aryloxyl groups, acetyl groups, substituents linked to silicon (the R/Si ratio) as well as certain functional groups incorporated in the organic substituents.

Silicon

A variety of methods are available for determination of silicon in organosilicon compounds. The majority of these methods are based on conversion of silicon into its inorganic forms (SiO_2, H_2SiO_3 etc.) followed by determination of the resulting inorganic compounds by methods of quantitative inorganic analysis.

Silicon may, however, be determined directly and without preliminary chemical treatment of the polymer. Specifically, Motoyama and Hashizume [23] determined silicon in silicone rubbers in the presence of silicon as finely powdered silica by X-ray emission spectroscopy.

Polymer Decomposition Techniques

The techniques of decomposition of silicone polymers may be classed in the following four groups.

1. Combustion in a combustion tube
2. Wet ashing
3. Fusion (with solid oxidising agents and/or alkaline agents)
4. Hydrolysis

Combustion in a Combustion Tube

Combustion analysis for quantitative determination of silicon in organic compounds was used for the first time by Rochow [24,25]. Carbon and hydrogen are determined simultaneously by this method. The silicon content is found from the difference in weight of the combustion tube before and after the analysis, as the silicon remains in the tube as SiO_2 after the analysis.

Silicon, carbon and hydrogen were also determined in one operation by Klimova, Korshun, and Bereznitskaya [26,27] by their rapid "empty-tube" combustion analysis procedure. The silicon content was found from the difference in weight of the glass vessel accommodating the unknown prior to and after the analysis. To prevent formation of silicon carbide and to trap the silica fumes produced during the combustion process a catalyst, chromium trioxide supported on asbestos, was placed in the vessel. This procedure also allowed chlorine to be determined in the same run [26].

The combustion methods which enable simultaneous determination of several elements to be made, despite being convenient in terms of economy of time, fail to yield accurate results. This is due to the formation of volatile pyrolysis products which fail to undergo complete combustion either in the vessel or in subsequent sections of the tube.

Wet Ashing

A procedure involving wet decomposition of organosilicon compounds with a mixture of 60% fuming sulphuric acid and fuming nitric acid (density 1·50 g per cm^3), followed by gravimetric determination of silicon as SiO_2, was used by Smith [28]. The same oxidising agent was adopted by Šir and Komers [29]. The silicic acid produced was determined acidimetrically by the formation of K_2SiF_6 in an acidic medium.

Kreshkov *et al.* [30] modified these methods by carrying out the oxidation process in a quartz flask to prevent silicon losses.

Haslam and Willis [6] report a method that consists in a very careful decomposition of the unknown at the lowest temperature possible using fuming sulphuric acid only. After this procedure a 0·3–1·0 g sample is weighed into a tared platinum evaporating dish, the dish is chilled in dry ice, and 2–4 ml of the acid is added. The dish and its contents are allowed to warm up to room temperature, then gently heated over a flame, and finally ignited to constant weight in a muffle furnace at 850°C.

Other oxidants used for this purpose, are a sulphuric acid–potassium permanganate mixture, perchloric acid alone or in mixtures with nitric acid, nitric acid alone or fuming nitric acid with bromine [18].

Fusion

This method for the quantitative analysis of silicon, as well as of other elements, in solid and liquid organosilicon compounds, is based on fusion with sodium carbonate or alkalis and oxidising agents, and is of the widest practical interest [6,7,18].

The commonest oxidant used is sodium peroxide and the analysis is carried out in a Parr micro bomb. To decrease the loss of volatile products of decomposition of silicones, Wetters and Smith [31] treated the samples prior to the alkali fusion with sodium hydroxide in butanol or sodium butoxide solution. A disadvantage of the alkali peroxide fusion method is that the reagent is introduced in very large quantities compared to the sample weight, and the resulting errors are due to the considerable quantity of electrolytes and to impurities present in the reagents. Additional shortcomings are possible contaminations from the bomb metal and problems with adequate sealing of the bomb.

Hydrolysis

Silicon was determined in a range of low molecular weight organosilicon compounds and hydrolysable silicone polymers by Kreshkov *et al.* [32]. These authors used gravimetric, volumetric and photocolorimetric methods to identify silicon in hydrolysates produced by a variety of acid or alkaline agents such as 10% aqueous ammonia, conc. or dilute hydrochloric acid, 40% hydrofluoric acid, and 15% potassium hydroxide solution. The best results were obtained by the volumetric method, in which silicon is first converted into the SiF_6^{2-} ion which is then determined acidimetrically or iodometrically.

In solutions from wet ashing or after hydrolysis, silicon can be determined by a range of gravimetric, colorimetric or other techniques common to inorganic analysis [18,33].

The subject has been covered in numerous recent studies by Kreshkov and his school. Thus, Myshlyaeva and Krasnoshchekov [34] determined fluorosilicic acid formed from the hydrofluoric acid hydrolysis of silicones, using high-frequency titrimetry with benzidine (other similar aromatic amines can also be used) in methanol or acetone solution. Kreshkov *et al.* [35] followed the same method but by potentiometric titration in acetone solution using the platinum–tungsten or platinum–calomel electrode systems. Finally Kreshkov *et al.* [36] determined fluorosilicic acid by precipitating it with benzidine then titrating the complex formed acidimetrically.

The most accurate method was found to be wet oxidation in quartz flasks using fuming sulphuric and nitric acids together [30], followed by gravimetric determination [30,37].

According to Haslam and Willis [6] the best results for the determination of silicon in silicone polymers are achieved by fusion in the Parr bomb as described in Kline's monograph. In the silicate melt thus produced silicon is best determined gravimetrically using 8-hydroxyquinoline [7,38]. This procedure is described below. When the silicon content is below 10% satisfactory results are achieved by a colorimetric technique (likewise described below). This involves the formation of a yellow molybdosilicate complex, which is subsequently reduced with sodium sulphite to a blue heteropoly acid.

Determination of Silicon after Decomposition of the Polymer

Gravimetric Method [38]

REAGENTS

Sodium Peroxide. Analytical grade.

Ammonium Molybdate. Ammonium molybdate (20 g) is dissolved in 80 ml of water and 1–2 ml of 14% ammonia is added.

8-Hydroxyquinoline. 8-Hydroxyquinoline (15 g) is dissolved in 20 ml of dilute (1:1) hydrochloric acid, and diluted with water to 1000 ml. The dilute solution of the reagent is prepared by mixing 200 ml of the 1·5% 8-hydroxyquinoline solution with 50 ml of conc. hydrochloric acid and 750 ml of water.

PROCEDURE

Fusion. Sodium peroxide (2–3 g), placed in a Parr bomb of capacity 25–30 ml, is mixed with 0·05–0·1 g of sucrose, and a quantity of sample (in a gelatin capsule) corresponding to 15–40 mg of Si is added. The bomb is filled with the peroxide, closed, and rapidly heated to red heat. The bomb is cooled by immersion in cold water, then opened. The cover is washed with water and the washings are collected in a 400 ml nickel beaker. The solution in the beaker is made up to a volume of 100–125 ml with water. The open bomb is placed in the beaker and covered with a watch glass, and allowed to stand for some time. When the melt is dissolved, the bomb is removed and washed first with water then with conc. hydrochloric acid. The beaker contents are made up to a volume of 250–275 ml and acidified with conc. hydrochloric acid until nickel hydroxide is completely dissolved. The contents are transferred to a 500 ml volumetric flask and the volume is made up to the mark (for parallel determinations the same quantities of hydrochloric acid should always be added).

Precipitation of the Complex. A 50 ml aliquot of the solution prepared as described above, is transferred with a pipette to a 250 ml volumetric flask, 15 ml of 18% hydrochloric acid is added along with 50 ml of water, and the whole is mixed; 15 ml of 20% ammonium molybdate solution is now added, the flask is stoppered and heated for 10 min at 75±3°C. After cooling 20 ml of 18% hydrochloric acid and 25 ml of 1·5% 8-hydroxyquinoline solution are added, then the flask is stoppered and heated for 10 min at 65±3°C. The contents are cooled and the precipitate is filtered off, washed with the dilute 8-hydroxyquinoline solution, and dried first at 110–120°C for 1 hr, then heated at 500°C for another hour.

A blank should be run separately.

CALCULATION. Silicon content in the polymer $X(\%)$ is found from the formula

$$X = \frac{15{\cdot}8(a-b)}{m}$$

where:

a = weight of the silicon complex precipitate, g
b = weight of the precipitate in the blank, g
m = weight of polymer sample, g
15·8 = conversion factor

Colorimetric Method [39]

REAGENTS

Ammonium Molybdate. Ammonium heptamolybdate (20 g) is dissolved in 180 ml of water in a nickel beaker. The solution must be prepared immediately before the analysis.

Sodium Sulphite. A 17% solution.

Silicic Acid Reference Solution.

PROCEDURE. The polymer is fused with sodium peroxide, then brought into solution as described in the gravimetric method. The clear solution is neutralised with hydrochloric acid, transferred to a 1000 ml volumetric flask and made up to the mark with water.

An aliquot of the solution, containing not more than 0·125 mg of silicon, is transferred to a 100 ml measuring flask and made up to a volume of 50 ml; 2 ml of 1*N* hydrochloric acid and 2 ml of ammonium molybdate solution are added, and the mixture allowed to stand for exactly 2 min. Next, 5 ml of sodium sulphite is added, the volume is made up to the mark, and exactly 7 min later absorbance is measured at 715 nm against a reference not containing silicon but of the same composition as the solution of the unknown. The silicon content, in mg, is found from a calibration curve prepared from a silicic acid reference solution used in such quantities as to give concentration points on the curve corresponding to 0·02, 0·04, 0·06, 0·08, 0·10 and 0·12 mg of silicon.

CALCULATION. Silicon content $X(\%)$ in the polymer is found from the formula

$$X = \frac{100a}{vm}$$

where:

a = silicon content in the coloured solution found from the calibration curve, mg,

v = the volume (ml) of the aliquot of the solution withdrawn for use in the colour reaction,

m = weight of polymer sample, g.

Carbon and Hydrogen

In the determination of carbon and hydrogen in organosilicon compounds by conventional combustion analysis certain products may be formed which exhibit a remarkably high thermal stability and resistance to oxidation. These include silicon carbide SiC and certain volatile silanes. These products fail to undergo oxidation under the conditions of the combustion analysis. Silicon carbide is known not to react with oxygen even at temperatures as high as 1100°C.

The formation of non-combustible compounds is the main source of negative errors in the analysis of organosilicon compounds by this method. Another effect, which gives rise to positive errors, is the formation of finely divided silica which, entrained in the oxygen stream, may reach the absorption vessels.

The elimination of these sources of error was the goal of a wide variety of investigations concerned with the quantitative analysis of organosilicon compounds. According to the procedures reported in the monograph by Pregl and Roth [40], the complete oxidation of carbon and hydrogen present in such compounds in the conventional method of combustion analysis can be accomplished when the unknown is mixed with cryolite and red lead oxide Pb_3O_4 (1:5) or with vanadium pentoxide. Rochow [24, 25], the author of the first method of elemental analysis of organosilicon compounds, prevented the formation of silicon carbide by first evaporating or gently thermally decomposing the compounds under test. The volatile products formed were subsequently oxidised on a platinum catalyst heated to 850°C.

Penetration of silica into the absorption vessels was prevented [41] by placing a layer of powdered pumice in the combustion tube. For the same purpose [42] a silver-asbestos wool plug was provided in the tube. As a catalyst for the oxidation of volatile products the pyrolysate of silver permanganate was employed ($AgMnO_4$ ignited at 500°C).

The "empty tube" technique of combustion analysis may also be used to advantage [26]. To prevent the formation of compounds that undergo combustion with difficulty, the unknowns were placed in a quartz test tube, which was afterwards filled with cupric oxide. Tetraarylsilanes and unsaturated compounds were found to be most prone to form carbides. The formation of compounds difficult to oxidise may also be prevented by using CrO_3 supported on asbestos as catalyst. If the material under analysis contains less than 3% hydrogen, water added in a certain quantity prevents formation of silicon carbide. The method consists in decomposing the unknown in a quartz tube packed with the catalyst and placed in the combustion tube [26].

Asbestos present in the tube containing the unknown traps silica, thus making it practicable to determine silicon along with carbon and hydrogen.

In all of these methods emphasis is placed on a slow and gentle heating at the first stage, which makes the procedures laborious.

In an attempt to develop a rapid method suitable for a wide variety of organosilicon compounds, Gawargious and Macdonald [43] examined the available procedures and tried a number of ways to prevent the formation of those compounds that undergo combustion with difficulty and to avoid the production of finely divided silica. Combinations of a range of catalysts were found to be unsatisfactory. Among these are tungsten trioxide, vanadium pentoxide, chromic oxide, potassium dichromate or asbestos used in mixtures with the sample to be thermally oxidised. Nor are these disadvantages overcome with such measures as an asbestos plug provided between the sample and the furnace, increased temperature, a platinum catalyst, or prior pyrolysis of the unknown in a tube filled with CrO_3 supported on asbestos. Satisfactory results could not be obtained with cobalto-cobaltic oxide [44] (by the Večeřa method). In the opinion of the authors the procedures available in the literature are perhaps suitable only for certain compounds. However, satisfactory results were achieved for a wide variety of organosilicon compounds, when a granulated magnesia bed was used as catalyst at 900°C [43]. This packing has been used successfully in a rapid quantitative analysis of carbon and hydrogen in fluorine-containing compounds [45]. According to the authors cited, magnesia efficiently combines with silica to form magnesium silicate and under the conditions of a rapid combustion process mechanical collection of silica is not possible. This method is described below.

APPARATUS. A quartz tube 36 cm long and 9 mm in diameter, drawn at one end to form a capillary and equipped with a side arm (oxygen feed) at the other end, is packed in the following way. The tube is packed in the order (from the capillary end): rolled silver gauze 7 cm long, 1 cm quartz-wool plug, granulated magnesia bed 10 cm long, finally another 1 cm quartz-wool plug. The tube is positioned in the furnace so that the magnesia bed lies in the middle of the furnace [43].

PREPARATION OF THE CATALYST. Magnesia is mixed with water in the ratio 1:3, and the resulting paste is dried at 120°C, then crushed and screened through a sieve of 0·35–1·2 mm mesh. The granules thus obtained are ignited at 800°C for 2 hr [23].

PROCEDURE. The combustion furnace (ca. 15 cm long) is previously heated to 900°C. A sample of 3–6 mg is weighed out in a platinum boat and placed in the tube 3–4 mm from the furnace. The absorption vessels for carbon dioxide and water should be attached to the combustion train prior to the placing of the sample in the tube. The oxygen flow rate is adjusted to 12 ml per min. The preliminary combustion of a sample is carried out for about 7 min by heating the boat with the low flame of a Bunsen burner. The part of the tube at the site of the boat is then heated for 4 min with a strong flame, and the oxygen stream is passed for another 5 min to remove the combustion products completely. Finally the absorption vessels are detached from the train and weighed.

All the foregoing procedures for the determination of carbon and hydrogen in silicones involve combustion in a tube in an oxygen stream. Recently Celon and Bresadola [46] have reported on the technique of the quantitative analysis of these elements in silicon- and boron-containing polymers, in which the combustion is carried out in a stationary oxygen atmosphere at 1060°C in the presence of an oxide catalyst. The combustion products, carbon dioxide and water, are determined by gas chromatography.

Chlorine

Chlorine is occasionally encountered in organic substituents of silicone polymers. It is more usual to find chlorine linked to silicone in these polymers. This chlorine originates from chlorosilanes which did not undergo complete hydrolysis in the polycondensation process. Hence, chlorine may be present in silicones derived from chlorosilanes but only in minor quantities.

The total chlorine content in silicone polymers, that is chlorine both linked to silicon and other atoms, is commonly determined by fusion with sodium peroxide in the Parr micro bomb [7]. The fusion is performed in the usual manner as described above. The resulting melt is dissolved in water, acidified with nitric acid, and the chloride ions present in the solution are titrated potentiometrically with a standard silver nitrate solution.

Instead of the fusion method another procedure may be used, in which the unknown is heated under reflux with sodium butoxide in butanol. The reaction mixture is cooled and acidified with nitric acid, and the chlorides present are titrated with silver nitrate.

If the organosilicon compound under analysis contains chlorine linked directly to silicon only, its content may be determined by methods based on hydrolytic splitting with water or alkalis at room temperature. According to Šir and Komers [47] the process is run with 1–2*N* sodium butoxide in *n*-butanol. The reaction mixture is acidified with nitric acid and titrated potentiometrically with 0·05*N* silver nitrate using a silver–mercurous sulphate (Hg_2SO_4) electrode system. Hydrolysis of chlorine attached directly to silicon may also be performed using a (1:1) ethylene glycol-water mixture, and the resulting hydrogen chloride determined by neutralisation [48].

Chlorine can also be determined by combustion in a combustion tube—sometimes together with carbon, hydrogen, and silicon—or in a flask by the Schöniger method [18].

Hydrogen Bonded to Silicon (Silane Hydrogen)

The SiH grouping present in silicone polymers may originate from impurities formed in the synthesis of chlorosilanes, or this may be intentionally introduced into these polymers (from the reactants made by the Grignard method) to manufacture cold-setting materials. Similar materials find considerable use in industry as water-repellent finishes for textiles and in the papermaking industry.

Chemical Methods

The SiH grouping readily reacts with active hydrogen in water or alcohols in the presence of alkalis:

$$\gt SiH + HOR \xrightarrow{NaOH} H_2\uparrow + ROSi \lt$$

$$\gt SiH + HOH \xrightarrow{NaOH} H_2\uparrow + HOSi \lt$$

Gaseous hydrogen is produced in these reactions, the volume of which can be measured and the SiH content calculated.

The available methods for the determination of silane hydrogen (SiH) utilise both alkoxides, specifically sodium butoxide [7], and aqueous sodium hydroxide solution [18,49–51].

However the Si—Si groupings also react readily with alkaline hydrolytic agents and in much the same way as SiH groups, with evolution of hydrogen, following the reaction:

$$\rangle Si—Si\langle + 2HOR \xrightarrow{NaOH} 2SiOR + H_2\uparrow$$

Thus only the total content of SiH and Si—Si groups can be determined by the methods based on the foregoing reactions [7].

The SiH grouping can be oxidised quantitatively to the Si—OH group by treatment with mercuric chloride. In aqueous solution this reaction proceeds as follows:

$$\rangle SiH + 2HgCl_2 + H_2O \rightarrow \rangle SiOH + Hg_2Cl_2 + 2HCl$$

This reaction underlies an analytical method by which hydrogen linked to silicon can be determined when present in concentrations down to 0·002% [52,53]. The determination is carried out in methyl ethyl ketone. In such a chemically inert solvent the reaction proceeds as follows:

$$\rangle SiH + 2HgCl_2 \rightarrow \rangle SiCl + Hg_2Cl_2 + HCl$$

The precipitated calomel is filtered off and weighed. The determination may also be performed by an acidimetric method. Then, instead of mercuric chloride methanolic mercuric acetate reagent is used. Two equivalents of acetic acid per one SiH group are formed in the reaction:

$$\rangle SiH + 2Hg(OOCCH_3)_2 + CH_3OH \rightarrow \rangle SiOCH_3 + Hg_2(OOCCH_3)_2 + 2CH_3COOH$$

The determination is carried out in a similar manner as in the case of olefinic bonds.

C=C bonds interfere in determinations by the methods based on the reaction with mercuric salts as these bonds also react with the reagent. At the same time, Si—Si groupings are also determined by these methods.

Alternatively the silane hydrogen may be determined by bromination [54,55]:

$$\rangle SiH + Br_2 \rightarrow \rangle SiBr + HBr$$

Excess bromine is determined iodometrically. For the determination of SiH groups Harzdorf [56] recommends a method based on the reaction with *N*-bromosuccinimide.

Instrumental Methods

Electroanalytical Methods

Kreshkov *et al.* [57] developed a variant of the method based on the reaction with mercuric chloride, in which this reagent was used as titrant in an amperometric titration. The titration is carried out in the presence of lithium chloride supporting electrolyte in methanol-benzene solution. A polarograph was employed in the determination.

Čermák and Dostál [58] worked out a polarographic technique for the determination of SiH groups based on the same reaction. The principle of this technique consists in converting the excess unreduced Hg^{2+} ions into $[HgI_4]^{2-}$ ions, followed by polarographic determination of these ions by the procedure described by Schwarz [59]. In this reaction the complex iodide ions are also formed with univalent mercury. The analysis is carried out at pH 4·9–5·5, as under these conditions the chemical stability of SiH groups is greatest.

By this method SiH groups can be determined in the presence of Si—Si bonds. Under the reaction conditions no addition of mercuric chloride to C=C bonds occurs, therefore olefins do not interfere with the determination.

REAGENTS

Mercuric Chloride. A 0·2*N* solution in 96% ethanol.

Potassium Iodide. Potassium iodide (20 g) and sodium acetate monohydrate (35 g) are dissolved in 250 ml of water.

PROCEDURE. A 50–200 mg sample of the polymer is weighed into a 10 ml volumetric flask and dissolved in a suitable solvent (methyl ethyl ketone, benzene, ethanol). The volume is made up to the mark, and an aliquot of this solution (0·1–1·0 ml) is transferred to a 25 ml volumetric flask containing 5 ml of mercuric chloride. After 1 hr, 5 ml of potassium iodide solution is added and it is acidified with dilute (1:5) acetic acid. 1 ml of 1*N* sodium thiosulphate solution is then added, followed by 1 ml of 0·5% gelatin solution. The flask is filled to the mark and the contents are thoroughly mixed. An aliquot of this solution is transferred to a polarographic cell, a nitrogen stream is bubbled through the solution for 5 min, and a polarogram is taken recording the polarographic wave height due to tetraiodomercurate ions.

Infrared Spectroscopy

SiH groups may be determined by infrared spectrophotometry using the distinct, fairly strong bands in the range 4·5–4·8 μm [6,60]. No other groups present in the majority of the silicone polymers absorb in this range. The method is suitable for the determination of these groups in the presence of Si—Si bonds.

Other Methods

According to the phototurbidimetric procedure [52,53] which is also based on the reaction with mercuric chloride the degree of turbidity is measured. Turbidity results from mercurous chloride being formed as a result of the reaction of SiH groups. Since this procedure is suitable for the determination of the SiH hydrogen down to a concentration of 5×10^{-5}% with a relative error not higher than 10%, it can be used to advantage when these groups are present in very low concentrations.

A photocolorimetric method [52] for determination of SiH groups involves oxidation of these groups with potassium permanganate. The colour of potassium permanganate solution decreases in intensity as a result of the reaction.

Hydroxyl Groups Bonded to Silicon

The silanol group content in silicone polymers constitutes an important characteristic of their useful properties, as this is a factor controlling the reactivity of these polymers. Quite understandably therefore, the question of determination of these groups has been given considerable attention.

For the determination of hydroxyl groups bonded to silicon several methods are known which differ principally in their chemical nature. In broad terms they may be classed into non-specific and specific methods for SiOH groups.

Of the non-specific methods the most widely recognised is the procedure based on the reaction with methylmagnesium iodide. There are several variants of this method, which differ in the solvent (ethyl ether, *n*-butyl ether, pyridine) or in the apparatus and the inert gas used (nitrogen, methane) [18]. In the most exact of these methods, developed by Guenther [61] methane is used as the inert gas and *n*-butyl ether as the reaction medium. This method was later improved by Lees and Lobeck [62].

Similar to the Tserevitinov method is the procedure based on reaction with butyllithium [63]. Ranking in the same group of methods are those based on the reaction of the active hydrogen of silanol groups with agents such as metallic sodium [18] and lithium aluminium hydride [64], which evolve hydrogen. In the former method the volume of gas evolved is measured and in the latter the rise in pressure in the apparatus.

Damm and Noll [65] established that silanols react with phenyl isocyanate in the same way as alcohols and developed a method for determination of silanol groups in polyorganosiloxanes, based on the reaction:

$$2HO{-}Si{\lessgtr} + 2C_6H_5NCO \rightarrow {\gtrless}Si{-}O{-}Si{\lessgtr} + C_6H_5NHCONHC_6H_5 + CO_2$$

The unreacted excess of phenyl isocyanate is fixed with an aliphatic amine, the excess of which is titrated in turn with hydrochloric acid either visually, using bromothymol blue as indicator, or potentiometrically.

Finally Kellum and Uglum [66] determined low molecular silanols and silanol groups in silicones by direct titration with lithium aluminium dibutyl amide.

All the non-specific methods of quantitative analysis of silanol groups discussed here have in common the feature that the reagents used in the determinations also react with other compounds containing active hydrogen, e.g. alcohols, and water which is both formed in silanol polycondensation and is also present in solvents or as atmospheric moisture.

Among the methods specific for the SiOH group is one based on reaction with the Karl Fischer reagent, and another involving condensation of silanol groups. The former is based on the reaction with methanol present in the Karl Fischer reagent:

$$\rangle SiOH + CH_3OH \rightarrow \rangle SiOCH_3 + H_2O$$

The water formed in a quantity of 1 mole per 1 equivalent of silanol groups reacts with the Karl Fischer reagent. The excess reagent is titrated visually or potentiometrically in the usual manner. Originally this method was used by Grubb [67] for the determination of SiOH groups in silanol monomers. Noll *et al.* [65,68] successfully analysed silicones, using this procedure.

The method based on the condensation of silanol groups as in the following

$$R_xSi(OH)_{4-x} \rightarrow R_xSiO_{\frac{4-x}{2}} + \frac{4-x}{2}\,H_2O$$

is the best of all the conventional methods known for determination of these groups. In all other methods this condensation is also likely to occur.

For the reaction to proceed quantitatively older procedures involve the use of such catalysts as trace alkalis, acids or iodine. The water produced is continuously removed by azeotropic distillation with toluene or benzene, using a Dean–Stark trap. The quantity of water formed is either measured as the volume collected in the trap or titrated with the Karl Fischer reagent.

However, in the case of silicone resins the reaction fails to proceed to completion in the presence of those catalysts. Recently Smith and Kellum [69] have succeeded in developing a new catalyst composed of boron trifluoride, acetic acid and pyridine. With this catalyst the determination time is substantially shortened and the results obtained are accurate and reproducible in the determination of a wide range of silicone polymers.

REAGENTS. The boron trifluoride is prepared by dissolving 42 ml of ethereal BF_3 complex in 208 ml of glacial acetic acid; 10 ml of this mixture contains 0·012 mole of boron trifluoride and 0·14 mole of acetic acid.

PROCEDURE. A sample of 10–20 g of silicone fluid or resin, or 1–5 g of a low molecular silanol, is weighed into a 200 ml round-bottomed flask. Solid materials are dissolved in an equal amount of xylene. The contents of the flask are treated with 10 ml of dry pyridine (pyridine is not added in the analysis of low molecular silanols), then 10 ml of the catalyst is introduced and the whole is thoroughly mixed. After 5 min 50 ml of toluene is added and a few chips of silicon carbide to facilitate boiling. The flask is attached to a Dean–Stark apparatus, a Drierite guard-tube is attached to the top of the condenser, the contents are immediately heated to boiling and distillation is carried out for 30 min. On cooling, the contents of the Dean–Stark trap are transferred to a 100 ml volumetric flask containing anhydrous ethanol. The condenser and the receiver are flushed with three small portions of ethanol and the washings are collected in the volumetric flask. The volume in the flask is made up to the mark. A 10 ml aliquot of this solution is introduced to a volume of methanol, previously titrated to the end point with the Karl Fischer reagent, and titrated in the usual manner with biamperometric detection of the end point.

The silanol content in the sample is found in terms of the equation of the condensation reaction of these groups given above. In the presence of larger quantities of SiH groups a correction should be made, since these groups react with silanol groups as in the following reaction

$$\rangle SiH + HO{-}Si\langle \rightarrow \uparrow H_2 + \rangle Si{-}O{-}Si\langle$$

without formation of water.

The necessary correction to be added to the SiOH figure is established from previous determination of the SiH content by one of the methods described earlier.

As is seen from the foregoing descriptions all the conventional methods, including the specific ones, require a parallel determination of water content in the material under analysis. This is of particular significance in the case of silicones of high viscosity, from which the residual water from the condensation process is difficult to remove, and which at the same time contain SiOH groups in minor quantities.

Instrumental Methods

Infrared Spectroscopy

The silicon atom has no clear effect on the position of the infrared absorption band of the OH group. The spectra of dilute silanol solutions in carbon tetrachloride exhibit the band at 3690 cm^{-1} characteristic of the free hydroxyl group [69].

In the spectra of concentrated solutions of these compounds a broad strong band appears at ca. 3 μm [65,70], due to the group participating in hydrogen bonding. In this situation the SiOH group can hardly be identified by infrared absorption spectroscopy. The Si—O stretching band occurs as a broad band between 11·2 and 12·0 μm, but it may be overlapped by the Si—Me bands [6].

NMR Spectroscopy

Hampton *et al.* [71] examined NMR spectra of silanols in dimethylsulphoxide. In this solvent at concentrations below 20 mole % the lines due to the SiOH protons are clearly separated from the signals of the OH groups attached to carbon, and from those of the protons of water. Thus, silanol groups can be quantitatively analysed in the presence of alcohols and water.

Alkoxyl and Aryloxyl Groups

Alkoxyl groups are present in certain water-repellent compounds and silicone resins, as well as in certain grades of silicone rubbers. In these polymers the organic structural units usually contain up to four carbon atoms.

The commonest method used for the determination of alkoxyl groups in silicone polymers is the Zeisel method but the conventional technique, such as is followed for alkoxyl groups linked to carbon, yields too low results [33]. A satisfactory accuracy can be achieved by a modified procedure and altered temperature conditions. Nessonova and Pogosyants [72] in their determination of ethoxyl groups in organosilicon compounds made use of a mixture composed of hydrogen iodide, phenol, propionic acid and red phosphorus. Bournique [73] used fuming hydrogen iodide in place of the constant boiling acid. Syavtsillo and Bondarevskaya [74], who employed this method for the analysis of butoxyl groups, gradually heated the sample with hydriodic acid to 100°C to effect complete removal of butyl iodide distilling from the reaction mixture. The alkyl iodide from the reaction is usually determined by bromine oxidation combined with iodometric titration of the iodic acid produced [33,75]. Another procedure involves absorption of alkyl iodide in pyridine and titration with tetrabutylammonium hydroxide [76].

The Zeisel method, despite its being widely used in analysis, is laborious and for higher alkoxyl groups yields too low results on account of the poorer volatility of the corresponding alkyl iodides.

Alkoxyl groups bonded to silicon are fairly often determined by the method in which the unknown is hydrolytically split [77]. The alcohols formed are then analysed by standard techniques [1,78]. However, the method is laborious and inaccurate because of too low a hydrolysis rate in the final stages, and a prolonged process for isolation of the alcohol from the reaction mixture.

Aryloxyl groups are usually determined by bromination. These groups are first cleaved with hydrogen chloride in glacial acetic acid, then the phenol produced is brominated with bromide-bromate mixture:

$$R_3SiOC_6H_5 + HCl \xrightarrow{CH_3COOH} R_3SiCl + C_6H_5OH$$

$$KBrO_3 + 5KBr + 6HCl \longrightarrow 3Br_2 + 6KCl + 3H_2O$$

$$C_6H_5OH + 3Br_2 \longrightarrow C_6H_2Br_3OH + 3HBr$$

The excess reagent is determined iodometrically.

When the phenyl groups are substituted bromine enters the available *ortho* and *para* positions. The number of these available positions may be less than three, which should be allowed for in calculations.

Magnuson [79], and independently Dostál and Čermák [80] developed a versatile procedure by which both aryloxyl groups and higher alkoxyl groups can be quantitatively determined. The procedure is based on the reaction with acetic anhydride in ethyl acetate solution in the presence of perchloric acid as a catalyst:

$$\gt SiOR + Ac_2O \rightarrow \gt SiOAc + AcOR$$

As a result of the reaction the acetoxysilane esters and alkyl or aryl acetates are produced. The unreacted excess of acetic anhydride is titrated with standard alkali solution. Acetoxysilane does not interfere here, as it hydrolyses quantitatively to acetic acid and silanol:

$$\gt SiOAc + H_2O \rightarrow \gt SiOH + AcOH$$

In the determination aromatic hydrocarbons such as benzene, toluene, xylene, and others do not interfere [80]. Alcohols and phenols acetylate quantitatively, but these can be determined along with alkoxy- and aryloxysilanes by one of the colorimetric procedures described below. Organic esters and ethers such as dioxan, tetrahydrofuran or anisole also partly acetylate. For this reason aldehydes and ketones (acetaldehyde, formaldehyde, acetone) interfere.

INSTRUMENTAL METHODS

Gas Chromatography

Alkoxyl groups, including vinyl groups, are determined in polyorganosiloxanes by fusion with alkalis by Hanson and Smith [81]. The liberated alcohols (and ethylene) are determined by gas chromatography.

Colorimetry

The colorimetric determination of alkoxyl groups in organosilicon compounds as described by Kreshkov and Bork [33] consists in cleavage of these groups with saturated sodium hydroxide solution to form the corresponding alcohol. The alcohol is distilled with benzene and determined with methylene blue. This dye readily dissolves in alcohol and turns the solution blue, but it is insoluble in benzene. The procedure allows methoxyl and ethoxyl groups to be determined when present in a concentration of at least 0·05%.

A similar principle underlies the method for determination of ethoxyl groups developed by Sprung and Guenther [82] with this difference, that the alcohol resulting from hydrolysis is determined by using a colour reaction with ammonium ceric nitrate, $(NH_4)_2Ce(NO_3)_6$.

Ultraviolet Spectroscopy

Phenoxyl groups, in contrast to alkoxyl groups which do not have a benzene ring, absorb radiation in the ultraviolet range of the spectrum. This distinction serves as a basis for a spectrophotometric method for determination of aryloxyl groups in polyorganosiloxanes. Errors in the determination do not exceed 0·1 mole % [33].

Infrared Spectroscopy

Alkoxyl groups linked to silicon exhibit characteristic bands in the range between 8·4 and 10·6 μm. Various alkoxyl groups absorb individually at other places in the infrared range [18]. Methoxyl groups in silicones can be identified by bands at 3·55 μm and 8·4 μm, whereas ethoxyl groups are identified by bands at ~8·5 μm and ~10·5 μm [6].

Brown and Smith [83] analysed methoxyl groups in silicones quantitatively by the 8·4 μm band, while Sidnev *et al.* [84] determined isopropoxyl and butoxyl groups by measuring absorbance at 1380 and 1450 cm^{-1}, respectively.

Aryloxysilanes exhibit characteristic absorption bands in the range 10·3–10·9 μm.

Hydrocarbon Substituents Bonded to Silicon

The ratio of the hydrocarbon substituent content to silicon content, R/Si, is a major index in the characterisation of silicone polymers, as this reflects the functionality of the structural units of these polymers.

In order to determine the R/Si ratio of siloxanes substituted only with aliphatic substituents, the method reported by Booth and Freedman [85] can be used. It is based on hydrofluoric acid cleavage in conc. sulphuric acid:

$$[R_2SiO]_x + 2xHF \xrightarrow{H_2SO_4} xR_2SiF_2 + xH_2O$$

Under suitable conditions the hydrocarbon substituents may be cleaved when conc. sulphuric and nitric acids are used.

The reactions mentioned are not, however, quantitative and can merely be employed for the qualitative characterisation of the materials under analysis, and for establishing the type of hydrocarbon substituents [7]. Of the methods available in the literature only a few are suitable for quantitative analysis of alkyl and aryl substituents and are outlined below.

The phenyl group content can be determined by bromination [75]. Bromine splits the Si—Ph bond

$$\geqslant Si{-}Ph + Br_2 \rightarrow \geqslant SiBr + PhBr$$

As opposed to the reaction with SiH groups the process runs quantitatively only at elevated temperatures, and thus these two groupings can be determined even when both are present in the same sample. The unreacted excess bromine, as already mentioned, is determined iodimetrically.

Phenyl groups in phenylchlorosilane and phenylmethylsiloxane can be analysed by ethylation with ethyl bromide in presence of anhydrous aluminium chloride [86], whereas alkyls (methyl, ethyl) in polyalkylsiloxane are determined [87] by cleavage with sodium hydroxide to yield the corresponding gaseous hydrocarbons which are measured by volume.

INSTRUMENTAL METHODS

The average value of the R/Si ratio can be found from the values of certain physical constants [1]. For instance the density of liquid polymethylsiloxanes unambiguously determines the R/Si ratio, because the density of these materials is independent of both the distribution of molecular weight in the polymer and the arrangement of structural units of diverse functionality. Alternatively this ratio may be found from the value of the molar refraction.

Ultraviolet Spectroscopy

Utraviolet spectroscopy is suitable for quantitative analysis of phenyl groups which, as opposed to alkyls, absorb in that spectral range. According to Uriu and Hakamada [88] total phenyl groups can be found by measuring the absorbance at 240 nm. Molar absorptivity at 264 nm for polyorganosiloxanes of high molecular weight was found to have a constant value for different polymers, equal to 335 l. cm^{-1} $mole^{-1}$. Phenyl substituents in methylphenylsilanes were determined by measuring absorbance at 260, 266 and 272 nm (in chloroform) [89].

Infrared Spectroscopy

For the determination of the Si—CH_3 groups in polymeric silicones, use is most often made of the band at 7·9–8·0 μm, which is well separated from other bands [90,91].

Phenyl groups in these polymers are detected by the band at 7·0 μm and determined by the bands of the —Si—Ph system, at 7·0, 19·5 and 20·6 μm, and the bands due to the benzene ring at 1595 cm^{-1} [90] and at 14·36 μm [91].

For the determination of the ratio

$$\frac{Ph_2Si}{PhSi + Ph_2Si}$$

advantage was taken of the phenyl bands in the range of 19–22 μm [92].

Determination of alkyl groups in presence of phenyl groups (that is, the R/Si ratio) is difficult. Grant and Smith [93] determined the ratio of phenyl to methyl groups by measuring the absorbance of bands at 3·2 and 3·4 μm. Lady *et al.* [94] used the absorbance ratio at 7·0 and ca. 8·0 μm for the same purpose. The latter procedure has the advantage of being characteristic of silicones, but here the measurements should be made in solution, as the film technique (cast on sodium chloride) leads to considerable errors unless the film thickness is strictly controlled.

The method based on the absorbance ratio at 3·2 and 3·4 μm is far more convenient, provided solvents are absent, as non-uniformity in the film thickness affects the measurement accuracy to a much lesser extent, due to the relatively close position of these bands.

If the 3·2 and 3·4 μm bands overlap for instrumental reasons, the methyl and phenyl groups are best determined by means of the bands at 8·0 and 14·36 μm respectively [91].

Olefinic Double Bonds

The C=C olefinic bonds in organosilicon compounds in the absence of SiH groups can be determined by methods based on the mercury salt addition reaction or on the reaction with Wijs reagent (see p. 51). The latter procedure is most used.

Instrumental Methods

Vinyl groups in silicone rubbers are determined by the method based on the reaction with phosphorus pentoxide and water [95,96].

$$2H_2C{=}CHSiO_{3/2} + H_3PO_4 \rightarrow 2C_2H_4 + 2SiO_2 + HPO_3$$

or by the procedure involving heating with 90% sulphuric acid [97]. The ethylene produced is determined by gas chromatography.

Vinyl and allyl groups may also be determined by infrared spectrophotometry [98]. The absorbance is measured in the bands at 980–940 and 3080–3020 cm^{-1}.

20.4 ADDITIVES AND CONTAMINANTS

WATER

Water present in silicone polymers is commonly determined by titration with the Karl Fischer reagent. As the silanol groups would interfere here, this analysis is best carried out using the procedure developed by Smith and Kellum who recommend the use of higher aliphatic alcohols as the reaction medium (see pp. 57, 447).

TRACE METALS

Fujiwara and Narasaki [99] developed a method for determination of trace quantities of iron and aluminium in silicone resins. The sample to be analysed was first wet-ashed by heating with conc. sulphuric acid in a Kjeldahl flask, with subsequent determination of the metals present by colorimetric methods: iron in the colour reaction with 2,2′-bipyridyl, and aluminium with 8-hydroxyquinoline. Dostál *et al.* [100] determined heavy metal cations e.g. Co, Fe, Mn, and Cu, in silicone resins directly, but Pb and Zn indirectly, by means of a colour reaction with piperidinium pentamethylenedithiocarbamate.

ACID NUMBER

The acid number in silicone polymers is determined by titration of an aqueous extract of the sample under analysis with standard potassium hydroxide solution.

SILICONES ON FABRICS

Silicone resins used as finishes in glass cloth, glass laminates, as well as in moulding powders are determined by extraction with diethylamine or piperidine [101]. In the analysis of other silicone-treated textiles the samples are first extracted with a suitable solvent such as toulene, benzene, carbon tetrachloride, propyl alcohol or tetrahydrofuran. Silicone resins, which contain SiH or SiOH groups, can only be extracted if they are uncured. For a cured resin only minor quantities of the plastic material can be extracted. However, that amount is sufficient if the analysis is made by infrared spectrophotometry. Best extractions are achieved with the amines specified above.

References

1. Bažant, V., Chvalkovský, V., Rathousky, J., *Organosilicon Compounds*, Czech. Acad. Sci., Prague, 1965.
2. Noll, W., *Chemie und Technologie der Silicone*, Verlag Chemie, Weinheim, 1968.
3. Rościszewski, P., *Zastosowanie silikonów* (*Application of Silicones*), WNT, Warsaw, 1964.
4. Burkhard, C. A., Rochow, E. G., Booth, H. S., Hartt, J., *Chem. Revs.*, **41**, 97 (1947).
5. Flood, E. A., *J. Am. Chem. Soc.*, **55**, 1735 (1933).
6. Haslam, J., Willis, H. A., *Identification and Analysis of Plastics*, 1st Ed., Iliffe Books, London, 1965.
7. McHard, J., A., in *High Polymers, Vol. XII, Analytical Chemistry of Polymers*, Kline, G. M., (Ed), *Part I, Analysis of Monomers and Polymeric Materials*, Interscience, New York, 1959.
8. Noll, W., *Naturwissenschaften*, **49**, 505 (1962).
9. Thompson, H. W., *J. Chem. Soc.*, **1947**, 289.
10. Wright, N., Hunter, M. J., *J. Am. Chem. Soc.*, **69**, 803 (1947).
11. Wurst, M., *Coll. Czech. Chem. Comm.* **29**, 1458 (1964).
12. Clark, H. A., Gordon, A. F., Young, C. W., Hunter, M. J., *J. Am. Chem. Soc.*, **73**, 3798 (1951).
13. Preisler, L., *Z. Anal. Chem.*, **240**, 389 (1968).
14. Luskina, B. M., Turkel'taub, G. N., Syavtsillo, S. V., *Zavodsk. Lab.*, **33**, 1496 (1967).
15. Franc, J., Ceeová, J., *Coll. Czech. Chem. Comm.*, **33**, 1570 (1968).
16. Madorsky, S. L., Strauss, S., *Modern Plastics*, **38**, [6], 134, 207 (1961).
17. Franc, J., Plaček, K., Mikeš, F., *Coll. Czech. Chem. Comm.*, **32**, 2242 (1967).
18. Kreshkov, A. P., Bork, V. A., Bondarevskaya, E. A., Myshlyaeva, L. V., Syavtsillo, S. V., Shemyatenkova, V. T., *Prakticheskoe rukovodstvo po analizu monomernykh i polimernykh kremniiorganicheskikh soedinenii* (*Analysis of Organosilicon Monomers and Polymers*), Goskhimizdat, Moscow, 1962.
19. Kreshkov, A. P., Bork, V. A., *Zh. Analit. Khim.*, **6**, 78 (1951).
20. Hummel, D. O., Scholl, F., *Atlas der Kunststoff-Analyse*, Carl Hanser Verlag, Munich; Verlag Chemie, Weinheim, 1968.
21. Kriegsmann, H., *Z. Elektrochem.*, **64**, 541, 848 (1960).
22. Kriegsmann, H., *ibid.*, **65**, 336, 342 (1961).
23. Motoyama, M., Hashizume, G., *Kogyo Kagaku Zasshi*, **74**, 88 (1971).
24. Rochow, E. G., *An Introduction to the Chemistry of the Silicones*, Wiley, New York, 1951.
25. Rochow, E. G., Gilliam, W. F., *J. Am. Chem. Soc.*, **63**, 798 (1941); **70**, 2170 (1948).

26. Klimova, V. A., Korshun, M. O., Bereznitskaya, E. G., *Dokl. Akad. Nauk SSSR*, **84**, 1175 (1952); **96**, 81 (1954).
27. Klimova, V. A., Korshun, M. O., Bereznitskaya, E. G., *Zh. Analit. Khim.*, **11**, 223 (1956).
28. Smith, B., *Acta Chem. Scand.*, **11**, 579 (1957).
29. Šir, Z., Komers, R., *Chem. Listy*, **50**, 88 (1956).
30. Kreshkov, A. P., Myshlyaeva, L. V., Krasnoshchekov, V. V., *Plast. Massy*, **1964**, [1], 65.
31. Wetters, J. H., Smith, R. C., *Anal. Chem.*, **41**, 379 (1969).
32. Kreshkov, A. P., Myshlyaeva, L. V., Krasnoshchekov, V. V., *Plast. Massy*, **1962**, [12], 51.
33. Kreshkov, A. P., Bork, V. A., *Usp. Khim.*, **28**, 5576 (1959).
34. Myshlyaeva, L. V., Krasnoshchekov, V. V., *Tr. Mosk. Khim-Tekhnol. Inst. im. D. I. Mendeleeva*, **1963**, 178.
35. Kreshkov, A. P., Myshlyaeva, L. V., Khachaturyan, O. V., Krasnoshchekov, V. V., *Izv. Vyssh. Ucheb. Zavedenii, Khim. i Khim. Tekhnol.*, **7**, 198 (1964).
36. Kreshkov, A. P., Myshlyaeva, L. V., Krasnoshchekov, V. V., *Zavodsk. Lab.*, **29**, 924, (1963).
37. Kreshkov, A. P., Syavtsillo, S. V., Shemyatenkova, V. T., *Zavodsk. Lab.*, **22**, 1425 (1956).
38. McHard, J. A., Servais, P. C., Clark, H. A., *Anal. Chem.*, **20**, 325 (1948).
39. Kahler, H. L., *Ind. Eng. Chem., Anal. Ed.*, **13**, 536 (1941).
40. Pregl, F., Roth, H., *Quantitative Organische Mikroanalyse*, Springer Verlag, Vienna, 1958.
41. Lämmer, H., *Chem. Tech.*, **4**, 491 (1952).
42. Körbl, J., Komers, R., *Chem. Listy*, **50**, 1120 (1956).
43. Gawargious, Y. A., Macdonald, A. M. G., *Anal. Chim. Acta*, **27**, 300 (1962).
44. Gawargious, Y. A., Macdonald, A. M. G., *ibid.*, **27**, 119 (1962).
45. Campbell, A. D., Macdonald, A. M. G., *Anal. Chim. Acta*, **26**, 275 (1962).
46. Celon, E., Bresadola, S., *Anal. Chem.*, **40**, 972 (1968).
47. Šir, Z., Komers, R., *Chem. Listy*, **50**, 162 (1956).
48. Luskina, B. M., Terent'ev, A. P., Syavtsillo, S. V., *Plast. Massy*, **1964**, [2], 62.
49. Schumb, W. C., Lefever, R. A., *J. Am. Chem. Soc.*, **76**, 2091 (1954).
50. West, R., *J. Am. Chem. Soc.*, **76**, 6015 (1954).
51. Behmel, K., Schulze, M., Wannagat, U., *Z. Anal. Chem.*, **241**, 1 (1968).
52. Bork, V. A., Shvyrkova, L. A., *Trudy Komiss. Anal. Khim., Akad. Nauk USSR*, **13**, 148 (1963).
53. Kreshkov, A. P., Bork, V. A., Shvyrkova, L. A., *Zavodsk. Lab.*, **28**, 151 (1962).
54. Fritz, G., Burdt, H., *Z. Anorg. Allgem. Chem.*, **317**, 35 (1962).
55. Terent'ev, A. P., Bondarevskaya, E. A., Kirillova, T. V., *Zavodsk. Lab.*, **33**, 156 (1967).
56. Harzdorf, C., *Z. Anal. Chem.*, **256**, 192 (1971).
57. Kreshkov, A. P., Bork, V. A., Shvyrkova, L. A., *Zh. Analit. Khim.*, **17**, 359 (1962).
58. Čermák, J., Dostál, P., *Collect. Czech. Chem. Comm.*, **28**, 1384 (1963).
59. Schwarz, J., *Z. Anal. Chem.*, **115**, 161 (1939).
60. Kaye, S., Tannenbaum, S., *J. Org. Chem.*, **18**, 1750 (1953).
61. Guenther, F. O., *Anal. Chem.*, **30**, 1118 (1958).
62. Lees, J. F., Lobeck, R. T., *Analyst*, **88**, 782 (1963).
63. Gilman, H., Benkeser, R. A., Dunn, G. E., *J. Am. Chem. Soc.*, **72**, 1689 (1950).
64. Stenmark, G. A., Weiss, F. T., *Anal. Chem.*, **28**, 1784 (1956).
65. Damm, K., Noll, W., *Kolloid-Z. Z. Polym.*, **158**, 97 (1958).
66. Kellum, G. E., Uglum, K. L., *Anal. Chem.*, **39**, 1623 (1967).
67. Grubb, W. T., *J. Am. Chem. Soc.*, **76**, 3408 (1954).
68. Damm, K., Gölitz, D., Noll, W., *Angew. Chem.*, **76**, 273 (1964).
69. Smith, R. C., Kellum, G. E., *Anal. Chem.*, **39**, 338 (1967).

70. Shull, E. R., *Anal. Chem.*, **32**, 1627 (1960).
71. Hampton, J. F., Lacefield, C. W., Hyde, J. F., *Inorg. Chem.*, **4**, 1659 (1965).
72. Nessonova, G. D., Pogosyants, E. K., *Zavodsk. Lab.*, **24**, 953 (1958).
73. Bournique, R. A., *Chemist-Analyst*, **43**, 40 (1954).
74. Syavtsillo, S. V., Bondarevskaya, E. A., *Zh. Analit. Khim.*, **11**, 613 (1956).
75. Fritz, G., Burdt, H., *Z. Anorg. Allgem. Chem.*, **307**, 12 (1961).
76. Cundiff, R. H., Markunas, P. C., *Anal. Chem.*, **33**, 1028 (1961).
77. Kantor, S. W., *J. Am. Chem. Soc.*, **75**, 2712 (1953).
78. Bondarevskaya, E. A., Syavtsillo, S. V., Potsepkina, R. N., *Zh. Analit. Khim.*, **14**, 501 (1959).
79. Magnuson, J. A., *Anal. Chem.* **35**, 1487 (1963).
80. Dostál, P., Čermák, J., *Coll. Czech. Chem. Comm.*, **30**, 34 (1965).
81. Hanson, C. L., Smith, R. C., *Anal. Chem.*, **44**, 1571 (1972).
82. Sprung, M. M., Guenther, F. O., *J. Am. Chem. Soc.*, **77**, 3990 (1955).
83. Brown, P., Smith, A. L., *Anal. Chem.*, **30**, 549 (1958).
84. Sidnev, A. I., Khvashchevskaya, Yu. V., Pravednikov, A. N., *Vysokomol. Soedin., Ser. A.*, **12**, 1233 (1970).
85. Booth, H. S., Freedman, M. L., *J. Am. Chem. Soc.*, **72**, 2847 (1950).
86. Kreshkov, A. P., Shemyatenkova, V. T., Syavtsillo, S. V., Palamarchuk, N. A., *Zh. Analit. Khim.*, **15**, 635 (1960).
87. Voronkov, M. G., Shemyatenkova, V. T., *Izv. Akad. Nauk USSR, Ser. Khim.*, **1961**, 178.
88. Uriu, T., Hakamada, T., *Kogyo Kagaku Zasshi.* **62**, 1421 (1959).
89. Walędziak, H., *Chem. Anal.* (*Warsaw*), **10**, 579 (1965).
90. Biernacka, T., Wokroj, A., *Chem. Anal.* (*Warsaw*), **10**, 1233 (1965).
91. Snowacka-Wokroj, A., Biernacka, T., *Chem. Anal.* (*Warsaw*), **9**, 303 (1964).
92. Yamamoto, H., *Kogyo Kagaku Zasshi*, **64**, 1464 (1961).
93. Grant, G. D., Smith, A. L., *Anal. Chem.*, **30**, 1016 (1958).
94. Lady, J. H., Bower, G. M., Adams, R. E., Byrne, F. P., *Anal. Chem.*, **31** 1100 (1959).
95. Heylmun, G. W., Bujalski, R. L., Bradley, H. B., *J. Gas Chromatogr.*, **2**, 300, (1964).
96. Krasikova, V. M., Kaganova, A. N., *Zh. Analit. Khim.*, **25**, 1409 (1970).
97. Bissel, E. R., Fields, D. B., *J. Chromatogr. Sci.*, **10**, 164 (1972).
98. Kreshkov, A. P., Mikhailenko, Yu. Ya., Senetskaya, L. P., *Plast. Massy*, **1965**, [8], 48.
99. Fujiwara, S., Narasaki, H., *Anal. Chem.*, **36**, 206 (1964).
100. Dostál, P., Čermák, J., Kartous, J., *Coll. Czech. Chem. Comm.*, **33**, 1539 (1968).
101. Grantham, R. L., Hastings, A. G., *J. Polymer Sci.*, **27**, 115 (1958).

INDEX